ANIMAL BEHAVIOR

ANIMAL BEHAVIOR

EIGHTH EDITION

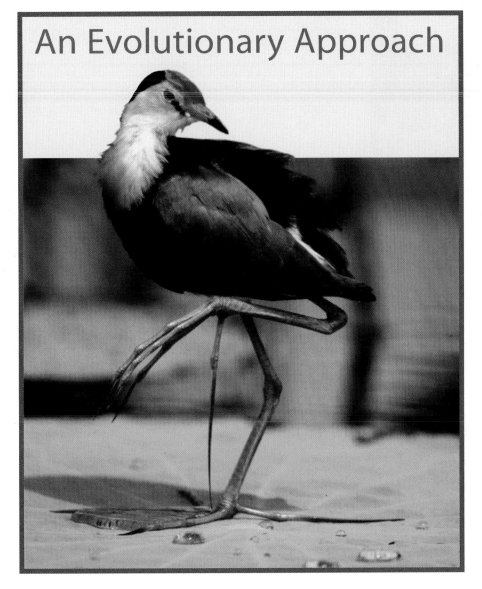

An Evolutionary Approach

JOHN ALCOCK
Arizona State University

 Sinauer Associates, Inc. • Publishers
Sunderland, Massachusetts U.S.A.

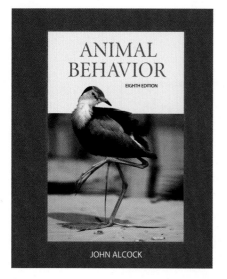

About the Cover

The comb-crested jacana is notable for its elongate toes, which enable it to walk on lily pads, and its unusual mating system, which features highly aggressive females that attempt to acquire a "harem" of male partners who will incubate their eggs. Explaining the mating system of jacanas is one of many challenges for behavioral biologists. Photograph by Bruce Lyon.

Animal Behavior: *An Evolutionary Approach,* Eighth Edition

Copyright © 2005 by Sinauer Associates, Inc.

For information, address:

Sinauer Associates, Inc., 23 Plumtree Road, Sunderland, MA 01375 U.S.A.

FAX: 413-549-1118

E-mail: publish@sinauer.com

Internet: www.sinauer.com

Library of Congress Cataloging-in-Publication Data

Alcock, John, 1942-
    Animal behavior: an evolutionary approach / John Alcock. –8th ed.
      p. cm.
    Includes bibliographical references.
    ISBN-10: 0-87893-005-1
    ISBN-13: 978-0-87893-005-0
    1. Animal Behavior — Evolution. I. Title.
    QL751.A58 2005
    591.5—dc22   2005006804

Printed in U.S.A.

10   9   8   7   6   5   4   3

With admiration for persons who are trying to save
what is left of the natural world

Contents in Brief

Contents

Preface

The first edition of my textbook was published in 1975. Now it is thirty years later, and I have just survived my 62nd birthday, something I had to do before I could receive my very own Golden Age Passport, which entitles me to free entry into our superb National Parks and Monuments for the rest of my life. Although I am not thrilled at being old enough to be a Golden Age Passport holder, I do feel privileged to have been around for the three decades when the discipline of animal behavior really came into its own. These days a host of behavioral journals provide dozens of excellent research reports every month from biologists and psychologists whose understanding of theory and technology lets them explore topics that were barely considered, if at all, when I first began doing research myself in the 1970s. The effect of all this activity is the production of far more new material than any one person can absorb, let alone squeeze into a textbook of manageable length. So what I have done here, as in previous revisions, is to provide a sampler of recent and older research findings organized around a few durable themes that help provide structure for it all. Paramount among these big ideas is the notion that the theory of natural selection provides a wonderful tool for understanding the evolutionary causes of behavior whether we are talking of roundworms or rhinoceroses. I would like to think that my textbook is imbued with Darwinian theory to such an extent that you, my readers, might come to adopt an evolutionary approach yourselves whenever you are exercising your curiosity about some aspect of the natural world. I also hope that by the time you have finished the course for which this textbook is required you will have a better understanding of the scientific logic employed by the researchers whose discoveries are reviewed here. If you really understand the theory of natural selection and what makes science special, I believe that you have made real progress toward becoming a truly educated person with all the pleasures that a good university education can provide.

Every time that I am called upon to revise this textbook, I am reminded just how fortunate I am to have certain kinds of assistance at hand. My debt to the persons responsible for the Web of Science (http://www.webof science.com) is enormous. If your institution of higher learning provides this research tool to you for free, you should immediately learn how to use it because it will come in handy whenever a professor decides to give you an assignment that requires a little library research.

In addition, I am also very grateful for the reviewers of the previous edition, who have taken the time to tell me what I have done wrong and what I can do better this time around. You might think that after seven tries, I would have the book pretty much where it should be by now but my reviewers have never been shy about pointing out assorted defects, and after I overcome my initial impulse to reject their advice, I find that I often agree that revision is in order. So to the extent that I have made improvements in this edition, I owe thanks to Tim DeVoogd (Chapters 1 and 2), Marla Sokolowski with aid from Doug Wahlsten and Dave Pfennig (Chapter 3), Andrew Bass and James Fullard (Chapter 4), Tom Hahn (Chapter 5), Larry Dill (Chapter 6), Phil Starks with aid from Caroline Blackie, Nina Fefferman, and Aferdita Spahillari (Chapters 7, 13, and 14), Steve Shuster (Chapter 8), Greg Grether (Chapter 9), Bob Montgomerie (Chapters 10 and 11), Doug Mock (Chapter 12), Peter Nonacs (Chapter 13), and Steve Gaulin (Chapter 14). Steve Shuster was especially generous with his time, trying repeatedly, albeit without complete success, to get me to see his side of an argument on the matter of conditional strategies, an idea that plays a moderately important role in some chapters of this textbook. Mike Beecher accepted the onerous task of reading the entire manuscript, which he did with good grace, generating a great many useful suggestions for revision. I owe him very special thanks, even though he made a considerable amount of extra work for me, which was not always appreciated at the time.

Many other colleagues have generously provided illustrations of various sorts; I have acknowledged the suppliers of photographs and drawings in the caption to the relevant figures in the text. Bruce Lyon has once again come up with the striking photograph on the cover of the book as well as some other excellent photos that appear in the text itself. Acknowledgements to the publishers who have kindly granted permission to use their copyrighted material appear on pages located between the Bibliography and the Index.

My editor at Sinauer Associates, Graig Donini, has supervised the process of putting the book together with calm efficiency and I thank him for that. In addition, other members of the Sinauer team have all done their special tasks very well, helping keep the job on schedule and producing an attractive book in the end. I speak of Kathaleen Emerson, David McIntyre, Janice Holabird, Joan Gemme, Chris Small, Elizabeth Morales, Mara Silver, Gayle Sullivan, Jason Dirks, Marie Scavotto, Linda Vandendolder, and Dean Scudder. Thank you. I am especially blessed that Norma Roche is my copy editor once again because she has an extraordinary ability to detect and correct grammatical errors and infelicities of expression, saving me from myself time and again. I don't know what I did to deserve her assistance but I know enough to be grateful.

I am also fortunate in having a batch of elderly but still fairly lively colleagues at Arizona State University with whom I have lunch on most days. In the Memorial Union at ASU we discuss a wide range of matters, not all of them highly intellectual or esoteric. In fact, most of our topics of conversation revolve around such fairly mundane things as the inexplicable activities of our children, the equally inexplicable behavior of some of our colleagues, the truly inexplicable willingness of a slight majority of our countrymen to elect our current commander-in-chief to a second term, and many other puzzles of everyday life. On occasion a subset of the group discusses one or another sporting event. Although Dave Brown, Jim Collins, Stuart Fisher, Dave Pearson, Ron Rutowski, and I almost never resolve much of anything at lunch, I can assure you that having these conversations with persons of like mind adds a great deal to my existence.

Likewise, I am very fortunate in having a wife, Sue, who has managed to maintain our marriage for even longer than the life of my textbook, an achievement that I appreciate. Sue long ago provided us with two children, Nick and Joe, whom we still see fairly often. They not only give me material for my lunchtime conversations but also are genuinely nice guys, willing to overlook or at least tolerate certain of my idiosyncracies, which they have admittedly made me aware of on occasion, and able to teach Sue and me such valuable things as how to use the DVD player and how a twenty-first century telephone works. Sue, Nick, and Joe, like my academic friends at ASU, make life far more interesting than it would be otherwise and in so doing, they have in their own way contributed a great deal to the writing of this textbook.

Supplements to accompany *Animal Behavior*, Eighth Edition

Instructor's Resource CD (ISBN 0-87893-015-9)
Available to qualified adopters of *Animal Behavior*, Eighth Edition, the Instructor's Resource CD includes electronic versions of all the textbook figures (both art and photos) and tables. These are provided in both high-resolution and low-resolution JPEG format, and in PowerPoint® presentations. All have been formatted and optimized for excellent legibility and projection quality.

Teaching Animal Behavior
by John Alcock
Included on the Instructor's Resource CD, this revised and updated Instructor's Manual includes answers to the expanded discussion questions presented in the textbook, complete sample exams, and an extensive new set of sample quiz questions. Also included is a chapter-by-chapter list of film and video resources.

Learning the Skills of Research: Animal Behavior Exercises in the Laboratory and Field
Edited by Elizabeth M. Jakob and Margaret Hodge
New for the Eighth Edition, this electronic lab manual (included on the Instructor's Resource CD) consists of 18 complete lab exercises, all with accompanying instructor documentation. The exercises are geared toward helping students learn about all stages of the scientific process: hypothesis development, observing and quantifying animal behavior, statistical analysis, data presentation, and more.

ANIMAL BEHAVIOR

1 | An Evolutionary Approach to Animal Behavior

For hundreds of thousands of years, our ancestors observed animals hungrily, learning the fine details of their behavior in order to put the next meal on the table. Even today, the subject of animal behavior still has great practical significance. If you'd like to maximize the production of wood ducks for duck hunters, then you had best determine the consequences of putting out lots of conspicuous nest boxes in the ducks' breeding areas [1084]. If you wish to protect your cotton crop against the pestiferous pink bollworm, it might be good to know how adult females of this moth attract mates with special scents, the better to interfere with this communication system [187]. Likewise, if you think it desirable to reduce the incidence of date rape, then perhaps you should know something about the biological basis of human sexual behavior [1193]. Or if your goal is to minimize the ever-increasing environmental destruction caused by our species, then why not get to the evolutionary root of the problem [936]?

Although some people study animal behavior to help solve one of the many problems facing human societies today, others become behavioral biologists simply because they find the subject intrinsically fascinating. Among these researchers are those who want to find out why males of some, but not all, species of dragonflies fly around holding onto their

◀ **Charles Darwin's study** in Down House where he developed the theory of evolution by natural selection, the foundation for the modern study of animal behavior. Photograph by Mark Moffett mates long after they have copulated with them. Others would like to know how night-flying moths avoid getting nailed by bats, which maneuver so much faster than their prey. And why do some seabird parents sit back and let their two babies fight until one has killed the other? And how can male red-sided garter snakes copulate in spring, when they have almost no testosterone circulating through their bodies?

In the pages ahead, you will learn something about dragonfly mate guarding, moth evasive tactics, avian siblicide, and the relationship between testosterone and sexual activity in red-sided garter snakes. The practical applications of these particular findings are modest, but there is something fundamentally satisfying about the study of animal behavior, a satisfaction regularly experienced by the thousands of curious naturalists who try to figure out why animals do what they do. Those researchers have provided the material that appears in my textbook. I hope that you will find their discoveries worth learning about, even entertaining on occasion. But in addition, I would like you to understand how scientists reach conclusions of the sort that are presented here. I believe that the process of doing science is every bit as interesting as the findings that are its end product. Therefore, the focus throughout this textbook will be on how scientific logic promotes effective thinking and leads to convincing conclusions. So let's get started by examining what some scientists have learned about the behavior of the prairie vole.

Understanding Monogamy

The prairie vole (*Microtus ochrogaster*) is a mouselike mammal that lives in burrows in grasslands in the central parts of the United States and southern Canada. In almost every respect, the drab little prairie vole is nothing to write home about. But one feature of its behavior does stand out: the prairie vole is **monogamous** (Figure 1.1). In other words, male prairie voles often content themselves with a single sexual partner, with whom they stay for an entire reproductive cycle, sometimes for a lifetime [444]. In contrast, many other voles (of which there are dozens of species), and most other mammals, are **polygynous**. Unlike the monogamous (one male–one female) prairie vole, males of polygynous species, such as the meadow vole (*Microtus pennsylvanicus*), roam from one female to another and another, and some persuade several females to copulate with them in a single breeding season.

So why should the prairie vole be monogamous when most other mammals are polygynous? One research team, led by Larry Young at Emory University's School of Medicine, has come up with an answer [722]. The team found that cells in certain parts of male prairie vole brains are loaded with protein receptors that bond chemically with a hormone called vasopressin. Vasopressin is produced and released into the bloodstream by other brain cells when a vole copulates. Molecules of vasopressin are carried to the ventral pallium, an anatomically distinct brain "center" whose cells possess the vasopressin-binding receptors. (The word *pallium* is a generic term that applies across species and refers to a structure near the brain's base that responds to hormonal signals and communicates with other parts of the brain that control emotion and memory.) When the receptor proteins in the ventral pallium, called V1-a receptors, bond with vasopressin, the resulting chemical events trigger activity in the receptor-rich cells (Figure 1.2). This activity affects neural pathways in the brain that apparently provide the vole with positive feedback. Researchers believe that these rewards encourage the male vole to remain in the company of his mate, forming a long-term social bond with her. In contrast, the reward system in the brains of polygynous vole species is different, in part because V1-a

Figure 1.1   The monogamous prairie vole. In this species, males in at least some populations form long-term relationships with females, pairing off as couples that live together and coordinate their parental care activities. Photograph by Lowell Getz.

(A) Corpus callosum   Lateral septum
Anterior commissure   Ventral pallium
(B)

Figure 1.2   The brain of the prairie vole is a complex, highly organized machine. (A) A cross section of the brain with just a few of its anatomically distinct regions labeled. The ventral pallium contains many cells with receptor proteins that bond with the hormone vasopressin. (B) A brain section that has been treated in such a way that regions with large numbers of vasopressin receptors appear black. The only area with densely packed vasopressin receptors is the ventral pallium, which occurs in both the left and right halves of the brain; the left-hand portion of the ventral pallium is outlined in black. After Lim et al. [722].

receptors are less numerous in the ventral pallium of these species. As a result, when these males copulate, their brains do not provide the same kind of feedback that leads to the formation of a durable social attachment between a male and his mate. Therefore, these males move on after copulating instead of staying put.

But Young and his colleagues have also provided a somewhat different answer to why some prairie vole males live with one female for life, an answer that focuses on a possible genetic basis for the monogamous mating system rather than on the fine details of brain cell functioning [959]. Young's group knew that the V1-a receptor protein, which is so important in the prairie vole's vasopressin-based system of social bonding, is encoded by a specific gene, the *V1aR* gene. The prairie vole's *V1aR* gene has a specific chunk of DNA that is lacking in the polygynous montane vole's version of the same gene. The extra DNA of the prairie vole's gene has the effect of making it more active in cells within the ventral pallium.

Suspecting that the *V1aR* gene had something to do with the mating systems of voles, Young and his team reasoned that if they could transfer extra copies of the prairie vole's form of the *V1aR* gene into the right cells of the right part of the brain of male prairie voles, then they should be able to make these voles bond even more eagerly with a female than they would naturally. Using a safe viral vector, the researchers inserted extra copies of the gene in question directly into cells in the ventral pallium of their prairie vole subjects. Once in place, the "extra" genes enabled these particular cells to make even more V1-a receptor proteins than they would have otherwise. The genetically modified, receptor-rich males did indeed form especially strong social bonds with female companions, even if they had not mated with them (Figure 1.3). Apparently, by increasing the number of brain cells with an active form of one particular gene, the research team was able to boost the tendency of male prairie voles to remain

Figure 1.3   A gene that affects male pairing behavior in the prairie vole. Males that receive added copies of the *V1aR* gene in the ventral pallium (*V1aR*-VP males), when given a choice between spending time with a familiar female and with a strange one, spend significantly more time in contact with the familiar female over a 3-hour trial period. In contrast, familiar partners were not significantly preferred by males that received the *V1aR* gene in the caudate putamen of the brain (*V1aR*-CP males), or by males that received a different gene, the *lacZ* gene, in the ventral pallium or the caudate putamen. After Pitkow et al. [959].

close to a social partner. They concluded, therefore, that the *V1aR* gene contributes to the monogamous behavior of male prairie voles in nature [1348].

Other researchers have come up with still another explanation for why prairie voles are monogamous. Jerry Woolf and his co-workers at Memphis State University argue that monogamy occurs in this species because prairie vole males that formed close attachments to a mate in the past left more descendants than males with a tendency toward polygyny [1335]. Woolf and company think that males that bond with one mate are reproductively successful because they can keep their female partner from copulating with other males. They found that when they experimentally prevented male prairie voles from associating with their mates in a laboratory setting, 55 percent of the females copulated with more than one male. (The females in this experiment could choose among three males that had been tethered so that they could not interfere with one another, nor could they prevent the females from leaving them and moving on to another male.) In nature, males that left their mates might well lose them to other males, which would translate into reduced paternity for the less monogamous males.

So here is a very different answer to the monogamy puzzle, one that revolves around the possible reproductive benefits to males that prevent their mates from mating with other males. According to this view, prairie vole monogamy exists because in the past males that lived with their mates, and thereby kept them under surveillance, sired most or all of the offspring of their monogamous partners. This reproductive tactic apparently has resulted in more descendants for males than if they had employed the alternative tactic of mating and moving on. This outcome would be especially likely if prairie voles were sparsely distributed in the past, as in fact they often are today [445]. In low-density populations, a male that left one female would have difficulty finding another available partner, particularly if other males guarded their mates.

Thus, under some conditions, monogamy can actually enhance male reproductive success, even though monogamous males, by definition, forgo having offspring with more than one female. If mate-guarding males in the past tended to have more surviving descendants than males that behaved in other ways, then those monogamous individuals would have shaped the evolutionary history of their species. According to this argument, when we see monogamy in today's prairie voles, we are looking at the historical result of reproductive competition between males that differed in their mating tactics.

But in addition to explaining monogamy in terms of the reasons why this mating system spread through the vole species some time ago, we can also provide a different sort of historical explanation for the behavior. This other angle requires that we trace the sequence of events that took place during evolution as monogamy originated and spread among some vole lineages (see Box 1.1 on how this can be done). There was surely a time when the prairie vole, or, more likely, some now extinct species that eventually gave rise to the modern prairie vole, was *not* monogamous. As noted above, in the overwhelming majority of living mammalian species, males attempt to mate with more than one female, which suggests that this pattern probably also prevailed long, long ago in the lineage that eventually evolved into the first vole. Indeed, not all voles are monogamous; the red-backed vole, for example, appears to exhibit the ancestral pattern of polygyny even today (Figure 1.4). If polygyny was the original mating system, then at some time in the past, polygyny gave way to monogamy in a population ancestral to modern prairie voles. Judging from the data presented in Figure 1.4, the shift toward monogamy may have occurred in an ancestral species that gave rise to two modern genera, *Lemmiscus* and *Microtus*. A number of the species in these genera combine a tendency toward male parental care and toward monogamy. Within this group is the prairie vole.

Box 1.1   **How are phylogenetic trees constructed and what do they mean?**

The diagram in Figure 1.4 is intended to represent the evolutionary history of seven of the modern species of voles living today. To create **phylogenetic trees** of this sort, it is necessary to determine which species are most closely related to each other, and thus which are descended from a more recent common ancestor. Phylogenetic trees can be drawn up on the basis of anatomical, physiological, or behavioral comparisons among species, but more and more often molecular comparisons are used for their construction. The molecule DNA, for example, is very useful for this purpose because it contains so many "characters" on which comparisons can be based; namely, the specific sequences of nucleotide bases that are linked together to form an immensely long chain. Each of the two strands of that chain has a *base sequence* that can now be read automatically by the appropriate machine. Therefore, one can, in theory and in practice, extract a specific segment of DNA from either the nucleus or mitochondria in the cells of a cluster of species and identify the base sequence of that particular segment.

For the purposes of illustration, here are three made-up base sequences from one of the two strands of DNA that constitute part of a particular gene found in three made-up species of animals:

Species X	A T T G C A T A T G T T A A A
Species Y	A T T G C A T A T G G T A A A
Species Z	G T T G T A C A T G T T A A T

These data could be used to claim that species X and Y are more closely related to each other than either is to species Z. The basis for the claim is the fact that the base sequences of species X and Y are nearly identical (differing by a single change in position 11 of the chain), whereas species Z differs from the other two by four and five base changes, respectively.

The shared genetic similarity between X and Y can be explained in terms of their history, which must have featured a very recent common ancestor (B in the phylogenetic tree shown here). Species B must have split so recently into the two lineages leading to living species X and Y that there has not been sufficient time for multiple mutations to become incorporated into this segment of DNA. The lesser, but still substantial, similarities between all three species can be explained in terms of their more ancient common ancestor, species A. The interval between the time when A split into two lineages and the present was long enough for several genetic changes to accumulate in the different lineages, with the result that species Z differs substantially from both species X and species Y.

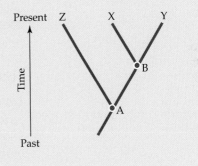

Discussion Question

1.1   Use the data in Figure 1.4 and information in Box 1.1 to outline the history of the polygynous mating system of the montane vole. Why does the meadow vole also exhibit polygyny? How similar genetically are the montane and meadow voles? The red-backed vole and the montane vole?

The change from polygyny to monogamy in an ancestor of the prairie vole may have originated in an ancestral species in which low-ranking males developed a novel tactic, infanticide, to combat the ability of a few dominant males to control many females. By killing the pups sired by other males, a subordinate male might have been able to mate with their mother and produce pups of his own (a topic covered in more detail later in this chapter). Females that

Figure 1.4   **The evolutionary relationships** of the prairie vole and six of its relatives. A tendency toward monogamy appears to have originated early in the history of the genus *Microtus* and to have been retained in the highly monogamous living species within this group, which include the prairie and pine voles. Polygyny was almost certainly the ancestral state that preceded the evolution of male parental care and mate guarding in *Microtus*, but a few modern species in the genus (e.g., the montane and meadow voles) have evidently shifted from monogamy back to polygyny. Data from McGuire and Bemis [811] and Conroy and Cook [236].

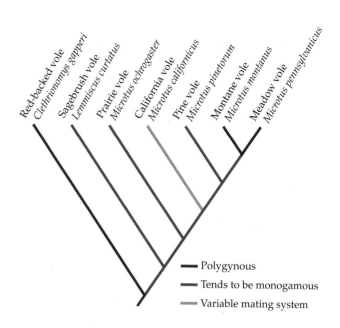

then mated with a number of males might have benefited if these males were loath to kill the pups of their mates. In response to female promiscuity, males that acquired a mate might have employed a countermeasure, guarding the female to protect her against such competitors. But mate guarding restricts a male's mobility, reducing his chances to acquire several mates in short order. If under these circumstances mate-guarding males also produced more surviving young by protecting their partner's pups, male parental behavior would have provided a further impetus for monogamy. Once the trait appeared in one ancestral species, it could have been retained in some or all of the descendant species that trace their history back to that ancestor. In other words, the unusual monogamous mating system of today's prairie voles may be the result of a series of changes in populations that preceded them, including the change from polygyny to monogamy (Figure 1.5).

Levels of Analysis

So which answer is correct? Is prairie vole monogamy the product of brain physiology, or is it due to a special form of a gene that is present in the prairie vole's brain cells, or is it the product of the reproductive benefits gained by mate-guarding males, or does it stem from a series of modifications that gradually converted a polygynous species into a monogamous one? Here we come to a critical point, central to everything that follows in this book: all four of these answers could be right, because none of these explanations for prairie vole monogamy precludes any of the others. In the jargon of behavioral biology, these four different explanations for monogamy represent different *levels of analysis* (Table 1.1). Each level of analysis potentially contributes one element that could be integrated into a satisfyingly complete explanation for the behavior [1096].

Let's illustrate the connections between the different levels of analysis of prairie vole monogamy in the following way. Imagine that at some point in evolutionary time, a male vole, perhaps a member of a species that no longer exists, happened to have a mutant (altered) gene that somehow changed the way he behaved toward his mate (perhaps by altering the number of vasopressin receptors in his brain). Instead of loving her and leaving her, this male remained with the female, and in so doing, he prevented her from mating with other males and sired all her offspring. Imagine also that because of his behavior, this novel

MODERN SPECIES

Monogamous pairs, as in the prairie vole

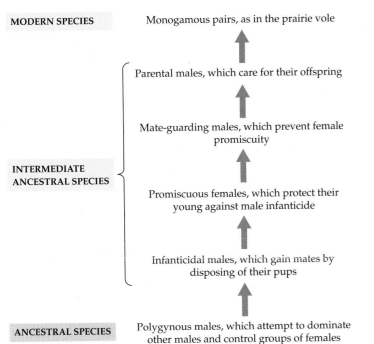

Parental males, which care for their offspring

Mate-guarding males, which prevent female promiscuity

INTERMEDIATE ANCESTRAL SPECIES

Promiscuous females, which protect their young against male infanticide

Infanticidal males, which gain mates by disposing of their pups

ANCESTRAL SPECIES

Polygynous males, which attempt to dominate other males and control groups of females

Figure 1.5   The possible history behind monogamy in the prairie vole. Evolution consists of one change layered on another. Monogamy in prairie voles may be the product of a series of behavioral shifts, with polygyny leading to infanticide, which may have favored females that tended to mate with several males to keep the males confused about the paternity of their young. But female promiscuity could have then favored males that guarded a single mate, and even cared for the offspring they produced, leading to the kind of monogamy exhibited by today's prairie voles. After Wolff and Macdonald [1334].

mate-guarding male was somewhat more successful at producing descendants than the other males of his species, which tended to seek out many copulatory partners instead of keeping an eye on just one. Because of the differences in the reproductive success of the two genetically different kinds of males, the genetic makeup of the next generation changed, with the special form of the gene associated with mate guarding monogamy becoming somewhat more common.

If this pattern were repeated generation after generation, the ancestors of today's prairie voles would have shifted from polygyny to monogamy as the frequencies of the different forms of various genes changed. Imagine that a particular form of the *V1aR* gene had a developmental effect that contributed to the kind of monogamy associated with higher reproductive success. Male parents with this gene would pass it on to their relatively numerous offspring, where it would be used during the development and operation of vole brains in the next generation.

TABLE 1.1 *Levels of analysis in the study of animal behavior*	
Proximate Causes	**Ultimate Causes**

1. Genetic–developmental mechanisms
    Effects of heredity on behavior
    Development of sensory–motor
      systems via gene–environment
      interactions
2. Sensory–motor mechanisms
    Nervous systems for the detection
      of environmental stimuli
    Hormone systems for adjusting
      responsiveness to environmental stimuli
    Skeletal–muscular systems for
      carrying out responses
	1. Historical pathways leading to a current behavioral trait
	    Events occurring over evolution
	      from the origin of the trait to the present
	2. Selective processes shaping the history of a behavioral trait
	    Past and current usefulness of the
	      behavior in promoting lifetime reproductive success

Sources: Holekamp and Sherman [550], Sherman [1096], and Tinbergen [1200]

Voles with the modern form of the *V1aR* gene have cells in the ventral pallium that acquire a certain amount of the protein encoded by this gene. The protein in question acts as a receptor for vasopressin, which is released when male prairie voles copulate with their mates. The chemical interaction between vasopressin and the V1-a receptors in cells of the ventral pallium causes a cascade of neural activity that rewards males for being close to a female. Thus, to secure additional positive reinforcement, males form a social attachment to one copulatory partner and remain with her over the long haul, guarding her against other males. If monogamous males in the current generation leave more descendants on average than any novel types that happen to have genetically different mating tendencies, then most males of the next generation will continue to exhibit the mating behavior that leads to monogamy, the mating system that has been more successful in the past than any other option that has arisen. But if in the future, the environment of the prairie vole were to change in certain ways, any males who happened to have a hereditary tendency to be polygynous might well begin to leave more descendants than their more monogamous conspecifics; if so, the population would evolve away from monogamy and toward polygyny. This capacity to change with changing environmental circumstances demonstrates that species have no preset endpoint, no evolutionary goal.

Proximate and Ultimate Explanations in Biology

In sketching this compound explanation for prairie vole monogamy, we have integrated the different levels of analysis pursued by different teams of behavioral researchers. These teams have discovered something about (1) how a gene contributes to the development of the behavior in male voles, (2) the physiological foundation for the behavior in terms of the operation of the male vole's brain, (3) the adaptive value of the behavior in terms of its contribution to male reproductive success, and (4) the transformation of a polygynous ancestor into the monogamous modern prairie vole (Figure 1.6). But we can consolidate the four levels of analysis into two larger categories, labeled proximate and ultimate [801, 897].

Proximate explanations about behavior deal with what is responsible for the building and operation of an animal that enables it to behave in a particular way. We can, for example, view the prairie vole as a machine with internal devices that make it capable of being monogamous. These devices include the sensory mechanisms that come into play during copulation, the brain cells that release vasopressin after mating takes place, the system that transports excreted vasopressin to other brain centers, cells in the ventral pallium with their receptor proteins that bond with vasopressin, the visual and olfactory systems that provide inputs that enable males to recognize a particular female with whom they have mated, and on and on.

All of these systems, mechanisms, cells, proteins, and the like were in turn assembled from simpler chemicals, thanks to events that took place during the development of the vole as it grew from a fertilized egg into an adult male. The development of the machine required genetic information, which each vole inherits from its parents. Every vole's bank of genes is capable of interacting with its chemical environment in particular ways. The interaction between heredity and chemical building blocks causes cells to grow, proliferate, and specialize, creating a complex battery of internal devices able to record information about a copulation, to release vasopressin under special circumstances, to motivate the male vole to stay close to a female, and more.

Therefore, we can explain how an animal is able to behave in particular ways if we understand, first, how that animal developed, and second, how its internal mechanisms work after they are constructed. In effect, by looking inside the

Evolution continues

Contribution of genes to next generation, the reproductive success of individuals

Behavior, including the mating system

INTERNAL MECHANISMS

Physiological system of adults, including the brain

Developmental system of young voles

The genes that have survived to the present in prairie voles

HISTORY

The spread of adaptations in previous generations of the vole

Previous evolutionary history in lineage leading to the prairie vole

Figure 1.6   The connection between evolutionary history and the mechanisms of behavior. Evolutionary processes determine which genes survive over time. The genes that animals possess influence the development of the mechanisms that make behavior possible. Behavior affects the genetic success of individuals in the current generation. Evolutionary change is ongoing.

creature, we may be able to figure out what causes it to behave one way instead of another. Because these internal developmental and physiological causes occur during one animal's lifetime, they are said to be the immediate, or **proximate**, causes of its behavior.

In contrast, another set of causes of an animal's behavior can be traced to events that took place over many generations. A long-term historical process has occurred, during which some forms of certain genes have made it to the present, while other forms have not. Voles that reproduced successfully in the past transmitted their special genes to future generations, with the result that these chunks of DNA are still around to influence the development of each living vole's nervous system, muscles, hormones, and skeleton. In contrast, voles that failed to reproduce took any different forms of genes they may have had to the grave with them. If we can determine why some voles were able to reproduce better than others in the past, then we have gained understanding about why some genes have survived over time, which provides a long-term explanation for the existence of particular proximate causes of behavior.

Furthermore, if we can secure information on the precise sequence of events that took place over the long term, we provide an extra historical dimension to the explanation for monogamy. We could in theory learn what mating system was replaced when monogamy first appeared in the prairie vole, when that original form of monogamy arose and what it was like, and how many modifications subsequently took place before the modern mating system of today's prairie voles was fully in place.

Therefore, if we want to know why prairie voles are monogamous, we need to know more than the proximate or immediate causes of male behavior. We need to know things about the history of the species, the long-term processes that gradually shaped vole attributes over time. Because these historical causes involve events that happened in previous generations, they are said to be the evolutionary, or **ultimate**, causes of behavior.

Discussion Questions

1.2   The great behavioral biologist Niko Tinbergen said that there were four main questions for behavioral researchers [1196]. I will paraphrase these questions as follows:

1. How does the behavior promote an animal's ability to survive and reproduce?
2. How does an animal use its sensory and motor abilities to activate and modify its behavior patterns?
3. How does an animal's behavior change during its growth, especially in response to the experiences that it has while maturing?
4. How does an animal's behavior compare with that of other closely related species, and what does this tell us about the origins of its behavior and the changes that have occurred during the history of the species?

Place these questions within the four levels of analysis framework and then assign each to the proximate or ultimate category. If you heard that because evolutionary questions are "ultimate" ones, they are therefore more important than questions about proximate causes, you would respectfully disagree. Why?

1.3   When a female baboon copulates, she vocalizes loudly, but her cries are longer and louder if her partner happens to be a high-ranking "alpha" male [1085]. A primate researcher has suggested that females cry out more vigorously when copulating with top males because this warns low-ranking baboons to stay clear. (Subordinate males sometimes harass mating pairs to such an extent that the copulation ends prematurely, but if that happens, they may be attacked by the dominant male whose mating has been so rudely interrupted.) The same researcher also says, however, that the more vigorous cries may simply reflect the fact that females are more strongly stimulated by the larger, more energetic, alpha males. Are the two explanations really in competition with each other? Explain why they could both be right and how they could be shown to be complementary.

How to Discover the Causes of Behavior—Scientifically

In describing the findings of various research teams on what causes prairie voles to be monogamous, I presented the conclusions that the different groups reached, emphasizing that these explanations were different, but complementary. But why did these researchers think that their discoveries were valid? And why should we take their claims seriously? The answer to these questions requires that we take a look at the logical foundations of scientific inquiry, an endeavor that employs a successful approach for sorting out possible explanations [965].

Let's take the proximate explanation for prairie vole monogamy that focuses on the presence of large numbers of V1-a receptors in the ventral pallium of the vole's brain (see Figure 1.2). Before the researchers reached the conclusion that brain cells with this protein had a major role to play in the formation of social attachments between a male and a female vole, they compared the brains of prairie voles with those of polygynous montane voles. The comparison revealed that receptors for vasopressin were numerous in the prairie vole, but much less so in the montane vole. This receptor difference suggested a possible explanation for the behavioral difference between the two kinds of voles, a working **hypothesis** that required verification before it could be accepted.

To determine whether vasopressin receptors in the ventral pallium really are essential to the formation of monogamous bonds in prairie voles, the researchers had to develop a way to test their idea. They did so by employing what can be called "if–then logic": they reasoned that *if* the vasopressin receptor explanation was correct, *then* a given experiment should generate specific results. For example, if the researchers were able to increase the number of V1-a receptors in the right part of the brain of living prairie voles, then these rodents should be even more inclined to monogamy than unaltered males. In other words, the researchers predicted what would happen if the receptor hypothesis was correct; namely, that the experimental males would be especially likely to form social partnerships with females.

Once the research team had a logical expectation or prediction to work with, they did the experiment to secure the data needed to test the prediction [722]. They found that the receptor-rich male voles did indeed attach themselves more strongly to females, so much so that they would bond with females they had not even mated with (see Figure 1.3). In this case, the actual results matched their expectations, providing solid evidence in support of the vasopressin receptor hypothesis. Had the experimental voles been no more monogamous than control groups with unaltered numbers of vasopressin receptors, then the researchers' confidence in the vasopressin receptor hypothesis would have fallen.

The vasopressin receptor hypothesis can be used to make many other predictions, one of which is that if you could transfer the prairie vole's vasopressin receptor gene to male meadow voles, you should be able convert them from polygyny to monogamy. This experiment has been done by using a safe viral vector to transfer the *V1aR* genes of prairie voles to the ventral pallium of male meadow voles. The genetically altered males formed much stronger attachments to a current mating partner than did unaltered control animals (Figure 1.7) [723]. This result provides another line of support for the vasopressin receptor explanation for monogamy in prairie voles.

To test an ultimate hypothesis, researchers use the same procedures as they do for proximate explanations. Take the hypothesis that prairie voles are monogamous because over evolutionary time, monogamous males have been able to force their mates to be "faithful," which has meant that the offspring produced almost always carry the genes of the mate-guarding male. *If* this hypothesis is true, and *if* we were to devise an experiment in which we prevented male prairie voles from guarding their mates, *then* we would expect to find at least some female voles willing to mate with more than one male. Jerry Woolf and his co-researchers conducted the necessary experiment, as described above. They found that, yes, many female prairie voles did indeed copulate with more than one partner when they had the opportunity to do so [1335]. The fact that this prediction proved to be correct provides support for one particular adaptive explanation for prairie vole monogamy.

One test is good; more tests are better, which is why most hypotheses are subjected to multiple tests. Those hypotheses that consistently lead to predic-

Figure 1.7 **Testing the hypothesis** that monogamy in prairie voles is linked to the possession of a specific gene. Male meadow voles that have received an infusion of the prairie vole *V1aR* gene in the ventral pallium (*V1aR*-VP males) huddle more closely with a female sexual partner after mating than do males that lack the foreign gene. In contrast, males that receive a ventral palladial infusion of the *lacZ* gene from prairie voles (Ctrl-VP males) are not significantly more likely to huddle with their mates than genetically unaltered meadow voles. After Lim et al. [723].

tions that are subsequently found to be correct are the ones we accept as probably true; hypotheses that regularly fail their tests are the ones we discard as probably false.

Discussion Questions

1.4   Imagine that it would be possible to inject prairie voles with a chemical that tied up the V1-a receptors in the ventral pallium. If that experiment were done, what outcome would result in the *rejection* of the vasopressin receptor hypothesis for monogamy in the prairie vole? Which is better, a tested hypothesis that is rejected, or a tested hypothesis that is accepted?

1.5   Take the material in the paragraph on page 5 in which work by Larry Young's team on the *V1aR* gene hypothesis for prairie vole monogamy is described. Rewrite this material so that you present a single sentence each for the *question about a cause* that motivated the research, the *hypothesis* about what the cause was, a *prediction* that was derived from the hypothesis, the *test evidence* that was gathered to check the prediction, and the *conclusion* that the researchers accepted on the basis of their research.

1.6   Here's an imaginary quote from a scientific paper: "We hypothesized that house mice that received the prairie vole *V1aR* gene should exhibit a greater tendency toward monogamy than others that were not genetically modified in this fashion." What's the problem here?

Darwinian Theory and Ultimate Hypotheses

When Jerry Woolf and his colleagues became interested in explaining prairie vole behavior from an evolutionary perspective, they based their work on Charles Darwin's theory of evolution by natural selection, which he presented in *The Origin of Species* in 1859 (Figure 1.8) [285]. From that time forward, biologists have used Darwin's theory to guide them whenever they have wanted to explain something in ultimate terms. Because of the importance of this theory to the study of animal behavior, and because it is so often misunderstood, we shall now review what the theory is—and is not.

Darwinian theory is based on the premise that evolutionary change is *inevitable* if just three conditions are met:

1. **Variation**, such that members of a species differ in some of their characteristics (Figure 1.9)

2. **Heredity**, with parents able to pass on some of their distinctive characteristics to their offspring

3. **Differences in reproductive success**, such that some individuals have more *surviving* offspring than others in their population, thanks to their distinctive characteristics

If there is hereditary variation within a species (and there almost always is), and if some hereditary variants consistently reproduce more successfully than others, then the increased abundance of living descendants of the more successful types will change the makeup of the species (Figure 1.10). The species will evolve as it becomes dominated by individuals that possess the traits associated with successful reproduction in the past. (You may ask yourself why, then, is there still so much genetic variation related to color pattern within *Harmonia axyridis*? For some possible answers to this question, see [802].)

Because the process that causes evolutionary change is natural, Darwin called it **natural selection**. Darwin not only laid out the logic of natural selection the-

Figure 1.8   Charles Darwin, shortly after returning from his around-the-world voyage on the *Beagle*, before he wrote *The Origin of Species.* Copyright Science Photo Library.

Figure 1.9 **A variable species.** The ladybird beetle *Harmonia axyridis* exhibits hereditary variation in its color pattern. Photographs by Mike Majerus.

ory clearly, but also provided abundant evidence that hereditary variation is common within species and that high rates of mortality are also the rule. Thus, alternative types within a species are forced into an unconscious competition for the limited spaces available for survivors. In other words, the conditions necessary and sufficient for evolutionary change are standard in living things.

When Darwin developed his theory of natural selection, he and his fellow scientists knew very little about the nature of heredity. Now, however, biologists know that considerable variation among individuals within a species arises because of differences in their **genes**, the segments of DNA that faithfully encode the information needed for the synthesis of proteins, such as the stretch of DNA that codes for the vasopressin receptor protein present in the prairie vole. Since genes can be copied and transmitted to offspring, parents can pass on the hereditary information needed for the development of critically important attributes, like those that affect the propensity for monogamy among male prairie voles.

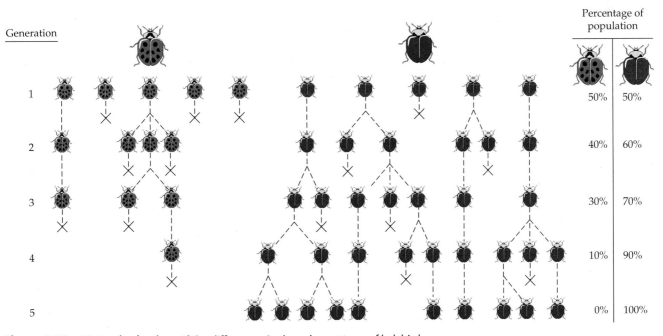

Figure 1.10 **Natural selection.** If the differences in the color patterns of ladybird beetles are hereditary, and if one type of beetle leaves more surviving offspring on average than another, then the population will evolve, becoming increasingly dominated by the reproductively successful type.

Genetic variation within a species occurs when a given gene exists in two or more forms, or **alleles**, within the species' gene pool. The different alleles sometimes affect the nature of the protein coded for by the gene, so that genetically different individuals transmit different instructions for protein manufacture to their offspring. If some alleles are superior to others in their ability to help make individuals reproductively successful, then those alleles will get themselves passed on from generation to generation and will become more common over time, gradually displacing their "competitors" over the course of evolution. We can summarize evolution at the genetic level with a simple equation:

Genetic variation + differential reproduction
= evolutionary change at the genetic level

The logical conclusion of this way of thinking about selection at the level of the gene is that alleles will spread in proportion to how well they help build bodies that are unusually good at reproducing. As E. O. Wilson puts it, a chicken is really the way that chicken genes make more copies of themselves [1315]. Given the way in which selection acts, we can assume that chickens (and all other organisms) are probably very good at reproducing and passing on their special genes.

Discussion Questions

1.7   Imagine that a mutation occurs this year in the gene that codes for the V1-a receptor protein in prairie voles. The altered protein increases the tendency of a male to form social bonds with a mate, in effect making him somewhat more monogamous than the typical male prairie vole. What is required if enhanced monogamy is to become more common in this species over time? If monogamous males have 3.7 pups on average during their lifetime while enhanced monogamous males have 4.1 on average, will the allele associated with enhanced monogamy necessarily make up an increasing proportion of the population? ("No" is the correct answer. Why? Hint: How does natural selection "measure" individual reproductive success?) Now imagine two types of prairie voles, one type that usually lives for 1.5 years and another for 0.8 years. Would it ever be possible for the one that died sooner to replace the longer-lived type over time?

1.8   If you wanted to invent the term "genetic success" for use in evolutionary studies based on natural selection theory, what definition would you come up with?

Darwinian Theory and the Study of Behavior

No matter how the concept of natural selection is presented, it is a blockbuster of a theory. Not only is the logic of the theory ironclad, but the conditions required for natural selection apply to nearly every organism, which means that almost every species has probably been shaped by natural selection in the past. In turn, this means that individuals of almost every species are endowed with alleles that have made it to the present because they were better than any alternative allele at getting their bearers to reproduce successfully. In other words, almost every hereditary attribute of almost every one of the millions of species of plants, animals, fungi, protozoans, and bacteria probably has something to do with reproductive success. The breadth of application of Darwinian theory is therefore most impressive. For all these reasons, when biologists wish to understand the ultimate reasons why an animal does something, they almost always try to come up with a working hypothesis that is consistent with natural selection theory. As a result, evolutionary hypotheses about trait X, Y, or Z (or the genes underlying those traits) usually involve suggestions about how the trait promotes reproduction.

Figure 1.11   A band of Hanuman langur females and their offspring. Males fight to monopolize sexual access to the females in groups like this one. Photograph by S. Nagendra.

Admittedly, some traits do not look especially likely to increase an individual's reproductive success. For example, male Hanuman langurs expend much energy and time trying to kill the infants of the females that they live with in bands in northern India [569]. These graceful, long-limbed primates form groups composed of one or more large adult males and several smaller females and their offspring (Figure 1.11). Young langurs are especially likely to be attacked when a new adult male becomes dominant, generally as a result of having chased off the previous top male. The new male on the block, which may have come from another band, may then try to separate infants from their mothers, harassing the mothers and biting their babies viciously when he gets a chance (Figure 1.12A).

(A)

Figure 1.12   Male langurs commit infanticide. (A) A nursing baby langur that has been paralyzed by a bite to the spine (note the open wound) by a male langur. This infant was attacked repeatedly over a period of weeks, losing an eye and finally its life at age 18 months. (B) An infant-killing male langur flees from a female belonging to the band he is attempting to join. A, photograph by Carola Borries; B, photograph by Volker Sommer, from Sommer [1140].

(B)

But killing baby langurs is not easy. Infanticidal males have to deal with female langurs, which join forces to fight back in defense of their infants (Figure 1.12B) [1139]. A male risks injury whenever he tries to destroy a youngster. Why should he take the risk, especially since his behavior would hardly seem to endear him to the very females that must become his mates if he is to ever produce offspring of his own? Given the apparent disadvantages to would-be infanticidal males, perhaps infanticide is not the product of natural selection, but rather an abnormal, pathological response to overcrowding. In fact, this nonevolutionary hypothesis for infanticidal behavior has been advocated by some langur watchers [269], who noted that langurs are often fed by Indian villagers and so perhaps now live at densities much higher than in the past. Under these novel conditions, male behavior could be thrown out of sync, leading to maladaptive, hyperaggressive reactions to youngsters.

But another langur observer, Sarah Hrdy, felt that Darwinian theory provides a legitimate solution to the puzzle of infanticide [570]. She thought that killer males could boost their reproductive chances by leaving the mothers of dead infants with no other adaptive option except to mate with them. With her nursing infant eliminated, the female might resume her reproductive cycling sooner than otherwise, and so could be impregnated sooner by the infant killer, who would leave more descendants as a result.

For the moment, let's put aside the question of whether Hrdy's hypothesis is right or wrong. The point here is that her answer to "Why infanticide?" derives from Darwinian theory. She thought of a way in which males could conceivably increase their reproductive success by practicing infanticide. If they did, then the proximate bases for infanticide could have spread in past generations of Hanuman langurs living under completely natural conditions.

So we have two potential explanations for infanticide by male langurs: the social pathology hypothesis and Hrdy's quicker reproduction hypothesis. But here's another idea. Perhaps male langurs commit infanticide as a means of population regulation. Langurs living at high densities may very well stress their food supplies, which could favor a mechanism for the prevention of overpopulation. Infanticide, although brutal and vicious, could nevertheless help keep langur bands from destroying the resources vital to their long-term survival.

The Problem with Group Selection

This population regulation hypothesis is certainly evolutionary because it explains infanticide in historical terms. It proposes that the trait evolved because in the past, groups (even entire species) that lacked the means to keep their population in reasonable harmony with the available food supply went extinct, while those that had some infanticidal males were more likely to keep their numbers in line with resources and so persist over the long haul. But note that although the hypothesis claims that infanticide spread because of its beneficial consequences, the beneficiary of infanticide is not the male who does the killing, but rather the entire group to which he belongs. Thus, the evolutionary process that causes infanticide to spread is not Darwinian selection, which is based on differences among *individuals* in their reproductive success, but rather a form of **group selection**, which is based on differences among groups in their ability to survive.

Group selection theory was laid out in detail in a book written in 1962 by V. C. Wynne-Edwards [1343]. According to Wynne-Edwards, species have such great potential to destroy the very things on which they depend that only those that happened to acquire the capacity for population regulation would have been able to avoid self-destruction. In contrast, species that were fortunate enough to have some self-sacrificing individuals would have been able to keep their numbers down and their essential resource base intact. Wynne-Edwards specifically identified infanticide as one of the mechanisms of "social mortality" that contributed to population stabilization in the animal kingdom (although in 1962 no one knew of its occurrence in langurs).

Group selection theory was challenged in 1966 by George C. Williams in *Adaptation and Natural Selection*, which many rank as the most important book on evolutionary theory written since *The Origin of Species*. Williams showed that the survival of alternative alleles was much more likely to be determined by differences in the reproductive success of genetically different individuals than by survival differences among genetically different groups [1307]. We can illustrate his point with reference to langurs. Imagine that in the past there really were male langurs prepared to risk serious injury, even death, by killing infants in order to regulate their band's size for the survival benefit of the group. In such a case, group selection would be said to favor the allele(s) for male infanticide because the group as a whole would benefit from the removal of excess infants.

But in this species, Darwinian natural selection would also be at work, provided that at any time there were two genetically different kinds of males, one that practiced infanticide for the good of the group and the other that permitted other males to pay the price for population reduction. If the non-killers lived longer and reproduced more, which of these two types would become more common in the next generation? Whose hereditary material would increase in frequency over time? Would infanticide long persist in such a population of langurs?

This kind of thought experiment told Williams and his readers that Darwinian selection acting on differences among individuals within a population or species will usually have a stronger evolutionary effect than group selection acting on differences among entire groups. Yes, group selection can occur, provided that groups can retain their integrity for long periods and differ genetically in ways that influence their chances of survival. But if group selection favors a trait that involves reproductive self-sacrifice while natural selection acts against it, natural selection seems likely to trump group selection, as we have just seen in our hypothetical langur example. Although research continues on group selection [1133], almost all behavioral biologists have been persuaded by Williams to distinguish between group selection and individual (or gene) selection hypotheses. The overwhelming majority of researchers exploring ultimate questions about behavior look first to Darwinian theory for guidance in the production of plausible hypotheses.

Figure 1.13   Variation in suicidal tendencies in a make-believe lemming-like species. Courtesy of Gary Larson.

Discussion Question

1.9 Lemmings are small rodents that live in the Arctic tundra. Their populations fluctuate wildly. At high population densities, large numbers leave their homes and travel long distances, during which time many die, some by drowning as they attempt to cross rivers and lakes. One popular explanation for their behavior is that the travelers are actually committing suicide to relieve overpopulation. By heading off to die, the suicidal lemmings leave shelter and food for those who have stayed behind. These surviving individuals will perpetuate their species, saving it from extinction. What theory was used to produce this hypothesis? How would George C. Williams use Gary Larson's cartoon (Figure 1.13) to evaluate the hypothesis critically?

Testing Alternative Hypotheses

We can argue that the group benefit hypothesis for infanticide has some logical deficiencies that reduce its plausibility, while Hrdy's quicker reproduction hypothesis has the virtue of resting on a logical and extremely powerful theory. But, as we have already noted, scientists do not call it a day after coming up with a hypothesis, no matter how reasonable it may seem. Hypotheses *have* to be tested. In the case of langur infanticide, we have three alternatives to consider: the nonevolutionary social pathology hypothesis and the two evolutionary possibilities outlined above. To test these alternatives, we first need to derive testable predictions from them.

If high population density really does cause male langurs to behave abnormally (the social pathology hypothesis), or if it truly threatens the survival of langur groups and so activates infanticide by self-sacrificing males (the population regulation hypothesis), then we would expect to see infanticide by males only in areas where Hanuman langurs live in exceptionally high densities. However, contrary to this prediction, infanticide regularly occurs in troops living at moderate or even low densities in areas where they are not fed by people [124, 873]. This finding weakens our confidence in both the social pathology and population regulation hypotheses for infanticide by langur males.

What about Hrdy's quicker reproduction hypothesis? It is not the only Darwinian explanation for infanticide. Another possibility is that males kill infants after taking over a band of females in order to eat the dead youngsters, thereby replenishing energy reserves that have been depleted in the takeover struggle. Cannibalism is widely practiced in the animal kingdom, even by some primates [818], so we really ought to consider it a possibility for langurs as well. As it happens, both the (1) cannibalism and (2) quicker reproduction hypotheses generate one prediction that is identical; namely, that infanticide should occur soon after takeovers, when the males are (1) presumably energetically stressed and (2) need to reproduce quickly if they are to have as many descendants as possible. The fact that infanticide is associated with the period soon after a takeover does not enable us to discriminate between the two hypotheses. Nonetheless, if the cannibalism hypothesis is true, we can also predict that male langurs will consume the infants they kill. Since no one has observed them doing so, it is unlikely that infanticide has evolved in Hanuman langurs because of its nutritional benefits for infanticidal males.

The quicker reproduction hypothesis leads to some additional predictions. For one thing, we would not expect to see males killing their own babies, which carry their genes, because that would obviously damage a male's reproductive success. In one group of langurs that contained several adult males, not just one dominant male, 16 cases of infanticide were recorded in which it was possible to match the presumptive killer's DNA with that of the dead infant. In every instance, the killer was *not* the father of the victim—as predicted [125].

The quicker reproduction hypothesis also yields the prediction that females who have lost their infants to a killer male will promptly resume sexual cycling and become receptive to the very male that disposed of their youngsters. In fact, langur females who have lost their young are quick to regain their sexual receptivity, which is suppressed when females are nursing young. The infant-deprived females soon become pregnant and give birth to babies whose DNA matches that of the infanticidal male [570, 1140]. Thus, tests of the quicker reproduction hypothesis have been positive.

But the more predictions and tests, the better. For example, if infanticide is adaptive for male langurs because it speeds up their chances of impregnating females who would otherwise be tied up in the care of some other male's progeny, then infanticide should also be adaptive in other species whose social systems resemble that of the Hanuman langur. We can check this prediction by

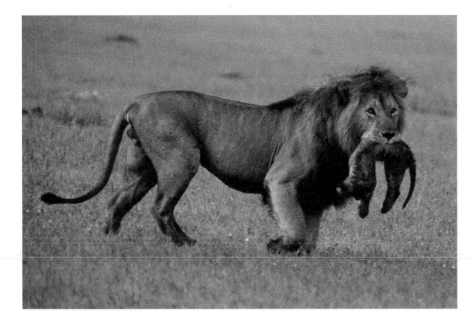

Figure 1.14   Infanticide by a male lion. This male carries a cub he has killed after displacing the adult males that once lived with his pride. Photograph by George Schaller.

observing what happens when male lions compete to control a pride, a group of females that live together. On occasion, new males oust the previous pride masters. Under these circumstances, Darwinian selection should have resulted in the spread of infanticidal behavior by newcomers, who, by eliminating the cubs of the previous residents, would leave the lionesses with little option except to come into heat at once in order to replace their youngsters with the new males as fathers. As predicted, incoming males hunt down youngsters less than 9 months old and try to kill them, often succeeding (Figure 1.14) despite resistance from the cubs' mothers [988]. Once a lioness has lost her cubs, she will resume sexual cycling and mate with her infanticidal companion. Had she retained her infant, she would not have become sexually receptive until her progeny were 2 years old. Since the average tenure of an adult male with a pride of females is only 2 years, the reproductive benefits of infanticide for the male after a takeover are obvious. Indeed, in some populations, male lions may kill a quarter of all the cubs that die in their first year of life [988].

The existence of male infanticide in both lions and langurs, despite the fact that the two species are not closely related, provides strong evidence that the trait permits reproductive gains for killer males because it speeds up their access to fertile females. Infanticide by males has now been recorded in over 50 species of mammals [1334], as well as in birds [1231] and spiders [1070]. Each case potentially provides evidence for testing the hypothesis that infanticide has evolved in langurs because it increases reproductive opportunities for successful infant killers.

Other kinds of infanticide can also provide relevant evidence for testing the quicker reproduction hypothesis. For example, in a species in which *females* compete for sexual access to males, we can predict that females will have evolved the capacity for infanticide if it leads males to accept them sooner than otherwise. This prediction has been confirmed for a giant water bug whose males take care of large masses of eggs deposited by their mates (Figure 1.15) [1128]. The male will not mate with additional females if the clutch of eggs under his care is sufficiently large. Therefore, a female that finds a brooding male sometimes attacks and destroys all his eggs. Under these circumstances, the now egg-less male may then mate with the killer female and care for her clutch of eggs after she lays them on a stick or branch.

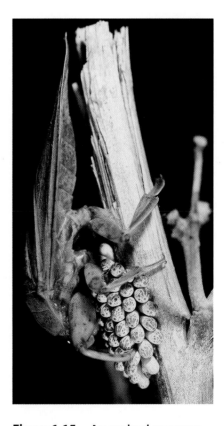

Figure 1.15   An evolved response to the risk of infanticide. This male water bug guards a clutch of eggs against infanticidal females that may destroy his current clutch in order to replace these eggs with their own. Photograph by Bob Smith.

Much the same sort of thing also occurs in the wattled jacana, a water bird in which males incubate the eggs of a partner and guard the chicks when they hatch. Here too females sometimes commit infanticide by killing the chicks of other females before mating with the males that have lost their broods. The killer females deposit their eggs in the care of these males, gaining caretakers for their offspring sooner than if they had waited for the males to finish rearing their current brood [367].

Discussion Questions

1.10 Male Hanuman langurs apparently use a rule of thumb when deciding which infants to kill: they attack youngsters of females that they did not copulate with prior to the birth of those offspring. In light of this finding, some observers have an explanation for why it is that after a takeover *pregnant* females may mate with newcomers. These researchers have suggested that the females' behavior may create confusion about the paternity of their babies, the better to limit male infanticidal behavior. If females also engage in paternity confusion when their band contains several adult males, what predictions can you make about (1) the duration of female receptivity each reproductive cycle, (2) the relationship between the period when females are in heat and the timing of ovulation, and (3) the occurrence of copulations by subordinate males as well as the dominant male langur, who attempts to monopolize sexual access to the female?

1.11 In an assassin bug, a predatory insect, males stand guard over clutches of eggs that females (usually their mates) have laid on leaves [1186]. The males protect the clutch against parasitic wasps that attempt to lay their eggs in assassin bug eggs. From time to time, a male stabs one of the eggs in his clutch with his beak, injecting it with digestive enzymes before drawing out the fluid. Needless to say, this act destroys the egg and so constitutes a kind of infanticide. If infanticide of this sort has evolved through sexual competition for quicker reproduction, as in Hanuman langurs and lions, how should males treat clutches laid by their copulatory partners versus those that the males "adopted" when they forced another male to leave an egg mass he had fertilized, something that happens occasionally? Once you have a prediction in hand, consider these two findings: (1) the parentage of the eggs has no effect on the number eaten, which increases steadily as the period of egg guarding lengthens, and (2) males do not lose weight while egg guarding, even though they cannot feed elsewhere. Are these findings relevant for the quicker reproduction hypothesis? Do they support another hypothesis for the tendency of males to eat some of the eggs under their care?

Certainty and Science

You must have deduced from my summary of the research on infanticide that I think the quicker reproduction hypothesis applies to langurs and lions, water bugs and jacanas. I do—but I could be wrong, and indeed, some other researchers surely think that I am wrong [68, 112, 272]. These differences of opinion remind us that all scientific conclusions must be considered tentative to some degree. In the past, conclusions have shifted dramatically when new evidence has thrown a once widely accepted hypothesis into doubt. For example, when I was a student at Amherst College, my paleontology professor convinced me that the earth's continents have always been where are now located. However, as new data came in, the old view was abandoned, and now the mainstream hypothesis is that the continents "float" around the planet on moving plates.

The rejection of established wisdom happens all the time in science. Scientists tend to be a skeptical lot, perhaps because special rewards go to those who can show that published conclusions are erroneous. Researchers constantly criticize their colleagues' ideas, in good humor or otherwise, sometimes causing their fellow scientists to change their minds. The uncertainty about Truth that scientists accept, at least when talking about someone else's ideas, often makes nonscientists nervous, in part because scientific results are usually presented to the public as if they were written in stone. But anyone who has taken a look at the history of science will learn that new ideas continually surface and old ones are regularly replaced or modified [680]. Indeed, a great strength of science lies in the willingness of at least some scientists to consider new ideas and to test old hypotheses repeatedly, even if some colleagues think that doing so is a waste of time.

Please keep this point in mind as we review some current scientific conclusions in the chapters ahead. We will first examine the proximate and ultimate aspects of bird song (Chapter 2) before looking more closely at different proximate analyses of behavior (in Chapters 3 through 5). Then we turn to ultimate questions about evolutionary history and adaptation (in Chapters 6 through 13). The book concludes with a chapter on the evolution of human behavior. Thanks to the many behavioral researchers who have explored these issues, there is much to write about, so let's get started.

Summary

1. The causes of any behavior can potentially be understood in terms of four different levels of analysis, which deal with (1) how the behavior develops, (2) how physiological mechanisms work to make the behavior possible, (3) how the behavior promotes the animal's reproductive success, and (4) how the behavior originated and has been changed over evolutionary time.

2. These four levels of analysis can be condensed into two: (1) those that deal with the proximate or immediate causes of behavior, which are linked to the operation of internal developmental and physiological systems, and (2) those that have to do with the ultimate or long-term evolutionary causes of behavior, which are linked to questions of adaptive value and historical modification.

3. Proximate and ultimate causes are interrelated. The genes present in today's animals have survived a historical process dominated by the effects of past differences among individuals in their reproductive success. A living animal's naturally selected genotype can then exert proximate influences over the course of that individual's development, affecting the nature of the mechanisms that the animal possesses, which in turn enables it to do certain things.

4. Both proximate and ultimate questions about behavior can be investigated scientifically in much the same way:

   a. We begin with a question about what causes an animal to do something.
   b. We devise a working hypothesis, a possible answer to the question.
   c. We use this potential explanation to make a prediction about what we expect to observe in an experiment or in nature if the hypothesis is true.
   d. We then collect the necessary data to determine if the prediction is correct or false.
   e. If the actual results do not correspond to the expected ones, we conclude that the underlying hypothesis is probably false; if the evidence does match the predicted results, we conclude that the hypothesis can be tentatively accepted as true.

5. The nature of our theories affects the kinds of working hypothesis that we generate. Darwinian evolutionary theory leads us to consider how a given behavior might advance the reproductive success of individuals; if the behavior has been shaped over time by Darwinian natural selection, then it must have been better than all the others that have appeared over the generations in terms of promoting individual reproductive success.

6. An alternative theory, the group selection theory of V. C. Wynne-Edwards, generates working hypotheses that focus on how a given behavior helps groups survive; if the behavior in question has been shaped over time by group selection, then it must be superior to all others in terms of helping entire groups avoid extinction.

7. Almost all behavioral biologists today use Darwinian theory rather than group selection theory as the foundation for their hypotheses because selection at the level of the individual is likely to be more powerful than group selection in causing evolutionary change. The logic of this conclusion can be seen if one imagines what would happen over time to a group-benefiting behavior that caused those who behaved in this manner to leave fewer descendants than others who did not sacrifice themselves (and their distinctive genes) for the benefit of their group.

8. The beauty of science lies in the ability of scientists to use logic and evidence to evaluate the validity of competing theories and alternative hypotheses. Persons who use the scientific approach can make progress by eliminating explanations that fail their tests while accepting others that pass their exams.

Suggested Reading

Books written by scientists that capture the sense of curiosity and excitement that biologists feel as they study the proximate basis of animal behavior include Vincent Dethier's *To Know a Fly* [314] and Kenneth Roeder's *Nerve Cells and Insect Behavior* [1027]. Niko Tinbergen's *Curious Naturalists* [1197] and Konrad Lorenz's *King Solomon's Ring* [742] bridge the gap between proximate and ultimate approaches. See also Howard Evans's *Life on a Little Known Planet* [372] and Michael Ryan's *The Túngara Frog* [1045].

For other books by scientists that capture the delight of field research, consider Evans's *Wasp Farm* [373], Bernd Heinrich's *In a Patch of Fireweed* [531], George Schaller's *The Year of the Gorilla* [1064], and Tinbergen's *The Herring Gull's World* [1199], as well as Jane Goodall's *In the Shadow of Man* [456], Shirley Strum's *Almost Human* [1158], Cynthia Moss's *Elephant Memories* [855], and *Journey to the Ants* by Bert Hölldobler and E. O. Wilson [554]. A realistic account of an ambitious field research project comes from Craig Packer [917]. *The Langurs of Abu* tells Sarah Hrdy's story of her work on infanticide [570]. Bernd Heinrich's superb *Ravens in Winter* offers an unusually clear picture of how scientists test alternative hypotheses [533]. For a provocative essay on the nature of science, read "Conduct, misconduct, and the structure of science" [1337]. For a scientific article that illustrates how to apply the four levels of analysis to a human behavior, see Debra Zeifman's paper on what causes babies to cry [1359].

Charles Darwin had something useful to say about the logic of natural selection in *The Origin of Species* [285], and so do Daniel Dennett [313] and Richard Dawkins [300, 303]. G. C. Williams's classic *Adaptation and Natural Selection* [1307] demolishes "for-the-good-of-the-species" arguments.

2 | *Understanding the Proximate and Ultimate Causes of Bird Song*

I began bird-watching avidly at age 6. Since then, my "life list" has grown to well over a thousand species, ranging from the tiny calliope hummingbird to the giant Andean condor. Although most birders, myself included, enjoy the challenge of identifying bird species by their appearance, we also get pleasure from learning to recognize species by their songs. So it was that long ago my father taught me that yellow warblers sing "sweet-sweet-sweet-sweeter-than-sweet," whereas the common yellowthroat (another warbler) goes "witchety-witchety-witchety," while the ovenbird (yet another warbler) belts out a loud "teacher-teacher-teacher-teacher-teach."

In fact, almost every bird species is associated with its own distinctive vocalization, a phenomenon that has long been explained in ultimate terms as an adaptation that helps a male attract a female of the "right" species. Thus, when a female ovenbird returns from her wintering ground in southern Mexico to the Ohio woodland where she will breed, she surely can hear male ovenbirds calling well before she ever lays eyes on a potential mate. By singing "teacher-teacher-teacher," the male ovenbird enables females to find him efficiently, which is in his reproductive interest. Females benefit from this system in part because it helps them track down ovenbird males while avoiding males of other species, which

Studies of bird song have relied heavily on male white-crowned sparrows, one of which is shown singing here. Photograph by Doug Nelson

decreases their risk of producing infertile or otherwise disadvantaged hybrid offspring.

But if males of different species sing different songs in order to attract females of their species, why should different individuals of the *same species* produce somewhat different versions of their species' song, as they often do? Yes, every yellow warbler sings something like "sweet-sweet-sweet-sweeter-than-sweet," but all sorts of variations on this theme exist within the species. So what's going on here? What happens during a bird's development that causes it to sing a little differently from some other members of its species? Does the operation of an individual male's brain differ from that of others in ways that affect his song? Do the song differences among individuals influence how many offspring they have? Are these behavioral differences rooted in their species' evolutionary history? This chapter attempts to answer these questions as a way to reinforce the foundation points presented in Chapter 1; namely, that if we want to understand what causes animals to behave a particular way, we have to tackle the problem from both proximate and ultimate angles.

Different Songs: Proximate Causes

In the short film *Why Do Birds Sing*? Peter Marler describes how he became interested in the proximate causes of song variation. In the 1950s, he was a fledgling limnologist studying mud deposits in British lakes. As Marler traveled about, he also casually kept an eye and ear open for the birds in the vicinity of his research sites. An experienced birder, he knew the chaffinch's song by ear, a song that has been described as a rattling series of chips terminated by a descending flourish. But he soon realized that the chaffinches living near one lake sang songs that were definitely different from the chaffinch songs he heard around other lakes not very far away. He came to believe that the birds lived in geographically discrete populations, each endowed with its own special version of their species' basic song.

When Marler moved to the University of California in the 1960s, he and his students investigated the same phenomenon in white-crowned sparrows, the males of which sing a complex whistled vocalization during the breeding season. Marler's team found that different populations of this bird often have distinctively different dialects. Thus, white-crowns living in Marin, north of San Francisco Bay, sing a song type that is easily distinguished from that produced by white-crowns that live in Berkeley, even though the two cities are only 50 miles apart (Figure 2.1) [786]. In at least some populations of white-crowns, the local dialect has persisted for decades with only modest changes [510], showing that bird dialects are stable (just like human ones).

Marler wanted to know what developmental factors were responsible for the sparrow's dialects. This proximate level of analysis required him to investigate how the birds' genes and environments interact to produce differences among individual males as they are growing up. Marler knew that either of two developmental explanations could potentially account for the dialects of Californian white-crowns. First, Marin birds might differ *genetically* from Berkeley white-crowns in ways that affect the development of the neural mechanisms that control singing behavior, so that birds in the two populations eventually come to sing different songs. Alternatively, the differences between birds in the two areas might not be hereditary, but rather *environmental*; in other words, differences in the experiences that males have early in life might be responsible for their later song differences. In particular, perhaps young males in Marin learn to sing the dialect of that region by listening to what adult Marin males are singing, while farther south in Berkeley, young male white-crowns have differ-

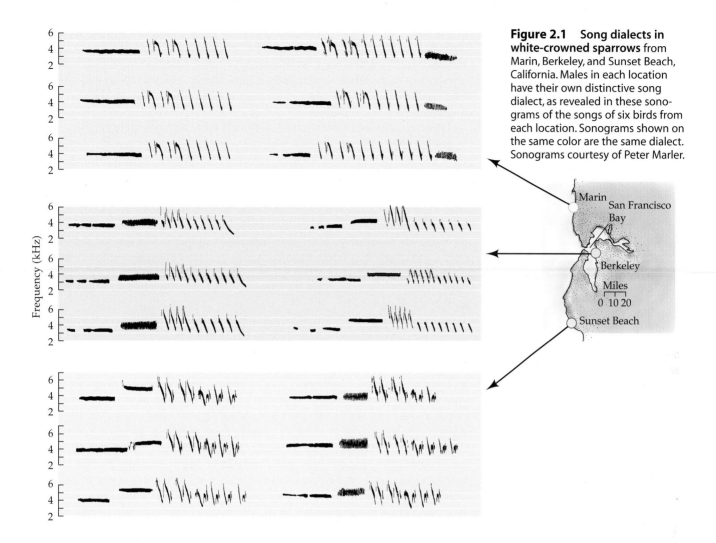

Figure 2.1   Song dialects in white-crowned sparrows from Marin, Berkeley, and Sunset Beach, California. Males in each location have their own distinctive song dialect, as revealed in these sonograms of the songs of six birds from each location. Sonograms shown on the same color are the same dialect. Sonograms courtesy of Peter Marler.

ent formative experiences as a result of hearing males singing the Berkeley dialect. After all, a young person growing up in Mobile, Alabama, acquires a different dialect from someone in Bangor, Maine, simply because the Alabaman child hears a different brand of English than a youngster reared in downeast Maine.

Marler and his colleagues explored the development of white-crown singing behavior by taking eggs from the nests of white-crowned sparrows and hatching them in the laboratory, where the baby birds were hand-reared. Even when these young birds were kept isolated from the sounds made by singing birds, they still started singing when they were about 150 days old, but the best they could ever do was a twittering call that never took on the rich character of the full song of a wild male white-crowned sparrow from Marin or Berkeley or anywhere else [785].

This result suggested that something critical was missing from the hand-reared birds' environment, perhaps the opportunity to hear the songs of adult male white-crowned sparrows. If this was the key factor, and if a young male were isolated in a soundproof chamber but exposed to tapes of white-crowned sparrow song, then the experimental subject ought to be able to sing a complete white-crown song in due course. And that is exactly what happened when 10- to 50-day-old white-crowns were allowed to listen to tapes of white-crowned sparrow song. These birds also started singing on schedule when they were

Father

Son 1 (intact)

Frequency (kHz)

Son 2 (deaf)

Time (seconds)

Figure 2.2 Hearing is critically important for song learning in the zebra finch (and many other songbirds). A sonogram of a male zebra finch's song is shown with those of two of his male offspring. The first son's hearing was intact, and he was able to copy his father's song accurately. The second son was experimentally deafened early in life, and as a consequence, he never sang a typical zebra finch song, let alone one that copied his father's song precisely. (Incidentally, zebra finches are the only known bird species in which sons learn what kind of song to sing from their fathers; in other species in which song learning occurs, as we will see, young males copy songs from older males other than their fathers.) Photograph courtesy of Atsuko Takahashi; sonograms from Wilbrecht et al. [1303].

about 150 days old. At first their songs were incomplete, but by the age of 200 days, the isolated birds not only sang the species-typical form of their song, they closely mimicked the exact version that they had heard on tape. Play a Berkeley song to an isolated young male white-crown, and that male will come to sing the Berkeley dialect. Play a Marin song to another male, and he will eventually sing the Marin dialect.

These results offer powerful support for the environmental differences hypothesis for white-crown dialects. Young birds that grow up near Marin hear only the Marin dialect as sung by older males in their neighborhood. They evidently store the acoustical information they acquire from their tutors and later match their own initially incomplete song versions against their memories of tutor song, gradually coming to duplicate a particular dialect. Along these lines, if a young hand-reared white-crown is unable to hear itself sing (as a result of being deafened after hearing others sing but before beginning to vocalize itself), then it never produces anything like a normal song, let alone a duplicate copy of the one it heard earlier in life [663]. Indeed, the ability to hear oneself sing appears to be critical for the development of a complete song in a host of songbirds (Figure 2.2).

Marler and others did many more experiments designed to determine just how song development takes place in white-crowns. For example, they wondered whether young males were more easily influenced by the stimuli provided by singing adults of their own species than by those of other species. In fact, young, isolated, hand-reared white-crowns that hear only songs of another species almost never come to sing that kind of song (although they may incorporate notes from the other species' song into their vocalizations). If 10- to 50-day-old birds listen only to tapes of song sparrows instead of white-crowns, they develop aberrant songs somewhat similar to the "songs" produced by males that never hear any bird song. But if an experimental bird has the chance to listen to tapes of the white-crowned sparrow along with songs of another sparrow species, then by 200 days of age it will sing the white-crowned sparrow dialect that it heard earlier [664]. The young bird's developmental system is such that listening to the other sparrow's song has no apparent effect on its later singing behavior.

Discussion Question

2.1 Work on the development of singing behavior in male white-crowned sparrows has demonstrated that the birds must learn to sing a particular dialect of the full song of their species. This finding eliminates the genetic differences hypothesis for the song differences between white-crowns in Marin and in Berkeley. Would we be right, therefore, to conclude that the genetic information present in the cells of white-crowned sparrows is irrelevant for the development of the bird's singing behavior? In this regard, what importance do you attach to the finding that white-crown males apparently can learn their species' song far more easily than the song of other sparrows? What about the finding that white-crown males that hear white-crown song only during a 40-day period early in life can nevertheless generate a complete song, although they do not start singing until 3 to 4 months after exposure to a tutor tape?

Social Experience and Song Development

The experiments with isolated white-crowns exposed to taped songs in laboratory cages led Marler to the following view of song development in this species (Figure 2.3): At a very early age, the white-crown's still immature brain is able to selectively store information about the sounds made by singing white-crowns

Figure 2.3   Song learning hypothesis based on laboratory experiments with white-crowned sparrows. According to this hypothesis, young white-crowns have a critical period 10 to 50 days after hatching when their neural systems can acquire information from listening to white-crown song, but not from any other species' song. Later in life, the bird matches his own subsong with his memory of the tutor's song, and eventually imitates it perfectly—unless he is deafened. Based on a diagram by Peter Marler.

while ignoring other species' songs. At this stage of life, it is as if the brain possesses a restricted computer file capable of recording only one kind of sound input. Then, months later, when the bird begins to sing, it accesses the file. By listening to its own subsongs (incomplete versions of the more complex full song that it will eventually sing) and comparing those sounds against its memories of the full songs it has heard, the maturing bird is able to shape its own songs to match its memory of the song it has on file. When it gets a good match, it then repeatedly practices this "right" song, and in so doing crystallizes a full song of its own, which it can then sing for the rest of its life.

The ability of male white-crowned sparrows to learn the songs of other male white-crowns in their birthplace just by listening to them sing provides a plausible proximate explanation for how males come to sing a particular dialect of their species' full song. However, very occasionally, observers have heard wild white-crowns singing songs like those of another species, including the song sparrow. These rare exceptions led Luis Baptista to wonder whether some other factor in addition to acoustical experience might influence song development in white-crowns. One such factor might be social experience, a variable excluded from Marler's famous experiments with isolated, hand-reared birds whose environments offered acoustical stimuli, but not the opportunity to interact with living, breathing companions.

To test whether social stimuli can influence song learning in white-crowns, Baptista and his colleague Lewis Petrinovich placed young hand-reared white-crowns in cages where they could see and hear living adult song sparrows or strawberry finches [65]. Under these circumstances, white-crowns learned their social tutor's song, even when they could hear, but not see, adult male white-crowned sparrows (Figure 2.4). In fact, social experience with another species can trump early tape tutor experience with white-crown song even after the young males are more than 50 days old, the end of the window of opportunity for song learning from tape tutors [64]. Clearly, social experience does have extremely powerful developmental effects on white-crowned sparrow singing behavior.

We now know that social factors have similar effects on song acquisition in many other birds as well [77]. Captive starlings, for example, are adept at mimicking human speech, producing such phrases as "see you soon baboon" or

Figure 2.4 Social experience influences song development. A white-crowned sparrow that has been caged next to a strawberry finch will learn the song of its social tutor. (A) The song of a tutor strawberry finch. (B) The song of a white-crowned sparrow caged nearby. The letters beneath the sonograms label the syllables of the finch song and their counterparts in the song learned by the sparrow. Sonograms courtesy of Luis Baptista.

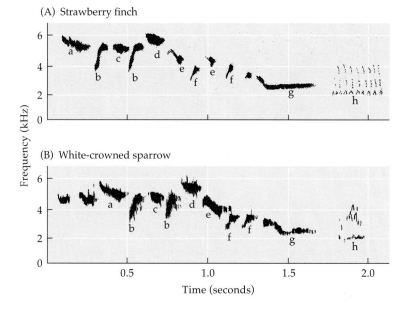

(A) Strawberry finch

(B) White-crowned sparrow

"basic research" as well as the sounds of laughter or kissing or coughing. But they will do so only if they are hand-reared in a human household where they literally become part of the family (Figure 2.5) [1274]. In nature, young starlings and white-crowned sparrows are strongly stimulated by interactions with vocalizing companions of their own species, and these experiences then influence song development.

Thus, in proximate developmental terms, male white-crowns sing different dialects because young males can be influenced by certain kinds of acoustical and social experiences. These experiences shape song development by providing young males with specific song memories that they will draw on later when they begin singing. The birds use their stored song memories as models to imitate on the road to producing their own full song. Males with different model song memories will therefore develop different songs as adults.

Figure 2.5 Social effects on song learning. Kuro the starling learned to include words in his vocalizations because he had a close relationship with the family of Keigo Iizuka. Photograph by Birgitte Nielsen, courtesy of Keigo Iizuka.

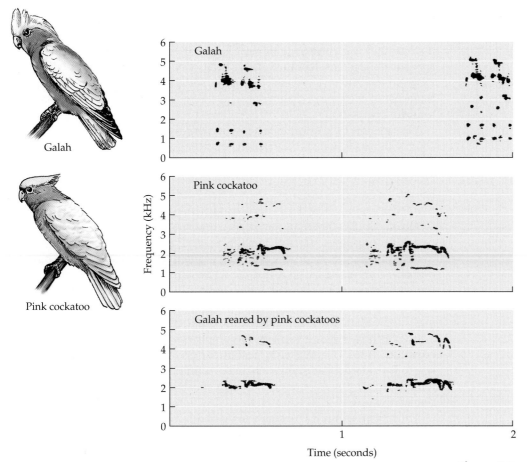

Figure 2.6   Sonograms of contact calls of galahs and pink cockatoos reared under different conditions. The top panel shows the contact call of a galah reared by galahs; the middle panel shows the call of a pink cockatoo reared by its genetic parents; the bottom panel shows the call of a galah reared by pink cockatoo foster parents. After Rowley and Chapman [1037].

Discussion Question

2.2   A natural experiment sometimes occurs in Australian woodlands when galahs (a species of parrot) lay eggs in nest holes in trees that are then stolen from them by pink cockatoos (another parrot). Thus, the cockatoos become foster parents for baby galahs. The young foster-reared galahs produce begging calls and alarm calls that are identical to those produced by galahs cared for by their genetic parents. However, the adopted galahs eventually give contact calls very much like those of their adoptive cockatoo parents, as you can see from Figure 2.6. (The birds produce these signals to help maintain contact with others when traveling in flocks [1037].) Someone claims that these observations show that galah begging and alarm calls are genetically determined, whereas contact calls are environmentally determined. Explain why this claim is wrong (you may find it helpful to read ahead in Chapter 3). Then defend the superficially similar statement that the differences between the alarm calls given by adopted galahs and by their cockatoo foster parents are the result of genetic differences between them. What other behavioral differences are the result of *differences* between the social environment of the adopted galahs and that of certain other individuals?

The Development of the Underlying Mechanisms of Singing Behavior

If we want to understand more about the development of a white-crowned sparrow's dialect, we have to go beyond identifying those elements of the young bird's social and acoustical environment that affect its later behavior. For one thing, where are a month-old white-crown's song memories stored? And what part of the brain controls the sounds that it produces at 5 months of age? And how does the young male know how to match his song memories with his own initially simpler songs? These questions require us to consider the internal devices that the young male possesses that are capable of using social and acoustical inputs to steer his singing behavior along a particular developmental pathway. The mechanisms that make song learning possible are located in the brain of the young male. How did the male's brain acquire the hardware for song acquisition?

We can start by exploring why male white-crowns sing a full and complex song, whereas female white-crowns do not. If singing behavior is dependent on the way in which the bird's brain is built, then it follows that male and female brains must differ in white-crowns. These brain differences could in theory arise because of genetic *or* environmental differences (or both) between the two sexes that influence how the birds' nervous systems develop.

No one doubts that male and female birds differ genetically. Male birds have two Z chromosomes, whereas females have a Z and a W chromosome. (This sex determination system differs from that of mammals, in which males have two different chromosomes, an X and a Y, while females have two X chromosomes.) Because chromosomes are where the genes are located, and because the avian W chromosome has many fewer genes than the Z chromosome, it follows that male and female birds are genetically distinct. These genetic differences are known to have large effects on development. An embryonic female white-crown, with her W and Z chromosomes, develops gonadal cells that will give rise to her ovaries. An embryonic male, with his different genes, develops differently, so that he eventually acquires sperm-producing testes, not egg-producing ovaries.

Thus, from a very early stage, male and female gonadal cells follow different developmental paths as their different genes interact with the compounds available for the construction of more cells and the assembly of the testes or ovaries. Moreover, gonadal cells in the two sexes also differ in the kinds of chemicals they produce for transport to other cells. In particular, cells in the young testes, but not in the pre-ovaries, manufacture the hormone testosterone. This hormone, as it travels through the male white-crown's developing body, becomes part of the chemical environment of other cells, including those in the male's developing brain. Testosterone can be picked up by cells that have the appropriate protein receptors in their nuclei, setting in motion a chain of chemical events that alters gene activity within the testosterone-sensitive cells. The result is the growth of special neural circuits that will eventually be used by singing males.

Note that the differences between the brains of male and female sparrows are the product of both genetic and environmental differences. A chromosomal (genetic) difference translates into a hormonal (environmental) difference, which leads to differences in the genetic activity of certain cells in the brains of males and females, which produce additional differences in the proteins manufactured by those cells, with follow-on (environmental) effects on other cells, and so on. When developmental biologists speak of development as an interactive process, they envision a cascade of alterations in genetic activity regulated by changes in the chemical environment of cells. (Remember that the genetic contribution to development is the information encoded in the organism's DNA; everything else constitutes the environmental contribution to development, including the cellular chemical products of preceding gene–environment interactions.)

As it turns out, the gonads are not the only, or even the most important, source of chemical signals needed for male brain development. Another key substance is the hormone estrogen. Although we usually think of estrogen as a female hormone (because it is produced by the ovaries), cells in male bird

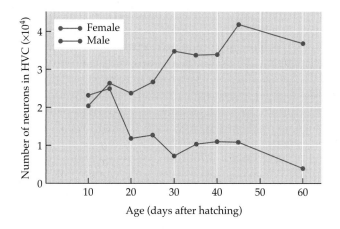

Figure 2.7   Changes in the song system of young male and female zebra finches. Between 10 and 40 days after hatching, the number of neurons in the female's HVC, a component of the song system, declines rapidly, whereas the number of HVC cells in males increases greatly. After Kirn and DeVoogd [644].

brains produce estrogen on their own for transport to other cells. There, this self-manufactured environmental signal activates the development of specific neural pathways found in the male's brain. These circuits link the elaborate network of neural structures, known as the song control system, whose integrated operation is required if the male is to sing his songs correctly [556].

The importance of estrogen for development of a male song control system has been demonstrated by testing the prediction that insertion of small estrogen pellets under the skin of a nestling female should result in the masculinization of her song control system. This experiment has been done with zebra finches, whose brains are anatomically similar to those of white-crowned sparrows. Normally, certain special clusters of brain cells that grow rapidly in size in young male zebra finches shrink over time in young females (Figure 2.7) [319, 644]. In immature females subjected to estrogen treatment, however, these song control units increase in size, as predicted, providing support for the hypothesis that a critical chromosomal–hormonal connection sets off a series of gene–environment interactions underlying the development of a male song control system in a zebra finch's brain [489].

Although in zebra finches both sexes listen to and memorize their species' song in preference to the songs of other species [129], only the male's brain eventually acquires the structures needed to produce zebra finch full song. To do so requires a whole series of highly coordinated changes in gene activity in the brain structures that make song learning and song production possible. And indeed, when geneticists have examined when particular genes are turned on and then off during the early stages of brain development in the zebra finch, they have found a special pattern that almost all males share (Figure 2.8).

Figure 2.8   The timing of gene activity in different components of the avian song control system in males. Different genes follow different schedules of activity over the course of development of the young zebra finch. The more intense the color, the more active the gene. Thus, for example, the gene that encodes the enzyme nitrous oxide (NO) synthase is moderately active in area X of the nestling's brain early in life, but production of that protein soon declines at this site. After Clayton [217].

Nor do changes in gene activity cease when the brain achieves its mature form. If a male zebra finch's or white-crowned sparrow's song control system is to serve as a song learning mechanism, certain cells within the system must in effect anticipate certain events. Thus, when a young bird is bombarded with sounds produced by singing adults of its own species, these sounds activate special sensory signals that get relayed to particular parts of the brain. In response to these distinctive inputs, some cells in these locations must then alter their biochemistry in order to change the bird's behavior. Biochemical changes typically require changes in gene expression (in which the information encoded in a gene is actually used to produce a product, such as an enzyme). So, for example, when a young white-crown hears its species' song, certain patterns of sensory signals generated by acoustical receptors in the bird's ears are relayed to song control centers in the brain where learning occurs. These sensory inputs are believed to alter the activity of certain genes in the set of responding cells, leading to new patterns of protein production that reshape those cells biochemically. Once they have been altered, the modified song control system can do things that it could not do before the bird was exposed to the song of other males of its species.

This theory of learning has been tested by examining cells in song control centers for changes in gene expression after a bird has heard a relevant song. You will recall that one acoustical stimulus that apparently affects the acquisition of song by white-crowned sparrows is the young bird's own songs. The sparrow's ability to hear itself sing when it is 150 to 200 days old is essential for the crystallization of a normal full song from the variable subsongs it initially sings. Likewise, male zebra finches go through a period early in life in which they appear to be matching elements in their initial subsongs against stored memories of the full songs they heard others singing previously. During this process, certain neurons in the finch's anterior forebrain become more and more responsive to the bird's own song as opposed to tutor songs; by the time the finch is an adult, with a fully crystallized song, certain of its auditory neurons have become highly selective, responding much more strongly to this song than to any other [340]. Presumably this developmental process involves a biochemical restructuring of the neurons in question, which in turn demands genetic information that can be turned on or off by specific environmental events.

We now know that as zebra finches attempt to match a tutor's song, the activity of a gene called *ZENK* rapidly increases in certain song control neurons, resulting in corresponding increases in cellular amounts of the protein encoded by that gene [817]. In other words, when a zebra finch listens to itself sing, it generates sensory feedback that activates a particular gene in certain cells. This gene's activity translates into the production of a specific protein, which is believed to have something to do with subsequent alterations in the neural circuits that control the finch's song (Figure 2.9). This hypothesis has received support from the observation that as the zebra finch gets closer and closer to singing an accurate copy of a tutor song, *ZENK* gene activity falls in a particular part of the brain [599]. As the finch gains a crystallized full song, further changes in cell architecture or biochemistry in the song control system cannot improve its ability to sing that song, and therefore the *ZENK* gene shuts down in the appropriate set of cells.

Studies like this one illustrate the tight link between the developmental and physiological levels of analysis of bird song. Changes in genetic activity in response to key environmental stimuli translate into changes in neurophysiological mechanisms that control the learning process. Therefore, an understanding of both developmental and neurophysiological systems is required to give us a full picture of the proximate causes of behavior.

Figure 2.9   Gene expression in a component of the zebra finch song system. The white and yellow area corresponds to area X. The brightness of this area is related to the high level of activity of the *ZENK* gene, which has resulted in the production of relatively large amounts of the protein coded for by this gene. The absence of white and yellow elsewhere in the brain indicates that the *ZENK* gene was not active in those areas. Photograph courtesy of David Clayton.

Discussion Question

2.3   Naturally, female starlings possess the same *ZENK* gene males do. In the female brain, the ventral caudomedial neostriatum, or NCMv, responds to signals sent to it from auditory neurons that fire when the bird is exposed to sounds, such as those made by singing male starlings. When captive female starlings are given a choice between perching next to a nest box where they can hear a long starling song versus perching next to another nest box where a shorter starling song is played, they spend more time at the long song site (Figure 2.10). What proximate hypothesis could account for the song preferences of female starlings? What prediction can you make about the activity of the *ZENK* gene in the NCMv of female starlings exposed to long versus short songs? How might you check your prediction? What would be the scientific point of collecting the data necessary to evaluate your prediction?

How the Avian Song Control System Works

Having identified some of the many influences of both genetic and environmental factors on the development of white-crowned sparrow singing behavior, let's move to another proximate level of analysis, one that focuses explicitly on the operating rules that characterize the brain of the bird. White-crowns, and other songbirds, have a brain with many anatomically distinct clusters of neurons, or **nuclei**, such as the just-mentioned NCMv, as well as neural connections that link one nucleus to another. The various components of the brain are made up of cells (neurons) that communicate with one another via bioelectric messages (action potentials) that travel from one neuron to another via elongate extensions of the neurons (axons) (see Figure 4.12). Some components of the brain are deeply involved in song learning, while others are necessary for the production of memorized song patterns. Figuring out which anatomical unit does what is a task for neurophysiologists more interested in the operational mechanics of the nervous system than in its development.

These researchers have long been especially interested in the higher vocal center (or HVC) in the brains of white-crowned sparrows and other songbirds. This dense collection of neurons connects to the robust nucleus of the

European starling

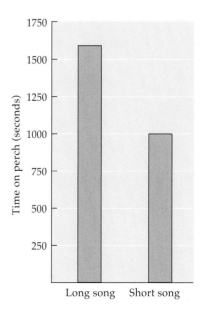

Figure 2.10   The song preferences of female starlings, as measured by their willingness to perch near a nest box from which a long starling song is played versus one from which a shorter song is played. After Gentner and Hulse [441].

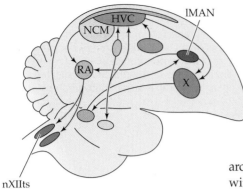

Figure 2.11   The song system of a typical songbird. The major components, or nuclei, involved in song production include the robust nucleus of the arcopallium (RA), the higher vocal center (HVC), the lateral portion of the magnocellular nucleus of the anterior nidopallium (lMAN), the caudomedial neostriatum (NCM), and area X (X). Neural pathways carry signals from the HVC to nXIIts to the muscles of the song-producing syrinx. Other pathways connect the nuclei, such as lMAN and area X, that are involved in song learning rather than song production. After Brenowitz et al. [143].

arcopallium (mercifully shortened to RA by anatomists), which in turn is linked with the tracheosyringeal portion of the hypoglossal nucleus (whose less successful acronym is nXIIts). This bit of brain anatomy sends messages to the syrinx, the sound-producing structure of birds that is analogous to the larynx in humans. The fact that the HVC and RA can communicate with the nXIIts, which connects to the syrinx, immediately suggests that these brain elements exert control over singing behavior (Figure 2.11).

This hypothesis about the neural control of bird song has been tested. For example, if neural messages from the RA cause songs to be produced, then the experimental destruction of this center, or surgical cuts through the neural pathway leading from the RA to the nXIIts, should have devastating effects on a bird's ability to sing. Experiments designed to test these predictions have been done with a variety of songbirds [204], with the result that we can now say with confidence that the RA does indeed play a critical role in song production. If this is true, then in bird species, like the white-crowned sparrow, in which males sing and females do not, the RA should be larger in male brains than in female brains, and it is (Figure 2.12) [52, 865, 886].

Other brain nuclei appear to be essential for song learning, rather than for the production of vocal signals. Destruction of the lateral magnocellular nucleus of the anterior nidopallium (lMAN), for example, does not strongly interfere with an adult zebra finch's ability to sing the song it has learned earlier in life, but if the operation is performed on a juvenile bird *before* it has acquired a mature song, then the bird will fail to sing a normal song in adulthood. Further evidence of the importance of this component of the control system for song learning came from checking the prediction that the lMAN should be much reduced or absent in bird species that sing, but do not learn their songs. There are many such species, and, as predicted, they lack the well-defined lMANs (and certain other forebrain nuclei) of white-crowned sparrows and other vocal learners [204].

Evidence that the HVC, as well as the lMAN, plays a role in song learning comes from a study of a group of closely related European warbler species that differ in the extent to which males learn a number of different songs. Among these species, the larger the song repertoire, the larger the HVC. In addition, individuals of the same species can differ in the size of their song repertoire. An analysis of all currently available studies on the relation between the volume of the HVC and the size of the male song repertoire within a species shows that overall, the two variables are significantly correlated: a male of a given species with a relatively large repertoire usually has a larger HVC than other less gifted members of his species [432].

In the studies just outlined, however, an assumption has been made that large amounts of neural tissue are required if the male is to learn a complex song or songs. But one could argue that the experience of singing a complex song or acquiring a large repertoire causes the HVC to expand in response to stimulation of this region of the brain. One way to see whether the HVC must be large if the bird is to learn or whether learning makes the HVC grow would be to create two groups of male songbirds, one that is permitted to shape its song reper-

Figure 2.12   Differences in the size of one nucleus of the song system, the robust nucleus of the arcopallium (RA), in (left) the male and (right) the female zebra finch. Photographs courtesy of Art Arnold; from Nottebohm and Arnold [886].

toire through learning and the other that is raised in acoustical isolation. If it is the learning experience that causes certain parts of brains to grow, then isolated males should have smaller HVCs than those that learn songs. This experiment has been done with sedge warblers; isolated males had brains that were in no way different from those of males that had learned their songs by listening to the songs of other males [705]. Thus, at least in this species, the male brain develops independently of the learning experiences of its owner, suggesting that the production of a large HVC is required for learning, rather than the other way around.

Although the neurophysiology of song learning has been explored at the level of entire brain nuclei, neuroscientists could in theory find out how a given neuron contributes to communication between birds. In fact, this kind of research has been done by Richard Mooney and his co-workers in a study of song perception by swamp sparrows [846]. Swamp sparrows sing two to five song types, each type consisting of a "syllable" of sound that is repeated over and over in a trill that lasts for a couple of seconds. If a young male swamp sparrow is to learn a set of song types, he must be able to discriminate between the types being sung by males around him, and then later must be able to tell the difference between his own song types as he listens to himself sing.

One mechanism that could help a young male control his song type output would be a set of specialized neurons in the HVC that respond selectively to a specific song type. Activity in these cells could contribute to his ability to monitor what he is singing so that he could adjust his repertoire in a strategic manner (by, for example, selecting a song type that would be particularly effective in communicating with a neighboring male, as we will see below). Of course, the contribution that any one cell makes to a behavioral decision is dependent on a host of other cells with which that neuron communicates. But the existence of song type specialists would help us understand how the component parts of various neural mechanisms contribute to song discrimination.

The technology exists to permit researchers to record the responses of single cells in a swamp sparrow's HVC to playbacks of that bird's own songs. In so

Figure 2.13   **Single cells and song learning** in the swamp sparrow. (Top) Sonograms of three song types: A, B, and C. (Bottom) When the three songs are presented to the sparrow, one of its HVC relay neurons reacts substantially only to song type B. Other cells not shown here respond strongly only to song type A, while still others fire rapidly only when song type C is the stimulus. After Mooney et al. [846].

doing, Mooney and his associates discovered a number of HVC relay neurons that generated intense volleys of action potentials when receiving neural signals from other cells upon exposure to one song type only [846]. Thus, one relay neuron, whose responses to three different song types are shown in Figure 2.13, produces large numbers of action potentials in a short period when the song stimulus is song type B. This same cell, however, is relatively unresponsive when the stimulus is song type A or C. So here we have a special kind of cell that could help the sparrow identify which song type it is hearing, the better to select the best response for a particular situation.

Different Songs: Ultimate Causes

Although a great deal has been learned about how the song control system of songbirds develops and operates, we still do not know everything about the underlying mechanisms of singing behavior. But even if we had all the proximate questions answered, our understanding of singing by white-crowned sparrows would still be incomplete until we dealt with the ultimate causes of the behavior. Because the proximate mechanisms underlying bird song did not emerge out of thin air, we can ask questions such as, when in the distant past did an ancestral bird species start learning its species-specific song, thereby setting in motion the events that led to dialects in birds like the white-crowned sparrow? Evidence relevant to this question includes the finding that song learning occurs in members of just three of the twenty-three avian orders: the parrots, the hummingbirds, and the "songbirds," which belong to that portion of the Passeriformes that includes the sparrows and warblers, among others (Figure 2.14) [142]. Members of the remaining twenty orders of birds produce complex vocalizations, but they do not have to learn how to do so, as shown in some cases by experiments in which young birds that were never permitted to hear a song tutor, or were deafened early in life prior to the onset of song practice, nevertheless came to sing normally (Figure 2.15).

One evolutionary question is, did the song learning that occurs within each of the three orders evolve independently of the other groups? The fact that the nearest relatives of each of the three song-learning orders do not learn their

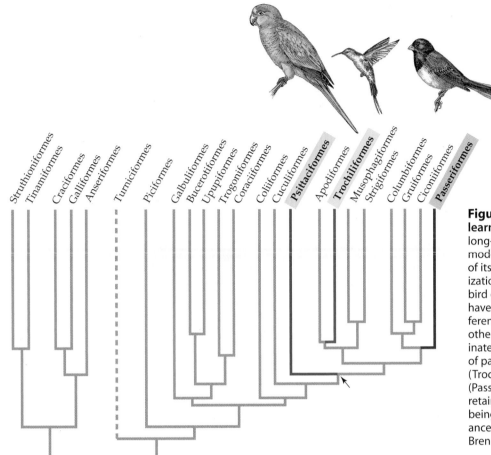

Figure 2.14   The phylogeny of song learning in birds. If we assume that the long-extinct bird that gave rise to all modern species did not learn elements of its songs, but instead produced vocalizations instinctively, as do many modern bird groups, then song learning must have evolved independently in three different lineages of modern birds. On the other hand, song learning may have originated in a common ancestor (see arrow) of parrots (Psittaciformes), hummingbirds (Trochiliformes), and passerine songbirds (Passeriformes), and may have been retained in these three lineages while being lost in other descendants of that ancestral song-learning species. After Brenowitz [142].

songs suggests one of two things: either song learning originated three different times in the approximately 65 million years since the first truly modern bird appeared, or song learning began in the ancestor of all the lineages from Psittaciformes to Passeriformes, but was lost subsequently in all these groups except the parrots, hummingbirds, and songbirds. The second scenario requires only three losses, if they occurred far enough back in time, at the base of evolutionary lines leading to (1) the Apodiformes, (2) the combined Musophagiformes and Strigiformes, and (3) the combined Columbiformes, Gruiformes, and Ciconiformes. In other words, either there have been three independent origins of song learning or three separate losses of the trait (see Figure 2.14). Because neither scenario requires fewer changes than the other, we cannot invoke the principle of parsimony to resolve this issue (the principle is used to give preference to the simplest scenario, namely the one that requires the fewest changes to move from an ancestral pattern to a more recent one.)

If the independent origins scenario is true, however, we would expect to find major differences among the song control systems of the three vocal learning groups. If we are to check this prediction, we need to identify the brain structures that endow parrots, hummingbirds, and songbirds with their singing abilities. One way to do so is to capture birds that have just been singing (or listening to others sing) and kill them immediately in order to examine their brains for regions in which a product of the *ZENK* gene can be found [590]. As noted above, this gene's activity is stimulated in particular parts of the brain when birds produce songs and in different parts of the brain when birds hear songs. Other stimuli and activities do not turn the gene on, so its product is absent or

Figure 2.15   The song of a vocal non-learner, the eastern phoebe, a member of part of the Passeriformes that does not need to hear others sing in order to produce a complete song of its own. Birds 1 and 2 were normal males, capable of hearing; their songs were recorded in the field. Birds 12 and 13 were males that were deafened early in life and so had not heard other phoebes calling, nor were they able to hear themselves sing. Learning apparently does not figure in the development of song in this species. After Kroodsma and Konishi [672].

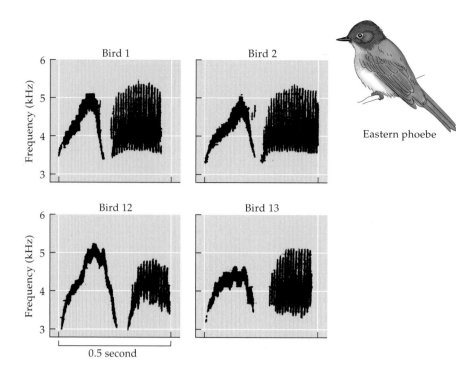

Eastern phoebe

very scarce when the birds are not communicating. By comparing how much ZENK protein appears in different parts of the brains of birds that were engaged in singing or listening to songs immediately before their deaths, one can effectively map those brain regions that are associated with bird song.

The *ZENK* activity maps of the brains of selected parrots, hummingbirds, and songbirds reveal strong similarities in the number and organization of discrete centers devoted to song production and processing. In all three groups, for example, cells that form the caudomedial neostriatum (NCM) activate their *ZENK* genes when individuals are exposed to the songs of others. The NCM is located in roughly the same part of the brain in all three groups, where it constitutes part of a larger aggregation of anatomically distinct elements that contribute to the processing of song stimuli. When parrots, hummingbirds, and songbirds vocalize, other brain centers, many in the anterior part of the forebrain, respond with heightened *ZENK* gene activity. Once again, there is considerable (but not perfect) correspondence in the locations of these distinctive song production centers in parrots, hummingbirds, and sparrows (Figure 2.16). The many similarities in brain anatomy among the three groups of vocal learners would seem to rule against the hypothesis that song learning abilities evolved independently in the three groups of birds. Thus, we have to take seriously the possibility that the ancestor of all those birds in the orders sandwiched between parrots and songbirds (see Figure 2.14) was a song learner, and that the mechanism for vocal learning possessed by that bird was then retained in some lineages while dropping out early on in others—a hypothesis that continues to generate debate [384, 590].

The Reproductive Benefits of Song Learning

Let us accept the possibility that in three bird lineages, neural mechanisms that promote song learning have been retained over evolutionary history, while this ability has been selected against in related groups. What fitness benefits resulted in the retention of song learning in parrots, hummingbirds, and songbirds? We can begin to tackle this question by asking another: What do birds gain by having a distinctive vocal message, whether it is learned or unlearned? One hypoth-

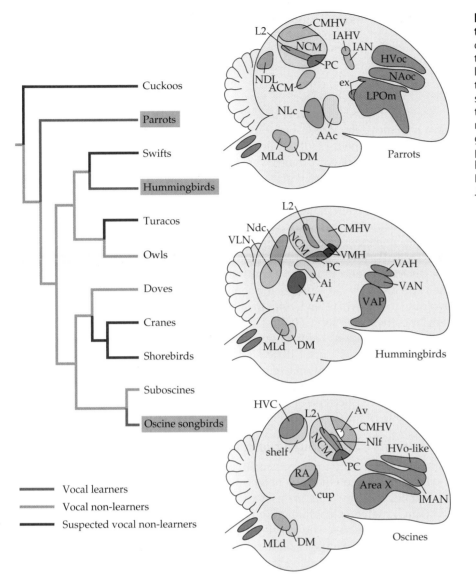

Figure 2.16 **The song control systems** of parrots, hummingbirds, and oscine songbirds are distributed throughout the brain in remarkably similar patterns. On the left is a diagram of the evolutionary relationships among some major groups of birds, including the three orders of vocal learners. On the right are diagrams of the brains of these groups, with the various equivalent components of the song control system labeled (e.g., HVC, higher vocal center; NCM, caudomedial neostriatum). After Jarvis et al. [590].

esis is that a distinctive vocalization conveys information about species membership, as we noted at the outset of this chapter. After all, if humans can tell a yellow warbler from an ovenbird by their songs, it is not unreasonable to assume that members of the two species of warblers can also make this acoustical discrimination. One benefit that might flow to males that sang differently from members of other species would be enhanced deterrence of competitors of their own species for territories and mates. Rival males of the same species pose the greatest threat to a singer's reproductive success because they may want to oust him from good breeding habitat or to lure his mate away from him. By broadcasting a clear message that a particular site or a particular female is defended by a physiologically fit singer, a male might encourage male conspecifics (members of the same species) to move on rather than engage in time-consuming conflicts with the songster.

Evidence in support of the species identity announcement hypothesis comes from experiments in which resident male white-throated sparrows (a relative of white-crowns) were removed from their territories, after which speakers were placed in the vacant territories. In half of the territories, the taped songs of the

White-throated sparrow

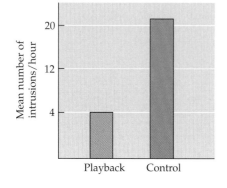

Figure 2.17   Does bird song repel territorial intruders? Territories from which resident male white-throated sparrows were experimentally removed attracted fewer intruders when the taped song of the removed male was broadcast from his vacant territory. After Falls [380].

former territory holders were broadcast from the speakers. New males were slower to move into territories with broadcast songs than those in which the speakers were silent [380] (Figure 2.17). Likewise, in paired experiments in which two song sparrow males were simultaneously removed from their territories, but only one was replaced with a taped song broadcast, the silent territory was always the first to be invaded by an intruder [889]. These results support the hypothesis that singers can gain by repelling other males of their species.

In much the same way, male singers able to communicate their species identity to females might attract mates more readily than those whose songs were less distinctive, and so less recognizable. Females that rapidly locate males of their species can begin reproducing sooner with a member of their own species, avoiding the risks of mating with members of another species. The mate attraction hypothesis for song distinctiveness correctly predicts that females will respond to taped songs of their own species more strongly than to taped songs of another species (Figure 2.18) [869]. If in the past female songbirds have been particularly strongly drawn to easily identified members of their species, then natural selection driven by mate choice would have resulted in the spread of the favored songs because of the clarity with which they announced the species membership of the singer.

The Benefits of Learning a Dialect

Although much more could be said about the reasons why so many bird species can be recognized on the basis of their song alone, this chapter revolves around why males *of the same species* may differ in the songs they sing. So let's shift our focus to why male white-crowns sing dialects that they have learned from others of their species. The puzzle here is that song learning has some obvious reproductive disadvantages for individual sparrows. Song learning takes time, energy, and special neural mechanisms, all of which could be devoted to other reproduction-enhancing activities. The fact that many—indeed, most—birds do not make these costly investments, but instead produce perfectly good songs instinctively, tells us that song learning is not essential for the acquisition of a species-distinctive signal. So we must identify some reproductive advantages for learning that are large enough to outweigh its costs if we are to account for the spread of the trait in a given songbird species.

Figure 2.18   White-crowned sparrow females are attracted to the songs of male white-crowned sparrows, but not to the songs of other species. In an experiment with female white-crowns, researchers temporarily removed their mates from their territories, then played two digital versions of white-crowned sparrow song (composed of elements taken from two different males) and vocalizations of song sparrows and dark-eyed juncos. The females (A) came closer to the speakers and (B) trilled more in response to the playbacks when white-crown songs were played. After Nelson and Soha [869].

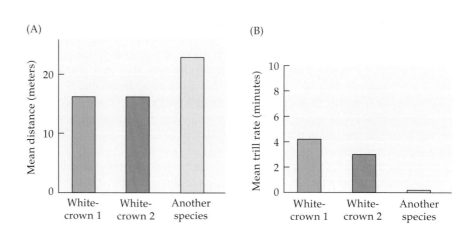

One possible benefit of song learning would be the ability of a young male to fine-tune his song so that it resembled the songs of one or more other individuals in a particular region. Fine-tuning could conceivably help the young male in several ways. First, imagine that males in a particular location have acquired a dialect that can be transmitted unusually effectively in that habitat. A young male that learns his song from his elders might then generate calls that travel farther and with less degradation than if he sang another dialect better suited to a different acoustical environment [204]. And in fact, males of the great tit do sing different versions of their species' song in different woodlands characterized by different foliage densities, yielding dialects that overcome the sound interference problems associated with particular habitats (Figure 2.19) [578].

A second hypothesis on the benefits of song learning centers on the advantages of matching songs to the singer's social environment. The idea here is that males able to learn the local version of their species-specific song can communicate better with rivals that will also be singing that particular learned song variant. If this is so, then we can predict that young males should learn directly from their territorial neighbors, using those individuals as social tutors. By adopting the song of a neighbor, a new boy on the block could signal his recognition of that male as an individual *and* demonstrate his capacity to learn a new song, which could be based on his health, body condition, or other indicators of competitive ability. Males that are evenly matched competitors have something to gain by accepting each other's presence rather than engaging in fruitless and expensive aggression. Thus, a mutual tolerance pact based on learned song signals could benefit both an established resident and a rival newcomer by helping them conserve time and energy.

Note that this hypothesis leads to the prediction that males should be able to fine-tune their song even after settling on an initial version relatively early in life. This prediction has been checked by examining the songs sung by young male white-crowns before and after they acquired their first territory. As predicted, males in at least some populations did change what had appeared to be their crystallized full song in order to more closely match the song of a neighbor [63, 87] (but see [867]).

Figure 2.19 **Songs match habitats.** Great tits from dense forests produce pure whistles of relatively low frequency, whereas males of the same species that live in more open woodlands use more and higher sound frequencies in their more complex songs. After Hunter and Krebs [578].

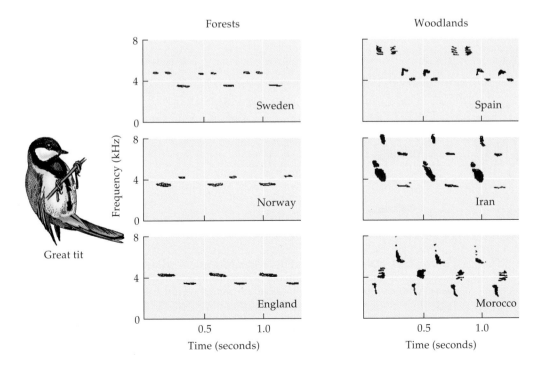

2.4   In the greater spear-nosed bat, unrelated females roost together in groups. The members of any given group tend to produce similar calls, but different groups have different calls. If you transfer some young bats into a new group and, at the same time, move some of their siblings into still another group, the shifted bats eventually change their calls to resemble those in their new groups [128]. What proximate hypothesis and what ultimate hypothesis on the acquisition of acoustical communication are supported by these results?

2.5   At the proximate level, two hypothesis have been advanced to account for the ability of some yearling white-crowned sparrows to match the dialect of neighboring males. The late acquisition hypothesis states that yearlings have a developmental window similar to that of very young birds, which enables them to listen to and learn directly from their immediate neighbors, overriding (if need be) any dialect learning that took place earlier in life. In contrast, the selective attrition hypothesis argues that fledglings can memorize a number of dialect versions of their species song early in life; then, after settling next to some older males, these birds gradually discard certain of their learned variants until they are left with the one that best matches the dialect of their neighbors. In light of these two alternative hypotheses, what significance do you attach to Figure 2.20? Outline the entire scientific process here, from the question that motivated the work to the scientific conclusion.

The ability of males to produce the songs, or at least song elements, of rival neighbors is also characteristic of the aptly named song sparrow, a bird that has a repertoire of many different, distinctive song types. (White-crowned sparrows typically sing just one song type, with different populations having their own dialects. Song sparrows sing many song types, with less sharply defined dialects in different regions.) Young song sparrows usually learn their songs from tutors that are their neighbors in their first breeding season, with their final repertoire tending

Figure 2.20   Dialect selection by male white-crowned sparrows. (A) After arriving on their first territories, yearling males FCN33 and PPM initially sang two different dialects, but after a time they each settled on one song type (the upper of the two song types shown for each male in the left-hand part of the figure). The song type they came to use matched that of their neighbor(s) (MWR for FCN33; FST17 and MBY for PPM). (B) Initially, male FCN33 mostly sang one of his two song dialects, but soon after his arrival on a territory near MWR, he stopped singing that song in favor of the one that matched his neighbor's song. After Nelson [868].

to be similar to that of an immediate neighbor [884]. In fact, Michael Beecher and his colleagues found through playback experiments that when a male heard a tape of a neighbor's song coming from the neighbor's territory, he tended to reply to that tape by singing a song from his own repertoire that matched one in the repertoire of that particular neighbor. This kind of *type matching* occurred when, for example, male BGMG heard the taped version of male MBGB's song type A (Figure 2.21), which led him to answer with his own song type A.

Another kind of response involves *repertoire matching*, in which a bird exposed to a song type from a neighbor responds not with the exact same type, but with a song type drawn from their shared repertoires. Repertoire matching would occur if male BGMG answered back with song type B or C when he heard song type A from male MBGB's territory (see Figure 2.21).

A third option for a male song sparrow is to reply to a song from a neighbor with an unshared song. This kind of mismatched response would be illustrated by male BGMG singing song type D, F, or H (see Figure 2.21) upon hearing male MGBG's song type A.

Song sparrow

Figure 2.21 Song type matching in the song sparrow. Males BGMG and MBGB occupy neighboring territories and share three song types (A, B, and C: the top three rows of spectrograms); six unshared song types (D, E, F, G, H, and I) appear on the bottom three rows. After Beecher et al. [81].

The fact that song sparrows type-match and repertoire-match so often indicates that they recognize their neighbors and know what songs they sing, and that they use this information to shape their replies [80]. But how do male sparrows benefit from their selection of a song type to reply to a neighbor? Perhaps their choice enables them to send graded threat signals to their neighbors, with type matches telling the targeted receiver that the singer is highly aggressive, whereas replying with an unshared song could signal a desire to back off, while a repertoire match might signal an intermediate level of aggressiveness. If so, then playback experiments should reveal differences in the responses of territorial birds to taped songs that contain a type match, a repertoire match, or neither. When Michael Beecher and various colleagues checked this prediction, they found that tapes containing a type match did indeed elicit the most aggressive response from a listening neighbor, while repertoire match songs generated an intermediate reaction, and nonshared song types were treated less aggressively still (Figure 2.22) [78, 176]. Thus, being able to learn to match a neighbor's song is apparently a way in which a male can convey information about his willingness to stand and fight. Males that are really willing to keep a contest going match the song types of their opponents; those that wish to avoid conflict sing a nonshared song type at a neighbor.

If this ability to modulate challenges is truly adaptive, then we can predict that male territorial success, as measured by the duration of territorial tenure, should be a function of the number of song types that a male shares with his neighbors. Song sparrows can hold onto their territories for up to eight years, which tends to create a relatively stable community of neighbors. And, in fact, the number of years that an individual male song sparrow holds his territory increases about fourfold as the number of song types shared with his neighbors goes from fewer than five to more than twenty [79]. This finding suggests strongly that the potential for neighbors to form long-term associations in song sparrow communities selects for males able to advertise exactly what their aggressive intentions are toward particular individuals. If so, then in the fluid social settings of certain other species in which territorial males come and go quickly, song learning and sharing of the sort exhibited by song sparrows should not occur. Instead, under these different ecological circumstances, males should acquire songs characteristic of their species as a whole to facilitate communication with any and all conspecifics, rather than developing a dialect or a set of shared song types characteristic of a small, stable population.

Donald Kroodsma and his colleagues have tested this proposition by taking advantage of the existence of two very different populations of the sedge wren. In the Great Plains of North America, sedge wren males are highly nomadic, moving from breeding site to breeding site throughout the summer. In this population, song matching and dialects are absent; instead, the birds improvise variations on songs heard early in life as well as inventing entirely novel songs of their own, albeit employing the general pattern characteristic of their species

Figure 2.22 Song matching and communication of aggressive intent in the song sparrow. Song sparrow males can control the level of conflict with a neighbor by their selection of songs. When a focal male sings a shared song at a rival, the neighbor has three options, one that will escalate the contest, one that keeps it at the same level, and another that de-escalates the interaction. Likewise, the initiator of the contest can use his ability to select a matching song type, a repertoire match, or an unshared song type. The three different kinds of songs convey information about the readiness of the singer to escalate or defuse an aggressive encounter. After Beecher and Campbell [79].

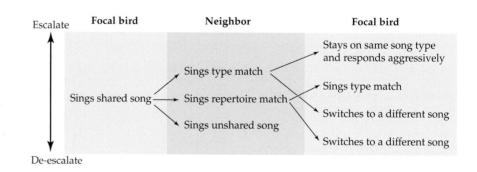

[673]. In contrast, sedge wren males living in Costa Rica and Brazil remain on their territories year-round. In these areas, dialects and song matching are predicted to occur, and these learning-proficient males do have dialects and do sing like their neighbors [675]. The differences in song learning that occur within this one species provide support for the general argument that learning and dialects arise when males gain by communicating specifically with other males that are their long-standing neighbors.

Female Preferences and Song Learning

Still other ultimate hypotheses on song learning take aim at the social environment provided by females. There can be little doubt that females are the intended recipients of at least some bird songs. When, for example, a male Cassin's finch has been close to a female for a day and she disappears, the number of songs he sings and the time he spends singing increase dramatically compared with when she was around (Figure 2.23). It seems certain that the male is attempting to communicate with a potential mate in an effort to get her back.

But could the fact that male Cassin's finches learn their songs contribute to a male's ability to attract a female back to him? Consider a species that is divided into stable subpopulations. In such a species, males in each of these groups are likely to have genes that have been passed down for generations by those males' successful ancestors. Therefore, by learning to sing the dialect associated with their place of birth, males could announce their possession of traits (and underlying genes) well adapted for that particular area. Females hatched in that area might gain by having a preference for males that sang the local dialect because they would endow their offspring with genetic information that promotes the development of locally adapted characteristics [53].

This hypothesis received some support from the discovery that male white-crowned sparrows that sang the local dialect around Tioga Pass, California, were less infected with a blood parasite than non-dialect singers. This finding means that female white-crowns in this population could potentially use song information to secure healthier mates. And, in fact, males with songs not matched to the local dialect fathered fewer offspring than local dialect singers, a result consistent with the hypothesis that females prefer to mate with males that sing the local dialect [761].

On the other hand, if female preferences are designed to get them to mate with males produced in their natal area, then female white-crowns should prefer males with the dialect that they heard while they were nestlings—namely, their father's dialect—and they do not, at least in one Canadian population [210]. Furthermore, young male white-crowns are not locked into their natal dialects, but can change them [321] if they happen to move from one dialect zone to another; therefore, female white-crowns cannot rely on a male's dialect to identify his birthplace with complete certainty, all of which casts doubt on the proposition that female song preferences enable them to endow their offspring with locally adapted gene complexes from locally hatched partners.

Another version of the female choice hypothesis for male song learning is based on the possibility that females look to the learned details of a potential mate's song for information about his developmental history. If a female could tell just by listening to a male that he was unusually healthy, she could acquire a mate whose genes had worked well during his development and so should be worth passing on to her offspring. For example, in zebra finches, males with more complex songs have a larger than average HVC [5, 6], a finding reminiscent of one mentioned earlier—the correlation between song repertoire size and HVC size among related European warblers. Since HVC size is affected by

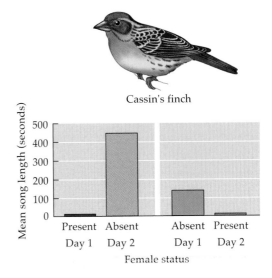

Cassin's finch

Figure 2.23 Evidence that male Cassin's finches direct their songs at females. When a male is paired off with a female on day 1, he sings relatively little. But when the female is removed on day 2, the male invests considerable time in singing, presumably to call her back. If the male is caged without a female on day 1, however, he allocates little time to singing, and he sings even less when a female is introduced into his cage, indicating that it is the loss of a potential mate that stimulates the male to sing. After Sockman et al. [1134].

heredity, female zebra finches that prefer a complex song may mate with males whose genes could give their male offspring a competitive edge.

Alternatively, a would-be mate that had grown up under especially good conditions might be physiologically advantaged, and so able to provide his offspring with above-average parental care. In the great reed warbler, well-fed nestlings have larger learned song repertoires as first-year adults than individuals that have been nutritionally stressed in the nest [887]. Thus, superior songsters may be stronger, healthier individuals capable of offering their offspring better paternal care than the average male in their population. This pattern is also seen in the sedge warbler, a reed warbler relative, whose males learn their song repertoires and whose females prefer males with larger repertoires. Katherine Buchanan and Clive Catchpole have shown that these song-rich males bring more food to their offspring, which grow bigger—a result that almost certainly raises the reproductive success of females that find large song repertoires sexually appealing, which in turn selects for males that are able to sing in the favored manner [167].

But the question still remains, why might female choice favor males that *learn* their complex songs, rather than simply manufacturing an innately complex song with no learned component? One hypothesis is that the *quality* of vocal learning could give female songbirds a valuable clue about the quality of the singers as potential mates. The key point here is that song learning occurs when males are very young and growing rapidly. If rapid growth is difficult to sustain, then young males that are handicapped by genetic defects or nutritional stress should be unable to keep up, resulting in suboptimal brain development. According to this hypothesis, individuals with even slightly deficient brains may be less able to meet the demands of learning a complex species-specific song [888].

These predictions have been tested in experiments with swamp sparrows. By bringing very young nestling males into the laboratory, Steve Nowicki and his co-workers were able to control what they were fed. One group of nine males (the controls) received as much food as they could eat, while another group of seven (the experimentals) received 70 percent of the food volume consumed by the control males. During the first 2 weeks or so, when the sparrows were totally dependent on their handlers for their meals, the controls came to weigh about a third more than the experimentals, a difference that was then gradually eliminated over the next 2 weeks as the sparrows came to feed themselves on abundant seeds and mealworms. Even though the period of nutritional stress was brief, this handicap had large effects on both brain development and song learning (Figure 2.24). Components of the song system in the food-deprived experimental group were significantly smaller than the equivalent regions in the control sparrows. Likewise, the deprived birds came to sing poorer copies, compared with the controls, of the taped song that both groups listened to during their early weeks of captivity [890]. These results, and similar ones from other studies [1144], confirm the possibility that adult female songbirds might learn something about the developmental history, and thus the quality, of potential mates by listening to their songs.

But do female birds actually pay attention to the information about mate quality potentially encoded in male songs? Nowicki's team again supplied supporting evidence by playing male song sparrow songs to female sparrows in the laboratory [891]. Some of the songs had been more accurately copied from tape tutors than others by males of diverse learning abilities. Female sparrows can be primed to respond with a tail-up precopulatory display to male songs they find sexually stimulating by giving them hormone implants. The more accurately copied songs elicited significantly more precopulatory displays from the females than those that were copied less perfectly (Figure 2.25).

(A)

Swamp sparrow

(B)

Figure 2.24 Nutritional stress early in life has large effects on both (A) brain development, as expressed in the volume of the HVC, and (B) song learning, as measured by the match between learned songs and tutor tapes. Nestling swamp sparrows were hand-reared and exposed to tutor tapes. The control group was fed all they could eat, whereas the experimental group received only 70 percent of that amount for about 3 weeks. After Nowicki et al. [890].

(A) Tape tutor song

(B) Good copy of (A)

(C) Tape tutor song

(D) Poor copy of (C)

1 second

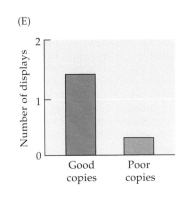

(E)

Figure 2.25   The mean number of precopulatory displays given by female song sparrows in response to playback of the songs of males that had been able to copy their tutor songs very accurately and of males of lower copying skill. The upper panel shows (A) a tape tutor song and (B) a copy sung by a male able to copy a high proportion of the notes in the taped song it listened to; the bottom panel shows (C) a tape tutor song that was (D) poorly copied by another individual. (E) The females' response to good and poor song copies. After Nowicki et al. [891].

Many studies of the hypothesis that female choice drives the evolution of song learning by male songbirds have involved laboratory manipulations. One always has to worry that these experiments are somewhat artificial and so not relevant to what happens in real life. However, one can say that the laboratory results gathered thus far do provide support for the hypothesis that those young male song sparrows able to copy the songs of others gain a reproductive advantage because receptive females can and do analyze a male's capacity for song learning when they listen to his vocalizations.

Discussion Question

2.6   William Searcy and a team of researchers played taped songs to captive female song sparrows that had been given hormone implants shortly after being taken to the laboratory [1078]. The recorded songs came from male song sparrows that lived in the population from which the females had been taken as well as from males living various distances (18, 34, 68, 135, and 540 kilometers) from the female subjects. Songs from males living 34 or more kilometers from the populations from which the females came were not nearly as effective in eliciting the precopulatory display as songs from local males; in contrast, songs from males living only 18 kilometers away were about as sexually stimulating as local songs. These data have relevance for more than one ultimate hypothesis on song learning by male sparrows. What are the hypotheses, and what importance do these findings have for them?

Proximate and Ultimate Causes Are Complementary

We have reviewed only a small portion of what is known about learning and dialects in bird song, but we have covered enough to illustrate the links between the proximate and ultimate causes of behavior. If we want to understand why male white-crowns from Marin and Berkeley sing somewhat different songs, we must understand how the song control mechanisms develop. Male white-crowns differ genetically from females, a fact that guarantees that different gene–environment interactions will take place within embryonic male and female birds. The cascading effects of differences in gene activity and in the protein products that result from gene–environment interactions affect the assembly of all parts of the bird's body, including his brain and nervous system. The way in which a male white-crown's brain, with its special subsystems, can respond to experience is a function of the way in which the male has developed.

Figure 2.26   What causes differences among individuals? The development of an individual's traits depends on the interaction between the genetic information inherited from its parents and the nutrients, chemical substances, and experiences derived from its environment. Therefore, differences among individuals can be caused by differences in either their genes or their environments, or both. The individual differences resulting from these proximate causes have the potential to affect the evolution of a species.

Process repeats itself

↑

Differences in gene–environment interactions in individuals

↑

Genes transmitted to the next generation

↑

Differences in reproductive success = Natural selection

↑

Song differences among individuals
Species differences in song (which may affect mate choice)
Sex differences in song (which may affect mate choice)
Dialect and repertoire differences (which may affect territory defense)

ULTIMATE CAUSES

↑

Differences in song system operation

↑

Differences in song system construction

↑

Hormonal differences

↑

PROXIMATE CAUSES

Differences in gene–environment interactions

By the time a male sparrow has become an adult, he possesses a large, highly organized song control system whose physiological properties permit him to produce a very particular kind of song. The learned dialect he sings is a manifestation of both his developmental history and the operating rules of his brain.

The proximate differences among white-crowns that generate differences in their dialects have an evolutionary basis. In the past, in the lineage leading to today's white-crowns, males have differed in their song control systems and thus in their behavior. Some of these differences have been hereditary, which means that natural selection has had an opportunity to work. Some males have surely been better than others at announcing their desirability as mates or their capacity to cope with rival males. These males' superior behavioral skills have translated into greater contributions of their genes to the next generation, where those genes have been available to participate in interactive developmental processes within members of that generation. The hereditary attributes of those animals will in turn be measured against one another today in terms of their ability to promote genetic success in the current round of selection. This selective process links the proximate bases of behavior with their ultimate causes in a never-ending spiral through time (Figure 2.26).

Discussion Question

2.7   What features of language learning in humans are similar to song learning in birds? Do these similarities suggest certain hypotheses on the proximate bases of human language learning, especially the genetic and developmental components?

Do comparisons with birds also suggest some interesting hypotheses on the adaptive value of learned language for members of our species? *After* you have attempted to answer these questions, go to the ISI Web of Science, if it is available at your college or university library, and try to find references on the shared proximate and ultimate causes of vocal learning by humans and birds. You may find it useful to know that those who have written knowledgeably on this subject include Peter Marler and Fernando Nottebohm.

Summary

1. Different species of birds sing different songs, and even within a single species, individuals may vary in how they sing their species' special vocalization. These features of bird song can be subjected to both proximate and ultimate analyses.

2. Some birds learn their species' song, which may lead to geographic differences in the songs sung by members of the same species. The underlying developmental processes of song learning and dialect formation are dependent on both genetic information and environmental inputs, which include the bird's acoustical and social experiences as well as the proteins and other chemical constituents of its brain and other body parts.

3. In addition to focusing on how the mechanisms of behavior are assembled during development, proximate analyses include attempts to understand the operating rules of these mechanisms. Birds that learn their songs possess an elaborate song control system whose component parts contribute to early learning of a song or songs, which are then copied when the learners begin to practice singing.

4. The ultimate causes of behavior can also be explored on two levels. Thus, understanding why some birds learn a regional dialect can lead us to explore the historical changes that occurred during the evolution of vocal learning in hummingbirds, parrots, and some passeriform songbirds. Another layer of evolutionary analysis of learned bird song involves studying why this means of song acquisition was favored by natural selection. Much evidence now exists that learned songs may enable individual males of some species to target signals to conspecific rival males as well as to potential mates, which may acquire information on the developmental history of the singer by listening to how well he sings.

5. In the past, any heritable variant mechanism of song learning that helped individuals leave more copies of their genes would have become more common over time, thanks to the action of natural selection. As certain forms of genes became more common, certain developmental outcomes would likewise have predominated, leading to the transformation of the song control systems in the brains of a songbird species under selection. Changes in the song control mechanisms would have consequences for how a songbird acquired its song, which in turn could affect its ability to reproduce successfully. Proximate and ultimate causes of behavior are thus linked together in a never-ending cycle.

Suggested Reading

Bird Song: Biological Themes and Variations [204] is a very readable review of the proximate and ultimate causes of bird song. See also the more technical *Ecology and Evolution of Acoustic Communication in Birds* [674]. A full analysis of all forms of animal communication has been written by Jack Bradbury and Sandra Vehrencamp [133]. Many excellent papers have been written on bird song, among them [80, 868, 890].

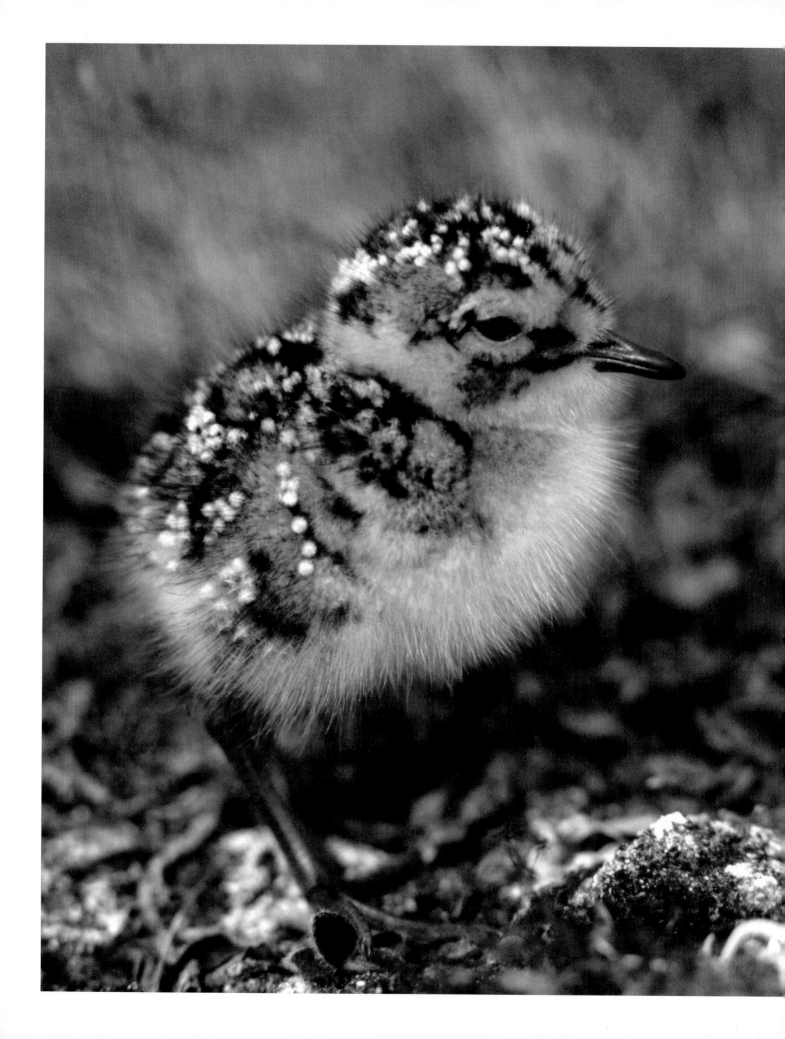

3 | The Development of Behavior

When you and I listen to a white-crowned sparrow's song, we are hearing the product of an extraordinarily long and complex developmental process. As the previous chapter indicated, behavioral development is influenced both by the genetic information the bird possesses in its DNA and by a host of environmental influences, ranging from the nongenetic materials in the egg yolk, to the hormones that certain of the bird's cells manufacture and transport to other cells, to the sensory signals generated when a baby sparrow hears its species' song, as well as the neural activity that occurs when a young adult male interacts with a neighboring territorial male.

The sparrow example highlights an extremely important concept; namely, that *development is an interactive process* in which genetic information interacts with changing internal and external environments in ways that assemble an organism with special properties and abilities. The process occurs because some of the genes in the organism's cell nuclei can be turned on or off by the appropriate signals, which are ultimately derived from the external environment. As genetic activity changes within an organism, the chemical reactions within its cells change, building (or modifying) the proximate mechanisms that underlie an organism's characteristics and capacities.

◀ *This recently hatched sanderling's growth and behavioral development will be influenced both by its genes and by its environment. Photograph by Bruce Lyon*

The fact that the development of any attribute of any living multicellular organism is dependent on both genes and environment means that no trait—not one—is "genetic" as opposed to "environmental," nor is any attribute environmentally determined in the sense of developing without genetic input. This claim is counterintuitive to many who want to divide the features of living things into those that are caused by "nature," the so-called genetically determined traits, versus those that are caused by "nurture," the so-called environmentally determined traits. The nature-versus-nurture approach is strongly embedded in the popular view of animal behavior, in which instincts are often said to be genetic, or more strongly affected by hereditary factors, than is learning, which is commonly believed to be largely or entirely due to the animal's environment. A main goal of this chapter is to address this misconception head on. Having stressed the importance of gene–environment interactions for development, we will then examine evidence that both genetic and environmental *differences* among individuals can lead to differences in development, which in turn can produce differences in how individuals behave. This point has great significance for an understanding of behavioral evolution, a subject that will be covered here with examples of how aspects of the developmental process promote the reproductive success of individuals.

The Interactive Theory of Development

Although most people think of insects as fairly boring automatons with only a limited set of basic instincts, in fact many insect species possess extremely sophisticated behavioral abilities. Take the worker honey bee, for example. These little creatures spend their lives helping their fellow hivemates. Although workers leave the egg laying to their mother queen bee (because workers are largely sterile), they are responsible for care of the larvae that hatch from their mother's eggs, the construction of honeycomb, the regulation of the hive's temperature, the defense of the colony against parasites and predators, and, of course, the collection of the pollen and nectar that their colony needs to survive and grow.

One of the many fascinating things about bee behavior is that workers go through a regular progression of occupations. When a worker emerges from a brood cell in the waxy comb tended by other workers, her first job is a humble one, the cleaning of comb cells. She then becomes a nurse bee that feeds honey to larvae in the brood comb before making the transition to a distributor of food to her fellow workers. The last phase of her life, which begins when she is about 3 weeks old, is spent foraging for pollen and nectar outside the hive (Figure 3.1) [728].

So what causes a worker to go through these different developmental stages? According to the interactionist approach that we outlined in the preceding chapter, the information in some of the bee's many thousands of genes (the bee's **genotype**) must respond to the environment in ways that influence the development of her measurable characteristics (the bee's **phenotype**), which include the proximate mechanisms underlying her behavior, such as her nervous system, and her behavioral traits as well. One result of these genotype–environment interactions is the standard shift in the behavioral phenotype of workers, with 3-week-old adults no longer feeding stored honey and pollen to larval bees, but instead flying from the hive to gather nectar and pollen from surrounding fields and woods. A developmental biologist might ask, does the active component of a worker bee's genotype change in predictable ways between her nurse phase and her forager phase (just as gene activity changes sequentially as a young nestling zebra finch grows older and its song system develops—see Figure 2.8)?

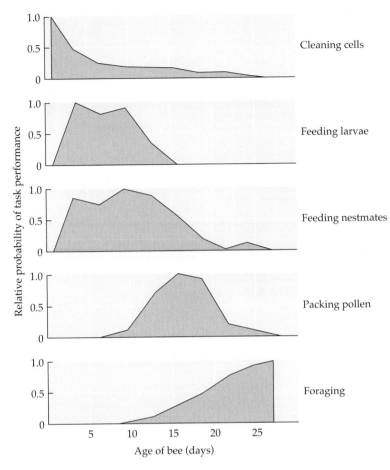

Figure 3.1 Development of worker behavior in honey bees. The tasks adopted by worker bees are linked to their age, as demonstrated by following marked individuals over their lifetimes. After Seeley [1080].

This question can now be answered, thanks to a fairly recent development in molecular biology called microarray technology, which makes it possible to look for relative levels of activity in a large set of genes by detecting the products (messenger RNAs) made when those genes have been "turned on." The microarray method uses a fluorescent labeling trick that enables researchers to scan a sheet on which tissue extracts have been spread and to see which gene products are abundant and which are not. When Charles Whitfield and his coworkers ran brain extracts from nurses and foragers through a variety of microarrays, they were able to compare the activity of about 5500 genes (of the roughly 14,000 in the bee genome) for these two kinds of individuals [1282]. Some of the genes that were turned on in nurse bees' brains differed substantially and consistently from those active in foragers' brains, and vice versa. Indeed, about 2000 of the surveyed genes showed different levels of product output in the two kinds of bees. The expression of the 17 genes whose activity differed most between nurses and foragers is shown in Figure 3.2: changes in genetic activity are correlated with the transformation of a nurse into a forager.

To make this claim is not to downplay the developmental role of the environment. For one thing, molecular building blocks are essential for making gene products; namely, the messenger RNAs and the proteins coded for by the DNA. The requisite building blocks for these critically important constituents of living things are chemicals present in the cellular environment. These compounds ultimately come from substances consumed by the queen prior to making her eggs, as well as from the honey and pollen eaten by the larvae and adults that develop from those eggs. Some of the resultant gene–environment products

Equivalent *Drosophila* gene	Expression ratio (F/N)	Typical colonies		Single-cohort colonies				Putative function
		YN	OF	YN	YF	ON	OF	
fax	0.63							Axonogenesis
fax	0.65							
BM-40-SPARC	0.52							Cell adhesion
GlyP	0.65							Glycogen phosphorylase
GlyP	0.72							
CG7322	0.70							Oxidoreductase
Rab10	0.77							Rho small monomeric GTPase
CG32703	2.45							MAP kinase
Eip71CD	1.86							Methionine sulfoxide reductase
Tps1	1.74							Trehalose-6-phosphate synthase
Tps1	1.55							
CG11334	1.80							Translation regulator
CAH1	2.35							Carbonic anhydrase
Inos	1.98							Inositol-3-phosphate synthase
CG5966	1.34							Triacylglycerol lipase
HLH3B	1.43							Transcription factor
U2af50	1.18							Pre-mRNA splicing factor

Reference code (upper right): >2, 1.5, 1.25, 1, 0.8, 0.67, <0.5

Figure 3.2 Gene activity varies in the brains of nurse bees and foragers. Shown here are individual records (each bar represents one bee's gene expression record for a particular gene) of bees from typical and manipulated colonies. The 17 genes surveyed here were the ones that showed the largest difference in activity between the brains of nurses and of foragers. In addition, these 17 genes are highly similar to genes found in fruit flies (*Drosophila*); the function of the corresponding gene in fruit flies has been established and is noted here. The activity level of a gene in forager relative to nurse bee brains is color-coded from high (>2) to low (<0.5); see the reference code at the upper right. The left-hand portion of the figure shows gene activity for young nurses (YN) and old foragers (OF) from unmanipulated colonies; the right-hand portion shows those for young foragers (YF) paired with nurses of the same age and old nurses (ON) paired with foragers of the same age taken from colonies manipulated to contain either all young or all old workers. The expression ratio (F/N) is calculated by dividing the activity score of a given gene in foragers by that for nurses. After Whitfield et al. [1282].

may play a special role in changing the activity of one or more key genes in an individual, initiating a cascading series of gene–environment outputs that eventually alters the behavior of the bee.

One particularly important developmental product appears to be a substance called juvenile hormone, which is found in low concentrations in the blood of young nurse workers, but in much higher concentrations in older foragers. As one might predict, if young bees are treated with juvenile hormone, they become precocious foragers [1023], but if one removes a bee's corpora allata (the glands that produce juvenile hormone), the bee delays its transition to foraging. Moreover, bees without corpora allata that receive hormone treatment regain the normal timing of the switch to foraging [1162].

So it appears that changes in juvenile hormone production have something to do with getting the ball rolling in ways that lead to a shift in behavior. But what causes the gene that codes for juvenile hormone to boost its output when the worker is about 3 weeks old? Part of the answer surely lies in the preceding sequence of gene–environment interactions that have taken place since the worker bee metamorphosed into an adult. But another part has to do with the current social environment of the worker, as shown by research in which experimental colonies were formed with a worker force consisting of uniformly young bees of the same age. Under these unusual conditions, a division of labor still manifested itself, with some individuals remaining nurses much longer than usual, while others began foraging as much as 2 weeks sooner than average.

What enabled the bees to make these developmental adjustments? One hypothesis is that a deficit in social encounters with older foragers may have stimulated an early developmental transition from nurse to forager behavior. This possibility has been tested by adding groups of older foragers to experimental colonies made up of only young workers. The higher the proportion of added older bees, the lower the proportion of young nurse bees that undergo an early transformation into foragers (Figure 3.3) [572]. As it turns out, the behavioral interactions between the young residents and the *older* transplants inhibit the development of foraging behavior, because transplants of *young* bees have no such effect on young resident bees. The inhibiting agent has been traced to a fatty acid compound called ethyl oleate, which only foragers manufacture and store in a chamber (the crop) off the digestive tract [711]. When returning foragers pass nectar contained in the crop to nurses back at the hive, ethyl oleate

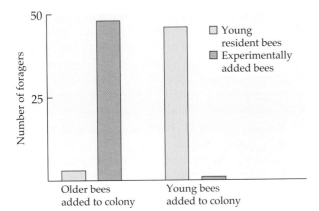

Figure 3.3 **Social environment and task specialization** by worker honey bees. In experimental colonies composed exclusively of young workers (residents), the young bees do not forage if older forager bees are added to their hive. But if young bees are added instead, the young residents develop into foragers very rapidly. After Huang and Robinson [572].

is probably transferred as well. The more foragers in a hive, the more likely nurses are to receive quantities of this chemical, which slows their transition to foraging status.

Discussion Question

3.1 Honey bees possess a gene (labeled *for*) that contains information for the production of a particular enzyme called PKG. If this genetic information is important for foraging activity in worker honey bees, what prediction follows about the levels of PKG enzyme present in the heads of foragers versus nonforagers taken from a typical bee colony composed of workers of many different ages? Figure 3.4 presents data on the quantity of the messenger RNA coded by *for* that is present in the brains of nurses and foragers in three typical honey bee colonies; this RNA is required for the production of PKG. Are these results consistent with your prediction? But since foragers are older than nurses in typical colonies, perhaps the greater activity of *for* is simply an age-related change that has nothing to do with foraging. What additional prediction and experiment are required to reach a solid conclusion about the causal role of *for* in regulating foraging in the honey bee? Suggestion: Take advantage of the ability to create experimental colonies of same-age workers. If the *for* gene contributes to the switch to foraging by worker bees [88], how can this change be influenced by the workers' social environment as well? Can you produce a hypothesis that integrates the genetic and environmental contributions to the switch in bee behavior?

The Nature or Nurture Fallacy

The honey bee example clearly illustrates why it would be a mistake to say that some behavioral phenotypes are more genetic than others. Bee foraging behavior cannot be purely "genetically determined" because the behavior is the product of literally thousands of gene–*environment* interactions, all of which are required to construct the bee's brain and the rest of its body. Indeed, the infor-

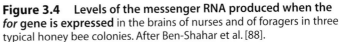

Figure 3.4 **Levels of the messenger RNA produced when the** *for* **gene is expressed** in the brains of nurses and of foragers in three typical honey bee colonies. After Ben-Shahar et al. [88].

mation in the DNA that makes up a gene is expressed only when the gene is in the appropriate environment. As Gene Robinson puts it, "DNA is both inherited and environmentally responsive [1025]." Environmental signals, such as those provided by juvenile hormone and ethyl oleate, influence gene activity. When a gene is turned on or off by changes in the environment, changes in protein output can directly or indirectly alter the activity of other genes within affected cells. A multitude of precisely timed and well-integrated changes in gene–environment interactions are responsible for the construction of every trait, and therefore no trait can be purely "genetic."

By the same token, it would be flat wrong to say that a given phenotype is "environmentally determined." The development of *every* attribute of every living thing requires the information contained in large numbers of genes and expressed in a multitude of *gene*–environment interactions. Neither genotype nor environment can be said to be more important than the other, just as no one would say that a chocolate cake owed more to the recipe used by the cook than to the ingredients that actually went into the finished product.

Discussion Question

3.2 The nature–nurture controversy involves those who believe that our nature (essentially our genes) dominates our behavioral development and others who argue just as forcefully that our nurture (especially our upbringing as children) is what shapes our personalities. Some have dismissed the controversy by saying that the two sides might as well be fighting about whether a rectangle's area is primarily a matter of its height or mostly a function of its width. What's the point of the rectangle analogy? Does the analogy have any weaknesses?

Behavioral Development Requires Both Genes and Environment

The contribution that DNA makes to behavioral development can be demonstrated by examining the development of the ability to learn, since learned behaviors are so often said to be environmentally determined, even though they are not. Of course the environment is involved when an animal learns something, but since learning takes place within a brain whose properties have been shaped by gene–environment interactions, the genetic influence on development simply cannot be ignored. This point is evident when we consider just how circumscribed and focused learned behaviors actually are. These constraints on learning are a consequence of specialized features of the brain, which in turn arise through the interplay between information-rich *genes* and the environment.

A classic example of the circumscribed nature of learning is provided by **imprinting**, in which a young animal's early social interactions, usually with its parents, lead to its learning such things as what constitutes an appropriate sexual partner. Thus, a group of greylag goslings, having imprinted on behavioral biologist Konrad Lorenz (Figure 3.5) rather than a mother goose, formed both a learned attachment to Lorenz [742] and, in the case of the male greylags when they reached adulthood, a preference for humans as mates. The experience of following a particular individual early in life must have somehow altered those regions of the male goose's nervous system responsible for sexual recognition and courtship. The special effects of imprinting could not have occurred without a "prepared" brain, one whose genetically influenced development enabled it to respond to the special kinds of information available from its social environment.

The fact that different species exhibit different imprinting tendencies provides further circumstantial evidence for a genetic contribution to learning. A group of Norwegian researchers provided this kind of evidence when they

Figure 3.5 Imprinting in greylag geese. These goslings have imprinted on the behavioral biologist Konrad Lorenz and are following him wherever he goes. Photograph by Nina Leen.

switched broods from blue tit nests into the nests of breeding great tits, and vice versa. Some of the cross-fostered youngsters grew up and survived to court and form pair bonds with members of the opposite sex (Figure 3.6A). Of the surviving fostered great tits, only 3 of 11 found mates—all of which were blue tit females that had been fostered by great tits. Of the surviving fostered blue tits, all 17 found mates, although 3 of these were the females that mated with male great tits reared by blue tits. Thus, some individuals of both species paired off with a member of another species as a result of the experience of having been reared by that other species [1121].

But the degree to which individuals imprinted on their foster parents clearly differed between the two songbirds. None of the cross-fostered great tits mated with a member of their own species, whereas most of the cross-fostered blue tits did. Moreover, even those few blue tit females that had a great tit as a social partner must have mated with blue tit males on the side, because all 33 offspring produced by those females were blue tits, not hybrids (Figure 3.6B). Thus, although mis-imprinting occurred in both species, the developmental effect of being reared by members of another species was far greater for great tits than for blue tits, an indication that the hereditary basis of the imprinting mechanism was not the same for these two species.

Figure 3.6 Cross-fostering has different imprinting effects in two related songbirds. (A) Males of the great tit (GT) that have been reared by blue tit (BT) foster parents try to pair with blue tit females but only a fraction succeed. In contrast, cross-fostered blue tits always find mates, generally of their own species. Control birds, which were reared by their own species, always find mates and pair with members of their own species. (B) When blue tit females pair with great tit males they also copulate with male blue tits. Here a female blue tit (far left) paired with a male great tit (upper left) rear a brood together that consists entirely of blue tit nestlings. After Slagsvold [1121]; photograph by Tore Slagsvold.

(A)

(B)

In addition to imprinting, bird species possess other specialized learning abilities, including the ability to remember where they have hidden food. The black-capped chickadee is especially good at this task. This bird's spatial memory enables it to relocate large numbers of seeds or small insects that it has hidden in bark crevices or patches of moss scattered throughout its environment. To establish just how good chickadees are at relocating their food caches, David Sherry provided captive chickadees with a chance to store food in holes drilled in small trees placed in an aviary. After the chickadees had placed a sunflower seed in 4 or 5 of 72 possible storage sites, they were shooed into a holding cage for 24 hours. During this time, Sherry removed the seeds and closed each of the 72 storage sites with a Velcro cover. When the birds were released back into the aviary, they spent much more time inspecting and pulling at the covers at their hoard sites than at sites where they had not stored food 24 hours before (Figure 3.7). Because the storage sites were all empty and covered, there were no olfactory or visual cues provided by stored food to guide the birds in their search; they had to rely solely on their memory of where they had hidden seeds [1099]. In nature, these birds store only one food item per hiding spot and never use the same location twice, yet they can relocate their caches as much as 28 days later [540].

Clark's nutcrackers may have an even more impressive memory, for they scatter as many as 33,000 seeds in up to 2500 caches that may be as many as 25 kilometers from the harvest site (Figure 3.8). The bird digs a little hole for each store of seeds, then completely covers the cache. A nutcracker does this work in the fall, and then relies on its stores through the winter and into the spring, recovering an estimated two-thirds of the caches, often months after making them [56, 824].

It could be that nutcrackers do not really remember where each and every seed cache is, but instead rely on a simple rule of thumb, such as "look near little tufts of grass." Or they might remember only the general location where food was stored, and once there, look around until they saw disturbed soil or some other indicator of a cache. But experiments similar to those performed with chickadees show that the birds do remember exactly where they hid their food. In one such test, a nutcracker was given a chance to store seeds in a large outdoor aviary, after which it was moved to another cage. The observer, Russ Balda, mapped the location of each cache, then removed the buried seeds and swept the cage floor, removing any signs of cache-making. No visual or olfactory cues were available to the bird when it was permitted to go back to the

Black-capped chickadee

Figure 3.7 Spatial learning by chickadees. (A) Black-capped chickadees spent much more time at sites in an aviary where they had stored food 24 hours previously (hoard sites) than they did during their initial exposure to those sites, even though experimenters had removed the stored food. (B) The chickadees also made many more visits to hoard sites than to other sites, evidently because they remembered having stored food there. After Sherry [1099].

Figure 3.8   A Clark's nutcracker holding a seed in its bill that the bird is about to cache underground. Photograph by Russ Balda.

aviary a week later and hunt for the food. Balda mapped the locations where the nutcracker probed with its bill, searching for the nonexistent caches. The bird's spatial memory served it well, for it dug into as many as 80 percent of its ex-cache sites, while only very rarely digging in other places [56]. Other long-term experiments on nutcracker memory have demonstrated that nutcrackers can remember where they have hidden food for at least 6, and perhaps as long as 9, months [57]. Indeed, when Balda tested one of his graduate students as if he were a food-storing bird, the student did only about half as well as a typical nutcracker when tested a month after making his caches [824].

The extraordinary ability of nutcrackers and chickadees to store spatial information in their brains is surely related to the ability of certain brain mechanisms to change biochemically and structurally in response to the kinds of sensory stimulation associated with hiding food. These changes could not occur without the genes needed to construct the learning system and the genes that are responsive to key sensory stimuli relevant to the learning task. The more general point is that even learned behaviors, which are obviously environment-dependent, are gene-dependent as well.

What Causes Individuals to Develop Differently?

One of the facts of development is that members of the same species frequently differ in their behavior. Thus, for example, members of an Alaskan population of black-capped chickadees store food more often, and retrieve caches more efficiently, than black-caps that live in the lowlands of Colorado (Figure 3.9) [975]. The superior food-storing ability of Alaskan chickadees may be related to their larger hippocampus, a brain structure associated with memory and successful cache retrieval in several avian species [219] (but see [762]). If the behavioral differences observed among these chickadees are caused by differences in the size of the hippocampus, the differences among individual birds in *either* their genetic information *or* environmental inputs could be responsible (because the construction of the hippocampus is dependent on both genes and environment).

Figure 3.9   Differences within a species in learned behavior. Black-capped chickadees that live in the severe cold of Alaska not only store more food, but remember better where they have put their caches, than chickadees of the same species that live in Colorado, where the climate is less demanding. The mean number of site inspections per food item found is much less for Alaskan chickadees than for their Coloradan cousins. After Pravosudov and Clayton [975].

3.3   Return to the chocolate cake analogy (see p. 60), and use it to illustrate how a change in either genes or environment could lead to developmental differences between individuals.

Although, in theory, the differences between an Alaskan chickadee and a member of the same species from Colorado could be due to genetic differences between them, or environmental differences, or both, no direct evidence is available to resolve this case. However, it is known for some other species that an environmental difference can generate differences in the size of the hippocampus. When Nicky Clayton and John Krebs hand-reared some marsh tits—close relatives of the black-capped chickadee—in the laboratory, they gave some of the young birds opportunities to store whole sunflower seeds, while others were always fed powdered sunflower seeds, which the birds ate, but never stored [218]. The individuals that had opportunities to store food gained more cells in the hippocampus than those birds lacking food-storing experience, a reflection of the fact that genes in hippocampal cells responded in different ways to different kinds of experiential inputs.

Environmental Differences and Behavioral Differences

The marsh tit case is merely one of thousands in which certain phenotypic differences among individuals have been linked to differences in their environment, rather than genetic differences. You will remember that the dialect differences among male white-crowned sparrows are also a product of differences in the birds' acoustical and social environments, which affect what young sparrows learn when they listen to males singing around them. Environmental differences are important whenever members of a species differ in a learned behavior. Consider why one spiny mouse huddles with a companion while another refuses to do so. The difference between these mice could stem entirely from differences in the experiences they had as pups; baby spiny mice learn who their littermates are, and later in life, they prefer to cluster together with familiar individuals rather than unfamiliar ones. Typically, members of this species huddle preferentially with their siblings with whom they have grown up, but if one experimentally creates a litter composed of nonsiblings that are cared for by the same female, these unrelated littermates will treat one another as if they were siblings [970].

David Pfennig and his colleagues suspected that a similar learned tendency to associate with familiar individuals occurs in paper wasp colonies because young adult females quickly learn what their natal nest and its other occupants smell like (Figure 3.10) [946]. Learning of this sort would, under natural conditions, enable females to recognize and respond positively to sisters, who share the natal nest odor, and to behave less tolerantly toward adults that do not smell like the familiar nest. If this hypothesis is correct, then it should be possible to fool paper wasps into tolerating nonkin by transferring newly emerged females to a portion of another nest for a few hours to expose them to the odor of that nest. As predicted, the transferred queens were less likely to fight with unrelated females from the nest they had smelled as young adults than with their genetic relatives from what was, to them, an unfamiliar nest. Thus, it is as if paper wasps inherit a behavioral command that reads, "Learn what your nest smells like and respond nonaggressively to those individuals that share this odor." Differences in the olfactory environments surrounding wasps can therefore cause them to differ in their social responses to others.

I want to emphasize again that the ability and motivation to record information about the scents associated with a nest require genetic information, which

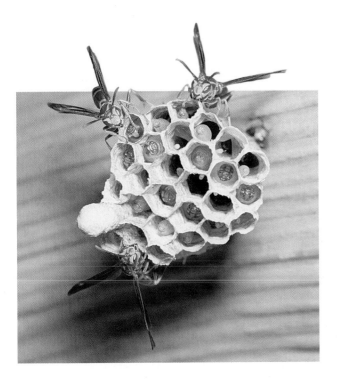

Figure 3.10   Nests of *Polistes* **paper wasps** contain odors that adhere to the bodies of the wasps reared in them, providing a proximate cue for the learned recognition of nestmates in these insects. Note the larvae with their bodies pressed up against the cell walls. Photograph by the author.

is needed for the construction of a nervous system with the capacity for this kind of learning. But the *difference* in the social behavior of two wasps reared in different nests arises because of a difference between them based on early olfactory experiences. The same applies to some mammals, judging from the results of a study of Belding's ground squirrels in which the newborn offspring of captive females were switched around at birth, creating four classes of individuals: (1) siblings reared apart, (2) siblings reared together, (3) nonsiblings reared apart, and (4) nonsiblings reared together. After having been reared and weaned, the juvenile ground squirrels were placed in an arena in pairs and given a chance to interact. In most cases, animals that were reared together, whether actual siblings or not, treated each other nicely, whereas animals that had been reared apart tended to react aggressively to each other [558].

The results described thus far are similar to those gathered from comparable studies of spiny mice and paper wasps. But an additional finding from the ground squirrel research was that biological sisters *reared apart* engaged in fewer aggressive interactions than nonsiblings reared apart (Figure 3.11). In other words, the squirrels had some way of recognizing their sibs that was not dependent on living with them as youngsters [557, 558]. Instead, a different kind of learning was probably involved, one that goes informally by the indelicate label of the "armpit effect" [796]. That is, if individuals can learn what they themselves smell like, then they can use this information as a reference against which to compare the odors of other individuals [559].

This hypothesis has been examined by Jill Mateo [794], who knew that Belding's ground squirrels possess two scent-producing glands, one around the mouth and the other on the back. Moreover, these animals regularly sniff the oral glands of other individuals, as if they were acquiring odor information that could conceivably be compared with the sniffers' own scents. By capturing pregnant ground squirrels and moving them to laboratory enclosures, Mateo was able to observe their juvenile offspring investigating objects (plastic cubes) that had been rubbed on the dorsal glands of other squirrels of varying degrees of relatedness to the youngsters. Because the captive squirrels had been separated

Belding's ground squirrel

Figure 3.11   Kin discrimination in Belding's ground squirrels. Sisters reared apart display significantly less aggression toward each other than other combinations of siblings reared apart, which are as aggressive to each other when they meet in an experimental arena as nonsiblings reared apart. After Holmes and Sherman [558].

from some of their relatives, they had never met them, and so had no prior experience with their odors. If, however, the test animals had learned what they themselves smelled like, and if close relatives produce scents more similar to their own than do distant relatives, an inexperienced youngster could in theory discriminate between unfamiliar relatives and nonrelatives on the basis of odor cues alone. As it turns out, the length of time a Belding's ground squirrel sniffs an object is an indicator of its interest in that object, which depends on the object's odor similarity to the squirrel's own scent. Thus, cubes that have been rubbed on a fairly close relative, genetically speaking, receive only a cursory inspection. Items smeared with odors of a more distant relative are given a significantly longer sniff, and the inspection time increases again for cubes daubed with a nonrelative's odor. Belding's ground squirrels are odor analyzers, spending less time with scents similar to their own and increasingly more time with odors less like their own (Figure 3.12). They therefore have the capacity to treat individuals differently on the basis of this learned label of relatedness [794].

Belding's ground squirrels possess a highly specific form of learning that enables them to remember what they themselves smell like in order to use this information to make decisions about which individuals to mate with and which individuals to ignore or reject. Differences among individuals in their odor "environment" translate into learned differences in their behavior—one exam-

Figure 3.12   Evidence for the ability of Belding's ground squirrels to learn their own odor, which they may match against that of other individuals. Juvenile squirrels were first given three trials during which they could investigate their own odors applied to plastic cubes. Note the squirrels' decline in responsiveness to their own dorsal gland scents over these initial trials. Then the squirrels were provided with plastic cubes daubed with dorsal gland odors from four categories of individuals. Cubes with scents from close relatives received less attention than those with scents from distant relatives or nonkin. Numbers in parentheses show the genetic similarity of the individual to the tested ground squirrel. After Mateo [794].

ple among many of how environmental differences can lead to the development of behavioral differences within a species.

3.4 A good predictor of a young person's vocabulary is the amount of time parents spent talking to their child when it was very young. Some have concluded that family environment is therefore the essential factor in determining a person's language skills. What's a logical problem with this conclusion?

Genetic Differences and Behavioral Differences

Although a great many differences in behavioral phenotypes have been traced to differences in the environment, others have been linked to genetic differences among individuals, which makes sense given the interactive, dual-factor nature of development. To see if a genetic difference underlies why some blackcap warblers spend the winter in southern Great Britain while most other members of this species migrate to Africa (Figure 3.13), Peter Berthold checked whether the offspring of "winter-in-Britain" birds would inherit their parents' behavior. To conduct this research, he and his colleagues captured some wild blackcaps in Britain during the winter and took them to a laboratory in Germany, where the birds spent the rest of the winter indoors. Then, with the advent of spring, pairs of warblers were released into outdoor aviaries, where they bred, providing Berthold with a crop of youngsters that had never migrated [97].

Figure 3.13   Different wintering sites of blackcap warblers. Blackcaps living in southern Germany and Scandinavia first go southwest to Spain before turning south to western Africa. Blackcaps living in eastern Europe migrate southeast before turning south to fly to eastern Africa. Where do the birds that winter in Great Britain come from?

Blackcap warbler

Figure 3.14 Funnel cage for recording the migratory orientation of captive birds. The bird can see the night sky through the wire mesh ceiling of the cage. As the bird jumps up onto the surface of the funnel, it leaves marks that show the direction in which it intended to fly. Photograph by Jonathan Blair.

Once the young birds were several months old, Berthold's team placed some in special cages that had been electronically wired to record the number of times a bird hopped from one perch to another. The electronic data revealed that when fall arrived, the young warblers became increasingly restless at night, exhibiting the kind of heightened activity characteristic of songbirds preparing to migrate. The immature blackcaps' parents also became nocturnally restless when placed in the same kind of cages in the fall. These observations showed that the British wintering population is *not* composed of birds that have simply lost their ability to migrate. Instead, the birds wintering in Britain must be migrants that flew to Britain from somewhere else.

But just where does the British wintering population come from? To answer this question, the researchers put some warblers in cages shaped like funnels and lined with typewriter correction paper. Whenever the bird leaped up from the base of the funnel in an attempt to take off, it landed on the paper and left scratch marks, which indicated the direction in which the bird was trying to go (Figure 3.14). Berthold's subjects, experienced adults and young novices alike, oriented due west, jumping up in that direction over and over, judging from the footmarks left on the paper. These data showed that the adults, which had been captured in wintery Britain, must have traveled there by flying west from Belgium or central Germany, a point eventually confirmed by the discovery of some blackcaps in Britain that had been banded earlier in Germany.

Discussion Questions

3.5 The black redstart is a bird species that migrates a relatively short distance from Germany to the Mediterranean region of Europe, whereas the common redstart travels as much as 5000 kilometers from Germany to central Africa. The scale in Figure 3.15 shows the duration of migratory restlessness in three groups of captive birds all hand-raised under identical conditions: black redstarts, hybrids created by crossing black and common redstarts, and common redstarts. Why do black redstarts exhibit migratory restlessness at night for fewer days than common redstarts? What does the behavior of the hybrids tell us about the genetic differences hypothesis for the difference in the duration of migratory restlessness in the two parental species?

3.6   Robert Plomin and his colleagues have compared the cognitive abilities of children with those of their parents (genetic or adoptive) and twin siblings [968]. What significance do you attach to these data (Figure 3.16) in the context of determining whether genetic or environmental differences are responsible for the differences between humans in their spatial and verbal abilities? If environmental differences are the key to understanding differences in these human phenotypes, what is the predicted relationship between the number of years a child has spent in an adoptive home and the degree of difference between its spatial and verbal attributes and those of its genetic parents?

Single-Gene Effects on Development

The blackcap warbler breeding experiments done by Berthold do not tell us how many genetic differences are responsible for the behavioral differences between the birds that winter in Great Britain and those that head for Africa. In theory, a single genetic difference could be responsible for the migratory differences observed in these birds. Please note that no one believes that migratory choices are encoded in a single gene in the warbler genome. However, a difference in even a single gene product can lead to many divergences in the gene–environment interactions occurring in different individuals, which may translate into large behavioral differences between them.

Single-gene effects of this sort have been documented in several different ways, but perhaps most dramatically via gene knockout experiments. Researchers are now able to inactivate a given gene in an animal's genome in order to determine how that particular gene contributes to development in a particular environment. Sometimes the developmental effect of knocking out a gene is spectacular, as demonstrated by the effects of scrambling the genetic

Figure 3.15   Differences in the migratory behavior of two closely related birds, the black redstart and the common redstart. The scale at the bottom shows the periods after hatching (day 0) when the young birds exhibit migratory restlessness at night. After Berthold and Querner [98].

Black redstart

Common redstart

Figure 3.16 Why do people differ in their test scores? The graphs show correlations in (A) verbal ability scores and (B) spatial ability scores for parents and offspring (P–O), monozygotic (identical) twins (MZ), and dizygotic (fraternal) twins (DZ) either living together or apart. The data are composite measures based on a number of different studies. After Plomin et al. [967].

code of the *fosB* gene of laboratory mice. Females with the experimental "mutation" are normal in most respects, but are totally indifferent to their newborn pups, which they fail to retrieve should they wriggle away from the nest. In contrast, normal females with two copies of the active *fosB* gene invariably gather displaced pups together and crouch over them, keeping them warm and permitting them to nurse (Figure 3.17) [161].

Figure 3.17   A single gene affects maternal behavior in laboratory mice. Wild-type female mice gather their pups together and crouch over them (above), but females with inactivated *fosB* genes (below) do not exhibit these behaviors (the pups can be seen scattered in the foreground). Photographs courtesy of Michael Greenberg; from Brown et al. [161].

Discussion Questions

3.7   Does the *fosB* knockout experiment demonstrate that a single gene determines the maternal behavior of a female mouse? You should know by now that the answer is no. Use this example to illustrate the difference between claiming that maternal behavior is genetically determined and claiming that certain differences among individuals in their maternal behavior phenotype are genetically determined. How is the idea that genes are responsive to particular kinds of environmental inputs illustrated by the fact that when female mice inspect their pups after birth, they receive olfactory stimulation, which affects the mouse's brain and triggers *fosB* gene activity in mice with the typical genotype? How might this gene's activity initiate additional changes in other genes, leading to a specific pattern of biochemical events?

3.8   In 1998, a team of geneticists claimed that men with two copies of the *I* allele of the *ACE* gene were much less able to improve their physical condition through exercise than people with a different allele (*D*) of this gene. According to these researchers, individuals with two copies of the *I* allele were very much underrepresented in groups of mountain climbers who climb to 7000 meters without oxygen [845]. Dispute the claim that the *D* allele is a "gene for mountain climbing." What role does the environment play in the development of the biochemical product of the *ACE* gene? Why are many genes almost certainly involved in the development of a person who enjoys mountain climbing? How, then, could a single difference in the *ACE* gene potentially result in a difference in mountain-climbing behavior between two people?

Other knockout mutations also have highly damaging, specific developmental consequences for mice. Males whose *Oxt* gene has been knocked out cannot produce oxytocin, an important brain hormone [385], with the correlated effect that these males cannot remember females with whom they have recently interacted. Each time a given female is removed from and then returned to the cage she shares with an *Oxt* mutant male, the male gives her a thorough and lengthy sniffing that is no different from his response the very first time they met (Figure 3.18). In contrast, if a female is placed in the cage of a normal male with a functional *Oxt* gene, he remembers what she smells like, so that if she is taken from his cage but then later returned to it, he will spend less time sniffing her on this occasion than he did the first time they met. The presence of a functional *Oxt* gene seems critical, therefore, if the male is to remember that he has interacted with a familiar female.

In yet another experiment, the *TRP2* gene was knocked out. When a male mouse without this gene encounters another male in his laboratory cage, he attempts to mate with the intruder, who rarely responds enthusiastically [1156]. In contrast, a male with functional copies of the *TRP2* gene makes war, not love, when he finds another male in his cage. Apparently the knockout males cannot identify their fellow males by their distinctive scents, and so treat every mouse as a potential copulatory partner, perhaps because of an alteration in their olfactory apparatus. The vomeronasal organ, an olfactory device in the mouse nose, contains a cluster of neurons that responds to sex-identifying mouse scents. However, if they lack the *TRP2* gene, these cells are incapable of reacting to male odors, so that the knockout male never receives signals from the vomeronasal organ that a male is in the vicinity. Because the genetically altered mouse does not detect the key cue of maleness, he responds to the other mouse as if it were a female.

Although knockout experiments have generated important evidence on the genetic differences hypothesis, other methods have also shown how important variation in even a single gene can be in causing developmental differences. For example, persons studying the wormlike larvae of *Drosophila melanogaster*, the humble fruit fly, have found two naturally occurring, distinctively different phenotypes. Some larvae (called "rovers") travel about four times farther when feeding on a yeast-coated petri dish during a 5-minute period than larvae of the other type (which are labeled "sitters") [308]. When adults reared from rover larvae are bred with adults reared from sitter larvae, these pairs of flies produce larval offspring (the F_1 generation) that are all rovers. When these larvae mature and interbreed, they produce an F_2 generation with three times as many rovers as sitters (Figure 3.19). Persons familiar with Mendelian genetics will recognize that rovers must have at least one copy of the dominant allele of a gene affecting larval foraging behavior, whereas sitters must have two copies of the recessive allele. If this analysis is correct, then if one could transfer the dominant allele associated with rover behavior to an individual of the sitter genotype, the

Figure 3.18   Social amnesia is related to the loss of a single gene.
(A) Male mouse inspecting a female. (B) Knockout male mice that lack a functional *Oxt* gene carefully inspect the same female every time she is reintroduced into their cage, whereas males with the typical genotype show less and less interest in a female that they have inspected previously. A, photograph by Larry J. Young; B, after Ferguson et al. [385].

(A)

(B)

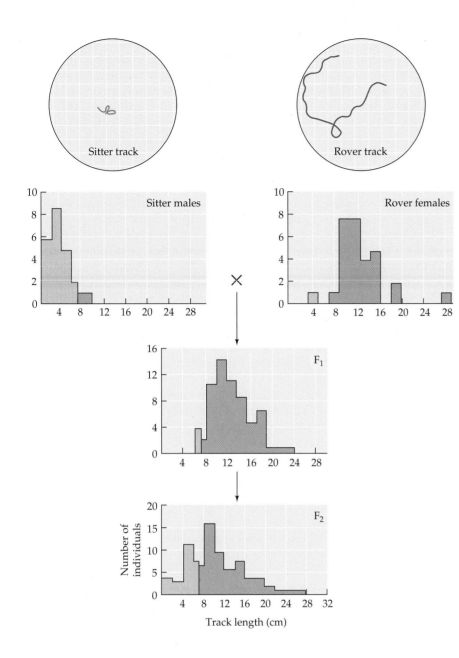

Figure 3.19   Genetic differences cause behavioral differences in fruit fly larvae. Representative tracks made by sitter and rover phenotypes feeding in a petri dish appear at the top of the figure. When adult male flies of the sitter strain mate with adult females of the rover strain, their larval offspring (the F_1 generation) almost all exhibit the rover phenotype (that is, they move more than 7.6 cm in 5 minutes). When flies from the F_1 generation interbreed, their offspring (the F_2 generation) are composed of rovers (purple) and sitters (tan) in the ratio of 3:1. After de Belle and Sokolowski [308].

genetically altered larva should exhibit rover behavior. This experiment has been done, with positive results [905]. Thus, the difference in foraging behavior between rovers and sitters stems from a difference in the information contained in a single gene—just one of the 13,061 genes [937] located on the four chromosomes of *Drosophila melanogaster* [307].

The techniques now available to molecular biologists have enabled researchers to identify the gene in question. This gene, which has been given the label *for*, is the same one mentioned earlier in the context of honey bee behavior. In both species, the gene codes for a "cGMP-dependent protein kinase" [905]. This enzyme is produced in certain cells of the larval fruit fly's brain, leading to the still untested hypothesis that the gene and its protein may have a special role to play in the larva's ability to analyze its chemical environment. Since the two naturally occurring variants of the gene produce different forms of a key protein, with different efficacies in promoting a particular reaction, sitters and rovers acquire different neural mechanisms, which may provide them with the physiological foundation for their behavioral differences.

Humans have a nervous system and a genome too, of course, and so we can expect that some of the behavioral differences among us have a genetic component. The search for genetic variation that affects human behavioral development has led some researchers to one particular part of the brain, the lateral frontal cortex, which is known to contribute to human intelligence [350]. Cells in this part of the brain become especially active when people try to solve either spatial or verbal problems. (Brain cell activity can be visualized by positron emission tomography, a technology that measures the extent of blood flow to particular parts of the brain.) It is possible, therefore, that a portion of the differences in intelligence among people in some populations will eventually be traced to differences among them in the gene–environment interplay that contributes to the development or activity of the lateral frontal cortex.

The lateral frontal cortex is such a complex chunk of brain tissue that we can be certain that literally thousands of genes are necessary for its complete development and effective operation [231]. In fact, half of the human genome, perhaps 10,000 genes or thereabouts, is active at some time in some part of the brain [736]. Thus, variation in any one of these thousands of genes could conceivably contribute to variation in brain phenotypes, and thus variation in the cognitive ability or behavior of human beings. Indeed, one variable gene (*COMT*) that codes for an enzyme called catechol-O-methyltransferase has been shown to affect performance on at least one intelligence test [1327]. The difference between two common alleles of *COMT* translates into a single difference in the long chain of amino acids that make up the enzyme in question. One variant form of the enzyme is four times as active at body temperature as the other type, which means that persons possessing the "fast" enzyme carry out one particular biochemical reaction at a relatively high rate. The reaction that is mediated by *COMT* breaks down a substance called dopamine, which is an important chemical communicator between certain brain cells. The rate at which dopamine is removed therefore affects signal transmission between cells in the prefrontal cortex, which in turn evidently affects a person's ability to perform certain cognitive tasks.

Another example of a relationship between variation in a gene, a neurotransmitter, and human behavior involves a segment of DNA found in a particular portion of chromosome 17 (humans have 23 pairs of chromosomes in all). The gene in question produces a protein that regulates the uptake of serotonin—another chemical, like dopamine, that relays messages between neurons in certain parts of the human brain. The activity of this serotonin transporter gene (labeled *5-HTT*) is controlled by a segment of DNA some distance from *5-HTT*. This regulatory chunk of DNA comes in two forms, one longer than the other; the shorter form causes the *5-HTT* gene to produce about a third less protein per unit of time than the longer form. As a result, a person's genotype affects how much protein is available to remove serotonin from the spaces between certain brain neurons, thus affecting the nature of neural activity in these cells, which depend on serotonin to communicate with one another. The regions of the brain that rely heavily on serotonin as a neurotransmitter include structures thought to play major roles in controlling our emotions, mood, and anxiety levels. Indeed, a small part of the difference among people in just how anxious they are has been linked to variation in the *5-HTT* regulator genotype [712].

Evolution and Behavioral Development

We have seen that in fruit flies and human beings, a genetic difference, even one as apparently minor as a single allelic difference, can have a significant effect on behavioral development. This finding is hugely important for persons interested in the evolution of behavior. If there is genetic variation that leads to behavioral variation in animal populations today, then surely the same applied to populations in the past. Developmentally potent genetic variation had to have existed in the past, if natural selection had anything to do with the evolution of a species, because selection occurs only when the members of a species differ hereditarily in ways that have both developmental effects and reproductive consequences (see Chapter 1). Under these conditions, the alleles that contribute to the development of reproductively superior phenotypes will become more common over time, while the other alleles will eventually disappear.

The potential for rapid evolutionary change has been realized in some populations under direct observation. For example, the spread and then decline of the blackish form of the moth *Biston betularia* (see Figure 6.16) has been documented and linked to specific changes in selection pressures over the last 150 years [466, 467]. Another case involves the blackcap warbler, whose use of southern Britain as a wintering grounds originated a mere 50 or so years ago [96].

In other species, we can infer that behavioral evolution has taken place over a somewhat longer period. For example, the garter snake, *Thamnophis elegans*, which occupies much of dry inland western North America, also currently lives in foggy coastal California, a region that it almost certainly invaded relatively recently [41]. The diets of snakes in the two areas, referred to hereafter as "coastal" and "inland" snakes, differ markedly. Whereas inland snakes feed primarily on the fish and frogs found in lakes and streams in the arid West, coastal snakes regularly eat the banana slugs that thrive in the wet forests of coastal California (Figure 3.20). I can only marvel at the ability of the snakes to consume these prey. When I once made the mistake of picking up a banana slug, the creature promptly covered my hand with massive amounts of an exceedingly sticky and repulsive mucus that greatly reduced my desire to touch these animals ever again.

If coastal garter snakes have evolved the special behaviors that permit them to identify and consume banana slugs, then they should differ genetically from inland snakes. To check this prediction, Steve Arnold took pregnant female snakes from the two populations into the laboratory, where they were held under identical conditions. When the females gave birth to a litter (garter snakes produce live young rather than laying eggs), each baby snake was placed in a separate cage, away from its littermates and its mother, to remove these possi-

Figure 3.20   A coastal Californian garter snake about to consume a banana slug, a favorite food of snakes in this region. Photograph by Steve Arnold.

Figure 3.21 **Response of newborn, naive garter snakes to slug cubes.** Young snakes from coastal populations tended to have high slug feeding scores (e.g., a score of 10 indicates that the snake ate a slug cube on each of the 10 days of the experiment). Inland garter snakes were much less likely than coastal snakes to eat even one slug cube (which would yield a score of 1). After Arnold [41].

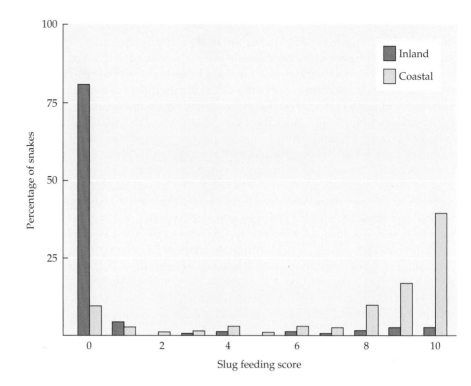

ble environmental influences on its behavior. Some days later Arnold offered each baby snake a chance to eat a small chunk of freshly thawed banana slug by placing it on the floor of the young snake's cage. Most naive young coastal snakes ate all the slug hors d'oeuvres they received; most of the inland snakes did not (Figure 3.21). In both populations, slug-refusing snakes ignored the slug cube completely.

Arnold took another group of isolated newborn snakes that had never fed on anything and offered them a chance to respond to the *odors* of different prey items. He took advantage of the readiness of newborn snakes to flick their tongues at, and even attack, cotton swabs that have been dipped in fluids from some species of prey (Figure 3.22). Chemical scents are carried by the tongue to the vomeronasal organ in the roof of the snake's mouth, where the odor molecules are analyzed as part of the process of detecting prey. By counting the number of tongue flicks that hit the swab during a 1-minute trial, Arnold measured the relative responsiveness of inexperienced baby snakes to different odors.

Populations of inland and coastal snakes reacted about the same to swabs dipped in toad tadpole solution (a prey of both groups), but behaved very differently toward swabs daubed with slug scent. Almost all inland snakes ignored the slug odor, whereas almost all coastal snakes rapidly flicked their tongues at it. Because all the young snakes had been reared in the same environment, the differences in their willingness to eat slugs and to tongue-flick in reaction to slug odor appear to have been caused by genetic differences among them.

If the feeding differences between the two populations arise because most coastal snakes have a different allele or alleles than most inland snakes, then crossing adults from the two populations should generate a great deal of genetic and phenotypic variation in the resulting group of hybrid offspring. Arnold conducted the appropriate experiment and found the expected result, confirming again that the behavioral differences *between* populations have a strong genetic component.

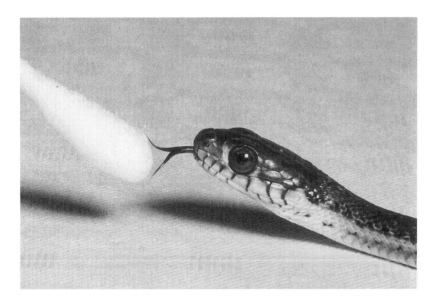

Figure 3.22   **A tongue-flicking newborn garter snake** senses odors from a cotton swab that has been dipped in slug extract. Photograph by Steve Arnold.

Having identified the predicted proximate basis for the different dietary preferences of coastal and inland garter snakes, Arnold proposed that among the original colonizers of the coastal habitat were a very few individuals that carried a rare allele or two for slug acceptance. These slug-eating snakes were able to exploit an abundant, if mucus-covered, food in their new habitat. If, as a result, their reproductive success was as little as 1 percent higher than that of their slug-ignoring fellows, the coastal population could have reached its present state of divergence from the inland population in less than 10,000 years [41]. This study illustrates once again that if there are genetic differences among individuals that affect their reproductive success, natural selection is a powerful agent for evolutionary change.

Discussion Questions

3.9   The typical effect of natural selection must be the elimination of genetic variation from populations, and yet in *Drosophila* fruit flies (and many other species), two or more alleles of certain genes occur at reasonably high frequency. The co-occurrence of the rover and sitter phenotypes in fruit fly larvae suggests that selection is actually maintaining both alleles of the *for* gene in this species. Figure 3.23 presents the results of experiments in which larvae were reared under low-density and high-density conditions. How do these experiments help us understand why genetic variation relevant to larval foraging behavior has persisted in *Drosophila melanogaster*?

3.10   Debi Fadool at Florida State University headed a research team that studied a strain of genetically modified mice that lacked the ability to make a protein called Kv1.3 [378]. In unaltered mice, this protein is found in regions of the brain that process olfactory information, leading Fadool and her team to predict that the two kinds of mice should differ in their ability to smell things. In fact, the genetically modified mice were able to smell scents at much lower concentrations than mice that possessed the protein; the knockout mice found odorous foods, such as peanut butter crackers, much faster than their wild-type cousins. What question about genetic evolution is raised by these findings? What ultimate explanation do you have for the fact that mice with Kv1.3 protein are actually *less* sensitive to food odors than mice without that protein?

Figure 3.23   Density effects on the foraging behavior of fruit fly larvae. Larvae reared in the laboratory for many generations under low-density conditions move shorter distances in the same period of time than larvae from populations reared under high-density conditions. Three different populations are presented for each of the two conditions, showing the consistent effect of density-dependent selection on larval behavior. After Sokolowski et al. [1136].

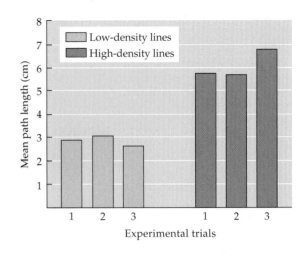

What Do Artificial Selection Experiments Tell Us about Behavioral Evolution?

We can test the proposition that behavior evolves readily not just by finding illustrative examples of behavioral evolution in the wild, but through **artificial selection** experiments, in which researchers get to determine which behavioral phenotypes will reproduce and which will not. If the phenotypic differences among the test subjects are hereditary, then over time, the particular allele(s) associated with the phenotypes that are allowed to reproduce should become more common in the experimental population, while the frequency of alternative alleles linked to less "successful" traits should decrease. In other words, we should be able to cause evolutionary change in the laboratory—provided that the variation in the study population at the outset is linked to genetic, not environmental, differences within the population.

Consider the work of Carol Lynch with house mice, which build nests of soft grasses and other plant materials in nature, but will happily accept cotton as nesting material in the laboratory [755]. The amount of cotton a mouse collects can be quantified as the number of grams it pulls into a nest cage over a 4-day period. In the starting generation of Lynch's experiment, individual mice moved between 13 and 18 grams of cotton into their cages from an external cotton supply. But was this variation in nest-building behavior caused by genetic or environmental factors?

To get an answer, Lynch attempted to create a "high line" by interbreeding males and females that collected a relatively large amount of cotton, as well as a "low line" (by crossing males and females that gathered relatively little cotton) and a "control line" (by crossing males and females chosen at random from each generation). The offspring produced by these crosses were reared under the same conditions their parents were, eliminating environmental variation as a possible cause for differences in their behavior. When the youngsters became adults, the amount of nest material they collected in 4 days was measured. The most avid cotton collectors in the high line were permitted to breed, creating a second selected generation, as were the least eager collectors in the low line.

Lynch repeated these procedures over 15 generations, with the eventual result that the high-line mice gathered about 50 grams of cotton for their nests on average, while the low-line mice brought in about 5 grams on average. The 15th-generation control mice brought in about 20 grams, roughly the same amount as their ancestors (Figure 3.24). This result shows that genetic differences among the mice in the original population caused some of the differences in nest-building behavior, which enabled Lynch to make her experimental populations evolve.

Figure 3.24   **Response to artificial selection on nest-building behavior by mice.** Artificial selection favoring mice that collect large amounts of cotton for their nests eventually led to the evolution of populations (the high lines) whose members collected much more cotton on average than the control lines, whose behavior was not selected and so did not evolve. Likewise, selective breeding of mice that gathered relatively little cotton resulted in the evolution of low lines, whose members made very small nests. The symbols indicate the average amounts of cotton collected over 4 days by each generation. After Lynch [755].

Discussion Questions

3.11   Analyze the science behind Carol Lynch's study of nest-building behavior in mice just described here. Formally lay out (in one sentence each) the question she asked, the hypothesis she proposed, her prediction(s) and test(s), and the conclusion she reached based on her research.

3.12   A few blackcap warblers live year-round in southern France, although 75 percent of the breeding population migrates from this area in winter. Perhaps the difference between the two behavioral phenotypes is environmentally induced. Make a prediction about the outcome of an artificial selection experiment in which the experimenter tries to select for both nonmigratory and migratory behavior in this species. Describe the procedure and present your predicted results graphically. Check your predictions against the actual results (see [94]).

One can have an interest in artificial selection experiments for practical as well as academic reasons. For example, let's say that you are worried that global warming will happen so fast that migratory birds won't be able to evolve rapidly enough to accommodate climatic changes as they occur. One way to check this possibility is to subject a migratory songbird, such as the blackcap warblers we met earlier, to a selection experiment designed to induce a change in the timing of fall migration. If artificial selection resulted in a delay in the departure time of migrating blackcaps, this finding would suggest that natural selection in the wild could rapidly alter the average departure time for the birds in response to changes in climate, quickly selecting for a new adaptive departure time. Were the artificial selection experiment to fail, however, this result would suggest that the birds would not be able to adjust rapidly to the challenges posed by rising temperatures—not a happy thought for persons concerned about bird conservation.

In fact, a research group was able to delay autumnal migration in blackcaps by more than a week in just two generations of artificial selection (Figure 3.25). Therefore, blackcaps must inherit genes from their parents that influence when they want to migrate. Indeed, genetic variation accounted for about one-third to one-half of the variation in the onset of migratory activity in samples of south-

Figure 3.25   **Response to artificial selection on the fall migration departure date** of blackcap warblers. After two generations of selection, the onset of migratory activity in captive birds had been shifted by nearly 8 days. After Pulido et al. [985].

Figure 3.26   Dogs are especially sensitive to signals from human beings. Dogs, but not hand-reared wolves, pay attention to the signals provided by humans about where food can be found. Dogs, but not wolves, use the information to go directly to a bowl with hidden food rather than to an empty bowl. The results are the mean number of correct choices made over a series of trials by seven dogs and seven wolves. The control represents a situation in which a human did not gaze at or point at the correct bowl. After Hare et al. [513].

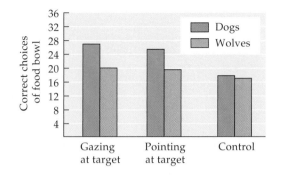

ern German blackcaps. By breeding males that started their migration late with females that shared a similar phenotype, the blackcap research team quickly created a late-departing line of birds [985]. This study gives us some hope that blackcaps will be able to evolve quickly on their own in response to climatic changes linked to global warming, although it would probably be better if they did not have to.

Discussion Question

3.13   Figure 3.26 shows the results of an experiment in which domestic dogs and hand-reared wolves, the closest relatives of dogs, were tested with respect to their ability to locate food hidden in one of two covered bowls. When the test subjects were allowed to watch humans intently gazing at the food-containing bowl or while pointing at it, the dogs went directly to the correct bowl significantly more often than the wolves. Even chimpanzees cannot match dogs at this task. Moreover, puppies that have been reared in a kennel with minimal human contact also use the cues provided by people to find food, and they do as well as puppies who have been reared in a household setting [513]. How is this case similar to, and how is it different from, the cases of artificial selection discussed in this section?

Adaptive Features of Behavioral Development

We have made the argument that changes in either genotype or environment (or both) could in theory generate changes in developmental sequences and thus changes in phenotypes. Given the fact that most organisms have thousands of genes and are subject to thousands of variable environmental factors, the number of possible gene–environment combinations is huge. It is not surprising, therefore, that considerable phenotypic variation characterizes most species (see Figure 1.9). On the other hand, the developmental process often appears to be capable of generating a typical or average outcome even in the face of shortfalls in genetic information, such as those caused by mutation, or deficits in environmental inputs of the sort that one would think would be important to normal development.

As a matter of fact, gene knockout experiments provide important evidence on this point. Although, as discussed earlier, the removal or inactivation of a given gene sometimes drastically affects the developmental trajectory of individuals lacking that gene, at other times knocking out a particular gene has little or no developmental effect, so that many knockout mice look and behave normally. These findings have led some geneticists to conclude that genetic systems must exhibit considerable information redundancy, which would explain why the loss of one gene–environment product is not fatal to the acquisition of one or more traits of importance to the individual [634, 949].

We also know that many animals overcome what you might think would constitute considerable environmental obstacles to normal development. For example, some young birds lack the opportunity to interact with their parents, and so cannot acquire the information that in other species is essential for normal social and sexual development (as discussed earlier in this chapter). So, when chicks of the Australian brush turkey hatch from eggs placed deep within an immense compost heap of a nest, they dig their way out and walk away, often without ever seeing a parent or sibling. So how do they manage to recognize other members of their species? Ann Göth and Christopher Evans studied captive young brush turkeys in an aviary in which they were exposed to feathered robots that looked like other youngsters. All that was required to elicit an approach from a naive youngster was a few pecks at the ground by the robot. Thus, young brush turkeys do not require extensive social experience in order for rudimentary social behavior to develop [457], and as adults, the birds are completely capable of normal sexual behavior despite having lived primarily by themselves beforehand.

Other experimenters have created genuinely abnormal rearing environments, only to find that various forms of sensory deprivation have little or no effect on the development of normal behavior. Bring up baby Belding's ground squirrels without their mothers and they still stop what they are doing to look around when they hear a tape of the alarm call of their species [795]. Male crickets that live in complete isolation sing a normal species-specific song despite their severely restricted social and acoustical environment [91]. Captive hand-reared female cowbirds that have never heard a male cowbird sing nevertheless adopt the appropriate precopulatory pose when they hear cowbird song for the first time, if they have mature eggs to be fertilized [641].

Developmental Homeostasis: Protecting Development against Disruption

The ability of many animals to develop more or less normally, despite defective genes and deficient environments, has been attributed to a process called **developmental homeostasis**. This property of developmental systems reduces the variation around a mean value for a phenotype (see Figure 3.32 below), and reflects the ability of developmental processes to suppress some outcomes in order to generate an adaptive phenotype more reliably. A clear demonstration of this ability comes from a classic experiment on the development of social behavior in young rhesus monkeys deprived of contact with others of their species by Margaret and Harry Harlow [514, 515]. (The Harlows' experiments were conducted nearly four decades ago, when animal rights were not the issue they are today; readers can decide for themselves whether their harsh treatment of infant monkeys yielded information of sufficient importance to justify the research.) In one such study, the Harlows separated a young rhesus from its mother shortly after birth. The baby was placed in a cage with an artificial surrogate mother (Figure 3.27), which might be a wire cylinder or a terry cloth figure with a nursing bottle. The baby rhesus gained weight normally and developed physically in the same way that nonisolated rhesus infants do. However, it soon began to spend its days

Figure 3.27   Surrogate mothers used in social deprivation experiments. This isolated rhesus infant was reared with wire cylinder and terry cloth dummies as substitutes for its mother. Photograph by Nina Leen.

Figure 3.28 Socially isolated rhesus infants that are permitted to interact with one another for short periods each day at first cling to each other during the contact period. Photograph by Nina Leen.

crouched in a corner, rocking back and forth, biting itself. If confronted with a strange object or another monkey, the isolated baby withdrew in apparent terror.

The isolation experiment demonstrated that a young rhesus needs social experience to develop normal social behavior. But what kind of social experience—and how much—is necessary? Interactions with a mother are insufficient for full social development of rhesus monkeys, since infants reared alone with their mothers failed to develop truly normal sexual, play, and aggressive behavior. Perhaps normal social development in rhesus monkeys requires the young animals to interact with one another. To test this hypothesis, the Harlows isolated some infants from their mothers, but gave these infants a chance to be with three other such infants for just *15 minutes* each day [515]. At first, the young rhesus monkeys simply clung to one another (Figure 3.28), but later they began to play. In their natural habitat, rhesus babies start to play when they are about 1 month old, and by 6 months they spend practically every waking moment in the company of their peers. Even so, the 15-minute play group developed nearly normal social behavior. As adolescents and adults, they were capable of interacting sexually and socially with other rhesus monkeys without exhibiting the intense aggression or withdrawal of animals that had no social contacts as infants.

Naturally one wonders about the relevance of these studies for another primate species, *Homo sapiens*, whose intellectual development is often said to be dependent on the early experiences that children have with their parents and peers. But is this true? We cannot, of course, do social isolation experiments with human babies but we can examine evidence of another sort on the resilience of intellectual development in the face of nutritional deprivation. Consider, for example, the results of a study of young Dutch men who were born or conceived during the Nazi transport embargo during the winter of 1944–1945, which caused many deaths from starvation in the larger Dutch cities [1150]. For most of the winter, the average caloric intake of city people was about 750 calories per day. As a result, urban women living under famine conditions produced babies of very low birth weights. In contrast, rural women were less dependent on food transported to them, and the babies they had that were conceived at the same time were born at more or less normal birth weights.

One would think that normal brain development would depend on adequate nutrition during pregnancy, when much of brain growth occurs. However, Dutch boys who were born in urban famine areas did not exhibit a higher incidence of mental retardation at age 19 than rural boys whose early nutrition was far superior (Figure 3.29A). Nor did those boys born to food-deprived mothers score more poorly than their relatively well-nourished rural counterparts when they took the Dutch intelligence test administered to draft-age men (Figure 3.29B) [1150]. These results are buttressed by the discovery that Finnish adults who experienced severe nutritional shortfalls in utero (during a 19th-century famine) lived just as long on average as those who were born after the famine was over [618].

No one believes that pregnant women or young children should be deprived of food [851], but the survival of a fetus and the intellectual development of a

(A)

(B)

Birth dates (months)

Figure 3.29   Developmental homeostasis in humans. Maternal starvation has surprisingly few effects on intellectual development in humans, judging from (A) rates of mild mental retardation and (B) intelligence test scores among 19-year-old Dutch men whose mothers lived under Nazi occupation while pregnant. (In this case, the lower the intelligence test score, the *greater* the intelligence of the subject.) The subjects were grouped according to the occupations of their fathers (Manual or Nonmanual) and whether their mothers lived or gave birth to them in a city subjected to food embargo by the Nazis (Embargo) or in a rural area unaffected by the embargo (No embargo). Individuals who were conceived or born under famine conditions exhibited the same rates of mental retardation and the same level of intelligence test scores as men conceived by or born to unstarved rural women. After Stein et al. [1150].

young person evidently can proceed successfully even under highly adverse conditions, perhaps because our developmental systems evolved in past environments in which episodes of starvation were not uncommon. The occurrence of normal development in stressful environments testifies to the adaptively guided, structured nature of the process, which can be thrown off course only by extremely unusual environmental shortfalls or severe genetic deficits.

The reproductive advantages of developmental homeostasis would be particularly great in species in which mate choice is based on body symmetry. In the barn swallow, for example, females have been reported to prefer males whose long outer tail feathers are the same length on each side [839], while females of the Iberian rock lizard associate preferentially with males endowed with a symmetrical distribution of pheromone-releasing pores on their thighs [789]. As for humans, some researchers have reported that both men and women find symmetry in facial features appealing (Figure 3.30) [574, 1016]. Perhaps prospective mates in these and other species respond positively to body or facial symmetry because these attributes announce the individual's capacity to overcome challenges to normal development [841]. Disruptions to development caused by mutations or by an inability to secure critical material resources early in life are thought to generate asymmetries in appearance. If body asymmetry reflects suboptimal development of the brain or other important organs, then a preference for symmetrical traits (or attributes closely allied with them) could enable the selective individual to acquire a partner with "good genes" to be transferred to their offspring. Alternatively, the benefit to the choosy individ-

Normal   High   Perfect

Figure 3.30 Facial symmetry and attractiveness. These images of human faces have been digitally manipulated to show varying degrees of symmetry. When asked to rate these faces, most persons find the image on the far right to be most attractive. After Rhodes et al. [1016].

ual in a parental species could be the acquisition of a mate in excellent physiological condition who could provide superior care for its young. In keeping with this prediction, the faces of human females with a history of health problems are judged less attractive, and the same is true of the faces of men whose caretakers had little income [574].

Debate exists on all aspects of this scenario, however [1171]. Although, as noted, some researchers report that asymmetrical individuals have indeed experienced developmental difficulties and deficits [49, 840], other researchers disagree [106, 349]. Moreover, although symmetrical individuals apparently enjoy a mating advantage in some species, in other species, no such advantage has been observed [720, 1206]. Finally, in some species in which a mate preference for symmetry has been reported, the differences between preferred symmetrical and rejected asymmetrical individuals are often so slight that it seems unlikely that the degree of symmetry per se provides the basis for making the choice. Starlings, for example, have been shown to be simply incapable of perceiving the kinds of very small differences that characterize most naturally occurring body asymmetries in their species [1170]. Furthermore, in our species, when women are asked to rate photographs of men's faces in terms of their attractiveness, their rankings do correlate with male facial symmetry, but the same rankings emerge when the women are provided with photographs of only the left or right side of the face, thereby eliminating information about facial symmetry. These results suggest that facial symmetry correlates with some other feature that women actually use to make their judgments [1067].

One species that meets all the criteria for visual mate choice based on body symmetry is the brush-legged wolf spider, whose males wave their hairy-tufted forelegs at females during courtship. Males that have larger tufts on one leg than the other (asymmetrical males) tend to be smaller and in poorer body condition than those whose tufts are symmetrical. To determine whether tuft symmetry was important, George Uetz and Elizabeth Smith took advantage of the willingness of female wolf spiders to signal their sexual receptivity while watching a video of courting males played on a tiny Sony Watchman microtelevision (Figure 3.31). Uetz and Smith recorded the reactions of females to digitally manipulated videotapes of a courting spider that were identical in every respect except for the degree to which the male's foreleg tufts were symmetrical. Female spiders signaled their readiness to mate (by raising the abdomen) more often when they saw the symmetrical male, showing that they found this kind of individual more sexually stimulating than the asymmetrical male the researchers had created digitally [1221]. At least in species of this sort, developmental homeostasis seems highly likely to confer a reproductive advantage by increasing the odds that individuals will be able to attract mates and leave descendants.

The Adaptive Value of Developmental Switch Mechanisms

The effect of developmental homeostasis is often a restriction in the degree of variation among individuals, which as a result have a greater likelihood of

Figure 3.31 **Testing mate choice in a female wolf spider.** The female (on the floor of the arena, to the left), responds to a moving image of a displaying male on the screen of a tiny television (to the right of the female). Photograph by George Uetz.

(A) Broadly variable
    phenotype

(B) Highly canalized
    phenotype

(C) Polyphenism

Phenotypic score

(D) Food-induced polyphenism

(E) Socially-induced polyphenism

(F) Predator-induced polyphenism

Figure 3.32 Developmental switch mechanisms can produce polyphenisms within the same species. Different phenotypes can arise when developmental switch mechanisms are activated in response to critical environmental cues. Top panel: Phenotypic variation within a species can range from (A) continuous, broad variation about a single mean value to (B) continuous but narrow variation about a single mean value to (C) discontinuous variation that generates several distinct peaks, each representing a different phenotype. Bottom panel: (D) In some cases, the amount or nature of the food eaten contributes to the production of certain polyphenisms, as in the castes of ants and other social insects. (E) In others, social interactions play a key role in switching phenotypes, as in the territorial and nonterritorial forms of the cichlid *Haplochromis burtoni*. (F) In still other instances, the presence or activity of predators contributes to the development of an antipredator phenotype, as in the soldier caste (left) of some aphids, which possess more powerful grasping legs and a larger, stabbing proboscis than nonsoldier forms (right). D, photograph by Mark Moffett; E, photograph by Russ Fernald; F, photograph by Takema Fukatsu; from Ijichi et al. [582].

acquiring an adaptive phenotype, such as a symmetrical body, rather than a less effective version of it. But there are a considerable number of species in which two or three quite distinct alternative phenotypes coexist comfortably, with the differences arising as a result of environmental differences among the individuals in question (Figure 3.32) [879, 1273]. At the proximate level, one challenge associated with such **polyphenisms** is to identify the environmental cues that activate the developmental mechanisms that steer development down one or another pathway (the process of "canalization"), so that an individual acquires one or another distinct phenotype, rather than any of a variety of intermediates between the alternative forms.

A representative example of polyphenism is provided by a tiger salamander in which there are two immature forms: (1) a typical aquatic larva, which eats small pond invertebrates such as dragonfly nymphs, and (2) a cannibal form, which grows much larger, has more powerful teeth, and feeds on other tiger salamander larvae unfortunate enough to live in its pond (Figure 3.33). The development of the cannibal type, with its distinctive form and behavior, depends on certain factors in the salamanders' social environment. For example, cannibals develop only when many salamander larvae live together [233]. Moreover, they appear more often when the larvae in a pond (or aquarium) differ greatly in size, with the largest individual much more likely to become a cannibal than its smaller companions [779]. In addition, the cannibal form is more likely to develop when the population consists largely of unrelated individuals than when many siblings live together [945]. If a larger-than-average salamander larva occupies a pond with many other young salamanders that do not smell like its close relatives [947], its development may well be switched from the typical track to the one that produces a giant, fierce-toothed cannibal. Thus, at the proximate level, any of several environmental cues can activate the developmental pathway leading to the cannibal form.

Figure 3.33   Tiger salamanders occur in two forms. The typical form (being eaten by its companion) feeds on small invertebrates and grows more slowly than the cannibal form (which is doing the eating). Cannibals have broader heads and larger teeth than their insect-eating companions. Photograph by David Pfennig, courtesy of James Collins.

What fitness benefits do tiger salamanders derive from having two possible developmental pathways and a switch mechanism that enables them to "choose" how to grow and behave? Individuals with some developmental flexibility may do better at coping with an environment with two or three distinct niches than individuals stuck with a one-size-fits-all phenotype. Larval salamanders are faced with two distinctly different sources of potential nutrients: insect prey and their fellow salamanders. If numerous salamander larvae occupy a pond, and if most are smaller than the individual that becomes a cannibal, then shifting to the cannibal phenotype gives that individual access to an abundant food source that is not being exploited by its fellows, so it can grow quickly. But a relatively small individual that was locked into becoming a big-jawed cannibal form would surely starve to death in a pond that lacked numerous potential victims of appropriate size. Because salamanders have no way of knowing in advance which of two food sources will be more available in the place where they happen to be developing, and because the two food sources are very different, selection appears to have favored individuals with the ability to develop in either one of two ways depending on the information they receive from their environments.

More generally, any time there are discrete ecological problems to be solved that require different developmental solutions, the stage may be set for the evolution of sophisticated developmental switch mechanisms that enable individuals to develop the phenotype best suited for their particular circumstances. The existence of two nonoverlapping categories of food (large versus small) or two levels of risk (predators present versus predators absent) or two kinds of social environments (superior competitors present or not) may select for the kind of developmental mechanism that can produce the one "right" specialist phenotype for the occasion, rather than a mechanism that generates a full range of intermediates.

So, for example, the fact that males of the cichlid fish *Haplochromis burtoni* (see Figure 3.32E) are either competitively superior or socially inferior to oth-

Figure 3.34   Activity of the gene that codes for gonadotropin-releasing hormone in the cichlid fish *Haplochromis burtoni*. After males switch from nonterritorial to territorial status, the GnRH gene becomes increasingly active over time in certain brain cells. Conversely, those males that switch from territorial to nonterritorial status show reduced activity in the GnRH gene. After White [1280].

ers helps explain why they have the capacity to shift between two different phenotypes. In this fish, males compete for a mate-attracting territory with winners holding sites until ousted by a stronger intruder. In such an either/or social environment, it pays to be either aggressively territorial (and to signal that state with bright colors) or nonaggressive (and to signal that state with dull colors) [386, 406]. Fish that behave in some intermediate fashion will almost certainly fail to hold a territory against motivated rivals, but will also fail to conserve their energy, which they can do only by dropping out (at least temporarily) from territorial competition. To this end, the fish respond to changes in their social status with changes in gene activity (Figure 3.34) within specific brain cells [1280]; as a result, the brain cells in question either grow in size, for fish that have acquired a territory, or decrease in size, for fish that have lost a territory, with attendant changes in behavior.

Discussion Question

3.14   Some marine fishes exhibit a spectacular polyphenism in that individuals can, under special circumstances, change their sex from female to male (in other species, the switch goes from male to female). This developmental change involves reproductive organs, hormones, and mating behavior [1251]. A key social cue for the switch in some species is a change in the makeup of the social unit in which the sex-altering individual lives; the removal of a dominant, breeding male from a cluster of females triggers a sex change in the largest female present. Here we have a case of socially induced polyphenism. Identify the apparent restrictions imposed on this system, starting with the most obvious one; namely, the ability to be transformed into a member of the opposite sex rather than some sort of intermediate sex. Speculate on the benefits associated with each restriction.

Although polyphenisms are common, they are far from universal, perhaps because many environmental features vary continuously rather than discontinuously. Under these conditions, individuals may not benefit from developmental systems that produce a particular phenotype targeted at a narrow part of the entire range of environmental variation. Moreover, the ability to adopt one of two or more phenotypes come with a price, such as the costs imposed by the negative side effects of the genes that promote flexible developmental responses. There may also be a cost associated with detecting and responding to the cues that signal one kind of environmental condition as opposed to another—a necessary step in the activation of the developmental switch mechanisms that steer individuals away from one phenotypic outcome and toward another [320].

The Adaptive Value of Learning

Learning is the adaptive modification of behavior based on experience. As such, it can be considered a polyphenism of sorts because it too confers a highly focused behavioral flexibility that requires developmental modifications in the nervous system. Learning does not produce behavioral change for the sake of change alone. Instead, selection favors investment in the mechanisms underlying learning only when there is environmental unpredictability that has reproductive relevance for individuals. What we have here is another cost–benefit argument, which presupposes that any proximate mechanisms that enable individuals to learn come with a price tag. We can check this assumption by predicting, for example, that the brains of male long-billed marsh wrens living in the western United States should be larger than those of their East Coast counterparts because young West Coast wrens learn nearly 100 songs by listening to others, whereas East Coast wrens have much smaller learned repertoires of about 40 songs [671]. When the birds' brains were examined, the song control systems of West Coast wrens weighed on average 25 percent more than the equivalent nuclei of East Coast wrens.

If learning mechanisms are costly, then we would expect learning to evolve only when there is some major counterbalancing benefit. As mentioned at the outset of this chapter, people tend to think of insects as instinct-driven automatons, but if we turn again to honey bees, we find that they are quite capable of learning where to search for food, what odors, shapes, and colors are associated with different pollen- or nectar-producing flowers, when during the day a particular plant species will open its flowers, how to get back to the hive after a foraging expedition, and much more [1240]. These abilities are all related to the fact that the conditions a worker bee will encounter cannot be precisely predicted before she goes out to forage. Instead, selection has favored a bee brain that can incorporate information about key variables in the bee's environment—information that alters genetic activity in brain cells, changing the structure of the brain, and ultimately modifying the behavior of the individual so that she can better exploit the particular pattern of food resources in her neighborhood.

Likewise, male thynnine wasps exhibit a special spatial learning ability that comes into play when a mimetic sex pheromone is released by a freshly opened orchid flower. Various orchids possess flowers with female decoy petals that smell and look (vaguely) like a female thynnine wasp. Males can be fooled into rushing to these flowers and attempting to mate with the petal (Figure 3.35) [1155], an activity that pollinates the orchid when an individual that has been tricked once comes to a second orchid and transfers pollen from orchid 1 to orchid 2. But having once been deceived by a particular flower, male wasps evidently learn to avoid the spot where that flower occurs, which explains why, when researchers move an orchid to a new spot, large numbers of males show

Figure 3.35 Male thynnine wasps can be deceived into "mating" with a flower. (A) A wingless female thynnine wasp that is releasing a sex pheromone to attract males. (B) Some Australian orchids possess flowers with a female decoy petal that can stimulate males into attempting to copulate with it. The orchid may secure a pollinator in this fashion. Note the yellow pollen sacs stuck to the male's back. Photographs by the author.

(A)

(B)

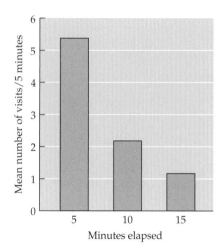

Figure 3.36   Male thynnine wasps can learn to avoid being deceived by an orchid. The number of visits to a deceptive orchid soon falls after the male wasps in an area have interacted with it and learned that an unrewarding source of sex pheromone is associated with that particular location. After Peakall [933].

up initially, but then fly away and do not return (Figure 3.36). As a result of grappling with a particular orchid's decoy petal, males evidently store information about the experience in their brains and thereafter avoid responding sexually to the scent *coming from that particular site* [933]. The reproductive benefits of the male wasp's behavioral flexibility are clear. Male wasps cannot be programmed in advance to know where female wasps and deceptive orchid flowers are on any given day. By using experience to learn where particular orchids are (in order to avoid them) while remaining responsive to novel sources of sex pheromone, the male wasp saves time and energy and improves his chance of encountering a receptive female that has begun to release sex pheromone.

That spatial learning evolves in response to particular ecological pressures can also be seen by comparing the learning abilities of four bird species, all members of the crow family (Corvidae), that vary in their predisposition to store food—a task that puts a premium on spatial memory. As we have seen, Clark's nutcracker is a food-storing specialist, and it has a large pouch for the transport of pine seeds to storage sites. The pinyon jay also has a special anatomical feature, an expandable esophagus, for carrying large quantities of seeds to hiding places. In contrast, the scrub jay and Mexican jay lack special seed transport devices and appear to hide substantially less food than their relatives.

Individuals from the four species were tested on two different learning tasks in which they had to peck a computer screen to receive rewards. One task required the birds to remember the *color* of a circle on the screen (a nonspatial learning task), and the other required them to remember the *location* of a circle on the screen (a spatial task). When it came to the nonspatial learning test, pinyon jays and Mexican jays did substantially better than scrub jays and nutcrackers. But in the spatial learning experiment, the nutcracker went to the head of the class, followed by the pinyon jay, then the Mexican jay, and, finally, the scrub jay (Figure 3.37). These results show that the birds have not evolved all-purpose learning abilities; instead, their learning skills are designed to promote success in solving the special problems that they face in their natural environments [895].

The logic of an evolutionary approach to learning leads us to expect that if males and females of the same species differed in the benefits derived from a particular learned task, then a sex difference in learning skills should evolve.

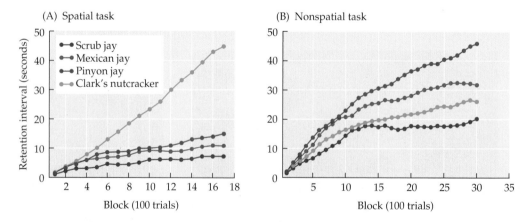

Figure 3.37   Spatial learning abilities differ among members of the crow family. (A) Captive Clark's nutcrackers performed much better than three other corvid species in experiments that required the birds to retain information about the location of a circle. (B) But when the birds' ability to remember the color of a circle was tested, the nutcrackers did not excel in this nonspatial learning task. After Olson et al. [895].

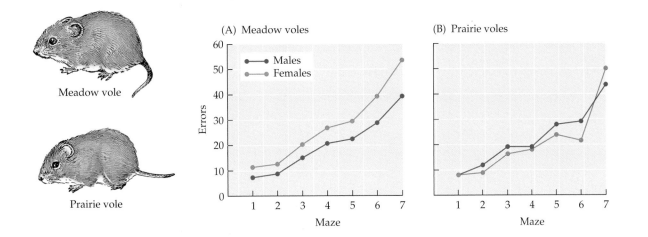

Steve Gaulin and Randall FitzGerald tested this prediction in studies of three species of voles, all members of the genus *Microtus*. Males of the polygynous meadow vole have home ranges more than four times as large as those occupied by each of their several mates. In contrast, males and females of the monogamous prairie vole (see Figure 1.1) and the monogamous pine vole share the same living space. When tested in a variety of mazes, which the animals had to solve in order to receive food rewards, males of the wide-ranging meadow vole consistently made fewer errors than females of their species. In both monogamous voles, however, males and females did equally well on these spatial learning tests (Figure 3.38); in these species, the two sexes usually have similar home ranges and so are confronted with equivalent spatial learning problems [437, 438]. Given these findings, you should not be surprised to learn that male meadow voles invest more heavily in the hippocampus than do females of their species, whereas no differences in hippocampal size exist between male and female pine voles [586].

Figure 3.38   Sex differences in spatial learning ability are linked to home range size. Spatial learning in voles was tested by giving individuals opportunities to travel through seven different mazes of increasing complexity in the laboratory and then letting them run through each maze again. (A) Polygynous male meadow voles, which roam over wide areas in nature, consistently made fewer errors (wrong turns) on average than the more sedentary females of their species. (B) In contrast, females matched male performance in the monogamous prairie vole, a species in which males and females live together on the same territory. After Gaulin and FitzGerald [437].

Discussion Question

3.15   In a study in which men and women were asked to sit at a computer and navigate through a virtual maze (Figure 3.39), the men were able to complete the task more quickly and with fewer errors over five trials than the women [835]. (Note, however, that in other tests involving language skills, women score higher on average than men.) What possible proximate developmental mechanisms might be responsible for this sex difference in navigational ability? Keeping in mind the evolutionary explanation for sex differences in spatial learning ability in voles, what prediction can you make about the nature of human mating systems over evolutionary time?

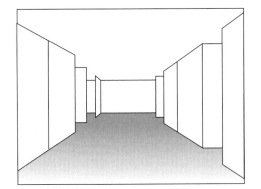

Figure 3.39   A virtual maze used for computer-based studies of navigational skills. After Moffat [835].

Figure 3.40   **Sex differences in the hippocampus.** Female brown-headed cowbirds have a larger hippocampus than males, as would be expected if this brain structure promotes spatial learning and if selection for spatial learning ability is greater on female than on male cowbirds. Red-winged blackbirds and common grackles do not exhibit this sex difference. After Sherry et al. [1100].

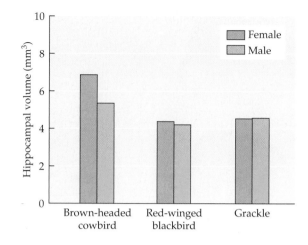

In a species in which females face greater spatial challenges than males, we would expect females to make larger investments in the expensive neural foundations of spatial learning. The brown-headed cowbird is such a species, because cowbirds are **brood parasites** that lay their eggs in other birds' nests. Females must search widely for nests to parasitize, and they must remember where potential victims have started their nests in order to return to them one to several days later when the time is ripe for the cowbird to add her egg to those already laid. In contrast, male cowbirds do not confront such difficult spatial problems. As predicted, the hippocampus (but no other brain structure) is considerably larger in female brown-headed cowbirds than in males. No such difference occurs in some nonparasitic relatives of this species [1100] (Figure 3.40).

Moreover, it is not just spatial learning that bears the clear imprint of natural selection. Consider **operant conditioning**, in which an animal learns to associate a voluntary action with the consequences that follow from that action [1118]. Operant conditioning (or trial-and-error learning) does occur outside psychology laboratories, but it has been studied extensively in Skinner boxes, named after the psychologist B. F. Skinner. After a white rat has been introduced into a Skinner box, it may accidentally press a bar on the wall of the cage (Figure 3.41), perhaps as it reaches up to look for a way out. When the bar is pressed down, a rat chow pellet pops into a food hopper. Some time may pass before the rat happens upon the pellet. After eating it, the rat may continue to explore its rather limited surroundings for a while before again happening to press the bar. Out comes another pellet. The rat may find it quickly this time, and then turn back to the bar and press it repeatedly, having learned to associate this particular activity with food. It is now operantly conditioned to press the bar.

Skinnerian psychologists once argued that one could condition with equal ease almost any operant (defined as any action that an animal could perform). Indeed, the successes of operant conditioning are legion, including the ability to get nutcrackers and jays to become computer users, as mentioned above. Certainly, one can readily condition a white rat to do a great many things, such as running in a running wheel, by sounding a warning noise and giving it an electric shock if it fails to get going. A rat that has had some experience with hearing the sound and receiving a shock while standing in the wheel, but not while running, will make the appropriate association and start running whenever it hears the warning signal [113]. However, Robert Bolles found that a rat cannot be conditioned to stand upright in the running wheel using the same procedures. Rats that happen to rear up just after the sound and are not shocked nev-

Figure 3.41   Operant conditioning exhibited by a rat in a Skinner box. The rat approaches the bar (top left) and then presses it (top right). The animal awaits the arrival of a pellet of rat chow (bottom left), which it consumes (bottom right), so that the bar-pressing behavior is reinforced. Photographs by Larry Stein.

ertheless fail to make the association between this behavior and avoidance of the shock. In fact, the frequency with which they stand up when they hear the warning signal actually *declines* over time (Figure 3.42).

Likewise, John Garcia and his colleagues found that white rats can learn to avoid some, but not all, sensory cues that are associated with certain consequences [433, 434]. For example, if a rat feeds on or drinks a novel, distinctively flavored food or liquid and then is exposed to X-rays, the animal becomes ill, because even tiny doses of radiation cause a buildup of toxic chemicals in irradiated tissues and body fluids. Subsequently, the rat often refuses to touch the particular substance that it consumed before becoming sick. The degree to which a food or fluid is avoided is proportional to (1) the intensity of the resulting illness, (2) the intensity of the taste of the substance, (3) the novelty of the substance, and (4) the shortness of the interval between consumption and illness [434]. But even if there is a long delay (up to 7 hours) between eating a distinctively flavored food and exposure to radiation and consequent illness, the rat still links the two events and uses the information to modify its behavior. In contrast, most kinds of associative learning will not take place when the interval between a cue and its consequence lasts more than a few seconds.

The specialized or biased nature of this taste aversion learning is further shown by the rat's complete failure to learn that a distinctive sound (a click) is a signal that always precedes an event associated with nausea. Nor can rats eas-

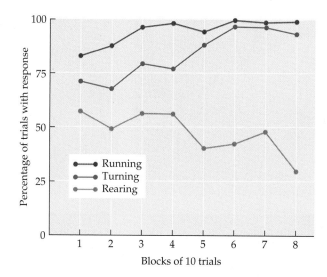

Figure 3.42   Biased learning. Learning curves of white rats for three operants (running, turning, rearing) that were equally rewarded (the rats were not shocked when they performed the appropriate operant). Rats failed to learn to rear up on their hind legs to avoid a shock. After Bolles [113].

ily make an association between a distinctive taste and shock punishment (Figure 3.43). If, after drinking a sweet-tasting fluid, the rat receives a shock on its feet, it often remains as fond of the fluid as it was before, as measured by the amount drunk per unit of time, no matter how often it is shocked after drinking this liquid. These failures surely relate to the fact that in nature, particular sounds are never associated with illness-inducing meals, any more than the consumption of certain fluids causes a rat's feet to hurt.

Understanding the natural environment of the ancestor of white rats, the Norway rat, also helps explain why white rats are so adept at learning to avoid novel foods with distinctive tastes that are associated with illness, even hours after ingesting the food. Under natural conditions, a Norway rat becomes completely familiar with the area around its burrow, foraging within that area for a wide variety of foods, plant and animal [741]. Some of these foods are edible

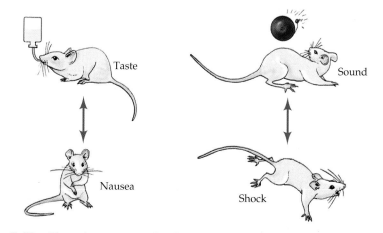

Figure 3.43   Biases in taste aversion learning. Although white rats can easily learn that certain taste cues will be followed by sensations of nausea and that certain sounds will be followed by skin pain caused by shock, they have great difficulty forming learned associations between taste and consequent skin pain or between sound and subsequent nausea. After Garcia et al. [434].

and nutritious; others are toxic and potentially lethal. A rat cannot clear its digestive system of toxic foods by vomiting. Therefore, the animal takes only a small bite of anything new. If it gets sick later, it avoids this food or liquid in the future, as it should because eating large amounts might kill it [434]. This case suggests that even what appears to be a general, all-purpose form of learning is actually a specialized response to particular kinds of biologically significant associations that occur in nature.

If this argument is correct, then other mammals that are dietary generalists, which also run the risk of consuming dangerous, toxic items, should behave like the Norway rat, which is to say that they should also quickly form taste aversions to bad-tasting, illness-inducing items. And they do. Three bat species that feed on a range of foods behaved in the predicted manner: they rapidly formed taste aversions when fed a meal laced with an unfamiliar flavor, cinnamon or citric acid, then injected with a chemical that made them vomit. When later offered a choice between food with and without cinnamon or citric acid, these three generalists avoided foods spiced with the novel additive.

We can also predict that dietary specialists, which concentrate exclusively on one or a very few safe foods, should be unable to acquire taste aversions in this manner. A study of the vampire bat, a blood-feeding specialist, has shown that this animal is quite incapable of learning that consumption of an unusual-tasting fluid will lead to gastrointestinal distress [1002] (Figure 3.44). The difference between the specialist vampire bat and its generalist relatives supports the hypothesis that taste aversion learning is an evolved response to the risk of food or fluid poisoning. Just as is true for all aspects of behavioral development, the changes associated with learned behavior are worth the cost only if they confer a net fitness benefit on individuals capable of modifying their behavior in a particular way.

Vampire bat

Figure 3.44 Vampire bats cannot form learned taste aversions. Vampire bats continued to consume a flavored fluid even if immediately after accepting this novel substance, they were injected with a toxin that caused gastrointestinal distress. In contrast, three insect-eating bat species completely rejected the novel dietary item when it was combined with injection of the toxin, no matter whether this was done immediately after feeding or after a delay. Two control groups were also used in the experiment, one in which the consumption of the novel food was paired with a harmless injection of saline solution, and another in which the toxin was injected, but not in conjunction with feeding on the fluid. After Ratcliffe et al. [1002].

Summary

1. The development of any trait is the result of an interaction between the genotype of a developing organism and its environment, which consists not only of the food it receives and the metabolic products produced by its cells (the material environment), but also its sensory experiences (the experiential environment). The value of genetic information lies in the ability of genes to respond to signals from the environment by altering their activity, leading to changes in the gene products available to the developing organism.

2. Because development is interactive, no measurable product of development (a phenotype) can be genetically determined. The statement, "In garter snakes, there is a gene for eating banana slugs," is shorthand for the following: "A particular allele in a garter snake's genotype codes for a distinctive protein; if the protein is actually made, which requires an interaction between the gene and its environment, the protein may influence the development or operation of the physiological mechanisms underlying the snake's ability to recognize slugs as food."

3. By the same token, the interactive nature of development means that no phenotype can be purely environmentally determined. The statement, "The rat's learned avoidance of rat poison is caused by its experience with this chemical," is shorthand for the following: "A specific experience with rat poison led to chemical changes in the rat's body, which were translated eventually into chemical changes in the rat's brain cells. These changes in turn altered the pattern of genetic activity in some parts of its nervous system and thereby modified the rat's response to the poison upon a second encounter with this stimulus."

4. Because development is interactive, changes in either the genetic information or the environmental inputs available to an individual can potentially alter the course of its development by changing the gene–environment interactions that take place within that individual. Therefore, the behavioral differences between two individuals can be genetically determined or environmentally determined, or both. Note that this claim is very different from the misconception that a given behavioral phenotype is caused either by an animal's genes or by its environment alone.

5. Because some differences between individuals are caused by genetic differences, populations have the potential to evolve by natural selection, which acts on genetic variation within groups. The ease with which behavior can evolve is demonstrated by the frequency with which artificial selection experiments produce behavioral evolution in the laboratory.

6. Because behavior can evolve, we expect to find that behavioral development has adaptive features. One such feature is developmental homeostasis, the capacity of the developmental process to ignore or overcome certain environmental or genetic shortfalls that might conceivably prevent animals from acquiring valuable traits that provide reproductive success. Indeed, normal physiological and behavioral phenotypes often develop in animals growing up in challenging, suboptimal environments and in animals burdened with potentially harmful mutations.

7. Other adaptive features include developmental switch mechanisms, which guide development into one of two or three alternative developmental pathways in response to specific environmental cues. Each of the resulting phenotypes can master the distinctive obstacles to success associated with its particular niche within the larger environment.

8. Another adaptive aspect of development involves learning mechanisms that can respond to particular environmental inputs related to individual experience and that generate functional changes in the behavior of animals. Learning, like other forms of developmental flexibility, reflects past selection for the capacity to make adaptive adjustments in behavior that match the environment of the individual.

Suggested Reading

Gene Robinson discusses how to integrate molecular biology, developmental biology, neurobiology, and evolutionary biology in the context of the nature–nurture debate [1024, 1025]. Environmental effects on development are nicely illustrated by studies of kin recognition, a topic that is the subject of an entire recent issue of the journal *Annales Zoologici Fennici* (2004, volume 41, number 6). The work on blackcap warbler migration provides a fine example of how the genetic effects on development can be examined [95, 96]. Ralph Greenspan provides an up-to-date review of some alternative approaches to behavior genetics [476]. You can also gain a sense of the controversy about how to analyze the development of human behavior, a topic not covered directly in this chapter, by reading papers by Thomas Bouchard [126, 127] in conjunction with a counterview from Marla Sokolowski and Doug Wahlsten [1135]. See also Jeremy Gray and Paul Thompson's review of a range of approaches to the genetics of human intelligence [472].

The Control of Behavior: Neural Mechanisms

Because nervous systems are the foundation for animal behavior, behavioral biologists have wanted to learn how these systems work. Although our current understanding of these systems has been helped by the use of all manner of impressive technological devices, even simple observations of animals in action can sometimes provide considerable information about the decision-making properties of neural mechanisms. For instance, I learned something about these mechanisms in the bee *Centris pallida* when I observed male bees trying to copulate with my thumb. This unseemly behavior occurred while I was studying whether large male bees have a mating advantage in this species. My research required that I capture coupled pairs of *Centris pallida* (Figure 4.1A), remove the male from the female, and measure his body size. While holding a male one day, I found more or less by accident that if I placed the bee whose copulation I had so unkindly terminated on my upturned thumb (Figure 4.1B), he would grasp it firmly and stroke it with his legs and antennae as if he were holding a female of his species. (Incidentally, male bees don't sting, so I was neither courageous nor stupid when I offered my thumb to the bee.) Despite the fact that my thumb has only the vaguest similarity to a female *Centris pallida*, it was evidently close enough for some males of this species.

(A)

(B)

Figure 4.1   A complex response to simple stimuli. (A) A male bee copulating with a female of his species. (B) A male bee attempts (unsuccessfully) to copulate with the author's thumb. Photographs by the author.

Even though I am not a neurophysiologist, I could tell that the bee's nervous system had some special operating rules. Apparently, when a sexually motivated male *Centris pallida* grasps any object approximately the size of a female of its species, the sensory signals generated by its touch receptors are sent to other parts of its nervous system, which produce messages that eventually translate into a complex series of muscle commands. The behavioral result is the sequence of movements that passes for courtship and mating in *Centris pallida*. That these activities can be stimulated by my thumb instead of a female bee indicates that the nervous system of a male *Centris pallida* is not terribly discriminating.

Nor is "my" bee at all unusual in this respect, given that some beetles will attempt to copulate with beer bottles or signs (Figure 4.2). And it is not just insects whose nervous systems employ rules of thumb as simple as "attempt to insert penis in yellow or orange objects." Although a toad has a much larger

(A)

(B)

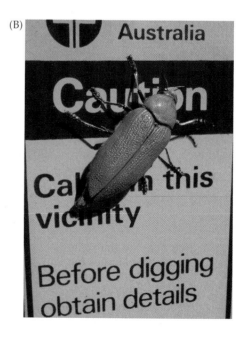

Figure 4.2   A simple rule of thumb governs this beetle's mating behavior. This large Australian beetle will attempt to mate with any object with approximately the same color as a female of the species, such as (A) a beer bottle or (B) a telecommunication sign. A, photograph courtesy of Darryl T. Gwynne; B, photograph by the author.

Figure 4.3   Pioneers in the study of animal behavior. The three men who won the Nobel Prize in Medicine in 1973 were Niko Tinbergen (left), Konrad Lorenz (middle), and Karl von Frisch (right). These ethologists received the prize in recognition of their contributions to an understanding of the proximate causes of behavior. Photographs by (left) B. Tschanz, (middle) Sybille Kalas, and (right) O. von Frisch.

brain than a bee or beetle, male toads have been known to embrace the toe of a boot, treating this rubbery object as if it were a sexually receptive female, which it is not.

The tendency of nervous systems to generate complex responses to simple stimuli is widespread. This phenomenon has attracted the attention of biologists who have wanted to figure out how neurons or neural networks acquire information from objects in the environment and then order the animal to respond in special ways. Thanks to research of this sort, we now know a great deal about such things as how moths flying at night can detect and avoid hungry bats rushing in for the kill, how the star-nosed mole uses its amazing nose to locate tasty worms in its underground tunnels, and how birds, butterflies, and sea turtles can accurately navigate over great distances to reach a particular destination. The proximate mechanisms that make these feats possible have adaptive properties that help individuals survive and reproduce in the environments utilized by their species. The interaction between proximate and ultimate analyses of behavioral control systems is the focus of this chapter.

How Neurons Control Behavior

When I was deceiving males of *Centris pallida*, I was not terribly surprised by the obtuse behavior of my insect friends, having read Niko Tinbergen's *The Study of Instinct* [1196]. There he describes how he and Konrad Lorenz (Figure 4.3) fooled many animals into providing elaborate behavioral responses to stimuli that barely resembled the natural object that normally triggers their behavior. For example, when gull chicks are hungry, they peck at the tip of a parent's bill (Figure 4.4). The parent typically responds by regurgitating a half-digested fish or other delicacy, which the chicks eagerly consume. You might think that a baby gull would not recognize its parent's bill unless the whole bird was present. How-

Figure 4.4   Begging behavior by a gull chick. A silver gull chick is being fed regurgitated food by its parent after pecking at the adult's bill. Photograph by the author.

Figure 4.5 **Effectiveness of different visual stimuli** in triggering the begging behavior of young herring gull chicks. Two-dimensional cardboard cutouts of the head of a gull with a red dot on its bill (A) are not much more effective in eliciting begging behavior in a gull chick than are models of the bill alone (B), provided the red dot is present. Moreover, a model of a gull head without the red dot (C) is a far less effective stimulus than is an unrealistically long "bill" with contrasting bars at the end (D). After Tinbergen and Perdeck [1201].

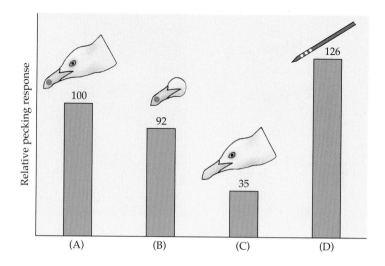

ever, two-dimensional cardboard cutouts of a herring gull head elicit pecking, and so do painted sticks with a high-contrast dot or band in the right place (Figure 4.5) [1201]. Tinbergen's experiments with these and other models revealed that herring gull chicks, at least very young ones, apparently ignore almost everything except the shape of the bill and its red dot. Tinbergen proposed that when a young gull sees certain very simple stimuli, sensory signals are relayed by neurons to its brain, where other neurons eventually generate the motor commands that cause the chick to peck at the stimulus—whether it is located on its mother's bill or a piece of cardboard or the end of a stick.

Tinbergen and Lorenz collaborated on another famous experiment that identified a simple stimulus capable of triggering a complex behavior. They found that if they removed an egg from under an incubating greylag goose and put it a half meter away, the goose would retrieve the egg by stretching its neck forward, tucking the egg under its lower bill, and rolling the egg carefully back into its nest. If they replaced the goose egg with almost any roughly egg-shaped object, the goose would invariably run through its egg retrieval routine. And if the researchers removed the object as it was being retrieved, the bird would continue pulling its head back just as if an egg were still balanced against the underside of its bill [1196]. From these results, Tinbergen and Lorenz concluded that the goose must have a special perceptual mechanism that is highly sensitive to a broad class of visual cues provided by eggs (and other objects of similar shape). Moreover, that sensory mechanism must send its information to a command center that automatically activates an invariant motor program for egg retrieval.

The gull chick's pecking response and the greylag goose's retrieval behavior are only two of dozens of instincts that Tinbergen, Lorenz, and other ethologists have studied. These founders of **ethology,** the discipline dedicated to the study of both the proximate and ultimate causes of animal behavior, were especially interested in the instincts exhibited by wild animals living under natural conditions. An **instinct** can be defined as a behavior pattern that appears in fully functional form the *first time* it is performed, even though the animal may have had no previous experience with the cues that elicit the behavior. You may recall a good example from Chapter 3: the tongue-flicking response of baby coastal garter snakes to extract of banana slug. You may also remember that tongue flicking cannot be "genetically determined," nor can any other instinct, because these behaviors are dependent on the gene–environment interactions that took place during development. In the case of a herring gull chick, these interactions

Figure 4.6   Instinct theory. Tinbergen and Lorenz proposed that simple stimuli, such as the red dot on a parent gull's bill, can activate or release complex behaviors, such as a gull chick's begging behavior. This effect is achieved, according to these ethologists, because certain sensory messages from the releaser are processed by innate releasing mechanisms higher in the nervous system, leading to motor commands that control a fixed action pattern, a preprogrammed series of movements that constitute an adaptive reaction to the releasing stimulus.

Releaser
(visual signal)

Innate releasing
mechanism

led to the construction of a nervous system that contains a network that enables the little bird to identify the key components of an adult gull's bill and to peck at the red dot on that bill. The neural network responsible for detecting the simple cue (the **sign stimulus** or **releaser**) and activating the instinct, or **fixed action pattern** (FAP), was given the label of an **innate releasing mechanism** (Figure 4.6) [1196].

The simple relationship between an innate releasing mechanism, sign stimulus, and FAP is highlighted by the ability of some species to exploit the FAPs of other species, a tactic known as code breaking [1291]. In the preceding chapter, we discussed the orchids whose flower petals provide the visual and olfactory releasers that elicit approach and attempted copulation by certain male wasps, much to the insects' disadvantage (see Figure 3.35). Likewise, the chicks of avian brood parasites, such as the European cuckoo and the North American cowbird, exploit adults of other bird species that use simple rules of thumb when they decide which of their nestlings to feed [507, 1291]. Typically, a parent songbird that has returned to its brood with food favors cheeping youngsters that are able to reach up high, with head moving and mouth gaping. Cuckoo and cowbird nestlings grow rapidly and become larger than their hosts' own offspring, and therefore can generate these releasers of parental feeding better than their smaller nestmates [717]. Because the parasite begs for food so effectively, it gets more than its fair share, eventually growing into a demanding youngster larger than its foster parents (Figure 4.7).

The trickery of instinct exploiters can be highly effective, as demonstrated by bolas spiders that release mimetic sex pheromones in order to lure moths to within attack range. When a sexually motivated male moth comes flying

Figure 4.7   A code breaker. This young cuckoo begs for food from its foster parent, a reed warbler, which provides for the cuckoo at great cost to itself and its own offspring. Photograph by Ian Wyllie.

Figure 4.8   Bolas spider deception. This spider "lassoes" moths that are attracted within range by the spider's deceptive use of mimetic sex phero- mones. The droplet of sticky glue dangles by a silken thread from the spider's fore- leg. Photograph by Ken Haynes and Ken Yeargan.

upwind toward the scent-releasing spider, the predator swings a sticky ball of glue at the end of a line of silk at the insect (Figure 4.8). If the spider hits the moth with the "bola" of glue, the predator reels him in, having exploited the moth's innate FAP, "Fly upwind along a gradient of chemical X," which would usually take him to a female of his species, rather than to his death.

One bolas spider, *Mastophora hutchinsoni*, is known to produce two differ- ent olfactory releasers that attract males of two species of moths. The spider releases only one mimetic sex pheromone at a time, sending out plumes of the odor favored by male bristly cutworm moths in the evening before switching around 11 p.m. to the pheromone that attracts male smoky tetanolitas. I trust that you can guess when during the night males of these two moth species typ- ically go hunting for females [526].

An equally remarkable case of deceptive use of releaser chemicals involves a wasp that lays its eggs in the bodies of caterpillars of a small butterfly species. These caterpillars grow up in colonies of red ants, which feed on the butterfly's secretions and in return protect the larvae against predatory insects, including other ants. Should an outsider ant come hunting for butterfly larvae, the resi- dent ants quickly determine by its odor that it does not belong to their colony. They then swing into action, releasing an assortment of volatile hydrocarbons that raise the alarm, attract other worker ants to the unwelcome visitor, and stimulate a wave of aggression against it, often resulting in the dismemberment of the invader. The butterfly-hunting wasp, however, overcomes these defenses by secreting a cocktail of these very same chemicals, which first attracts ants to the wasp, but then repels them, sending them scurrying off to assault their colonymates instead of the wasp. As ant attacks fellow ant, the wasp makes its way safely to its butterfly victim, into which it inserts its eggs (Figure 4.9A).

The team of entomologists who made this discovery did so by coating Teflon wasp-like dummies with synthetic copies of the chemicals secreted by the wasp. Dummies with a single chemical stimulated different responses depending on the compound used. The complete mixture had the integrated effects found when a living wasp invades a red ant nest (Figure 4.9B) [1185]. The wasp uncon- sciously takes advantage of a whole series of red ant reactions, each one of which is elicited more or less automatically when the ants come in contact with a particular chemical.

Discussion Questions

4.1   Suggest how a modern behavioral biologist might explore the effect of a re- leaser, such as a red dot on a moving gull bill, in terms of changes in gene expres- sion and neural activity in selected portions of the brain of the gull chick. It might be helpful to review the relationships between bird song, neural networks, and the *ZENK* gene as presented in Chapter 2 and in [58].

4.2   Fiddler crabs live on tidal flats, where male crabs court females and where predators, such as gulls, hunt the crabs. Crabs often approach or ignore one anoth- er, but a gull flying toward them sends them dashing for their burrows in the mud. Crabs keep their eyestalks perpendicular to the flat ground on which they roam when they are outside their burrows (Figure 4.10). How would you test the hypoth- esis that fiddler crabs hold their eyestalks in that position in order to use the follow- ing simple rule of thumb: Flee from any moving stimulus that appears above the level of the eye (at the top of the stalk) and approach or ignore any moving stimulus that appears below eye level? After answering, check [702]. Apply ethological ter- minology (see Figure 4.6) to this system by identifying the releaser, the fixed action pattern, and the innate releasing mechanism.

Figure 4.9   Complex code breaking by a wasp. (A) A parasitic wasp on top of a butterfly larva that is its victim. (B) To get to the larva, which was being protected by a colony of red ants, the wasp produced a bouquet of volatile hydrocarbons (see X-axis). These chemicals release a sequence of responses by red ants, whose reactions are listed on the left. The responses of ants to wasp dummies coated with a mixture of these chemicals and with each component individually are shown in the five graphs. Note that the result of the wasp's multiple mimicry of ant chemical signals is to induce ants to flee from the wasp and instead to assault their fellow colony members. The data are presented as the ratio of the number of ants responding to the hydrocarbon bouquet (or its individual components) relative to a control solution composed of the solvents used to extract the pheromones from the wasps. A, photograph by Jeremy Thomas; B, after Thomas et al. [1185].

(A)

(B)

Sensory Receptors and Survival

The behavior of red ants exposed to chemical secretions from the deceptive wasp can be described in ethological terms with reference to releasers, innate releasing mechanisms, and fixed action patterns. However, one need not employ ethological jargon while studying how simple cues trigger generally adaptive, but occasionally exploited, responses in animals of all sorts. Kenneth Roeder, for example, studied how night-flying moths manage to escape from bats without reference to innate releasing mechanisms and the like. His interest in the neural control of escape behavior was piqued when he spent some time outdoors at night in summer, armed with "a minimum amount of illumination, perhaps a 100-watt bulb with a reflector, and a fair amount of patience and mosquito repellent" [1027]. With these items in place, you too might see a moth that has been attracted to the light turn abruptly away right before a bat swoops into

Figure 4.10   The eyestalks of a fiddler crab point straight up, holding the eyes in a plane perpendicular to the ground. Photograph by John Christy.

view. Or you might see a moth dive straight down just as a bat appears. You can also sometimes get moths approaching a light to turn away abruptly, or even plummet earthward, if you jangle a set of keys. The moths' responses suggest that they can detect a simple acoustical cue, which provides the trigger for a particular behavior, just as minimal visual cues are sufficient to activate begging behavior in a baby gull.

As it turns out, the hypothesis that acoustical stimuli trigger the turning or diving behavior of moths is correct, but the sounds that the moth hears when the keys are rattled together are not the sounds that you and I hear. Instead, the moth detects the very high-frequency sounds produced by the clashing keys, which makes sense when you consider that most night-hunting bats vocalize ultrasonically using sound frequencies between 20 and 80 kilohertz (kHz)—well outside the hearing range of humans, but not moths.

Bats use ultrasonic calls to navigate at night—something that was not suspected until the 1930s, when researchers with ultrasound detectors were able to eavesdrop on the sounds produced by flying bats. Donald Griffin suggested that night-flying bats use high-frequency cries in order to listen for weak ultrasonic echoes reflected back from objects in their flight path [481]. This echolocation hypothesis for bat navigation initially had many skeptics, but they changed their minds when they read about Griffin's experiments with some captive little brown bats, a common North American species. When Griffin placed the bats in a dark room filled with fruit flies and wires strung from ceiling to floor, his subjects had no trouble catching the insects while negotiating the obstacle course—until Griffin turned on a machine that filled the room with high-frequency sound. As soon as the machine-produced ultrasound bombarded the bats, they began to collide with obstacles and crash to the floor, where they remained until Griffin turned off the jamming device. In contrast, loud sounds of 1–15 kHz (which humans can hear) had no effect on the bats because these stimuli did not mask the high-frequency echoes these predators listen to as they fly in the dark. Griffin rightly concluded that the little brown bat employs a sonar system to avoid obstacles and detect prey at night.

As Roeder watched moths evading echolocating bats, he felt sure that the insects were able to hear pulses of bat ultrasound. He knew that if he was right, then he should be able to find ears on some moths. Taking advantage of some earlier work on moth hearing organs, he located the structures that noctuid moths use to hear ultrasound (Figure 4.11). As it turns out, a noctuid moth has

Figure 4.11   Noctuid moth ears.
(A) The location of the ear. (B) The design of the ear, which features two auditory receptors (A1, A2) linked to a tympanum that vibrates when exposed to sounds. After Roeder [1027].

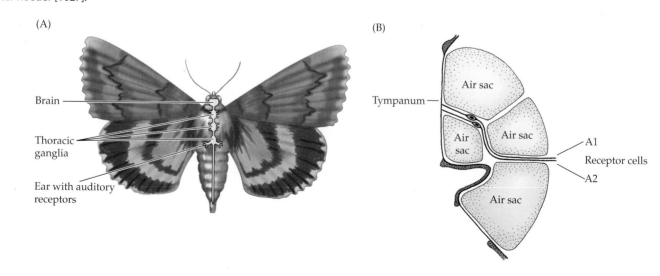

Figure 4.12   Neurons and their operation. This diagram illustrates the structure of a generalized neuron with its dendrites, cell body, axon, and synapses. Electrical activity in a neuron originates with the effects of certain stimuli on the dendrites. Electrical changes in a dendrite's cell membrane can, if sufficiently great, trigger an action potential, which begins near the cell body and travels along the axon toward the next cell in the network.

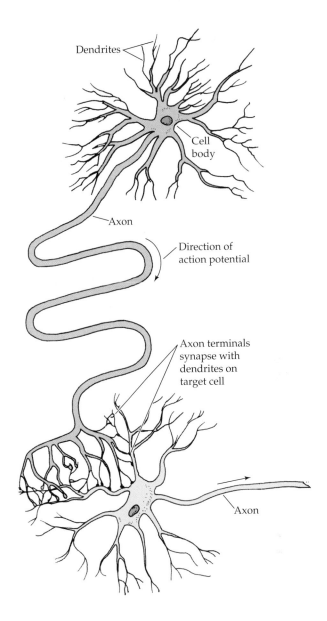

two ears, one on each side of the thorax. Each ear consists of a thin, flexible sheet of cuticle—the tympanic membrane, or tympanum—lying over a chamber on the side of the thorax. The tympanum is attached to two neurons, the A1 and A2 auditory receptors. These receptor cells are deformed when the tympanum vibrates, which it does when intense sound pressure waves sweep over the moth's body. Roeder decided to focus his attention on these auditory receptors, using a cellular approach to get at the proximate basis of moth behavior.

The moth's A1 and A2 receptors work in much the same way that most neurons do: they respond to the energy contained in selected stimuli by changing the permeability of their cell membranes to positively charged ions. The effective stimuli for a moth auditory receptor are provided by the movement of the tympanum, which mechanically stimulates the receptor cell, opening stretch-sensitive channels in the cell membrane. As positively-charged ions flow in, they change the electrical charge inside the cell relative to the charge on the other side of the membrane. If the inward movement of ions is sufficiently great, a substantial, abrupt, local change in the electrical charge difference across the membrane may occur and spread to neighboring portions of the membrane, sweeping around the cell body and down the axon—the "transmission line" of the cell (Figure 4.12). This brief, all-or-nothing change in electrical charge, called an **action potential**, is the signal that one neuron uses to communicate with another.

When an action potential arrives at the end of an axon, it may cause the release of a neurotransmitter at this point. This chemical signal diffuses across the narrow gap, or **synapse**, separating the axon tip of one cell from the surface of the next cell in the network. Neurotransmitters can affect the membrane permeability of the next cell in a chain of cells in ways that increase (or decrease) the probability that this neuron will produce its own action potential(s). If a neuron fires an action potential in response to stimulation provided by the preceding cell in the network, the message may be relayed on to the next cell, and on and on. Volleys of action potentials initiated by distant receptors may have excitatory (or inhibitory) effects that reach deep into the nervous system, eventually resulting in action potential outputs that reach the animal's muscles and cause them to contract.

In the case of the noctuid moth studied by Roeder, the A1 and A2 receptor cells are linked to relay cells called **interneurons**, whose action potentials can change the activity of other cells in one or more of the insect's thoracic ganglia

Figure 4.13 Neural network of a moth. Receptors in the ear relay information to interneurons in the thoracic ganglia, which communicate with motor neurons that control the wing muscles.

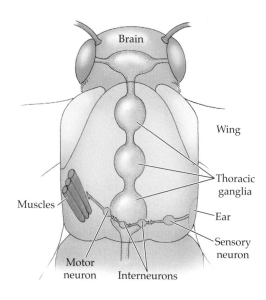

(a ganglion is a neural structure composed of a highly organized mass of neurons) that relays messages on to the moth's brain (Figure 4.13). As messages flow through these parts of the nervous system, certain patterns of action potentials produced by cells in the thoracic ganglia trigger other interneurons, whose action potentials in turn reach motor neurons that are connected with the wing muscles of the moth. When a motor neuron fires, the neurotransmitter it releases at the synapse with a muscle fiber changes the membrane permeability of the muscle cell. These changes initiate the contraction or relaxation of muscles, which drive the wings and thereby affect the moth's movements.

Thus, the moth's behavior, like that of any animal, is the product of an integrated series of chemical and biophysical changes in a network of cells. Because these changes occur with remarkable rapidity, a moth can react to certain acoustical stimuli in fractions of a second, an exceedingly useful ability if the insect is to avoid a bat zooming in for the kill.

Although the neurons of many animals are very much alike at a fundamental level, they differ greatly in their functions. Thus, the auditory receptors of noctuid moths are highly specialized for the detection of ultrasonic stimuli. Roeder demonstrated this point by attaching recording electrodes to the A1 and A2 receptors of living, but restrained, moths [1027]. When he projected a variety of sounds at the moths, the electrical responses of the receptors were relayed to an oscilloscope, which produced a visible record. These recordings revealed the following features (Figure 4.14):

1. The A1 receptor is sensitive to ultrasounds of low to moderate intensity, whereas the A2 receptor begins to produce action potentials only when an ultrasound is relatively loud.

2. As a sound increases in intensity, the A1 receptor fires more often, and with a shorter delay between the arrival of the stimulus at the tympanum and the onset of the first action potential.

3. The A1 receptor fires much more frequently to pulses of sound than to steady, uninterrupted sounds.

4. Neither receptor responds differently to sounds of different frequencies over a broad ultrasonic range. Thus, a burst of 30 kHz sound elicits much the same pattern of firing as an equally intense sound of 50 kHz.

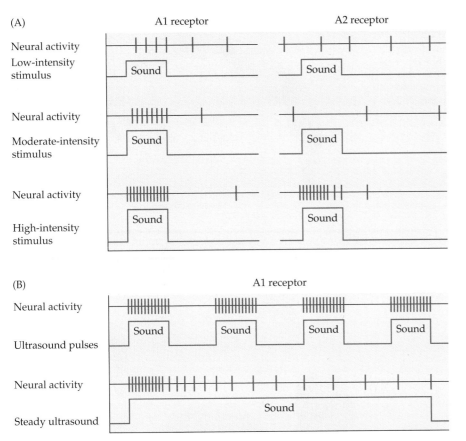

Figure 4.14 Properties of the ultrasound-detecting auditory receptors of a noctuid moth. (A) Sounds of low or moderate intensity do not generate action potentials in the A2 receptor. The A1 receptor fires sooner and more often as sound intensity increases. (B) The A1 receptor reacts strongly to pulses of ultrasound, but ceases to fire after a short time if the stimulus is a constant sound.

5. The receptor cells do not respond at all to low-frequency sounds, which means that moths are deaf to stimuli that we can easily hear, such as the chirping calls and trills of night-singing crickets. Of course, we are deaf to sounds that moths have no trouble hearing.

Although the moth's ears have just two receptors each, they could potentially provide an impressive amount of information to the moth's nervous system about echolocating bats. The key property of the A1 receptor is its great sensitivity to pulses of ultrasound, which enables it to begin generating action potentials in response to the faint cries of a little brown bat 30 meters away, long before the bat can detect the moth. In addition, because the rate of firing in the A1 cell is proportional to the loudness of the sound, the insect has a system for determining whether a bat is flying toward it (or not).

The moth's ears also gather information that could be used to locate the bat in space. If a hunting bat is on the moth's left, for example, the A1 receptor in its left ear will be stimulated a fraction of a second sooner and somewhat more strongly than the A1 receptor in its right ear, which is shielded from the sound by the moth's body. As a result, the left receptor will fire sooner and more often than the right receptor (Figure 4.15A). The moth's nervous system could also detect whether a bat is above it or below it. If the predator is higher than the moth, then with every up-and-down movement of the insect's wings, there will be a corresponding fluctuation in the rate of firing of the A1 receptors as they are exposed to, then shielded from, bat cries by the wings. If the bat is lower than the moth, there will be no such fluctuation in neural activity (Figure 4.15C).

As neural signals initiated by the receptors race through the moth's nervous system, they may ultimately generate motor messages that cause the moth to

Figure 4.15 How moths might locate bats in space.
(A) When a bat is to one side of the moth, the A1 receptor on the side closer to the predator fires sooner and more often than the shielded A1 receptor in the other ear. (B) When a bat is directly behind the moth, both A1 receptors fire at the same rate and time. (C) When a bat is above the moth, activity in the A1 receptors fluctuates in synchrony with the moth's wingbeats. (Figures not drawn to scale.)

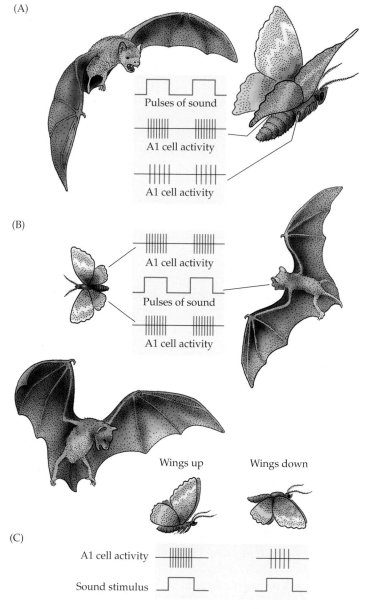

turn and fly directly away from the source of ultrasonic stimuli [1029]. When a moth is moving away from a bat, it exposes less echo-reflecting area than if it were presenting the full surface of its wings to the bat's vocalizations. If a bat receives no insect-related echoes from its calls, it cannot detect a meal. Bats rarely fly in a straight line for long, and therefore the odds are good that a moth will remain undetected if it can stay out of range for a few seconds. By then the bat will have found something else within its 3-meter moth detection range and will have veered off to pursue it.

In order to employ its antidetection response, a moth must orient itself so as to synchronize the activity of the two A1 receptors. Differences in the rate of firing by the A1 receptors in the two ears are probably monitored by the brain, which relays neural messages to the wing muscles via the thoracic ganglia and allied motor neurons. The resulting changes in muscular action steer the moth away from the side of its body with the ear that is more strongly stimulated. (You can imagine what would happen if the moth turned itself *toward* the side of its body with the more strongly stimulated ear!) As the moth turns away from the relatively intense ultrasound reaching one side of its body, it will reach a point at which both A1 cells are equally active; at this moment, it will be facing in the opposite direction from the bat and will be heading away from danger (Figure 4.15B).

Although this reaction is effective if the moth has not been detected, it is useless if a speedy bat has come within 3 meters of the moth. At this point, a moth has at most a second, and probably less, before the bat reaches it [615]. Therefore, moths in this situation do not try to outrun their enemies, but instead employ drastic evasive maneuvers, including wild loops and power dives, that make it harder for bats to intercept them. A moth that executes a successful power dive and reaches a bush or grassy spot is safe from further attack because echoes from the leaves or grass at the moth's crash-landing site mask those coming from the moth itself [1029]. Other nocturnal insects have independently evolved the capacity to sense ultrasound and to take similar kinds of evasive action against approaching bats (Figure 4.16) [827–829, 1347].

Roeder speculated that the physiological basis for this erratic escape flight lies in the neural circuitry leading from the A2 receptors to the brain and back to the thoracic ganglia [1028]. When a bat is about to collide with a moth, the intensity of the sound waves reaching the insect's ears is high. It is under these conditions that the A2 cells fire. Roeder believed that the A2 signals, once relayed to the brain, might shut down the central steering mechanism that regulates the activity of the flight motor neurons. If the steering mechanism was inhibited, the moth's wings would begin beating out of synchrony, irregularly, or not at all. As a result, the insect might not know where it was going—but neither

Figure 4.16   Bat ultrasonic cries trigger evasive behavior in a number of insects. In normal flight, a mantis holds its forelegs close to the body (top left), but when it detects ultrasound, it rapidly extends its forelegs (bottom left), which causes the insect to loop and dive erratically downward. Likewise, a lacewing may employ an anti-interception power dive (right) when approached by a hunting bat. The numbers superimposed on this multiple-exposure photograph show the relative positions of a lacewing and a bat over time. (The lacewing survived.) (Left, top and bottom) From Yager and May [1347], photographs courtesy of D. D. Yager and M. L. May; (right) photograph by Lee Miller.

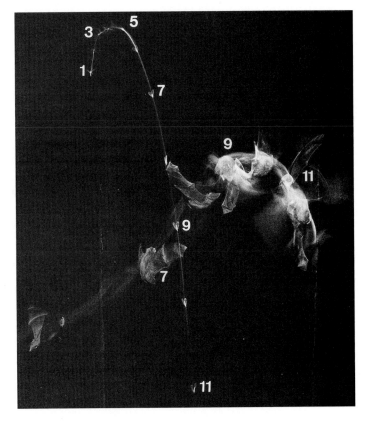

would the pursuing bat, whose inability to plot the path of its prey could permit the insect to escape.

Although Roeder's hypotheses about the functions of the A1 and A2 cells were plausible and supported by considerable evidence, especially with respect to the A1 receptor, other persons have continued to look at how these cells moderate the interactions between moths and bats. As a result, we now know that notodontid moths, even though they have just one auditory receptor per ear, still appear to exhibit a two-part response to approaching bats: the turn away from distant hunters and then the last-second erratic flight pattern when death is at hand. Thus, two cells may not be necessary for the double-barreled response of moths to their hunters [1165]. Even in moths with two receptors per ear, the A1 cell's activity changes greatly as a bat comes sailing in toward it because the bat's ultrasonic cries speed up greatly and become much more intense [421] (Figure 4.17). Presumably, higher-order neurons up the chain of command could analyze the correlated changes in the activity of the A1 receptor alone and make adaptive adjustments accordingly without involvement of the A2 receptor.

More doubts that the A2 cell is necessary to trigger erratic evasive behavior come from the finding that in some noctuid moths, both the A1 and A2 cells may more or less stop firing during the terminal buzz phase—the last 150 milliseconds—of a bat attack. One would think that these cells would keep signaling if either one was truly important in controlling the last-gasp evasive maneuvers of moths under attack. James Fullard and his co-workers suggest that perhaps these cells fail to signal at this late stage simply because the extremely loud and rapid attack vocalizations of a nearby bat incapacitate the cells. Thus, doubt exists about Roeder's hypothesis on the relation between the A2 cell and the close-range response to approaching bats [421].

This case study nicely illustrates several things, including the fact that scientists continually check one another's ideas and update their conclusions [1253]. Just as importantly, the moth–bat story also demonstrates how knowledge of natural history and the real-world problems confronting an animal can help biologists formulate productive hypotheses about a species' sensory mechanisms. Kenneth Roeder knew that noctuid moths live in a world filled with echolocating moth-killers; he searched for, and found, a specialized proximate mechanism that helps some moths cope with these enemies. Many other researchers have followed Roeder in exploring the relationship between the ecology of a species and its neurophysiological mechanisms.

Discussion Questions

4.3 An American cockroach can begin to turn away from approaching danger, such as a hungry toad lunging toward it or a flyswatter wielded by a cockroach-loathing human, in as little as a hundredth of a second after the air pushed in front of the toad's head or the descending flyswatter reaches the roach's body. Cockroaches have wind sensors that react to even slight air movements; these sensors are concentrated on its cerci, two thin projecting appendages at the end of its abdomen. One cercus points slightly to the right, the other to the left. Use what you know about moth orientation to bat cries to suggest how this simple system might provide the information the roach needs to turns away from the toad, rather than toward it. How might you test your hypothesis experimentally?

4.4 A bat's ability to hunt successfully for fruit flies or moths in total darkness is downright astonishing. What sorts of special sensory processing properties would you expect to find in the auditory system of a moth-hunting bat species? (Imagine how an engineer might overcome the problems inherent in designing a machine that generated loud ultrasonic sounds, but then had to detect faint echoes of those sounds coming back from an object moving through the air.)

(A) Ultrasonic cry of hunting bat

Approach

Terminal buzz

Start

100 milliseconds

Buzz   End of buzz

(B) Start of bat's cry and moth's response

- A1 cell
- A2 cell
- B cell

20 milliseconds

(C) Start of bat's buzz and moth's response

10 milliseconds

(D) End of bat's buzz and moth's response

10 milliseconds

Figure 4.17   Is the A2 cell necessary for anti-interception behavior by moths? (A) A sonogram of the ultrasonic pulses produced by a bat approaching and then attacking a prey. The bat's ultrasonic cries speed up greatly and become much more intense as it closes in on the prey (the terminal buzz phase). (B) A sonogram of part of the approach portion of the bat's cry is shown over the response of the A1 and B receptor cells in the noctuid moth's ear. (The B cell is a nonauditory neuron.) (C) A sonogram of the initial part of the bat's terminal buzz is shown over the response of the A1, A2, and B cells in the moth's ear. The A1 receptor fires more often in response to the terminal buzz than to the approach component of the bat's cry. (D) During the latter part of the bat's terminal buzz, only the noctuid's A1 cell is active, not the A2 receptor. After Fullard et al. [421].

Relaying and Responding to Sensory Input

We have focused thus far primarily on how the noctuid moth's auditory receptors collect information about ultrasound. Of course, if the insect is to act on that information, its receptors must be able to forward messages to those parts of its central nervous system that can process acoustical inputs and order appropriate reactions. Just how this is accomplished has not been worked out for noctuid moths, but we know something about the proximate mechanisms of information relay in another insect, the cricket *Teleogryllus oceanicus*, which also flees from ultrasound-producing bats [836]. As in noctuid moths, the cricket's ability to avoid bats begins with the firing of certain ultrasound-sensitive auditory receptors in its ears, which are found on the cricket's forelegs. Sensory messages from these receptors travel to other cells in the cricket's central nervous system. Among the receivers of these messages is a pair of sensory interneurons called *int-1*, also known as AN2, one of which is located on each side of the insect's body. Ron Hoy and his co-workers established that *int-1* plays a key role in the perception of ultrasound by playing sounds of different frequencies to a cricket and recording the resulting neural activity. Their recordings revealed that these cells became highly excited when the cricket's ears were bathed in ultrasound. The more intense a sound in the 40–50 kHz range, the more action potentials the cells produced, and the shorter the latency between stimulus and response—two properties that match those of the Al receptor in noctuid moths.

These results suggest that the *int-1* cell is part of a neural circuit that helps the cricket respond to ultrasound. If this is true, then it follows that if one could experimentally inactivate *int-1*, ultrasonic stimulation should not generate the typical reaction of a tethered cricket suspended in midair, which is to turn away from the source of the sound by bending its abdomen (Figure 4.18). As predicted, crickets with temporarily inactivated *int-1* cells do not attempt to steer away from ultrasound, even though their auditory receptors are firing. Thus, *int-1* is necessary for the steering response.

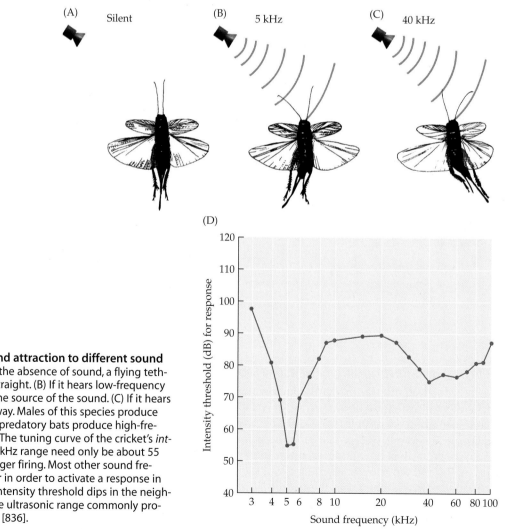

Figure 4.18   Avoidance of and attraction to different sound frequencies by crickets. (A) In the absence of sound, a flying tethered cricket holds its abdomen straight. (B) If it hears low-frequency sound, the cricket turns toward the source of the sound. (C) If it hears high-frequency sound, it turns away. Males of this species produce sounds in the 5 kHz range; some predatory bats produce high-frequency calls of about 40 kHz. (D) The tuning curve of the cricket's *int-1* interneuron. Sounds in the 5–6 kHz range need only be about 55 decibels (dB) loud in order to trigger firing. Most other sound frequencies have to be much louder in order to activate a response in this cell, although note that the intensity threshold dips in the neighborhood of 40 kHz, which is in the ultrasonic range commonly produced by bats. After Moiseff et al. [836].

The corollary prediction is that if one could activate *int-1* in a flying, tethered cricket (and one can, with the appropriate stimulating electrode), the cricket should change its body orientation as if it were being exposed to ultrasound, even when it was not. Experimental activation of *int-1* is sufficient to cause the cricket to bend its abdomen [881]. These experiments convincingly establish that *int-1* activity is both necessary and sufficient for the apparent bat evasion response of flying crickets; therefore, this interneuron can be considered a critical part of the relay apparatus between the receptors and the central nervous system that enables the cricket to behave in a particular way in response to an ultrasonic stimulus.

And just how does a flying cricket carry out orders from its brain to steer away from a source of ultrasound? This problem attracted the attention of Mike May, who began his study not by conducting a carefully designed experiment, but instead by "toying with the ultrasound stimulus and watching the responses of a tethered cricket" [799]. As May zapped the cricket with bursts of ultrasound, he noticed that the beating of one hindwing seemed to slow down with each application of the stimulus. Crickets have four wings, but only the two hindwings are directly involved in flight. If the hindwing opposite the source of ultrasound really did slow down, that would reduce power or thrust on one side of the cricket's body, with a corresponding turning (or yawing) of the cricket away from the stimulus.

On the basis of his informal observations, May proposed that the flight path of a cricket was controlled by the position of the insect's hind leg, which, when lifted into a hindwing, altered the beat of that wing and thereby changed the cricket's position in space (Figure 4.19). May went on to take a number of high-speed photographs of crickets with and without hind legs. Without the appropriate hind leg to act as a brake, both hindwings continued beating unimpeded when the cricket was exposed to ultrasound. As a result, crickets without hind legs required about 140 milliseconds to begin to turn, whereas intact crickets started their turns in about 100 milliseconds. These findings led May to assert that neurons in the ultrasound detection network order the appropriate motor neurons to induce muscle contractions in the opposite-side hind leg of the cricket. As these muscles contract, they lift the leg into the wing, interfering with its beating movement, thereby causing the cricket to veer rapidly away from an ultrasound-producing bat [799].

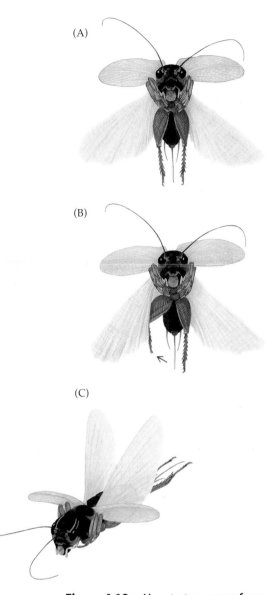

(A)

(B)

(C)

Figure 4.19   How to turn away from a bat—quickly. (A) A flying cricket typically holds its hind legs so as not to interfere with its beating wings. (B) Ultrasound coming from the right causes the leg on the opposite side of the body to be lifted up into the wing. (C) As a result, the beating of the left wing slows, and the cricket turns away from the source of ultrasound as it dives to the ground. Drawings by Virge Kask, from May [799].

Discussion Question

4.5   Outline Mike May's research in terms of the question that provoked his study, his hypothesis, prediction(s), evidence, and his scientific conclusion. In addition, what contribution to this research could come from learning that locusts, a group of insects unrelated to crickets, also possess a special mechanism for very rapidly altering leg position and wingbeat patterns in reaction to ultrasound, so that a flying individual banks sharply downward away from the side stimulated by the stimulus [306]?

Central Pattern Generators

In crickets and moths, ultrasound stimulation turns on auditory receptors that send sensory signals to interneurons, which may relay the message to other

Figure 4.20   Escape behavior by a sea slug. The slug in this photograph has just begun to swim away from its deadly enemy, a predatory sea star. The slug's dorsal muscles are maximally contracted, drawing the slug's head and tail together. Soon the ventral muscles will contract, helping the slug thrash away from the sea star. Photograph by William Frost.

neurons within the central nervous system. When these target cells respond, they can generate signals that turn on a set of motor neurons. The end result is that certain muscles contract or relax in ways that cause the moth to stop beating its wings or the cricket to lift one leg upward, braking the activity of one wing. The escape dives of moths and the evasive swerves of flying crickets pursued by bats are effective one-step responses triggered by simple releasing stimuli in these animals' environments. Most behaviors, however, involve a coordinated series of muscular responses, which cannot result from a single command from a neuron or neural network. Consider, for example, the escape behavior of the sea slug *Tritonia diomedea*, which is activated when the slug comes in contact with a predatory sea star (Figure 4.20). Stimuli associated with this event cause the slug to swim in the ungainly fashion of sea slugs, by bending its body up and down [1311]. If all goes well, it will move far enough away from the sea star to live another day.

How does *Tritonia* manage its multi-step swimming response, which requires from two to twenty alternating bends, each involving the contraction of a sheet of muscles on the slug's back followed by a contraction of the muscles on its belly? As it turns out, the dorsal and ventral muscles are under the control of a small number of motor neurons. The dorsal flexion neurons (DFN) are active when the animal is being bent into a U, and the ventral flexion neuron (VFN) produces a pulse of action potentials that turn the slug into an inverted U (Figure 4.21). But what controls the alternating pattern of DFN and VFN activity?

Dorsal flexion

Ventral flexion

DFN

VFN

Figure 4.21   Neural control of escape behavior in *Tritonia*. The dorsal and ventral muscles of the sea slug are under the control of two dorsal flexion neurons (DFN) and a ventral flexion neuron (VFN). The alternating pattern of activity in these two categories of motor neurons translates into alternating bouts of dorsal and ventral bending—the movements that cause this animal to swim. After Willows [1311].

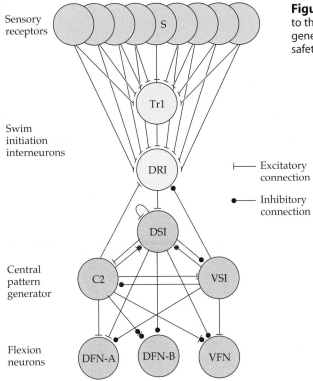

Sensory receptors

Swim initiation interneurons

Central pattern generator

Flexion neurons

Figure 4.22 **The central pattern generator of** *Tritonia* in relation to the dorsal ramp interneurons (DRI) that maintain activity in the cells that generate the sequence of signals necessary for the sea slug to swim to safety. After Frost et al. [415].

The escape reaction begins when sensory receptor cells in the skin of *Tritonia* detect certain chemicals on the tube feet of its sea star enemy. The receptors then relay messages to interneurons, among them the dorsal ramp interneurons (DRI), which, upon receipt of sufficiently strong stimulation, begin to fire steadily. This category of interneurons sends a stream of excitatory signals to a small group of interneurons (the dorsal swim interneurons, or DSI), which in turn is part of an assembly of interconnected cells, among them the ventral swim interneurons (VSI) and cerebral neuron 2 (C2), as well as the flexion neurons mentioned already (Figure 4.22) [442, 443]. A web of excitatory and inhibitory relations exists within this cluster of interneurons such that, for example, activity in DSI turns on C2, which leads to excitation of the DFN and the contraction of the dorsal flexion muscles. After a short period of excitation, however, C2 begins to block the DFN while sending excitatory messages to the VSI, which activate the VFN, leading to contraction of the ventral flexion muscles. The situation then reverses. Alternating bouts of activity in the interneurons regulating the DFN and VFN lead to alternating bouts of DFN and VFN firing, and thus alternation of dorsal and ventral bending [415].

The capacity of the simple neural network headed by the DRI to impose order on the activity of the motor neurons that control the dorsal and ventral flexion muscles means that this mechanism qualifies as a **central pattern generator**. Systems of this sort play a preprogrammed set of messages—a motor tape, if you will—that helps organize the motor output underlying movements of the sort that Tinbergen would have labeled fixed action patterns.

Central pattern generators are also found in vertebrates, as we can illustrate by looking at the plainfin midshipman, a fish that "sings" by contracting and relaxing certain muscles in a highly coordinated fashion [72]. Only the large males of this rather grotesque fish sing "humming" songs, which last more than a minute, and they do so only at night during spring and summer while guarding certain rocks. Their songs are so loud that they can annoy houseboat own-

Figure 4.23 Song-producing apparatus of the male plainfin midshipman fish. The sonic muscles control the movement of the swim bladder, thereby controlling the fish's ability to sing. After an illustration by Margaret Nelson, in Bass [72].

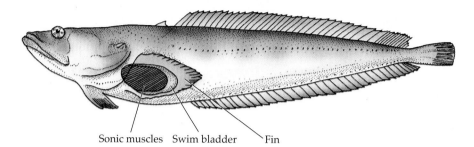

Sonic muscles   Swim bladder   Fin

ers in the Pacific Northwest. The male fish sings to attract females of his species; the fish spawn at the defended rocks, and the male guards the eggs his mates lay in his territory.

How do the male fish produce their songs? When Andrew Bass and his co-workers inspected the anatomy of the fish's abdomen, they found a large, air-filled swim bladder sandwiched between layers of muscles (Figure 4.23). The bladder serves as a drum; rhythmic contractions of the muscles "beat" the drum, generating vibrations that other fish can hear. Muscle contractions require signals from motor neurons, which Bass found connected to the sonic muscles. He applied a cellular dye called biocytin to the cut ends of these motor neurons, which absorbed the material, staining themselves brown. And the stain kept moving along, crossing the synapses between the first cells to receive it and the next ones in the circuit, and so on, through the whole network of cells connected to the sonic muscles. By cutting the brain into fine sections and searching for cells stained brown by biocytin, Bass and his colleagues mapped the fish's sonic control system. In so doing, they discovered two discrete collections of inter-related neurons that generate the signals controlling the coordinated muscle contractions required for midshipman humming. These two clusters, called the sonic motor nuclei, consist of about 2000 neurons each and are located in the upper part of the spinal cord near the base of the brain (Figure 4.24). The long axons of their neurons travel out from the brain, fusing together to form the occipital nerves, which reach the sonic muscles.

In addition to these components, two other anatomically distinct elements of the nervous system ally themselves with the sonic motor nuclei. First, next to each nucleus lies a sheet of pacemaker neurons—a multicellular central

Figure 4.24 Neural control of the sonic muscles in the plainfin midshipman fish. Signals from the central region of the brain (the mesencephalon) travel by way of the cerebellum and ventral medullary nuclei to the sonic motor nuclei in the upper part of the spinal cord. The firing of the pacemaker neurons regulates the frequency of firing by the neurons in the sonic motor nuclei; these signals, in turn, set the rate of contraction of the sonic muscles, and thus the frequency of the sounds produced by the fish. After an illustration by Margaret Nelson, in Bass [72].

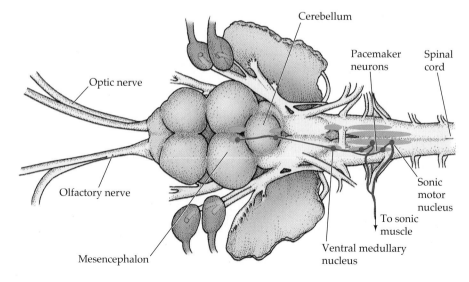

Cerebellum

Pacemaker neurons   Spinal cord

Optic nerve

Olfactory nerve

Sonic motor nucleus

To sonic muscle

Mesencephalon

Ventral medullary nucleus

pattern generator, if you will—that adjusts the activity of the sonic motor neurons so that the sequence of muscle contractions will yield a proper song. Second, in front of the pair of sonic motor nuclei are some special neurons that appear to connect the two nuclei, probably to coordinate the firing patterns coming from the left and the right nucleus, so that the muscle contractions and relaxations will be synchronous, the better to produce a humming sound [72].

The Proximate Basis of Stimulus Filtering

We have just examined the attributes of individual neurons and neural clusters that are involved in the detection of certain kinds of sensory information, the relaying of messages to other cells in the nervous system, and the control of motor commands that are sent to muscles. The effective performance of these basic functions is promoted by **stimulus filtering**, the ability of neurons and neural networks to ignore—to filter out—vast amounts of potential information in order to focus on biologically relevant elements within the diverse stimuli bombarding an animal.

Stimulus Filtering by Auditory Receptors

The noctuid moth's auditory system offers an object lesson on the operation and utility of stimulus filtering. First, the A1 receptors are activated only by acoustical stimuli, not by other forms of stimulation. Moreover, as noted, these cells completely ignore sounds of relatively low frequencies, which means that moths are not sensitive to the stimuli produced by chirping crickets or calling frogs—sounds they can safely ignore. Finally, even when the A1 receptors do fire in response to ultrasound, they do little to discriminate between different ultrasonic frequencies (whereas human auditory receptors produce distinct messages in response to sounds of different frequencies, which is why we can tell the difference between C and C-sharp). The noctuid moth's sensory apparatus appears to have just one task of paramount importance: the detection of cues associated with its echolocating predators. To this end, its auditory capabilities are tuned to pulsed ultrasound at the expense of all else. Upon detection of these critical inputs, the moth can take effective action.

Noctuid moths are not the only animals with biased auditory systems designed to filter out the irrelevant and focus on the important. The relationship between stimulus filtering and a species' special obstacles to reproductive success is evident in every animal whose sensory systems have been carefully examined. Consider the female midshipman fish that listen to the underwater humming "songs" of the males whose neural control systems we have just explored. These songs have components that range up to 400 kHz in frequency. During the summer breeding season, when males are singing lustily, female hearing is highly sensitive to the entire spectrum of sound frequencies present in male songs [1117]. But in the winter, recordings from the female's battery of auditory neurons reveal that very few individual cells are tuned to sound frequencies over 100 kHz. Males are not singing at this time, and therefore, females presumably have little to gain by detecting and analyzing the whole range of sound frequencies in male songs. As a result, their auditory system undergoes changes that enable the females to ignore many acoustical stimuli, which presumably do not provide useful information in the nonbreeding season.

Female midshipman fish employ stimulus filtering on a seasonal basis as they listen to sounds in their underwater world. The screening of acoustical stimuli also occurs in certain parasitoid flies, which use their hearing to locate singing male crickets, the better to place their larvae on these insects. The little maggots burrow into the unlucky crickets and proceed to devour them from

Figure 4.25   Tuning curves of a parasitoid fly.
Females, but not males, of the fly *Ormia ochracea* find their victims by listening for the calls of male crickets, which produce sound with a frequency–intensity spectrum that peaks between 4 and 5 kHz. The female fly, unlike the male fly, is maximally sensitive to sounds around 5 kHz. After Robert et al. [1020].

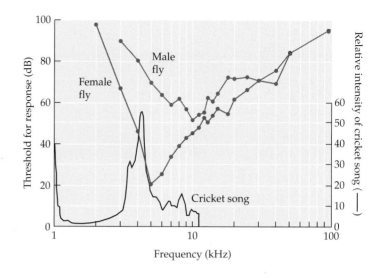

the inside out. Larvae-laden female flies of the species *Ormia ochracea* can find food for their offspring because they have ears tuned to cricket calls, as researchers discovered when they found *Ormia* coming to loudspeakers that were playing cricket song tapes at night.

The unique ears of the female fly consist of two air-filled structures with tympanic membranes and associated auditory receptors on the front of the thorax. Vibration of the fly's tympanic "eardrums" activates the receptors, just as in noctuid moth ears, and thus provides the fly with information about sound in its environment—but not every sound. As predicted by a trio of evolutionary biologists, Daniel Robert, John Amoroso, and Ronald Hoy, the female fly's auditory system is tuned to (i.e., most sensitive to) the dominant frequencies in cricket songs (Figure 4.25). That is, the female fly can hear sounds of 4–5 kHz (the sort produced by crickets) more easily than sounds of 7–10 kHz, which have to be much louder if they are to generate any response [1020]. In contrast, male *Ormia* are not especially sensitive to sounds of 4–5 kHz. Although males can hear those sounds, they do not depend on finding singing male crickets, and as a result, their auditory system has evolved its own distinctive properties. The differences in sensitivity to various sound frequencies within and between the sexes of this fly provide another example of stimulus filtering.

Discussion Question

4.6   Females of another parasitoid fly related to *Ormia* track down singing male katydids (*Poecilimon veluchianus*), whose ultrasonic mate-attracting calls fall largely in the 20–30 kHz range [1160]. What sound frequencies should elicit maximal response in the ears of this katydid-hunting parasitoid, if stimulus filtering enables the animal to achieve goals that are biologically relevant? What conclusion can you reach based on the data in Figure 4.26 [1020]?

Figure 4.26   Tuning curves of a katydid killer. Females of the fly *Therobia leonidei* parasitize male katydids, whose stridulatory calls contain most of their energy in the range of 20–30 kHz. The female fly is much more sensitive than the male fly to sounds in this range. After Stumpner and Lakes-Harlan [1160].

Cortical Magnification in the Tactile Mode

The sensory receptor system of every animal species is designed to screen out some kinds of sensory stimulation while reacting to other kinds. Much the same sort of biased treatment of different kinds of potential information occurs in the central nervous system, as we shall illustrate by examining the operating rules of the star-nosed mole's brain. This weird mammal lives in wet, marshy soil, where it burrows about in search of earthworms and other prey. In its dark tunnels, earthworms cannot be seen, and indeed, the mole's eyes are greatly reduced in size, so that it largely ignores visual information even when light is available. Instead, the mole relies heavily on touch to find its food, using its strange and wonderful nose to sweep the tunnel walls as it moves forward. Its two nostrils are ringed by 22 fleshy appendages, 11 on each side of the nose (Figure 4.27). These appendages cannot grasp or hold anything, but instead are covered with a thousand or so tiny sensory devices called Eimer's organs. Each of these organs contains several receptor cells that react to changes in pressure

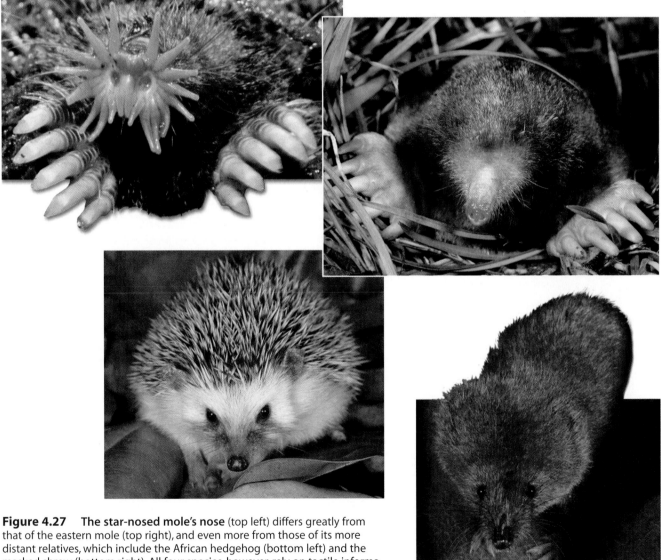

Figure 4.27   The star-nosed mole's nose (top left) differs greatly from that of the eastern mole (top right), and even more from those of its more distant relatives, which include the African hedgehog (bottom left) and the masked shrew (bottom right). All four species, however, rely on tactile information to a considerable degree in locating prey, which range from insects to earthworms. Photographs by Ken Catania.

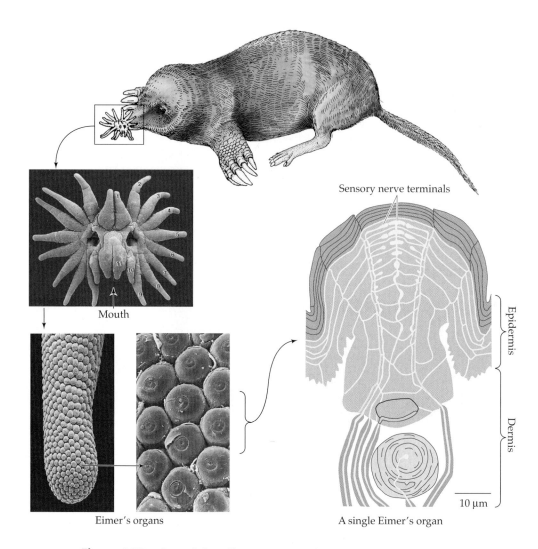

Figure 4.28   A special tactile apparatus. The 22 appendages of the star-nosed mole's nose are covered with thousands of Eimer's organs. Each organ contains a variety of specialized sensory cells that respond to mechanical deformation of the skin above them. After Catania and Kaas [201]; photographs by Ken Catania.

(Figure 4.28). With its vast array of receptors, the animal can collect extremely complex patterns of information about the surfaces touched by the nose [201].

Whenever the mole brushes an earthworm with, say, appendage 5, it instantly sweeps its nose over the prey so that the two projections closest to the mouth, labeled appendage 11 (Figure 4.29A), come in contact with the object of interest. The tactile receptors on each appendage 11 generate a volley of signals, which are carried by nerves to the brain of the animal. Although these two nose "fingers" contain only about 7 percent of the Eimer's organs on the star nose, more than 10 percent of all the nerve fibers relaying information from the nose's touch receptors to the brain come from these two appendages. In other words, the mole uses relatively more neurons to relay information from appendage 11 than from any other appendage.

Not only is the relay system biased toward inputs from appendage 11, but the animal's brain also gives extra weight to signals from this part of the nose. The information from the nose travels through nerves to the somatosensory cor-

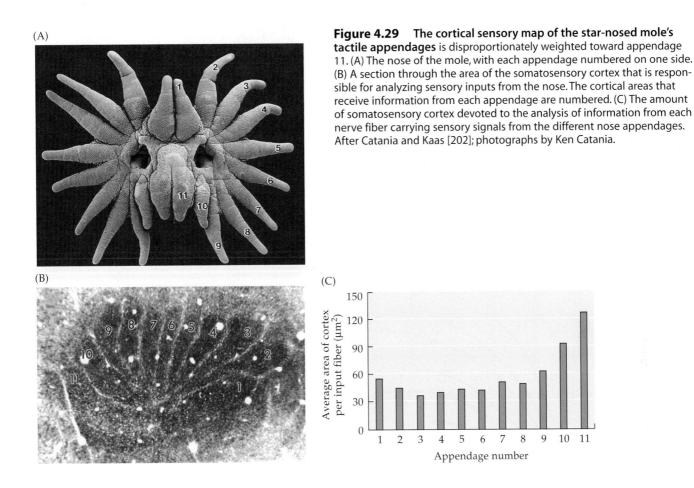

(A)

(B)

(C)

Figure 4.29   The cortical sensory map of the star-nosed mole's tactile appendages is disproportionately weighted toward appendage 11. (A) The nose of the mole, with each appendage numbered on one side. (B) A section through the area of the somatosensory cortex that is responsible for analyzing sensory inputs from the nose. The cortical areas that receive information from each appendage are numbered. (C) The amount of somatosensory cortex devoted to the analysis of information from each nerve fiber carrying sensory signals from the different nose appendages. After Catania and Kaas [202]; photographs by Ken Catania.

tex, the part of the brain that receives and decodes sensory signals from touch receptors all over the animal's body. Of the portion of the somatosensory cortex that is dedicated to decoding inputs from the 22 nose appendages, about 25 percent deals exclusively with messages from the two appendage 11s (see Figure 4.29) [202]. This discovery was made by Kenneth Catania and Jon Kaas when they exposed the cerebral cortex of an anesthetized mole and recorded the responses of cortical neurons as they touched different parts of the animal's nose. Perhaps the mole's brain is "more interested in" information from appendage 11 because of its location right above the mouth; should signals from this appendage activate a cortical order to capture a worm and consume it, the animal is in the right position to carry out the action immediately [201].

In the star-nosed mole, the disproportionate investment in brain tissue to decode tactile signals from one part of the nose is mirrored on a larger scale by the biases evident in the somatosensory cortex as a whole. The somatosensory cortex is given over largely to analyzing signals from the hands and nose at the expense of other parts of the body. This pattern of cortical magnification makes adaptive sense for this species because of the biological importance of the mole's hands for burrowing and the mole's nose for locating prey (Figure 4.30A).

If this argument is correct, then we can predict that the allocation of cortical tissue to somatosensory inputs will differ from species to species in ways that make sense given the special environmental problems each species has to solve. And it is true that other insectivores, even though they are related to the star-nosed mole, exhibit their own distinctive patterns of cortical magnification, all of which make ecological sense (Figure 4.30B–D).

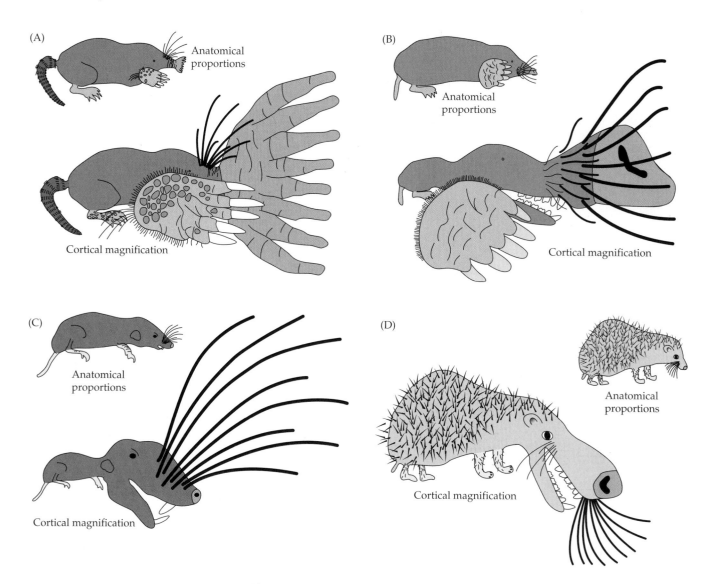

Figure 4.30 Sensory analysis in four insectivores. In each case, the smaller drawing shows the actual anatomical proportions of the animal; the larger drawing shows how the body is proportionally represented in the somatosensory cortex of the animal's brain. (A) The star-nosed mole devotes much more somatosensory cortex to processing inputs from its nose and forelimbs than it does to signals coming from receptors on other parts of its body. (B) Cortical magnification in the eastern mole focuses on sensory inputs from the feet, nose, and sensory hairs, or vibrissae, around the nose. (C) Cortical magnification in the masked shrew also reveals the importance of the vibrissae. (D) Sensory signals from the vibrissae are magnified cortically to a lesser degree in the African hedgehog. After Catania and Kaas [201] and Catania [200].

Discussion Question

4.7 Cortical magnification occurs in all mammals. The drawings at the right in Figure 4.31 are cortical maps based on the amount of brain tissue devoted to tactile inputs from different parts of the bodies of human beings and of the naked mole-rat (see Figure 13.34), a strange, nearly hairless mammal that uses its large front teeth to dig a vast network of underground tunnels while also excavating and processing tuberous roots for food. In what ways do these two maps support the argument that animal brains exhibit adaptive sensory biases?

Figure 4.31   Sensory analysis in humans and naked mole-rats. Brains evolve in response to selection pressures associated with particular physical and social environments, as shown by this comparison between two mammalian species. In each case, the drawing on the left shows the actual anatomical proportions; the larger drawing shows how the body is proportionally represented in the somatosensory cortex. Cortical map of naked mole-rat drawn by Lana Finch, from Catania [203].

Adaptation and the Proximate Mechanisms of Behavior

Our brief survey has revealed that both individual cells and neural networks have special operating rules that filter the information an animal receives, relays, and processes. These rules adaptively shape the ways in which members of different species perceive their environments. Many bees, for example, can see ultraviolet (UV) light (which we humans cannot see) reflected from flowers,

Figure 4.32 Ultraviolet-reflecting patterns have great biological significance for some species. In both sets of photographs, the image on the left shows the organism as it appears to humans, while the image on the right shows the organism's UV-reflecting (pale) surfaces. (Top) The ultraviolet pattern on this daisy advertises the central location of food for insect pollinators. (Bottom) Only males (top specimens) of this sulphur butterfly species have UV-reflecting patches on their wings, which helps signal their sex to other individuals of their species. Photographs by (top) Tom Eisner; (bottom) Randi Papke and Ron Rutowski.

which helps the bees find the nectar source within a flower quickly (Figure 4.32, top). Likewise, the ability of some butterfly species to see UV light helps them respond to the UV-reflecting patterns on the wings of other butterflies, which announce species membership and sex (Figure 4.32, bottom) [1044].

Nor is ultraviolet light perception limited to the insects [90]. Many, if not all, songbirds can see UV light, including the bluethroat of northern Europe and Asia. The bird's common name refers to a patch of blue feathers located, as you may have suspected, on the male's throat (Figure 4.33). These feathers reflect light that we perceive as blue, but they also reflect ultraviolet light that is invisible to us. If bluethroats can see UV light, then perhaps UV reflectance from the throat area, which is displayed during courtship, affects male mating success. This hypothesis generates the prediction that males whose blue throats have been altered to absorb, rather than reflect, ultraviolet radiation will become less attractive to females. To check this prediction, a Scandinavian research team placed captured male bluethroats of the same age in pairs in an aviary. One member of each pair had sunscreen plus a fatty substance from an avian preening gland rubbed onto its blue throat patch, while the other received an application of preening gland oil only. The chemicals in the sunscreen absorbed the UV wavelengths, whereas the glandular secretions did not. In 13 of 16 paired trials, a female bluethroat approached the UV-reflecting male more often than she approached the male whose ornament had been altered to become UV-absorbing [33]. Moreover, when similar manipulations were performed on free-living bluethroats early in the breeding season, the UV-reflecting males acquired a mate and began nesting sooner than their UV-absorbing rivals, and they were also more likely to secure copulations with females other than their primary partners, showing that UV signals influence a male's ability to attract mates (Figure 4.34) [604].

How might female bluethroats benefit from having the sensory equipment to see UV light? Perhaps a preference for bluethroat males whose throat patches reflect more UV light leads female bluethroats to choose older partners, since

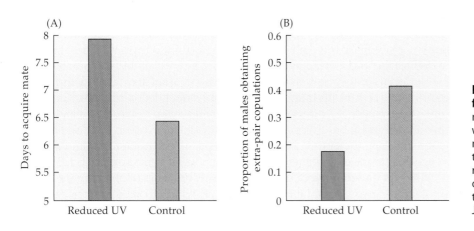

Figure 4.34   The ultraviolet reflectance of the throat patch affects mate choice in the bluethroat. (A) Males with experimentally reduced ultraviolet reflectance were slower to attract mates than control males. (B) Males with reduced ultraviolet reflectance also obtained fewer copulations with females that were not their primary mates. After Johnsen et al. [604].

older males typically have more UV-reflecting feathers than younger males [33]. Older male bluethroats have demonstrated their capacity to survive, and so may be able to pass on their survival-enhancing genes to their offspring.

Discussion Question

4.8   The ocean-dwelling crab *Bythograea thermydron* goes through three life stages, which live at three different depths and habitats. The minute larvae float in waters about 1000 meters below the surface, where only a faint blue light penetrates. The older, larger juvenile forms sink into deeper, darker waters, where the only sources of light are luminous fish and other deep-sea creatures that produce their own blue-green light. Finally, the adults sink to the ocean floor, where they live in the vicinity of deep-sea hydrothermal vents 2500 meters or thereabouts below the surface. Here only the faintest flickers of light are given off at the vents themselves. What would an evolutionary biologist predict about the properties of the eyes of the different life stages of this species? You can check your predictions with information in [600].

Adaptive Mechanisms of Human Perception

Even though the studies I have just described have not involved direct inspection of the nervous systems of bluethroats or butterflies, they tell us something about the properties of the sensory receptors and information decoders possessed by these animals. Bluethroats and butterflies have nervous systems with perceptual capabilities that help these animals manage their special environments. If we were to apply this principle to our own species, we could predict that natural selection should have endowed us with proximate mechanisms suited to the ecological problems we face. Indeed, much evidence suggests that our highly specialized auditory skills, for example, have evolved to match the human social environment, which is dominated by language [955]. Moreover, our visual perception complements our acoustical analysis of language in a most interesting way. As it turns out, a listener's perception of spoken language is heavily influenced by visual cues provided by the moving lips of a speaker. When someone sees a video of a person mouthing the nonsense sentence, "My gag kok me koo grive," while an audiotape synchronously plays the nonsense sentence, "My bab pop me poo brive," the viewer/listener will hear quite clearly, "My dad taught me to drive." This result demonstrates that our brains have circuits that integrate the visual and auditory stimuli associated with speech, using the visual component to alter our perception of the auditory channel [792]. Even though most people are totally unaware of their lip-reading skills, the ability increases the odds that they will understand what others say to them.

Lip reading depends on specific neural clusters within the visual cortex. In particular, a region of the brain called the superior temporal sulcus becomes especially active when we view moving mouths, hands, and eyes [22] (Figure 4.35). People are remarkably good at detecting even subtle movements of these body parts because this skill enables the viewer not only to read lips, but also to deduce the intentions of other individuals, a useful ability for members of a highly social species. Cells in the superior temporal sulcus are also activated by certain static visual stimuli, especially those associated with faces. They perceive, for example, the direction in which a companion's eyes are gazing, which says something about this person's current focus of interest. Because neurons within the superior temporal sulcus are "tuned" to particular facial stimuli, such as eye position, we can better predict the actions of those around us [22].

Nor is the superior temporal sulcus the only brain area devoted to the analysis of faces. An area called the fusiform gyrus, which is on the underside of the cerebral cortex, plays a critical role in our capacity to identify familiar faces quickly and accurately. And we are exceptionally good at this task. If an exper-

● Mouth movement
● Body movement
○ Eye gaze
○ Hand movement

imenter gives someone 5 seconds to inspect each of 50 photographs of unfamiliar faces, odds are that when the subject is tested later, he will be able to pick out about 90 percent of the previously glimpsed faces from a large collection of photographs of faces [190]. We know that face recognition is dependent on an intact fusiform gyrus because persons who have suffered an injury to this part of the brain are no longer able to recognize people by their faces [361], a phenomenon that gave Oliver Sacks the title for his book *The Man Who Mistook His Wife for a Hat* [1054]. Some persons with a damaged fusiform gyrus can see things perfectly well, they can name many objects, they can identify particular individuals by the sound of the voice or by familiar clothing, but when shown pictures of the faces of their friends, their spouses, even themselves, they are at a complete loss [309]. Interestingly, other slightly brain-damaged persons have just the opposite problem. They cannot identify ordinary objects when they see them, but have no difficulty recognizing particular faces, suggesting that the special neural mechanisms dedicated to face recognition are still intact in their brains [83].

Various techniques, including functional magnetic resonance imaging, can be used to identify exactly which parts of the cerebral cortex become active when a person looks at a picture of a face (Figure 4.36). Brain imaging reveals that neurons in a small part of the posterior fusiform gyrus fire only when a person looks at a face. This neural module, called the facial fusiform area, does not respond to pictures of inanimate objects, although another nearby region of the brain does, providing strong evidence of task specialization by modules that provide biologically relevant perceptions for their owners [501, 620].

Yet another approach to identifying which part of the brain does what has been to place electrodes directly on the surface of the cerebral cortex and record the electrical activity of neurons. This technique is used to map the brains of

Figure 4.35   Socially relevant movements of the lips, mouth, hands, and body activate neurons in different parts of the superior temporal sulcus in the human brain. The left and right hemispheres are shown on the left and right, respectively. Each circle represents findings by a particular research team. After Allison et al. [22].

Figure 4.36   A special-purpose module in the human brain: the face recognition center. Magnetic resonance images of the brains of two persons looking at a photograph of a human face. The most active part of the brain, the facial fusiform area, is shown in red. After Kanwisher et al. [619].

Figure 4.37   Specialization of function in different parts of the visual cortex of humans. Different neural circuits perform different analyses of images in our environment. The ability to remember faces, for example, is dependent on a specialized site in the fusiform gyrus. After Puce et al. [984].

epileptic patients prior to operations designed to remove the tissue responsible for their epileptic seizures. Needless to say, the idea is to destroy as little of the brain as possible, which means finding the dysfunctional tissue in order to leave the rest alone when the surgery is performed. In the course of recording electrical activity in different parts of the brains of these patients, who remain conscious, researchers have been able to locate those parts of the brain that react to different kinds of visual stimuli. This approach has confirmed that different sections on the underside of the brain are devoted to different kinds of visual analyses. The sites that respond strongly to images of entire faces are different from those that react primarily to parts of faces, such as eyes, which in turn are different from those that are activated when the subject is shown a picture of an object (Figure 4.37) [984]. Our brains are computers with stimulus-filtering circuits that enable us to perceive some things much more readily than others.

Discussion Question

4.9   Some evolutionary biologists have argued that the human cerebral cortex contains an assortment of regions specialized for the analysis of an assortment of biologically relevant stimuli. Thus, according to this view, our face recognition mechanism evolved because of the reproductive value for individuals of quickly recognizing the identity of others, given the highly social nature of our species. What problem is posed for this argument by the discovery that we also possess a "visual word form area" located in the left fusiform gyrus (Figure 4.38) [804]? You are using this particular region of your brain at this very moment as you read these words. In addition, what significance do you attach to the finding that when we read, we recognize each letter independently by its simple features? We never come to recognize words as wholes on the basis of their complexly distinctive patterns, even though we could read far more efficiently if we worked with entire word forms rather than moving from one letter to the next [934].

Adaptive Mechanisms of Navigation

The human visual system is biased toward gathering biologically significant information from our social environment. Persons in the past who were better

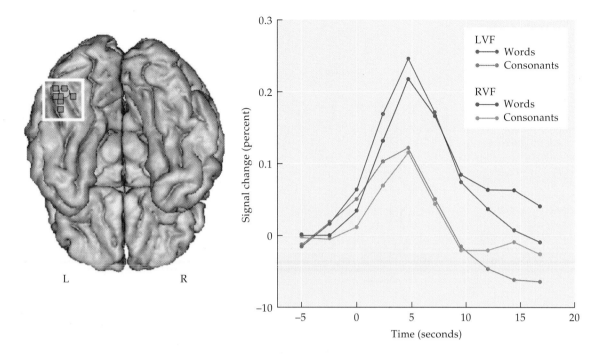

Figure 4.38   A cerebral word analysis center. (A) The fusiform gyrus, shown by the green squares and yellow circle, responds especially strongly to the visual stimulation provided by words. (B) Shown are the differences in neural activity over time of the fusiform gyrus on the left (LVF) and right (RVF) sides of the brains of human subjects that were permitted to see words or strings of consonants as opposed to checkerboards of black and white squares. Seeing words produces a greater change in activity in this part of the brain than seeing strings of consonants; both kinds of stimuli generate a greater response than seeing a checkerboard stimulus. After Cohen et al. [230].

than average at remembering who was friend and who was foe, and at deducing the intentions of others by examining their faces, have in effect left us a legacy in the form of our brains, with all their adaptive filters, oddities, and quirks. The point that nervous systems have been shaped by natural selection can be reinforced by considering the relationship between the brains of animals and their ability to travel to destinations of their choice. The human species is one in which our hunter–gatherer ancestors wandered widely from a home base in search of food and other resources. On this basis alone, we could predict that our brains should have features that promote efficient movement through space, just as behavioral ecologists predicted that food-storing birds should possess neural mechanisms—notably an enlarged hippocampus—that give them the spatial memory they need to relocate hidden food items (see p. 63). Does the human hippocampus play a part in getting us from point A to point B?

In order to answer this question, researchers have taken advantage of magnetic resonance imaging, which permits them to monitor hippocampal activity in fully conscious persons. In one such study, subjects were permitted to explore a virtual town with a complex maze of streets before being given the task of navigating from point A to point B as quickly as possible [769]. Individuals able to navigate more accurately evidently relied more heavily on the right hippocampus compared to other, less accurate subjects, judging from the higher levels of activity observed in this part of the brains of especially skillful navigators (Figure 4.39).

These and other research results suggest that our learned mental maps may be stored within the hippocampus. If true, then persons with complex mental

Figure 4.39 The hippocampus is essential for navigation by humans. (A) A map (aerial view) of the virtual town through which experimental subjects navigated. Three examples of attempts to get from A to B are illustrated. The shortest route (in yellow) between the two points was considered the most accurate. (B) Average magnetic resonance image for ten subjects while performing the navigation task, showing the location of peak neural activity (bright yellow), which lies in the right hippocampus. (C) The location of the right hippocampus in a brain viewed from the underside. (D) The accuracy of the ten navigators was a function of the intensity of the activity in their right hippocampus. After Maguire [769] and Carter [195].

(A)

(B)

(C)

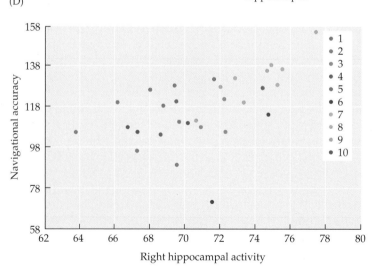

(D)

maps formed while navigating through varied environments, such as cities, should have more hippocampal tissue than persons who have less experience in acquiring spatial representations of their surroundings. We can check this prediction with the help of the taxi drivers of London, who must demonstrate that they can remember the locations of thousands of addresses before receiving their license. As predicted, the average posterior hippocampal size, as revealed by magnetic resonance imaging, in a sample of London taxi drivers was indeed larger than that in a comparable group of men who did not drive taxis for a living [770]. Navigational experience itself appears to be responsible for the development of a large hippocampus capable of storing a great deal of spatial information: the more years of taxi driving, the larger the posterior hippocampus.

Although the performance of these experienced taxi drivers in negotiating the streets of London is impressive, if you were to take a London cabbie and drop him off in Detroit with the requirement that he get to 16th and Ash without assistance from others, he would almost certainly be totally lost. Nor would he set off confidently if he were given a compass by a bystander. A compass would help only if he knew where he was and where he was headed, which would require a map of some sort. Although humans without a map and compass are handicapped when it comes to navigating across unfamiliar areas, many other animals show no such disability because they possess an internal map sense (knowing the location of home or some other goal) and an internal compass sense (knowing in what direction to move) (Figure 4.40).

Take the honey bee and the homing pigeon, for example. Both species are skilled navigators capable of crossing unfamiliar terrain on the way home, as demonstrated by a honey bee's ability to make a beeline back to its hive after a meandering outward journey in search of food, and by a homing pigeon's ability to make a pigeon-line back to its loft after having been released in a distant and strange location [1247, 1248]. Because honey bees and homing pigeons are both active during the daytime, they both, as one might suspect, use the sun's position in the sky as a directional guide [208, 1320]. Even we can do this to some extent, knowing that the sun rises in the east and sets in the west—provided that we also know approximately what time of day it is. Every hour the sun moves 15 degrees on its circular arc through the sky. Therefore, one has to adjust for the sun's movement if one is to use its position as a compass.

A honey bee leaving its hive notes the position of the sun in the sky relative to the hive as it flies off on a foraging trip. It might spend 15 to 30 minutes on its trip and might move into unfamiliar terrain in search of food. If on the homeward flight, it oriented as if the sun had not shifted, the bee would be lost. In reality, honey bees rarely get lost, in part because they can compensate for the sun's movement, thanks to an internal clock mechanism (see p. 150) [728]. This ability can be demonstrated by training some marked bees to fly to a sugar-water feeder some distance away from their hive (say, 300 meters due east of the hive). One can then trap the bees inside the hive and move it to a new location. After 3 hours have passed, the hive is unplugged, and the bees are free to go in search of food. They do not have familiar visual landmarks to guide them, and yet some marked individuals remember that food is found 300 meters due east. They fly 300 meters due east of the new hive location, to the spot where the food source "should be." They have compensated for the 45-degree shift in the position of the sun that has taken place during their 3-hour confinement.

Pigeons can also be tricked into demonstrating how important a clock sense is for their navigation system [1246]. You can reset a pigeon's biological clock by placing the bird in a closed room with artificial lighting and then shifting the light and dark periods out of phase with sunrise and sunset in the real world. If sunrise is at 6:00 a.m. and sunset at 6:00 p.m., for example, one might set the lights in the room to go on 6 hours earlier (at midnight) and off 6 hours earlier (at noon). A pigeon exposed to this routine for several days would become *clock-shifted* 6 hours out of phase with the natural day. If taken from the room and released at 6:00 a.m. at a spot some distance from its loft, the bird will behave as if the sun has been up for 6 hours (as if it is noon), which will cause it to orient improperly. For example, let's say that the pigeon is released at a place 30 kilometers due west of its loft. Its map sense somehow tells it this, and it attempts to orient itself to fly east (Figure 4.41). As Charles Walcott points out: "To fly east at 6:00 a.m., you fly roughly toward the sun, but because your clock tells you it is really noon, you know that the sun is in the south and that to fly east, you must fly 90 degrees to the left of the sun. And this is exactly what the birds do, although they presumably do not go through the reasoning process I have described."

Discussion Question

4.10   We put a homing pigeon on an experimental light–dark cycle in which the lights come on at noon and go off at midnight at a time of the year when actual sunrise is at roughly 6:00 a.m. and sunset is at 6:00 p.m. After several weeks on this schedule, we release the pigeon at noon on a clear day in unfamiliar territory due east of its home loft. First, in what direction should the pigeon fly? Second, are you surprised to learn that on a completely overcast day, the pigeon would fly directly home? What does this finding suggest about the homing mechanism(s) of this bird?

Wandering albatross

(A)

(B)

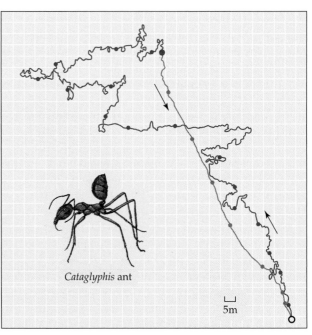

Cataglyphis ant

Figure 4.40   The ability to navigate over unfamiliar terrain requires both a compass sense (knowing in what direction to move) and a map sense (knowing the location of a goal). (A) The flight path taken by a wandering albatross on a foraging journey of over 4000 kilometers from its nest in the Crozet Islands in the southern Indian Ocean, north of Antarctica, and back again. (B) A trip of 592 meters by a foraging ant out from its nest (at the large open circle at the bottom) and then directly back home after capturing a prey (at the spot marked with a large red circle, 140 meters away from home). A, after Weimerskirch et al. [1265]; B, after Wehner and Wehner [1262].

**Figure 4.41   Clock shifting and altered naviga-
tion in homing pigeons.** The results of an experiment in which pigeons were released at Marathon,
New York (about 30 kilometers east of Ithaca, where
their home loft was located). On sunny days, birds in
the control group generally headed west, back toward
the home loft. But pigeons whose biological clocks
had been shifted by 6 hours usually misoriented so
that, on average, they headed north. After Keeton
[622].

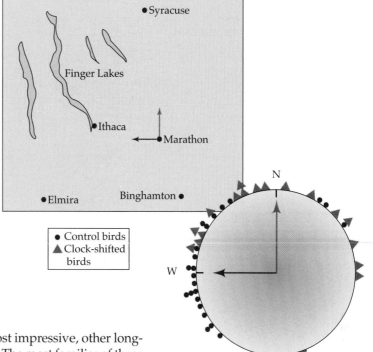

Adaptive Mechanisms of Migration

Although what honey bees and pigeons can do is most impressive, other long-
distance migrants are even more amazing navigators. The most familiar of these
are the many birds whose migratory journeys cover thousands of kilometers,
such as the white-crowned sparrows that move between Mexico and Alaska
twice a year (see also pp. 157–158). Although much has been learned about
the proximate mechanisms of bird migration [1321], here we will focus on two
other animals, the monarch butterfly and the green sea turtle.

In the fall, adult monarch butterflies in Canada head south, and they do not
stop until they reach one of a select group of forest groves high in the mountains of central Mexico, a trip that requires flying as far as 3600 kilometers
(Figure 4.42). Once in Mexico, the butterflies congregate by the millions in stands
of Oyamel firs, where they mostly sit and wait, day after day, through the cool
winter season until spring comes. Then the monarchs rouse themselves and
begin a return trip to the Gulf Coast of the United States, arriving in time to
lay their eggs on growing milkweed plants, which their caterpillar progeny will
consume.

Migrating monarchs fly during the daytime, which suggests that they might
use the sun as a compass to guide them in a southwesterly direction during the
fall migration (and in a northeasterly direction during the spring return flight).
If so, then it ought to be possible to do unto monarchs what has been done unto
pigeons; namely, to trick them into misorienting by manipulating their biological clock [416]. To test this prediction experimentally, a research team captured
migrating monarchs in the fall and held them in a laboratory under a regime of
12 hours of light and 12 hours of dark. They placed the butterflies in two groups,
one that received its 12 hours of light beginning daily at 7:00 a.m., and the other
that had the lights go on 6 hours earlier, at 1:00 a.m. After a number of days in
captivity, the monarchs were tethered and released in a flight cage under natural daylight conditions. There they could fly, and in so doing, indicate the direction in which they wanted to go (without actually leaving the test arena, thanks
to the tether).

The results showed that the monarchs attended to the sun's position in the
sky. The individuals that had had their biological clock set to the natural daily
cycle for the fall (lights on at 7:00 a.m.) struggled against their tethers to fly off
to the southwest. But those individuals whose clocks had been shifted 6 hours

Figure 4.42   The fall migration route of monarch butterflies takes some butterflies from Canada to Mexico, where the monarchs cluster in a few forest patches high in the central Mexican mountains. After Brower [153].

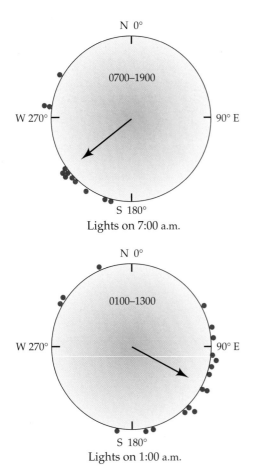

Lights on 7:00 a.m.

Lights on 1:00 a.m.

earlier oriented 90 degrees to the left, which meant that they were heading in a generally southeasterly direction (Figure 4.43) [416].

Sunlight is actually composed of many different wavelengths of light, and so the question arises, what wavelengths are critical for the navigational orientation of monarch butterflies? This question attracted the same team that had established that monarchs possess a time-compensated sun compass. The researchers examined whether ultraviolet light might be a key factor. As all sunscreen users know, the sun produces plenty of ultraviolet radiation, and although you and I cannot see this radiation, many other animals can, as we saw earlier in this chapter. To test the hypothesis that monarch navigation is dependent on ultraviolet radiation, the experimenters captured migrating monarchs in the fall, and then permitted tethered monarchs to start flying in the flight cage where they could orient in a direction of their choosing. The team then covered the flight cage with a UV interference filter, which screened out this component of sunlight. The monarchs quickly became confused, and many stopped flying altogether. Most individuals (11 out of 13) resumed flight, however, as soon as the filter was removed, evidence that the sun's UV radiation is indeed essential for monarch navigation [416].

Figure 4.43   Experimental manipulation of the biological clock changes the orientation of migrating monarchs. Individuals were tested in an outdoor flight cage after one group had been held indoors under an artificial light–dark cycle with lights on at 7:00 a.m. and lights off at 7:00 p.m. This group tried to fly in a southwesterly direction. A second group of butterflies that had been held with lights on at 1:00 a.m. and lights off at 1:00 p.m. flew in a southeasterly direction, evidence of the importance of a sun compass in helping monarchs stay on course. After Froy et al. [416].

Ultraviolet light appears to help monarchs get started flying in the right direction. But once airborne, the butterflies need to maintain their compass orientation and, as it turns out, they achieve this end with assistance from the cues provided by the polarized skylight pattern. Polarized light, which you and I cannot see, is produced when sunlight enters the Earth's atmosphere and the light scatters so that some light waves are vibrating perpendicular to the direction of actual sunrays. The three-dimentional pattern of polarized light in the sky created in this fashion depends upon the position of the sun relative to the Earth and so this pattern changes as the sun moves across the sky. Therefore, if an animal on the ground is able to perceive the pattern of polarized light in the sky, this stimulus can serve as a proxy for the position of the sun at any given time. As a result, creatures with a biological clock and the capacity to see polarized light could use the information in skylight as a compass in the same way that the sun's position in the sky can be used as a compass. Being able to make use of the directional information in polarized light has real advantages because this cue is available even when the sun is hidden behind clouds or mountains.

A team of researchers established that monarchs can orient to polarized light information by tethering butterflies captured on their fall migration in a small walled arena where the insects could not see the sun but could look at the sky overhead. Under these conditions, when the butterflies flew they were able to orient consistently to the southwest. Flying tethered monarchs retained this ability when a light filter was placed over the apparatus and aligned so as to permit the entry of linear polarized light waves in the same plane as occurs in the sky at the zenith (the highest point in the sky). However, when the filter was turned 90 degrees from its original orientation, changing the angle of entry of the polarized light visible to the monarchs at the zenith, they altered their flight orientation by 90 degrees as well, demonstrating their ability to use a polarized light compass (Figure 4.44) [1012].

Green sea turtles are the monarch butterflies of the turtle world. Like the monarch, these animals travel thousands of kilometers on their migrations, which take some turtles from their nesting beaches on Ascension, a tiny island in the middle of the South Atlantic Ocean, to their feeding grounds just off the

Figure 4.44   Polarized light affects the orientation of monarch butterflies. (A) Butterflies tethered in a flight cage received three treatments with respect to the angle of polarized light reaching them from the sky (the skylight e-vector) : no filter, which meant that they could see the natural pattern of polarized light in the sky; vertical filter, which did not interfere with the pattern of polarized light visible from the flight cage; and horizontal filter, which shifted the angle of polarized light arriving from the sky by 90 degrees. (B) Orientation choices made by two flying monarchs in the flight cage, one tested in the morning (AM) and the other in the afternoon (PM). During the period of flight, a computer automatically recorded the orientation of the butterfly every 200 milliseconds, yielding a record composed of many "events" during the minutes when the monarch was flying. The data show that when the insects could observe polarized light in the natural skylight pattern, they tended to fly to the southwest, but when the filter shifted the angle of polarized light, altering the skylight pattern by 90 degrees, the monarchs shifted their flight orientation accordingly. After Reppert et al. [1012].

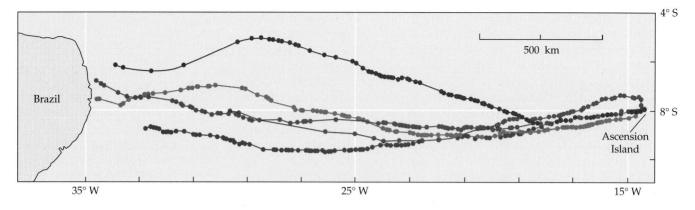

Figure 4.45   Migratory routes taken by five green sea turtles that nested on Ascension Island and then returned over 2000 kilometers to feeding areas in the South Atlantic Ocean near Brazil. After Luschi et al. [753]; photograph by Ursula Keuper-Bennett and Peter Bennett.

coast of Brazil, about 2000 kilometers away (Figure 4.45) [753]. Other green sea turtles shuttle between other feeding and breeding sites separated by vast distances, often spending several years at sea before making their way back to the beach where they nested before. Satellite transmitters have been attached to turtles so that they can be followed day and night. The transmitters revealed that the turtles travel long distances at night, suggesting that they do not need to track the sun's position in order to make their remarkable journeys. Therefore, Ken Lohmann and his co-workers hypothesized that the turtles must make use of some other navigational cue, of which there are several potential candidates, among them the earth's magnetic field [740]. These lines of magnetic force could in theory be used by the turtles to construct an internal map.

To test their hypothesis, Lohmann and company captured some young turtles at sea, brought them back to the mainland, tethered them in a cloth harness, and plunked them into an "ocean simulator"; namely, a plastic pool of seawater in the backyard of a house in Melbourne Beach, Florida. The pool was surrounded by a computer-driven magnetic coil system that the researchers used to alter the magnetic field around the pool, thereby simulating the conditions that a magnetic field detector would experience hundreds of kilometers to the north or south. If the turtles were capable of sensing the earth's magnetic field and using it as a map, then an individual that perceived conditions associated

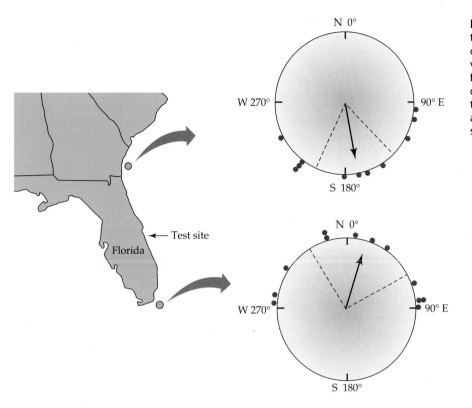

Figure 4.46   Experimental manipulation of the magnetic field affects the orientation of green sea turtles. Individuals that experience the magnetic field associated with an area to the north of their actual location swim south; turtles that sense the magnetic field of an area to the south of their actual location swim north. After Lohmann et al. [740].

with an area 340 kilometers to the north of the Melbourne Beach area should swim steadily south, rowing along in its tether, getting nowhere but establishing its orientation preference. If the magnetic field was one that a turtle would encounter 340 kilometers to the south, then the turtles should orient in such a way as to head north. The turtles did what was expected of them (Figure 4.46) [740], demonstrating that green sea turtles are indeed geomagnetic map navigators.

Although monarch butterflies and green sea turtles do not use the same navigational systems [856], they each possess perceptual skills absent in our own species. These two species provide additional evidence, as if any more were needed, that the sensory mechanisms and brains of animals have been shaped by natural selection arising from each animal's special requirements for survival and reproduction.

Summary

1. The operating rules of neural mechanisms constitute a proximate cause of behavior. Neurons acquire sensory information from the environment, relay and process that information, and order appropriate motor responses to environmental events. Different species have different neural mechanisms and therefore perform these tasks differently, providing proximate reasons why species differ in their behavior.

2. The classic ethological approach to the study of nervous systems showed that some animals possess neural elements shaped by natural selection to detect key stimuli and to order appropriate species-specific responses to these biologically relevant cues. Modern neurophysiological studies confirm that animals possess highly specialized sensory receptors whose design facilitates the acquisition of critical information from the environment, as well as interneurons and motor neurons with unique properties that con-

tribute to particular behavioral abilities. Moreover, central pattern generators within the central nervous system can produce a programmed series of messages to selected muscles, facilitating complex patterned responses to certain stimuli.

3. Stimulus filtering, the filtering out of irrelevant information, and other forms of biased sensory processing are fundamental properties of all animal nervous systems. Sensory receptors ignore some stimuli in favor of others, while interneurons relay some, but not all, of the messages they receive. Within the central nervous system, many cells and circuits are devoted to the analysis of certain categories of information, although this means that other inputs are not as thoroughly analyzed.

4. The proximate mechanisms of stimulus filtering and biased data analysis are the basis for adaptive perception and adaptive responses. By ignoring some potential information, animals are able to focus on the biologically relevant stimuli in their environment, increasing the odds of a prompt and effective reaction to these key stimuli. Because the obstacles to reproductive success differ among species, selection has resulted in the evolution of proximate mechanisms with different functional attributes. That is why some animals have perceptual abilities not possessed by humans, such as the capacity to hear ultrasound, or to see ultraviolet light, or to form geomagnetic maps useful for navigating to a particular destination across unfamiliar terrain.

Suggested Reading

Hans Kruuk's biography of Niko Tinbergen describes Tinbergen's life and science dispassionately and well [679]. Kenneth Roeder's *Nerve Cells and Insect Behavior* [1027] is a classic on how to conduct research on the physiology of behavior. Modern textbooks on the neurophysiology of behavior have been written by Jeffrey Camhi [185], Peter Simmons and David Young [1111], and Thomas Carew [188]. For a moderately advanced text on how human sensory cells convert the diverse kinds of sensory stimulation into electrical signals, read Gordon Fain's *Sensory Transduction* [379].

One of the main findings of neurobiologists highlighted in this chapter is that the sensory worlds of other species are remarkably diverse and not necessarily at all like our own (as illustrated by ultrasound-detecting bats), and that their special abilities can often be shown to be linked to special ecological factors acting on those species. Other examples include infrasound-detecting elephants [435], the electric sense of certain fishes (see [562, 616]), and the infrared perception of some snakes [872]. Another illustration of this point comes from a comparison of the vomeronasal organ of assorted vertebrates (see Michael Meredith's website http://www.neuro.fsu.edu/research/vomeronasal). This organ, which serves a special olfactory role, has evolved several very different functions, each appropriate to a task of biological significance for the species in question. Likewise, the different mechanisms underlying the extraordinary feats of animal navigation are fascinating (see [1261]), and if you have a special interest in sea turtle navigation, go to http://www.unc.edu/depts/oceanweb/turtles/.

Mapping the Brain by Rita Carter [195] is a beautiful book about the operation of the human brain. *Evolving Brains* by John Allman focuses on the evolution of the human brain, but contains a wealth of fascinating information on other species as well [23].

5 | The Organization of Behavior: Neurons and Hormones

*I*n the previous chapter, we talked about animals as if they were endowed with neural computers designed to detect key stimuli, discriminate among patterns of inputs, and order adaptive reactions. So, for example, when a flying moth is exposed to ultrasound from a predatory bat, its auditory receptors fire, setting in motion a chain of neural events, which help the insect respond adaptively to this acoustical stimulus. The capacity of neural mechanisms to filter out irrelevant information, to perceive some things very reliably, and to order effective reactions makes obvious biological sense.

But nervous systems do more than just activate response X in the presence of stimulus Y. Imagine a flying male moth hot on the scent trail of a distant pheromone-releasing female. If the male's nervous system operated simply by keeping the moth flying whenever female sex pheromone was in the air, a scent-tracking moth would be in real trouble if bats approached. But the moth's nervous system balances risk against reproductive opportunity. If the ultrasonic pulses a scent-tracking male hears are very loud, or if the pheromonal signal is very weak, he usually aborts his flight immediately [2, 1169]. Thanks to this feature of his nervous system, the male moth has the ability to give precedence to avoiding bats, and so may live to mate another day.

◀ *These male red-sided garter snakes* emerging from hibernation are ready to mate, despite the fact they have almost no testosterone in their blood. Photograph by François Gohier

(A)

(B)

Figure 5.1   **Different courtship displays of the male ring dove** are under the control of different hormones. (A) The aggressive strutting and bowing display, which is linked to testosterone. (The male is the bird at the left.) (B) The nest soliciting display, which is linked to estrogen. Photographs by Leonida Fusani.

Hormones often help animals keep their behavioral options straight. To pick just one example, when a male ring dove courts a female, he starts off by aggressively chasing his potential mate around, as well as strutting and bowing to her in a pushy manner. If that were his only courtship behavior, he would not get very far, because females will not mate unless this early aggressive display is followed by a calmer display called nest soliciting, in which the male stands on the nest with his tail raised (Figure 5.1). As it turns out, the hormone testosterone increases the likelihood of chase-strutting, whereas the hormone estrogen facilitates nest soliciting [428]. The male ring dove's brain gets its various sexual activities in the right order by producing an enzyme, aromatase, at the appropriate stage of courtship; aromatase catalyzes the conversion of testosterone to estrogen, thereby enabling the male ring dove to switch from strutting to nest soliciting in the proper sequence.

The ability of nervous and hormonal systems to perform these feats of behavioral organization is what this chapter is all about. The fundamental problem that we shall examine is how proximate mechanisms structure an individual's behavior—from moment to moment, over the course of a day, over a few weeks or a breeding season, or even over a whole year. We shall examine three classes of mechanisms that carry out these functions: neural command centers that communicate with one another, clocks that schedule the activity of these command centers, and hormonal systems that track changing physical and social environments and adjust the priorities of competing command centers.

How Neural Command Centers Organize Behavior

Because most animals have the capacity to do many different things in response to many different stimuli, at any given moment they face the question of which behavior to activate. At an ultimate level, it is easy to understand why animals rarely try to do two things at once. But at a proximate level, how are animals' nervous systems organized so that maladaptive conflicts do not occur?

One way to establish behavioral priorities is to have a nervous system endowed with **command centers**, which include the innate releasing mechanisms, central pattern generators, song control systems, and the like that we have met in preceding chapters. Let each command center be responsible for activating a particular response, but have the various centers communicate with one another in a hierarchical fashion so that an active center can suppress competing signals from another center (or vice versa). Note that a "command center" need not be a single bundle of neurons found in a particular part of the brain; alternatively, it may consist of any number of interconnected batteries of neurons that are capable of unified decision making.

Kenneth Roeder used command center theory to examine decision making in the praying mantis [1027]. A mantis can do many things: search for mates, sunbathe, copulate, fly, dive away from bats, and so on. Most of the time, however, the typical mantis remains motionless on a leaf until an unsuspecting bug wanders within striking distance. When the mantis's visual receptors alert it to the presence of the prey, the mantis makes very rapid, accurate, and powerful grasping movements with its front pair of legs.

Roeder proposed that the mantis's nervous system sorts out its options thanks to inhibitory relationships among an assortment of command centers within its neural network. The design of the mantis's nervous system (Figure 5.2) suggested that the control of muscles in each of the insect's segments is the responsibility of that segment's ganglion. Roeder tested this possibility by cutting one segmental ganglion's connections with the rest of the nervous system. Not surprisingly, the muscles within the neurally isolated segment subsequently failed to react when the mantis's nervous system became active elsewhere. However, if the isolated ganglion was stimulated electrically, the muscles and any limbs in that segment made vigorous, complete movements.

If the segmental ganglia are indeed responsible for telling the muscles in particular segments to carry out a given movement, what is the mantis's brain doing? Roeder suspected that certain brain cells are responsible for inhibiting (blocking) neural activity in the segmental ganglia, keeping cells in a ganglion quiet until they are specifically ordered into action by an excitatory command center in the brain. If so, then cutting the connection between the inhibitory

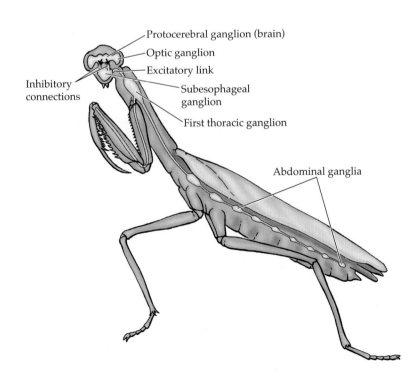

Figure 5.2   Nervous system of a praying mantis. If the connections between the protocerebral ganglion and the subesophageal ganglion are cut, the subesophageal ganglion sends a stream of excitatory messages to the segmental ganglia in the thorax and abdomen; the mantis then attempts to do several competing activities simultaneously.

brain cells and the segmental ganglia should have the effect of removing this inhibition and inducing inappropriate, conflicting responses. When Roeder severed the connections between the protocerebral ganglion (the mantis's brain) and the rest of its nervous system, he produced an insect that walked and grasped simultaneously, something that would be disastrous in nature. The protocerebral ganglion apparently makes certain that an intact mantis either walks *or* grasps, but does *not* do both at exactly the same time.

When Roeder removed the entire head, however—a procedure that eliminates the subesophageal ganglion as well as the protocerebral ganglion—the mantis became immobile. Roeder could induce single, irrelevant movements by poking the creature sharply, but nothing more. These results suggest that the protocerebral ganglion of an intact mantis typically sends out a stream of inhibitory messages to the subesophageal ganglion, preventing these neurons from communicating with the other ganglia. When certain sensory signals reach the protocerebral ganglion, however, neurons there stop inhibiting certain modules in the subesophageal ganglion. Freed from suppression, these subesophageal neurons send excitatory messages to various segmental ganglia, where new signals are generated that order muscles to take specific actions. Depending on what sections of the subesophageal ganglion are no longer inhibited, the mantis walks forward, or strikes out with its forelegs, or flies, or does something else.

Interestingly, mature male mantises do not always obey the "rule" that complete removal of the head eliminates behavior. Instead, a headless adult male performs a series of rotary movements that swing its body sideways in a circle. While this is happening, the mantis's abdomen is twisting around and down, movements that are normally blocked by signals coming from the protocerebral ganglion. This odd response to decapitation begins to make sense when you consider that a male mantis sometimes literally loses his head over a female, when she grabs him and consumes him, head first. Even under these difficult circumstances, the male can still copulate with his cannibalistic partner (Figure 5.3), thanks to the nature of the control system regulating his mating behavior. Headless, his legs carry what is left of him in a circular path until his body touches the female's, at which point he climbs onto her back and twists his abdomen down to copulate competently.

Figure 5.3   A no-brainer. Losing his head and brain did not extinguish the mating behavior of this male praying mantis mounted on the back of an intact female. Photograph by Mike Maxwell.

Whether male or female, adult or immature, the mantis, like many other animals, has a nervous system that appears to be functionally organized as a cluster of command centers, each with specific responsibilities. Some centers produce their own output, inhibiting the activities of other centers, which makes it possible for the mantis to do just one thing at a time. The importance of inhibitory relationships within nervous systems is also evident in the feeding blowfly, the subject of Vincent Dethier's classic studies [315]. These humble insects drink various exudates from plants, juices of liquefying animal corpses, and other savory fluids rich in sugars and proteins. During the night, the nutrients collected in the day's meals are metabolized to provide energy for the insect. In the morning, the blowfly flies off to seek additional food, which it locates in part by smell and in part by tasting with its feet when it happens to step into a fluid. The appropriate sensory inputs activate feeding commands: the fly extends its proboscis, spreads its labellum, and imbibes.

Dethier showed experimentally that the speed at which a fluid is sucked up and the duration of feeding are proportional to the concentration of sugar in the fluid. If the liquid is not very sugary, the fly's oral receptors cease firing quickly, and sucking stops. If sugar concentrations are high, however, the oral receptors may keep firing for 90 seconds or thereabouts before they quit, ending that bout of feeding. In short order, the sensory cells on the feet may become active again, leading to reextension of the proboscis and more drinking. Sooner or later, however, the fly stops drinking altogether, even when standing in the richest of sugary liquids. Dissection of a sated fly reveals that its crop, a storage sac off the digestive tract, is filled to overflowing, forcing fluid back into the foregut. Dethier hypothesized that distension of the foregut is detected by stretch receptors attached to this part of the digestive tract. He believed that those receptors might send messages to the brain, stimulating cells there to inhibit the feeding response.

As predicted from this hypothesis, receptors similar in design to stretch receptors in other organisms were found in the fly's foregut. As predicted, these receptors feed their sensory input into a pathway called the recurrent nerve, which runs from the foregut to the brain (Figure 5.4). As predicted, if one experimentally severs the recurrent nerve, the fly will not stop feeding. The insect continues on with bout after bout of drinking until its body literally explodes. Flies

Figure 5.4 Nervous system and digestive system of a blowfly. Severing the recurrent nerve eliminates feedback to the brain about the fullness of the crop and eliminates the signals that eventually block feeding in a blowfly whose stomach and crop are full.

Figure 5.5   Record of neural and behavioral activity of a calling cricket. (A) Male crickets call by rubbing their forewings together. (B) The record of a command neuron in the brain of a calling male that was subjected to a brief tactile stimulus, a puff of air on the tip of the abdomen. The cricket's wing movements and the sounds of its calls are shown on separate lines. A, photograph by Edward S. Ross; B, from Hedwig [528].

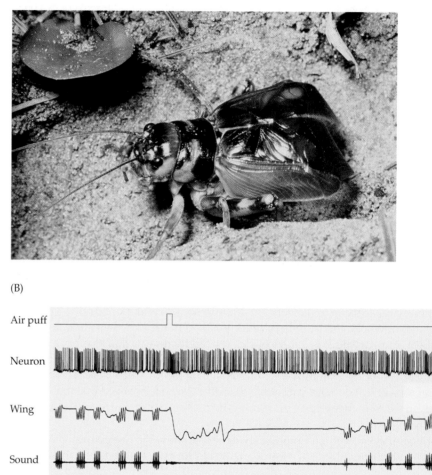

(A)

(B)

with intact recurrent nerves do not blow themselves up, thanks to inhibitory arrangements between elements of their nervous system [315].

Discussion Question

5.1   Figure 5.5 shows a record of the activity of a particular neuron in a cricket's brain that appears to act as a command cell controlling the male's chirping call, which is produced when the male rubs his wings together. About 3 seconds into the recording, the cricket was subjected to a sharp puff of air that struck its cerci, a pair of sensory appendages that project out from the rear of the abdomen. What is the relevance of this figure to the topic that we have just discussed? What adaptive value do you attach to the apparent proximate mechanisms that help male crickets make the behavioral decisions shown here?

Behavioral Schedules

The ability of neural command centers to communicate with and inhibit one another helps set an animal's behavioral priorities. But what if the hierarchical ordering of command centers has to change in order to meet the demands of a changing environment? Some activities are better done at some times of the day than others. Female *Teleogryllus* crickets, for example, usually hide in burrows or under leaf litter during the day and move about only after dusk, when it is relatively safe to search for mates [739]. Not surprisingly, male crickets wait for the evening to start calling for mates [738]. These observations suggest that the inhibitory relationships between the calling center in a male cricket's brain and other neural elements responsible for other behaviors must change cyclically over a 24-hour period. What mechanisms might regulate these changes?

Researchers interested in this question have developed two major competing theories. First, animals might change their priorities in response to a timing mechanism with a built-in schedule that acts independently of any cues from the animal's surroundings. That such environment-independent timing mechanisms might exist should be plausible to anyone who has flown across several time zones and then tried to adjust immediately to local conditions. The second theory is that animals alter the relationships among command centers in their nervous systems strictly on the basis of information gathered by mechanisms that monitor the surrounding environment. Such devices would enable individuals to modulate their behavior in response to certain changes in the world around them, such as a decrease in light intensity as evening comes on.

Let's consider these two possibilities in the context of the calling cycle of male *Teleogryllus* crickets. Each day's calling bout could begin at much the same time because the crickets possess an internal timer that measures how long it has been since the last bout began; they could use this environment-independent system to activate the onset of a new round of chirping each evening at dusk. Alternatively, the insect's neural mechanisms might be designed to initiate calling when light intensity falls to a particular threshold level. If this second hypothesis is correct, then crickets held under constant bright light should not call. But in fact, laboratory crickets held in rooms in which the temperature stays the same and the lights are on 24 hours a day still continue to call regularly for several hours each day. Under conditions of constant light, calling starts about 25 to 26 hours later than it did the previous day (Figure 5.6). A cycle of activity that is not matched to environmental cues is called a **free-running cycle**. Because the length, or **period**, of the free-running cycle of cricket calling deviates from

Figure 5.6   Circadian rhythms in cricket calling behavior. Each horizontal line on the grid represents one day; each vertical line represents a half hour on a 24-hour time scale. Dark marks indicate periods of activity—in this case, calling. The bars at the top and middle of the figure represent the lighting conditions; thus, for the first 12 days of this experiment, male crickets were kept in constant light (LL), and for the remainder, they were subjected to 12-hour cycles of light and dark (LD). Male crickets held under constant light exhibit a daily cycle of calling and noncalling, but the calling starts later each day. The onset of "nightfall" on day 13 acts as a cue that resets the calling rhythm, which soon stops shifting and eventually begins an hour or two before the lights are turned off each day. After Loher [738].

the many 24-hour environmental cycles caused by the earth's daily rotation about its axis, we can conclude that the cyclical pattern of cricket calling is caused in part by an environment-independent internal **circadian rhythm** (*circadian* means "about a day").

Now let's place our crickets in a regime of 12 hours of light and 12 hours of darkness. The switch from light to darkness offers an external environmental cue that the crickets could use to adjust their timing mechanism. In fact, they do, just as monarch butterflies and pigeons can reset their clocks when moved into a laboratory with artificial lighting (see Chapter 4). After a few days, the males start calling about 2 hours before the lights go off, accurately anticipating "nightfall," and they will continue until about 2.5 hours before the lights go on again in the "morning" (see Figure 5.6). This cycle of calling matches the natural one, which is synchronized with dusk; unlike the free-running cycle, it does not drift out of phase with the 24-hour day, but is reset, or **entrained**, each day so that it begins at the same time in relation to lights-out [738]. From these results, we can conclude that the complete control system for cricket calling has both environment-independent and environment-dependent components: an environment-independent timer, or **biological clock**, set on a cycle that is not exactly 24 hours long, and an environment-activated device that synchronizes the clock with local light conditions.

How Do Circadian Mechanisms Work?

Through experiments of the sort we have described, investigators have been able to infer something about the properties of the biological clocks of many animal species. Additional information on the clock mechanism has come from surgical manipulations similar to those done by Kenneth Roeder in his studies of the mantis nervous system. By cutting the connections between various neural components, researchers have been able to test predictions about the possible roles played by the different components. Thus, if one cuts the nerve carrying sensory information from the eyes of a male cricket to the optic lobes of his brain (Figure 5.7), the cricket enters a free-running cycle. Visual signals of

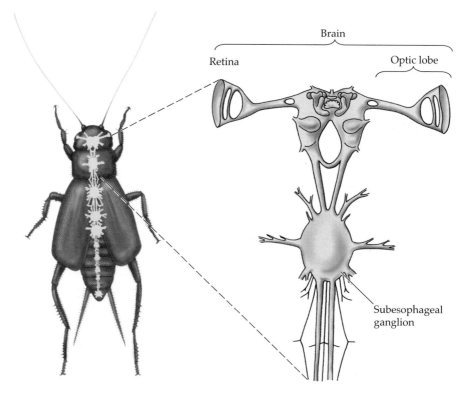

Figure 5.7   The cricket nervous system. Visual information from the eyes is relayed to the optic lobes of the cricket's brain. If the optic lobes are surgically disconnected from the rest of the brain, the cricket loses its capacity to maintain a circadian rhythm. Based on diagrams by F. Huber and W. F. Shurmann.

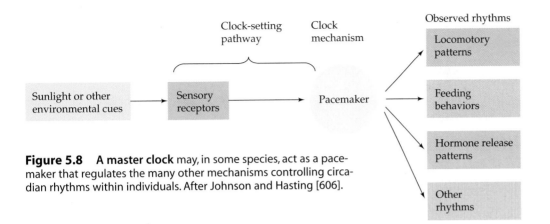

Figure 5.8   **A master clock** may, in some species, act as a pacemaker that regulates the many other mechanisms controlling circadian rhythms within individuals. After Johnson and Hasting [606].

some sort are evidently needed to entrain the daily rhythm to local conditions, but a rhythm persists in the absence of this information. If, however, one separates both optic lobes from the rest of the brain, the calling cycle breaks down completely; the cricket will now call with equal probability at any time of the day. These results are consistent with the hypothesis that a master clock mechanism (Figure 5.8) resides within the optic lobes, sending messages to other regions of the nervous system [606, 921].

Biologists interested in the control of circadian rhythms in mammals and other vertebrates have focused on the hypothalamus, an important regulatory structure in the brain, with special emphasis on the suprachiasmatic nucleus (SCN), a pair of hypothalamic neural clusters that receive inputs from nerves originating in the retina. The SCN is therefore a likely element of the mechanism that secures information about day and night length, information that could be used to adjust a master biological clock.

If the SCN contains a master clock or pacemaker that is critical for maintaining circadian rhythms, then damage to the SCN should cause individuals to lose those rhythms. Such an experiment has been done by selectively destroying SCN neurons in the brains of hamsters and white rats, which subsequently exhibit arrhythmic patterns of hormone secretion, locomotion, and feeding [1361]. If arrhythmic hamsters receive transplants of SCN tissue from fetal hamsters, they sometimes regain their circadian rhythms, but not if the tissue transplants come from other parts of the fetal hamster brain [310].

Perhaps the SCN clock of hamsters, rats, and other mammals operates via rhythmic changes in gene activity. A key candidate gene in this regard appears to be the *per* gene, which codes for a protein (PER) whose production varies over a 24-hour schedule in concert with that of the product of another mammalian gene called *tau* (Figure 5.9). The product of *tau* is an enzyme whose production is turned on when PER is at peak abundance in the cell. The enzyme degrades PER, contributing to a 24-hour cycle in which PER first increases in abundance and then falls [1349].

A striking feature of this system, whose complexity is daunting, is that the key clock genes regulating cellular circadian rhythms in mammals are also present in insects. *Drosophila* fruit flies and honey bees also have the *per* gene, a chain of DNA composed of somewhat over 3500 base pairs, which provides the information needed to produce the PER protein chain of nearly 1200 amino acids. Alterations in the base sequence involving as little as a single base substitution can result in dramatically different circadian rhythms in fruit flies (Figure 5.10), as well as in humans (carriers of one *per* mutation typically fall asleep at 7:30 in the evening and arise at 4:30 in the morning [1204]). These results strongly suggest that the gene's information plays a critical role in enabling circadian rhythms.

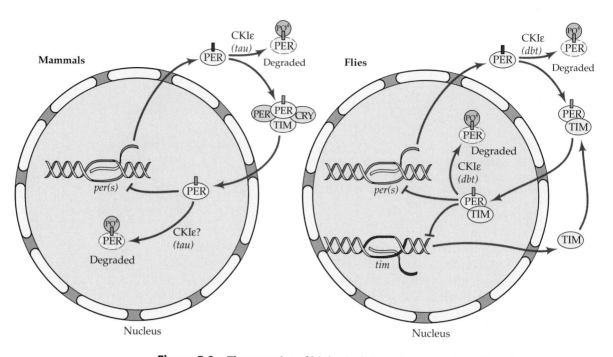

Figure 5.9 **The genetics of biological clocks** in mammals and fruit flies. In both groups, a set of three key genes produces proteins that interact with one another to regulate the activity of certain other genes on a cycle lasting approximately 24 hours. One of the genes (*per*) codes for a protein (PER) that gradually builds up inside and outside the cell nucleus over time. Another key gene, called *tau* in mammals and *dbt* in flies, codes for an enzyme, CKIε, that helps break down PER, slowing its rate of accumulation in the cell. But during peak periods of production of PER, more PER is available to bond with another protein (TIM), coded for by a third gene (*tim*). When the PER protein is bound in complexes with TIM (and another protein, CRY, in the case of mammals), it cannot be broken down as quickly by CKIε. Therefore, more intact PER is carried back into the nucleus where it blocks the activity of the very gene that produces it, though only temporarily. Then a new cycle of *per* gene activity and PER protein production begins. After Young [1349].

If expression of *per* does have this effect, then animals in which the *per* gene is relatively inactive should behave in an arrhythmic fashion. Very little PER protein is manufactured in young honey bees, which generally remain within the hive to care for eggs and larvae. And young honey bees are in fact just as likely to perform these nursing tasks at time X as at time Y over a 24-hour period. In contrast, older honey bees, which forage for food during the daytime only, exhibit well-defined circadian rhythms, leaving the hive to collect pollen and nectar only during that part of the day when the flowers they seek are most likely to be resource-rich. Foragers have almost three times as much PER protein in their brain cells as do young nurse bees, thanks to their heightened *per* gene activity [1205].

Discussion Question

5.2 You may recall that the transition to foraging in honey bees depends on the makeup of the colony, such that if there is a shortage of nurse workers within the hive, older bees will delay their shift to foraging role. What prediction follows about *per* gene expression in the brains of these socially delayed older nurses relative to foragers of the same age from other colonies with numerous young nurse bees? Provide proximate and ultimate hypotheses for the fact that social interactions can alter circadian rhythms in honey bees—and even fruit flies [714], which do not live in highly organized societies.

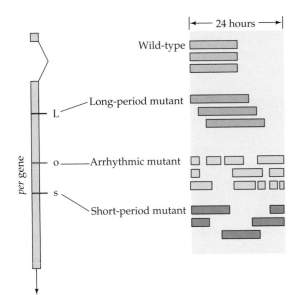

← 24 hours →

Wild-type

Long-period mutant

L

o —— Arrhythmic mutant

s

Short-period mutant

per gene

Figure 5.10   Mutations of the *per* gene affect the circadian rhythms of fruit flies. On the left is a diagram of the DNA sequence that constitutes the *per* gene. The locations of the base substitutions present in three mutant alleles found in fruit flies are indicated on the diagram. The activity patterns of wild-type flies, and those associated with each mutation, are shown on the right. After Baylies et al. [75].

Because the fruit fly, the hamster, the honey bee, and you and I all have the same gene serving much the same clock function, evolutionary biologists believe that we inherited this gene, as well as some others involved in the regulation of activity patterns, from a very ancient animal that lived perhaps 550 million years ago [1349]. In mammals and some other vertebrates, the *per* gene and certain others are expressed in the neurons of the SCN, which can be viewed as the home of a master clock or circadian pacemaker that communicates with a host of other brain components, thereby regulating many different behaviors on a daily schedule. In order to serve this role, one would imagine that the SCN master clock might send out chemical signals to the target systems it controls. If so, then we can predict that (1) the molecule that relays the clock's information ought to be secreted by the SCN in a manner regulated by the clock's genes, (2) there should be receptor proteins for that chemical messenger in cells of the target tissues, and (3) experimental administration of the chemical messenger should disrupt the normal timing of an animal's behavior.

A research team found a protein, prokineticin 2 (PK2), with precisely these properties [209]. In normal mice living under a cyclical regime of 12 hours of light and 12 hours of darkness, the PK2 protein is produced in a strongly circadian pattern by the SCN (Figure 5.11). Moreover, mice with certain mutations in key clock genes lack the circadian rhythm of PK2 production. Furthermore, as also predicted, only certain structures within the brain of the mouse produce a receptor protein that bonds with PK2. These regions are linked to the SCN by neural pathways and are believed to contain command centers that control var-

Figure 5.11   Expression of the gene that codes for PK2 in the SCN. Mice were exposed to a standard light–dark cycle of 12 hours of light and 12 hours of darkness (LD) prior to the start of the experiment. The graph shows hourly changes in the production of the messenger RNA encoded by the PK2 gene. The gene was expressed in a circadian rhythm whether the animals were held under the standard light–dark cycle, kept in complete darkness for 2 days (2DD), or held in darkness for 8 days (8DD). After Cheng et al. [209].

Figure 5.12 Circadian control of wheel-running by white rats changes when the brains of rats are injected with PK2. These rats are active primarily during the daylight hours whereas control rats injected with saline exhibit the standard preference for nighttime activity. After Cheng et al. [209].

ious behavioral activities in a circadian fashion. Finally, if one injects PK2 into the brain of white rats during the night, when the animals are normally active, the behavior of these animals changes dramatically. Instead of running in their running wheel, they sleep, shifting to daytime activity instead (Figure 5.12) [209].

All of these lines of evidence point to PK2 as the chemical messenger that the mammalian SCN uses to communicate with target centers in the brain so that they can control the animal's behavior in keeping with a 24-hour schedule. The SCN, in turn, receives information from the retina about the light–dark cycle of the animal's environment, so that it can fine-tune its autoregulated pattern of gene expression. The adaptive value of the environment-independent component of the clock system may be that a clock of this sort enables individuals to alter the timing of their behavioral and physiological cycles without having to constantly check the environment to see what time it is. The presence of an environment-dependent element, however, permits individuals to adjust their cycles in keeping with local conditions. As a result, a typical nocturnal mammal automatically becomes active at about the right time each night, while retaining the capacity to shift its activity period gradually to accommodate the changes in day length that occur as spring becomes summer, or summer becomes fall.

One way to test this hypothesis about the adaptive value of circadian rhythms would be to predict that if there were a mammal for which the day–night cycle was biologically irrelevant, then this species should lack a circadian pattern of activity. The naked mole-rat (see Figure 13.34) is such a species. This animal lives in groups that occupy an extensive network of underground tunnels. Because mole-rats almost never come to the surface, but instead live in total darkness, feeding on roots and tubers collected from their tunnels, what goes on above ground is not important to them. As predicted, naked mole-rats lack a circadian rhythm. Instead, individuals scatter generally brief episodes of activity among longer periods of inactivity, with the pattern changing irregularly from day to day (Figure 5.13) [299].

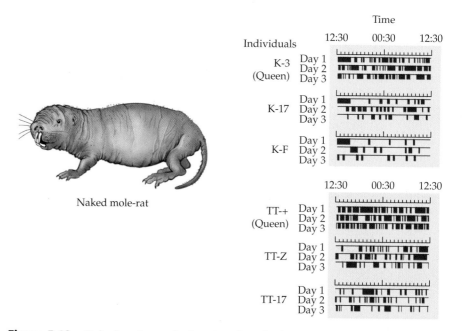

Naked mole-rat

Figure 5.13 Naked mole-rats lack a circadian rhythm. Patterns of activity are shown for six individuals from two captive colonies held under constant low light. Dark bars indicate periods when the individual was awake and active. After Davis-Walton and Sherman [299].

Long-Term Cycles of Behavior

Because of their unusual lifestyle, naked mole-rats do not have to deal with cyclically changing environments, and they have apparently lost their circadian rhythm as a result. But almost all other creatures confront not only daily changes in food availability or risk of predation, but also changes that cover periods longer than 24 hours, such as the seasonal changes that occur in many parts of the world. If circadian rhythms enable animals to prepare physiologically and behaviorally for certain predictable daily changes in the environment, might not some animals possess a **circannual rhythm** that runs on an approximately 365-day cycle [491]? A circannual clock mechanism might be similar to the circadian master clock, with an environment-independent timer capable of generating a circannual rhythm in conjunction with a mechanism that keeps the clock entrained to local conditions.

Testing the hypothesis that an animal has a circannual rhythm is technically difficult because individuals must be maintained under constant conditions for at least 2 years after being removed from their natural environment. One successful study of this sort involved the golden-mantled ground squirrel [935] of North America, which in nature spends the late fall and frigid winter hibernating in an underground chamber. Five members of this species were born in captivity, then blinded and held thereafter in constant darkness and at a constant temperature while supplied with an abundance of food. Year after year, these ground squirrels entered hibernation at about the same time as their fellows living in the wild (Figure 5.14).

In another study, several nestling stonechats were taken from Kenya to Germany to be reared in laboratory chambers in which the temperature and **photoperiod** (the number of hours of light in a 24-hour period) were always the same. Needless to say, these birds, and their offspring, never had a chance to encounter the spring rainy season in Kenya, which heralds a period of insect abundance and is the time when Kenyan stonechats must reproduce if they are to find sufficient food for their hungry nestlings. Thus, wild stonechats exhibit an annual cycle of reproductive physiology and behavior. The transplanted stonechats, despite their constant environment, also exhibited an annual reproductive cycle, but one that shifted out of phase with that of their Kenyan com-

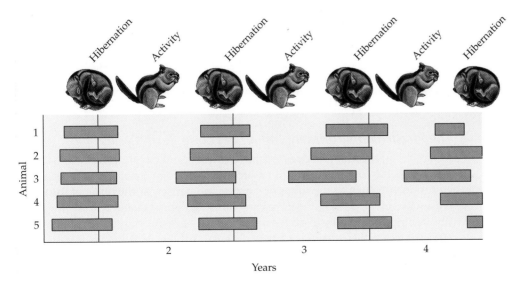

Figure 5.14 Circannual rhythm of the golden-mantled ground squirrel. Animals held in constant darkness and at a constant temperature nevertheless entered hibernation (green bars) at certain times year after year. After Pengelley and Asmundson [935].

Stonechat

Figure 5.15 Circannual rhythm in a stonechat. When transferred from Kenya to Germany and held under constant conditions, this male stonechat still underwent a regular long-term cycle of testicular growth and decline (purple lines) as well as regular feather molts (the two bars, which refer to wing feather and body molts). The cycle was not exactly 12 months long, however, so the timing of molting and testicular growth shifted over the years (see the dashed lines that angle upward from left to right). After Gwinner and Dittami [492].

patriots over time (Figure 5.15). One male, for example, went through nine cycles of testicular growth and decline during the 7.5 years of the experiment. Evidently, the stonechat's circannual rhythm is generated in part by an internal, environment-independent mechanism, just as is true for the golden-mantled ground squirrel [492].

The Physical Environment Influences Long-Term Cycles

In nature, environmental cues entrain circadian and circannual clocks so as to produce behavioral rhythms that match the particular features of the animal's environment, such as the times of sunrise and sunset, or the onset of the rainy season in a given year, or the increasing day lengths associated with spring. This fine-tuning of behavioral cycles involves mechanisms of great diversity that respond to a full spectrum of environmental influences, which can vary from species to species according to their ecological circumstances.

The banner-tailed kangaroo rat, for example, is an animal whose readiness to forage for food changes over the course of the lunar cycle [734, 735]. Robert Lockard and Donald Owings reached this conclusion by monitoring the activity of free-living kangaroo rats in a valley in southeastern Arizona. To measure kangaroo rat activity, Lockard invented an ingenious food dispenser/timer that released very small quantities of millet seed at hourly intervals. To retrieve the seeds, an animal had to walk through the dispenser, depressing a treadle in the process. The moving treadle caused a pen to make a mark on a paper disk that turned slowly throughout the night, driven by a clock mechanism. When the paper disk was collected in the morning, it carried a temporal record of all nocturnal visits to the dispenser.

Data collection was sometimes frustrated by ants that perversely drank all the ink or by Arizonan steers that trampled the recorders. Nevertheless, Lockard's surviving records showed that when the kangaroo rats had accumulated a large cache of seeds in the fall, they were selective about foraging, usually coming out of their underground burrows only at night when the moon was not shining (Figure 5.16). Because the predators of kangaroo rats (coyotes and owls) can see their prey more easily in moonlight, banner-tails that forage only in complete darkness are relatively safe. For this reason, the kangaroo rats apparently possess a mechanism that enables them to shift their foraging schedule in keeping with nightly moonlight conditions. Perhaps kangaroo rats simply avoid foraging under bright moonlight, or perhaps the rodents have an internally driven circalunar rhythm that approximates the lunar cycle. (How would you determine which hypothesis was correct?)

Whereas desert kangaroo rats match their foraging behavior to the lunar cycle and Kenyan stonechats employ a circannual rhythm to breed at the right time in their tropical environment, temperate-zone birds such as white-crowned sparrows have their own behavioral control system that is well suited for coping with the dramatic seasonal changes that occur in their environments. In the spring, males fly from their wintering grounds in Mexico or the southwestern United States to their distant summer headquarters in the northern United States, Canada, and Alaska. There they establish breeding territories, fight with rivals, and court sexually receptive females. In concert with these striking behavioral changes, the gonads of the birds grow with dramatic rapidity, regaining all the weight lost during the winter, when they shrink to 1 percent of their breeding-season weight. In order to properly time the regrowth of their gonads and the onset of their reproductive activities, the birds must somehow anticipate the spring breeding season. How do they manage this feat?

The sparrows' ability to change their physiology and behavior depends on their capacity to detect changes in the photoperiod, which grows longer as spring advances in temperate North America [383]. One hypothesis on how such a system might work proposes that the clock mechanism of white-crowns exhibits a daily change in sensitivity to light, with a cycle that is reset each morning at dawn. During the initial 12 hours or so after the clock is reset, this mechanism is highly insensitive to light; this insensitivity then steadily gives way

Banner-tailed kangaroo rat

Period when animals did not forage during moonlight

Period of continuous nocturnal foraging

Start of period of daytime activity in addition to nighttime foraging

Figure 5.16 Lunar cycle of banner-tailed kangaroo rats. Each thin black mark represents a visit made by a banner-tail to a feeding device with a timer. From November to March, the rats were active at night only when the moon was not shining. A shortage of seeds later in the year caused the animals to feed throughout the night, even when the moon was up, and later still to forage during all hours of the day. After Lockard [735].

Figure 5.17   A cycle of photosensitivity. An experiment with white-crowned sparrows tested the hypothesis that these birds possess a clock mechanism that is especially sensitive to light between hours 17 and 19 of each day. The lower line represents these hypothetical periods of photosensitivity. The white and black sections of the two upper horizontal bars show the light and dark periods of two different light–dark regimes. Only sparrows under the 8L:28D experimental regime were exposed to light during the supposed photosensitive phase of the cycle, and only they responded with testicular growth. After Farner [382].

to increasing sensitivity, which reaches a peak 16 to 20 hours after the starting point in the cycle. Photosensitivity then fades very rapidly to a low point 24 hours later, at the start of a new day and a new cycle. Therefore, if the days have fewer than 12 hours of light, the system will never become activated because no light is present during the photosensitive phase of the cycle. However, if the photoperiod is longer than 14 or 15 hours, light will reach the bird's brain during the photosensitive phase, initiating a series of hormonal changes that lead to the development of its reproductive equipment and the drive to reproduce.

If this model of the photoperiod-measuring system is correct, it should be possible to deceive the system. William Hamner, working with house finches [508], and Donald Farner, in similar studies with white-crowned sparrows [382], stimulated testicular growth by exposing captive birds to light during the hypothesized photosensitive phase of their circadian rhythms. In Farner's experiment, birds that had been on a regular schedule of 8 hours of light and 16 hours of darkness (8L:16D) were shifted to an 8L:28D schedule. Because the light periods were now out of phase with a 24-hour cycle, these birds sometimes experienced light during the time when their brains were predicted to be highly photosensitive. The male birds' testes grew under these conditions, even though there was a lower ratio of light to dark hours than under the 8L:16D cycle, which did not stimulate testicular growth (Figure 5.17) [382].

Discussion Question

5.3   In another experiment with white-crowned sparrows, groups of males that had been held on an 8L:16D schedule were housed in complete darkness for variable lengths of time (anywhere from 2 to 100 hours) before being exposed to an 8-hour block of light. A few hours later, the researchers measured the concentration in the birds' blood of luteinizing hormone (a hormone released by the anterior pituitary and carried to the testes, where it stimulates the growth of these tissues). Do the collected data shown in Figure 5.18 provide a test for the photosensitivity hypothesis just described?

Crossbills exhibit a much less constrained pattern of seasonal breeding than white-crowned sparrows do. Craig Benkman has shown that food intake, not photoperiod length, appears to be the primary determinant of breeding in two of these finch species, the white-winged crossbill and the red crossbill [89]. There are years when conifer seed production is so high that the birds can find ample food for themselves and a brood in almost any month, and they accordingly take advantage of these occasions to breed. This flexibility, however, does not mean that crossbills ignore all environmental cues except seed abundance. If you take captive nonbreeding white-winged crossbills that have been held under the short-day, long-night conditions of winter and shift them to a long photoperiod (20L:4D), they respond physiologically with increases in the size of their gonads and in the concentration of luteinizing hormone in their blood [499].

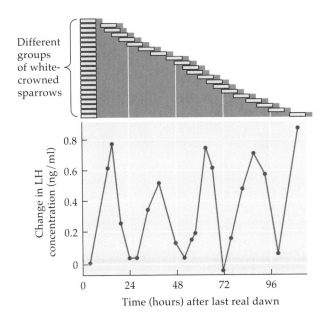

Figure 5.18   A hormonal response to light. Different groups of white-crowned sparrows were kept in complete darkness for different lengths of time, then exposed to a single 8-hour period of light. The upper diagram illustrates when the 8-hour light exposures occurred. The lower graph shows the changes from the start of the experiment in luteinizing hormone (LH) concentrations in the blood in each group of birds. LH is a hormone known to stimulate the growth of the gonads. After Follett et al. [394].

In studies of crossbills in the wild, Thomas Hahn noticed a break in crossbill breeding in December and January (Figure 5.19), even where food was plentiful. Therefore, even though these birds are more flexible and opportunistic than the average songbird, perhaps they too have an underlying reproductive cycle dependent on photoperiod. When Hahn held red crossbills at a constant temperature with unlimited access to food, while letting the birds experience natural photoperiod changes, he found that male testis length fluctuated in a cyclical fashion (Figure 5.20), becoming smaller during October through December of each year, even when the birds had all the seeds a crossbill could hope to eat [500]. In addition, free-living red crossbills undergo declines in the concentrations of sex hormones in their blood at this time of year, even in areas where

White-winged crossbill

Figure 5.19   Food intake and reproductive timing in the white-winged crossbill. Breeding populations usually occur in areas with relatively high food availability. Nonbreeding populations generally occur in areas where the birds have low food intake. Note, however, the absence of breeding populations in all locations in December and January. After Benkman [89].

Red crossbill

Figure 5.20   Photoperiod affects testis size in the red crossbill. Six captive birds were held under natural photoperiods, which changed over the seasons, but temperature and food supply were held constant. The curve represents the average testis length among these birds at different times during the year. After Hahn [497].

their food is abundant [498]. Therefore, the reproductive opportunism of these birds is not absolute, but rather is superimposed on the photoperiod-driven timing mechanism characteristic of temperate-zone songbirds in general.

The persistence of the standard timing system in the flexibly breeding crossbills could be explained at the evolutionary level in at least two ways: (1) the photoperiod-driven mechanism might be a nonadaptive holdover from the past, or (2) crossbills might derive reproductive benefits from the retention of a physiological system that reduces the likelihood that they will attempt to reproduce at times when factors other than food supply make successful reproduction unlikely (such as the need to molt and replace their feathers in the fall, which requires a large caloric investment) [318].

Changing Priorities in Changing Social Environments

As we have just seen, different features of the physical environment, such as moonlight, day length, or food supply, are used by different species to establish their behavioral priorities. In addition, animals can use changes in their social environment to make adaptive adjustments in their physiology and behavior. Thus, for example, when Hahn and several co-workers performed an experiment on crossbills in which some captive males were caged with their mates, while others were forced into bachelorhood but were kept within sight and sound of the paired crossbills in a neighboring aviary, the bachelor males experienced a slower return to reproductive condition after the winter break than did the paired males [500]. Similarly, just one 60-minute exposure to a female resulted in elevated testosterone concentrations in captive male starlings [956], which may have contributed to subsequent changes in the males' behavior, such as increased courtship singing.

The social environment also affects the behavioral priorities of house mice. After a male house mouse mounts a female and ejaculates, he immediately becomes highly aggressive toward mouse pups, killing any he finds. For almost 3 weeks after mating, he is likely to commit infanticide, but after that time, he becomes more and more likely to protect any young pups he encounters. When about 7 weeks have passed since ejaculation, he becomes infanticidal once again [938].

This remarkable cycle has clear adaptive value. After a male transfers sperm to a partner, 3 weeks pass before she gives birth. Attacks on pups during these 3 weeks will invariably be directed against a rival male's offspring, with all the benefits attendant on their elimination (see p. 20). After 3 weeks, a male that switches to paternal behavior will almost always care for his own neonatal offspring. After 7 weeks, his weaned pups will have dispersed, so that once again he can practice infanticide advantageously.

At the proximate level, what kind of mechanism could enable a male to switch from infanticidal Mr. Hyde to paternal Dr. Jekyll 3 weeks after a mating? One possible explanation involves an internal timing device that records the number of days since the male last copulated. If such a sexually activated timing mechanism exists, then an experimental manipulation that either increases or decreases the length of a "day," as perceived by the mouse, ought to have an effect on the absolute amount of time that passes before the male makes the transition from killer to caregiver.

Glenn Perrigo and his co-workers manipulated day length by placing groups of mice under two different laboratory conditions, one with "fast days," in which 11 hours of light were followed by 11 hours of darkness (11L:11D) to make a 22-hour "day," and another with "slow days" (13.5L:13.5D) that lasted for 27 hours. As predicted, the total number of light–dark cycles, not the number of 24-hour periods, controlled the infanticidal tendencies of males (Figure 5.21). Thus, when male mice in the fast-day group were exposed to mouse pups 20 real days after mating, only a small minority committed infanticide, because these males had experienced 22 light–dark cycles during this period. In contrast, more than 50 percent of the males in the slow-day group attacked newborn pups at 20 real days after mating because these males had experienced only 18 light–dark cycles during this time. These results demonstrate that a timing device registers the number of light–dark cycles that have occurred since mating, and that this information provides the proximate basis for the control of the infanticidal response [938].

Perhaps the timing device that controls the male mouse's treatment of mouse pups influences the hormonal state of the male. If so, then you might expect that during their infanticidal phase, male mice would have high concentrations of testosterone in their blood, given the well-established relationship between testosterone and male aggression. An alternative hypothesis is that extreme male aggression toward mouse pups occurs when the males are under the influ-

Figure 5.21   Regulation of infanticide by male house mice. (A) Male mice were held under artificial "slow-day" and "fast-day" experimental conditions. (B) Most of the males held under fast-day conditions had stopped being infanticidal by 20 real days (22 fast days) after mating; males experiencing slow days did not show the same decline in infanticidal behavior until nearly 25 real days had passed. After Perrigo et al. [938].

Figure 5.22   A hormonal effect on infanticidal behavior in laboratory mice. Males of the strain C57BL/6 are highly likely to attack their own first or second litters, rather than caring for their young. If, however, the progesterone receptor gene is removed from the genome, the resulting progesterone knockout (PRKO) mice cannot detect the progesterone in their bodies, and they do not exhibit infanticidal behavior toward their offspring. After Schneider et al. [1071].

ence of progesterone, because this hormone is known to suppress parental behavior in female rodents. To test these alternatives, Jon Levine and his co-workers used genetic knockout techniques to create a population of laboratory mice that lacked progesterone receptors, which meant that males of this line could not detect the progesterone in their bodies. When the knockout males were exposed to a pup, they never attacked it, whereas over half of the males from a genetically unmodified strain of laboratory mice assaulted a test infant (Figure 5.22).(This strain is an unusually aggressive one in which untreated males may kill their own pups.) There were no significant differences in testosterone or progesterone levels between the two groups of mice, only a difference in the ability of their brain cells to detect progesterone [1071]. The results of this experiment indicate that when certain levels of progesterone are present in an intact male, he is primed to be an infanticidal killer. As progesterone levels slowly fall (or sensitivity to the hormone declines in brain cells) after a copulation, the paternal capacity of the male slowly increases in time for him to avoid attacking his own brood.

Discussion Question

5.4   In the California mouse, *Peromyscus californicus*, males are highly paternal. Explanations for this behavior include the reduced progesterone hypothesis that we have just explored for laboratory mice. Alternatively, it is possible that increased levels of estrogen are responsible for male parental behavior. Make some predictions derived from these two hypotheses, including at least one that has to do with *testosterone* (if necessary, reread the introduction to this chapter to refresh your memory on the relation between testosterone and estrogen). Then examine Figure 5.23 and evaluate your hypotheses in the light of these data.

Mating alters the behavioral priorities of many animals, not just house mice (and prairie voles—see Chapter 1). Take the male Japanese quail, for example. When a male copulates with a female, even once, his behavior changes dramatically. Prior to mating, a male housed next to a female in a two-compartment cage will spend relatively little time peering through a window between the compartments in order to look at her. But after mating, a male appears positively fascinated by his partner, to the extent that he will stare at her for hours on end (Figure 5.24A) [62]. This behavior, which presumably has the function in nature of bringing the male together with a receptive female for yet another copulation, is heavily influenced by the presence of testosterone in particular regions of the male's brain. This conclusion is based in part on the finding that

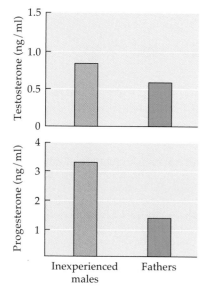

Figure 5.23   Testosterone and progesterone levels in two categories of male California mice: inexperienced males with neither a mate nor offspring and father mice with a mate and offspring. The differences in testosterone levels between the two categories of males are not statistically significant; in contrast, progesterone occurs at far higher levels in the blood of inexperienced males than in males with offspring. After Trainor et al. [1209].

(A)

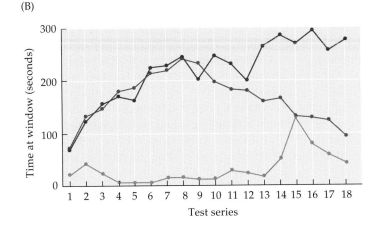

Figure 5.24 **Testosterone and the control of sexual motivation** in male Japanese quail. (A) Hormonal changes contribute to changes in how sexually experienced males respond to the opportunity to view a mature female quail. Testosterone is released from the testes and travels to certain parts of the brain. When testosterone enters target cells, chemical reactions catalyzed by the enzyme aromatase covert it to 17β-estradiol. That substance bonds with an estrogen receptor to form an estrogen–receptor complex, which is transferred to the target cell nucleus, where it promotes chemical changes that ultimately cause male quail to stare at females through a window in their two-compartment cage. (B) Castrated males that have received testosterone implants exhibit the staring-at-female response, unlike controls that have not been given a testosterone implant. But when implanted males are injected with an aromatase inhibitor, they lose the response, presumably because testosterone can no longer be converted to 17β-estradiol, the essential hormone for modulating male sexual motivation. After Balthazart et al. [62].

removal of the male's testes eliminates all sexual behavior—unless the male receives an implant containing testosterone, in which case he regains his motivation to seek out mates.

Interestingly, testosterone itself is not the signal that turns a male Japanese quail into an apparently lovelorn individual when separated from his partner. Instead, testosterone is converted into 17β-estradiol, an estrogen, in target cells within the preoptic area of the brain. The conversion requires an enzyme, aromatase, which is coded for by a gene that becomes active only after a male quail has mated. The estrogen produced in the presence of aromatase bonds with an estrogen receptor protein, and the resulting estrogen–receptor complex relays the signal to the nucleus of the cell, leading to further biochemical events. These events ultimately translate into the neural signals that cause males to stare intently at a currently unreachable sexual partner.

The importance of the enzyme aromatase to this process has been established experimentally by creating three groups of castrated male quail. The controls received no replacement testosterone, whereas males in two other groups received testosterone implants. The implanted males spent much more time at the window than the controls. After eight trials, one of the two groups of implanted males was given daily injections of an aromatase inhibitor; these birds' interest in the female in the next compartment fell steadily as seen in the reduction in time they spent looking through the window over the next 10 trials (Figure 5.24B).

Hormones and the Organization of Reproductive Behavior

Many of the hormonally mediated adjustments that animals make directly affect their reproductive behavior. In the Japanese quail, 17β-estradiol plays a major role in adjusting the degree to which a male becomes attached to one mate, whereas in the prairie vole, vasopressin is the key to the formation of sexual bonds (see p. 4). Males of many species not only have the capacity to form an attachment to a female, but can also modulate their aggressiveness toward other males. This shift in behavior is also under hormonal regulation. Thus, when a male cichlid fish is ousted from his mate-attracting territory, a change in gonadotropin-releasing hormone production contributes to reductions in his aggressiveness, the size of his testes, and his efforts to reproduce (see pp. 87–88) [1280]. On the other side of the coin, when a male plainfin midshipman fish makes the transition from the nonreproductive state to breeding condition in the summer, when he begins to attract females with his humming vocalization (see p. 118), a hormone, arginine vasotocin, helps link growth of the fish's testes with activation of his singing behavior [73]. In a host of other species, hormones also underlie the integrated shifts in sexual and aggressive behavior that facilitate reproduction at times when external environmental and internal physiological conditions are most favorable.

Hormonal organization of reproduction often involves coordinated changes in gamete production and in sexual activity—the so-called **associated reproductive pattern** (Figure 5.25). In the green anole, for example, when males first become active after a winter dormant phase, their circulating concentrations of the hormone testosterone are very low. As this hormone begins to be produced in greater amounts, however, the males' testes grow in size, and mature sperm are made. As their testosterone levels rise, the males begin to defend territories and court females [597]. The same pattern applies to red deer stags living in Great Britain, where they mate during September and October. Stags that have been living peacefully with one another all summer become aggressive as they begin to court females. At the same time, their testes generate sperm and testosterone. Adult males that have been castrated prior to the

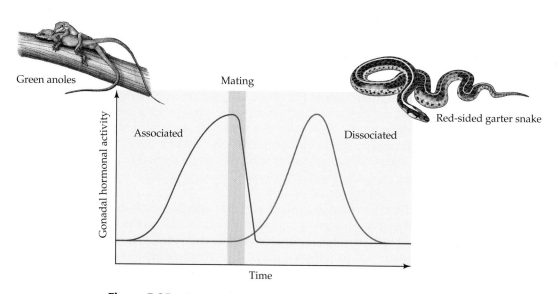

Figure 5.25   An associated reproductive pattern is one in which environmental cues trigger internal hormonal changes, which then activate behavioral responses within a relatively short time (as illustrated by the green anole). In a disassociated reproductive pattern, mating may be dissociated in time from gonadal (and hormonal) activity (as in the red-sided garter snake). After Crews [254].

Figure 5.26   Digitally produced variation in male facial features. The masculinized face appears on the right. Images courtesy of Tony Little; from Little et al. [729].

fall rutting season show little aggression and do not try to mate with sexually receptive females. If the behavioral differences between castrated and intact males stem from an absence of circulating testosterone in the castrated stags, then testosterone implants should restore aggressive and sexual behavior during the rut. Testosterone implants do have these effects, showing that changes in testosterone concentrations promote seasonally adaptive shifts in physiology and behavior [726].

Discussion Question

5.5   In women, the menstrual cycle involves hormonally mediated changes that regulate the production of a mature egg. Might the link between hormones and ovulatory physiology extend to female reproductive behavior, producing an associated reproductive pattern? What would evolutionary biologists predict about the relationship between the menstrual cycle and sexual desire? Why would they predict a difference in this relationship for married women versus those without a steady partner? Why might evolutionary biologists also predict that at the time of ovulation, women should find males with masculinized facial features especially attractive? (See Figure 5.26 for an example of variation in male facial features.) Check your responses against [431, 765, 954].

Although many species studied to date appear to possess associated reproductive patterns, the theory that reproductive behavior is under this kind of hormonal control continues to be tested, as it should be [258]. If testosterone is required for sexual behavior in birds, then castration should eliminate male sexual behavior, because removal of the testes eliminates a major source of testosterone. This prediction has been confirmed for some species, including the Japanese quail, which, if castrated, stops reproducing, but is back in business if given a testosterone implant [913], just like the red deer. But the prediction fails for white-crowned sparrows. Even without his testes, a male white-crown will mount females that solicit copulations, provided that he has been exposed to long photoperiods [848]. Moreover, some populations of white-crowned sparrows raise more than one brood per breeding season. Those male white-crowns that mate with females to produce a second or third clutch of fertilized eggs in the summer have relatively low testosterone concentrations in their blood at

Figure 5.27 **Hormonal and behavioral cycles in single-brood and multiple-brood populations** of white-crowned sparrows. Testosterone concentrations in the blood of male white-crowns peak shortly before the time when the males mate with females (M) in their first breeding cycle of the season. In populations that breed twice in one season, however, copulation also occurs during a second breeding cycle, at a time when testosterone concentrations are declining. After Wingfield and Moore [1324].

that time (Figure 5.27), further evidence that high levels of testosterone are not essential for male sexual behavior in this species.

Discussion Questions

5.6   In studying the hormonal control of behavior, it is common to remove an animal's ovaries or testes and then inject the creature with assorted hormones to see what behavioral effects they have. What advantage does this technique have over another approach, which is simply to measure the concentrations of specific hormones in the blood of animal subjects from time to time? The far less invasive direct measurement approach would show, for example, whether testosterone or estrogen concentrations were elevated when mating was occurring.

5.7   In the guinea pig, individual males vary in their sex drive, as measured (for example) by the number of times a male ejaculates when given access to receptive females for a standard period of time. One hypothesis for this variation is that male sex drive correlates with circulating testosterone concentrations. What prediction follows from this hypothesis? Figure 5.28 presents data from an experiment with three guinea pigs in which male sex drive was measured. All three males were castrated, after which their sex drive continued to be monitored, until finally, after some weeks, the three males were all given the same amount of supplemental testosterone, and their sex drive measured again at intervals. What is the relevance of these data for the hypothesis in question? What scientific conclusion can you reach based on these results?

The Role of Testosterone

The white-crowned sparrow is not the only bird in which high testosterone concentrations are not necessary for courtship and mating [1300]. This observation suggests that testosterone might have other functions in addition to stimulating sexual behavior in some species. One possibility is that testosterone is a hormonal facilitator of aggression. If so, then we can predict that in seasonally territorial birds, testosterone concentrations should be especially high early in the breeding season, when males are aggressively defending a territory against

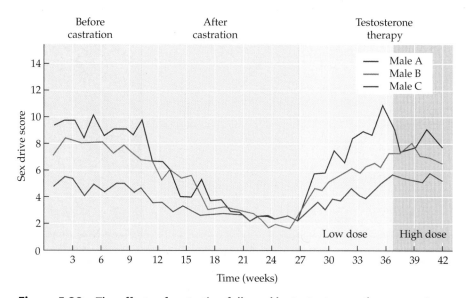

Figure 5.28   The effects of castration followed by testosterone therapy on three male guinea pigs. "Sex drive score" is a measure of copulatory activity. After Grunt and Young [487].

rivals. This phenomenon has been observed in white-crowned sparrows (see Figure 5.27) and some other birds [1324]. Even territorial spotted antbirds, which live in the tropics and do not exhibit elevated testosterone concentrations in any particular seasonal pattern, respond to recorded songs of their species with a rapid buildup of circulating testosterone [1300].

The hypothesis that testosterone has a role to play in regulating aggression leads to another prediction, which is that when competition among *females* is a regular feature of a bird's natural history, then aggressive females should possess relatively high testosterone levels. One bird species with competitive females is the dunnock (see Figure 10.28). Female dunnocks often wind up breeding in small groups that share one or a couple of males. In these situations, a female's reproductive success depends on how much assistance she can secure from the male or males with whom she lives. Therefore, females try to keep other females away from "their" partner(s), chasing them, giving aggressive "tseep" calls, even singing to announce their readiness to fight. When the testosterone levels of females in multi-female groups were compared with those of females who were paired with a single male, who did not have to fight with others of their sex, the results supported the hypothesis that testosterone is a hormonal facilitator of aggression (Figure 5.29) [695].

Even in species in which testosterone promotes adaptive sexual or aggressive behavior, it is the rule that testosterone concentrations fall to almost nothing outside the breeding season or after the likelihood of territorial challenges subsides. Why should this be? Perhaps the hormone comes with a price, reducing fitness at certain times or in certain situations. The hormone does have multiple effects (Figure 5.30), not all of them positive, including its interference with the immune system, which

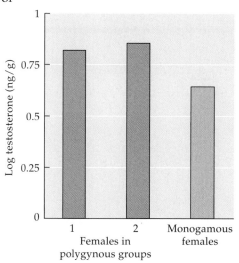

Figure 5.29   Testosterone and female aggression in the dunnock. Testosterone concentrations were higher in females competing for males in polygynous groups (orange bars) than in females in monogamous relationships. After Langmore et al. [695].

Figure 5.30   The chemical structure of testosterone and its diverse effects on physiology and behavior. After Wingfield et al. [1323].

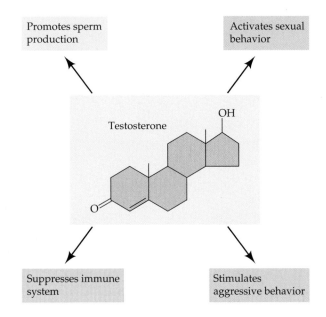

Figure 5.31   A survival cost of testosterone. (A) The three different forms of the lizard *Uta stansburiana* have different throat colors and different amounts of testosterone circulating through their blood on average. (B) The form with the highest testosterone levels also has a very high annual mortality rate. After Sinervo et al. [1112]; photograph by Paul Hamilton.

may explain why males of so many vertebrates are more likely than females to be infected by viruses, bacteria, and parasites [646]. High testosterone concentrations may also contribute to high glucocorticosteroid concentrations, which may contribute to or be reflective of physiological stress [1325], and thus greater vulnerability to disease-causing organisms.

In addition, the direct behavioral effects of testosterone can be costly. Males under the influence of testosterone, for example, often become so focused on mating with females or fighting with rivals that they become sitting ducks for predators or parasites. The bottom line is that in some species, males with higher testosterone levels are less likely to survive than those with more modest concentrations of the hormone in their blood (Figure 5.31) [1113] (see also Figure 8.25). Even if a testosterone-saturated male survives, he may neglect his young in favor of fighting with other males or trying to attract an additional mate. Thus, male dark-eyed juncos that have received testosterone implants do not experience higher mortality than control birds, but these songbirds do feed their young less often, lowering the survival chances of the offspring under their care [636].

Given the potential disadvantages of testosterone, which surely vary from species to species, it is not surprising that the pattern of hormonal control of aggression and reproduction is not the same for every species (Figure 5.32). In

Figure 5.32 Testosterone and territorial behavior. No one pattern exists for the relationship between testosterone concentrations and the duration of male territoriality. In some bird species (top three panels), a surge of testosterone occurs at the onset of territoriality and breeding, but in other birds (bottom two panels), males are territorial at times when they have little or no circulating testosterone. After Wingfield et al. [1323].

the song sparrow, for example, males defend territories long after the breeding season is over, when testosterone has essentially disappeared from their blood. But these birds produce nongonadal sex hormones, especially estrogen, which are thought to play a critical role in the post-breeding territorial phase. If estrogen does boost male territoriality, then males treated with fadrozole (FAD), which blocks the manufacture of estrogen, should sing less often and stay farther away from a simulated intruding rival than untreated controls or males that have received both FAD and replacement estrogen. You can evaluate the data for yourself (Figure 5.33) [1138].

Figure 5.33 Estrogen and territorial behavior. Male song sparrows treated with fadrozole (FAD), a chemical that blocks the production of estrogen, were significantly less likely (A) to sing or (B) to come within 5 meters of a simulated intruder than were control males or males given both FAD and estrogen. After Soma et al. [1138].

Figure 5.34   Spring mating aggregation of red-sided garter snakes. The males search for and avidly court females emerging from hibernation. Photograph by Nic Bishop.

The red-sided garter snake offers another demonstration that hormone–behavior relationships differ among species. This reptile lives as far north as southern Canada, where it spends much of the year dormant in a sheltered underground hibernaculum. On warm days in the late spring, the snakes begin to stir, and soon emerge, sometimes by the thousands, from their hibernaculum (Figure 5.34). Before going their separate ways, they engage in an orgy of sexual activity, with males slithering after females and attempting to copulate with them. Although males compete for females by trying to contact receptive partners before their fellow males do, they do not fight with one another for the privilege of mating.

Examination of the sex hormone concentrations in their blood reveals that these nonaggressive snakes have almost no circulating testosterone or any other equivalent substance. Yet they have no trouble mating, so red-sided garter snakes are animals with a **dissociated reproductive pattern** (see Figure 5.25) [257]. Indeed, various hormonal manipulations have been performed on adult male garter snakes without any effect on their sexual behavior. Removal of the pineal gland prior to hibernation, however, produces male snakes that almost always fail to court females the following spring [258]. The pineal gland provides a critical mechanism for detecting temperature increases following a period of hibernation, and warmer weather suffices to activate sexual behavior in male garter snakes independently of their testosterone concentrations.

This does not mean, however, that testosterone has no role at all to play in the reproductive cycle of the garter snake. In the fall, male snakes have high concentrations of testosterone, which contribute to the production of sperm, which are stored internally over the winter in anticipation of the spring mating frenzy.

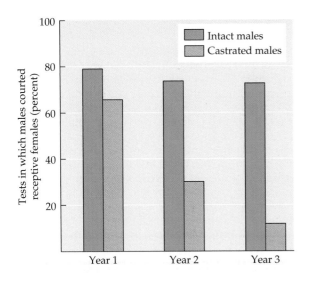

Figure 5.35   Testosterone and the long-term maintenance of mating behavior. Male garter snakes whose testes were removed shortly before the breeding season in year 1 remained sexually active during that breeding season, despite the absence of testosterone. But in years 2 and 3, these males became less and less likely to court receptive females compared with males that still possessed their testes. After Crews [255].

Furthermore, although temperature increases may be the cue for the *activation* of sexual activity, testosterone may play an *organizational* role in the development of the mechanisms underlying reproductive behavior in the red-sided garter snake, as it does in so many other vertebrates. Evidence on this point comes from experiments on adult male snakes that were castrated shortly before the breeding season. Without their testes, these individuals couldnot produce testosterone, but they still exhibited courtship behavior after a period of hibernation in the laboratory. If, however, the castrated snakes were tested again after a second hibernation period, their sexual activity fell sharply. These results suggest that the surge of testosterone that occurs prior to hibernation primes the neural systems that are needed for sexual behavior the following spring (Figure 5.35).

This hypothesis receives more support from the finding that testosterone implants given in the summer before the male snake's first hibernation turn 1-year-old males into sexually active animals, although snakes of this age are normally sexually immature. Therefore, testosterone production, which normally begins in the snake's second or third year, appears to be necessary for the full development and maintenance of the mechanisms that control sexual behavior [255]. However, high concentrations of testosterone are not necessary for copulation to occur in the spring—another example of the rule that there is no one hormonal mechanism that regulates the sexual behavior of all animals, or even all vertebrates, in exactly the same way [3, 256].

Despite the fact that hormonal systems for organizing behavior are highly diverse, you will have noted that the same hormones (especially testosterone and estrogen) have appeared repeatedly in this chapter. Likewise, in many vertebrates, much the same set of neural structures, including specific regions of the hypothalamus, interacts with certain hormones in ways that make behavioral organization possible. Thus, the diversity of neural and hormonal systems consists mainly of variations on a theme, rather than involving arrays of uniquely different mechanisms. Even the red-sided garter snake, whose reproductive behavior appears to be relatively independent of sex hormones, has neurons with receptors that bond with sex hormones in the neural pathways that control its mating behavior [670]. Evolutionary history is clearly responsible for observations of this sort. As clusters of species have arisen and diverged from a common ancestor, each descendant species has retained elements of its hormonal past, which have been modified and rearranged over time, but not replaced by something totally new. The imprint of evolution is very apparent in the proximate mechanisms that organize the behavior of today's animals.

Discussion Question

5.8   The endangered Amargosa River pupfish lives in Death Valley, where different populations of the species live in total isolation from one another in tiny permanent pools and short flowing stream segments [706]. Researchers have estimated that some populations have been cut off from others for only 400 to 4000 years. Yet the reproductive behavior of the different populations can be quite different. In some populations, males aggressively defend territories and court females drawn to those sites, but in others males are not aggressive toward one another and do not form territories, but instead pursue and court females as they become receptive. In other words, dramatic shifts in behavior have occurred in this species in just a few hundred to a few thousand years. Imagine that you wish to explain how these changes could occur so rapidly, knowing what you now know about the ability of hormones to organize sexual and aggressive behavior. You should be aware that arginine vasotocin (AVT) is a brain hormone that in some fish has been linked to decreased aggressive behavior. How might you establish experimentally that the hormone had this effect in the pupfish? If the hormone did lower aggression in this species, what predictions could you make about AVT or about AVT receptor protein differences between populations and the differences among territorial and nonterritorial males of the same population? If your results supported a particular proximate hypothesis about AVT's effect on aggression, describe how selection could lead to rapid changes in territorial behavior in this species.

Summary

1. Because an animal's environment provides various stimuli that could trigger contradictory responses, and because its physical and social environment often change over time, animals gain by having mechanisms that set priorities for their different behavioral options. One such proximate system includes behavioral command centers with the capacity to inhibit one another, so that animals do not try to do several things simultaneously.

2. As the environment changes, the nature of the inhibitory relationships between neural command centers may also change. Devices to achieve this end include the various pacemaker or clock mechanisms that regulate nervous system functioning and hormonal output in cycles that typically last either 24 hours or 365 days. Circadian and circannual clocks have environment-independent components, but they can also adjust their performance by acquiring information from the environment about local conditions, such as the time of sunrise or sunset.

3. Hormones function within those mechanisms that establish behavioral priorities. In many animals, changes in the physical environment (such as seasonal changes in photoperiod) and in the social environment (such as the presence of potential mates) are detected by neural mechanisms and translated into hormonal messages. These chemical signals often set in motion a cascading series of physiological and behavioral changes that make reproductive activity the top priority at times when it is most likely to translate into the production of surviving offspring.

4. The precise roles played by hormones in effecting behavioral change vary from species to species. Male sexual behavior, for example, may or may not be dependent on high testosterone concentrations in the blood. Nevertheless, the proximate mechanisms of behavioral organization of many different species show similarities, such as a reliance of one sort or another on testosterone and certain other widely distributed hormones for organizing male reproductive behavior. This pattern reflects the nature of evolutionary change, in which the attributes of today's species are modified versions of previous ones, not inventions that have arisen out of thin air.

Suggested Reading

Kenneth Roeder's classic *Nerve Cells and Insect Behavior* [1027] and Vincent Dethier's *The Hungry Fly* [315] discuss how some animals avoid conflicting responses and structure their behavior over the short haul. Randy Nelson has written a fine textbook that covers all the topics in this chapter in much more detail [870]. In just a few pages, Michael Young explains the exceedingly complex molecular basis of circadian rhythms about as clearly as possible [1349]. The disassociated mating pattern of the red-sided garter snake has attracted a great deal of attention, which has resulted in many interesting papers on topics only briefly touched on in this chapter, including [670, 707, 819]. These papers deal with some of the factors that control male and female behavior in this species; more articles can be tracked down via the Web of Science. The snake's proximate mechanisms for organizing sexual behavior can be contrasted with those of the green anole, about which there has been some debate (e.g., see [597] as well as [253], and then [746, 1326]).

Behavioral Adaptations for Survival

*I*t is hard to pass on your genes when you are dead. Not surprisingly, therefore, most animals are strongly motivated to stay alive long enough to reproduce. But surviving can be a challenge in most environments, which are crawling with deadly predators. You will remember that noctuid moths, for example, may be snatched out of the night sky by bats armed with sophisticated sonar systems that enable them to detect and intercept prey in complete darkness. During the day, those moths that have managed to make it through the preceding night, perhaps because of their ability to hear ultrasound, are at grave risk of being found, captured, and eaten by keen-eyed birds. Life is short for most noctuid moths, despite their ears and superb camouflage.

Because predators are so good at finding food, they place their prey under intense selection pressure, favoring those individuals with attributes that postpone death until they have reproduced at least once. The hereditary life-lengthening traits of these survivors can then spread through their populations by natural selection—an outcome that creates reciprocal pressure on predators, favoring those individuals who manage to overcome their prey's improved defenses. For example, the hearing abilities of certain moths and other night-flying insects may have favored individuals in some species of bats with unusually high-frequency sonar outside the

*◀ **The leafy sea dragon** is a fish that swims very slowly so as to make the most of its survival-enhancing resemblance to seagrass. Photograph by Jeffrey Jeffords*

detection range of their prey [419, 1053]. But there are some night-flying mantises, for example, that can now hear unusually high-frequency sounds, presumably as a counter to the evolutionary innovation of their hunters [263]. These interactions can therefore be considered evolutionary arms races between predator and prey.

With this chapter, which looks at the results of this ongoing evolutionary contest between the hunters and the hunted, the focus of the book shifts from the proximate to the ultimate causes of behavior. The main goals of the chapter are to establish what is meant by an adaptation and to show how one can use a cost–benefit approach to produce hypotheses on the possible adaptive value of a behavioral trait while using the comparative method to test those hypotheses.

Mobbing Behavior and the Evolution of Adaptations

I was in New Zealand not long ago, visiting a coastal nature reserve rich in wildlife. As I walked along the shore, I approached a place where hundreds of pairs of silver gulls had built their nests on the stony ground. While I was watching the gulls from a distance, a young researcher came down to the coast carrying a scale for weighing gull chicks and a clipboard for recording data. As she walked toward the colony, the gulls took notice, and soon those closest to her began to fly up, calling raucously. By the time she came within a few meters of the first nests, the colony was in an uproar, and many of adult gulls were swooping about, some diving at the intruder, others calling loudly (Figure 6.1).

Figure 6.1   Mobbing behavior of colonial, ground-nesting gulls. Silver gulls reacting to a trespasser in their breeding colony in New Zealand. Photograph by the author.

In gull colonies around the world, whenever a potential predator, such as a human, a hawk, a crow, or some other potential consumer of eggs or chicks, comes close to nesting birds, the gulls usually react strongly. Among the masses of screaming gulls, some may launch what appear to be kamikaze attacks on the unwanted visitor. No one likes being hit on the head by a gull's trailing foot as the bird pulls out of a wing-roaring dive; nor do visitors to gull rookeries enjoy being splattered by liquid excrement released by agitated gulls flying overhead.

We can easily guess why the gulls become upset when potential predators get close to their nests. The parents' assaults probably keep hungry intruders away from their youngsters, helping them survive. If so, then mobbing by gulls could increase their reproductive success, passing on the hereditary basis for their willingness to join groups of other gulls in screaming at, defecating on, and hitting those who might eat their eggs and youngsters. Indeed, this hypothesis was the one that quickly came to mind when Hans Kruuk, a student of Niko Tinbergen, decided to investigate mobbing behavior in the black-headed gull, another ground-nesting, colony-forming species (Figure 6.2), but one that lives in Europe rather than New Zealand [677].

Kruuk was interested in studying the evolutionary, ultimate causes of mobbing, not the proximate ones, which would have required examination of the genetic, developmental, hormonal, and neural bases of the behavior—all interesting and valuable things to know about, but not on Kruuk's agenda. To get at the evolutionary foundation of the behavior, Kruuk employed what is now called the **adaptationist approach**. In effect, he wanted to know whether the mobbing response of black-headed gulls was an adaptive product of natural selection. His working hypothesis was that mobbing behavior distracted certain predators, reducing the chance that they would find the mobbers' offspring, which would boost the fitness of mobbing parent gulls.

Kruuk used his hypothesis to make testable predictions about gull mobbing behavior [677]. He knew that natural selection occurs when individuals vary in

Figure 6.2 Two black-headed gulls on their nests in a large colony of this ground-nesting species. Photograph by Philippe Clément.

their hereditary attributes in ways that affect how many surviving offspring they contribute to the next generation (see Chapter 1). Imagine any given population of gulls, some of whose members mob nest predators while others do not. This population will surely evolve if the difference between the two types of individuals is hereditary and if one type consistently leaves more surviving progeny than the other. If, for example, the mobbing phenotype outreproduces the non-mobbing type generation after generation, then mobbing behavior will eventually become the norm, and non-mobbing will disappear (always assuming that the differences between the two types are hereditary). Thereafter, any hereditary change in the nature of the mobbing response that enhances individual success in passing on genes will also spread through the species, given enough time.

So, what kind of mobbing behavior did Kruuk expect to observe in black-headed gulls? Absolutely perfect, 100 percent effective mobbing behavior that always saved the mobber's eggs and chicks? No, for any number of reasons (Table 6.1) [252]. Selection cannot create the best of all possible genes for a particular task, but has to wait for mutations to occur by chance; only then can it winnow out the less effective alleles, leaving the one that does better at promoting reproduction in place. If a "better" allele does not appear, there is nothing selection can do about it, because selection is a consequence of the confluence of certain conditions (hereditary variation that causes differences in individual reproductive success), not a controlling force that acts for the good of the

---

TABLE 6.1 *Constraints on adaptive perfection*

Constraint 1: Failure of appropriate mutations to occur

Evolutionary constraints on adaptive perfection can arise from the failure of appropriate mutations to occur, which will prevent selection from keeping up with environmental change. Thus, maladaptive or nonadaptive traits can persist, especially in environments only recently invaded by a species. So, for example, some arctic moths live in regions where bats are absent, but still cease flying upon exposure to an experimental ultrasonic stimulus [1052]. Likewise, arctic ground squirrels respond to snakes in the manner of other ground squirrels, even though snakes do not occur where arctic ground squirrels live [244].

Man-made changes in the environment are especially likely to elicit inappropriate expression of traits that have served adaptive functions in the past under natural conditions [1068]. Thus, some moths are so strongly attracted to artificial lights that bats can easily capture them there [423]. Likewise, male Australian buprestid beetles suffer when they try to copulate with beer bottles [494] (see Figure 4.2A); some male frogs and toads die on roads when they attempt unsuccessfully to mate with previously road-killed females of their species [820]; and sea turtles regularly die of blocked digestive tracts after having consumed plastic bags [168, 690], which resemble jellyfish, the turtle's natural prey.

Constraint 2: Pleiotropy

Developmental constraints on adaptive perfection occur as a result of *pleiotropy* (the multiple developmental effects that most genes have). Not all the effects of a given gene are positive. If the negative consequences of a gene outweigh the positive ones, that gene will be selected against. The converse is that because some genetic effects are so valuable, the less significant, somewhat negative consequences of an otherwise adaptive proximate mechanism may be maintained in the population by selection. Thus, female rodents living in a communal nest may sometimes give milk to offspring other than their own as a by-product of their strong parental drive, which usually results in adaptive care of their own genetic offspring [524]. In humans, the parental tendency to want to care for one's own offspring rather than someone else's leads a few people under some circumstances to commit criminal child abuse [277] (see Chapter 14). Rules of thumb occasionally misfire.

Constraint 3: Coevolution

Coevolution (the kind of evolution that occurs when species interact in ways that affect the fitness of each other's members) means that evolutionary stability may never be reached. Instead, each species changes in response to selection pressure imposed by the other, so that first one species gains the upper hand, and then the other one does, as in the arms race between predator and prey. Once again, the inability of selective processes to generate an immediate solution to an environmental obstacle means that less than perfect traits can persist in species.

species. Moreover, even if a mutant allele happened along with a particularly positive developmental effect on mobbing, the new allele might very well damage the construction of other traits. Almost all genes do have multiple developmental consequences. An allele with a *net* negative effect on reproductive success would never spread, no matter how useful its effect on the mobbing trait. Finally, gulls have been around for a long time, but so have their predators, and selection works on both halves of the predator–prey equation. As we noted at the outset, any improvement in a gull's ability to thwart a certain predator would tend to select for improvement in that predator's ability to circumvent the bird's defense. As a result of this arms race, we could find that neither prey nor predator has complete control of the situation at any given moment [370].

Given the constraints that prevent adaptive perfection from evolving, what does evolve? As just noted, if there are hereditary alternative phenotypes in a population, the "better" one will spread; namely, the one that confers higher **fitness** (defined as higher reproductive success or higher genetic success) on the individuals that happen to have this superior alternative. In other words, an **adaptation** is a hereditary trait that either (1) spread because of natural selection in the past and has been maintained by selection to the present or (2) is currently spreading relative to alternative traits because of natural selection. In all such cases, the trait in question has conferred and continues to confer (or is just beginning to confer) higher genetic success, on average, on individuals that have it compared with other individuals with alternative traits. There are other definitions of adaptation [518, 701], but the one we shall use permits us to test hypotheses about possible adaptations by focusing on the *current* benefits of a given trait, which has some major practical advantages over trying to test whether certain traits provided genetic benefits to individuals in the past [1006].

Discussion Questions

6.1   Many people think that an adaptation is a trait that improves the survival chances of an organism. Under what circumstances would such a trait be an adaptation? Under what other circumstances would a survival-enhancing attribute actually be selected against?

6.2   Stephen J. Gould and Richard Lewontin claimed that adaptationists make the elementary mistake of believing that every characteristic of living things is a perfected product of natural selection [463], when in reality many attributes of living things are not adaptations (see Table 6.1). Moreover, in their eagerness to explain everything as an adaptation, adaptationists have, according to Gould and Lewontin, invented fables as absurd as the fictional "just-so" stories of Rudyard Kipling, who made up amusingly silly explanations for the leopard's spots and the camel's hump. How might adaptationists defend themselves against these charges? Do adaptationists have the means to discover whether their tentative explanations for trait X are wrong?

With our definition of adaptation in place, we can identify a naturally selected adaptation if we can establish that it was better than other alternatives that appeared in the past. But how do we do that? Adaptationists tackle this problem with a very useful tool borrowed from economics: namely, the **cost–benefit approach**, which enables them to analyze phenotypes in terms of their fitness benefits and fitness costs. (In evolutionary biology, **fitness benefit** refers to the positive effect of a trait on the number of surviving offspring produced by an individual or the number of copies of its alleles that it contributes to the next generation, while **fitness cost** refers to the damaging effects of the trait on these measures of individual genetic success.) So, for example, Kruuk knew that mobbing behavior comes with significant disadvantages for individuals in terms of their reproductive chances. These costs include the time and energy that mobbers expend when they are screaming, diving, and flapping about in response to an unwanted visitor near their nests. In addition, mobbers can lose not only calories, but also their lives. More than one black-headed gull has made a lethal miscalculation when dive-bombing an unusually agile fox that was able to wheel about, leap up, and catch the gull as it passed by. Given the very real fitness costs associated with mobbing, Kruuk knew that the trait could not be an adaptation unless there were some equally obvious, substantial benefits for the mobbers, so that, *on average*, the benefits (B) would exceed the costs (C) of the trait. The minimum requirement for an adaptation is, of course, that $B > C$.

Therefore, Kruuk predicted that mobbing gulls should force nest-robbing predators to expend more searching effort than they would otherwise, a simple prediction that can be tested just by watching gull–predator interactions [677]. Kruuk saw that egg-eating carrion crows have to continually face swooping gulls and so, while being mobbed, they cannot look around comfortably for nests and eggs. Because distracted crows are probably less likely to find their prey, Kruuk established that a probable benefit exists for mobbing. Moreover, the benefit of mobbing crows plausibly exceeds the costs, given that these predators do not attack or injure adult gulls.

However, the predator distraction hypothesis yields much more demanding predictions as well. Because adaptations are better than the traits they replace, we can predict that the benefit experienced by mobbing gulls in protecting their eggs should be directly proportional to the extent that predators are actually mobbed. Kruuk used an experimental approach to test this prediction [677]. He placed ten hen eggs, one every 10 meters, along a line running from the outside to the inside of a black-headed gull nesting colony. The eggs placed outside the colony, where mobbing pressure was low, were much more likely to be found and gobbled up by carrion crows and herring gulls than were eggs placed inside the colony, where the predators were communally harassed by the many parents whose offspring were threatened by them (Figure 6.3).

Thus, Kruuk assembled some observational and experimental evidence in support of the hypothesis that mobbing is an adaptation that helps adult black-headed gulls protect their eggs and youngsters. Note that these tests did *not* involve measuring gull reproductive success by counting the number of surviving offspring produced by individuals in their lifetimes. Kruuk looked instead at the number of hen eggs that were uneaten, on the reasonable assumption that had they been gull eggs in a gull nest, they would have had a chance to become

Figure 6.3 **Does mobbing protect eggs?** When hen eggs were placed outside and inside a black-headed gull nesting colony, crows searching for the eggs within the colony were subject to more attacks by mobbing gulls (red circles), and as a result, they discovered fewer hen eggs (blue circles). After Kruuk [677].

Black-headed gull

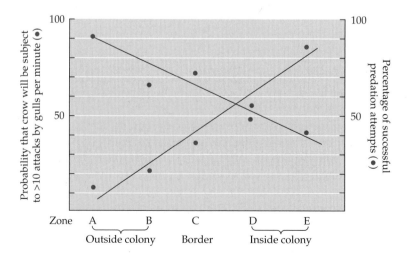

surviving offspring for the gull parent, in which case they would constitute part of the parent's lifetime genetic contribution to the next generation.

Behavioral ecologists often have to settle for an indicator or correlate of reproductive success or genetic success when they attempt to measure fitness. In the chapters that follow, *fitness* or *reproductive success* is often used more or less interchangeably with such things as egg survival (Kruuk's measure), young that survive to fledging, or number of mates inseminated, or even more indirectly, the quantity of food ingested per unit of time, the ability to acquire a breeding territory, and so on. Keep in mind, however, that the bottom line with respect to an individual's reproductive success is the total number of his or her offspring (or grandoffspring) that reach the age of reproduction. Proxies for this measure can only approximate actual reproductive success.

Discussion Questions

6.3   For many evolutionary biologists, the term "adaptation" must be reserved for a characteristic that provides "current utility to the organism and [has] been generated historically through the action of natural selection for its current biological role" [74]. What *could* "current utility" mean, and what do you think it *should* mean? Make use of the terms "fitness benefits" and "fitness costs" in your answer. If a trait originated for function X and later took on a different, but still adaptive, biological role Y, does that mean it is not an adaptation? Track down the evolutionary history of the flight feathers on the wings of modern birds (see, for example, [982]). Where did these feathers come from, and what function did their predecessor feathers exhibit? If you go back far enough in time, will the ancestral form of any current trait have the same function that it does now?

6.4   The arctic skua is a close relative of gulls that also nests on the ground and mobs colony intruders, including the great skua, a larger predator that eats many arctic skua eggs and chicks. In one study, hatching success and post-fledging survival were greater for arctic skuas that nested in dense colonies than in low-density groups (Figure 6.4); the number of near neighbors was, however, negatively correlated with the growth rate of their chicks [948]. Rephrase these findings in terms of the fitness costs and benefits of communal mobbing by the arctic skua. If the term *adaptation* meant a perfect trait, would communal mobbing by arctic skuas be labeled an "adaptation?"

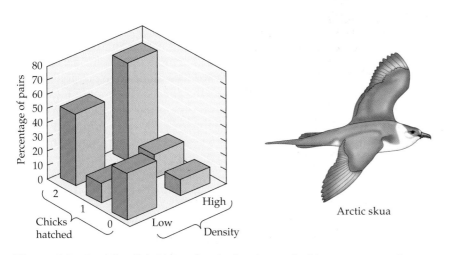

Figure 6.4   Benefit of high nest density for the arctic skua. In one population studied in 1994, skuas nesting with relatively many nearby neighbors were more likely to rear two chicks than were individuals nesting in areas with a lower density of breeding pairs. After Phillips et al. [948].

The Comparative Method for Testing Adaptationist Hypotheses

Experiments are highly valued in science, so much so that most persons believe that scientific research can be performed only in high-tech laboratories by officious white-coated researchers. As Kruuk's work shows, however, good experimental science can be done in the field. Moreover, the manipulative experiment is only one of several ways in which predictions from hypotheses can be tested. Another powerful technique for testing adaptationist hypotheses is via the **comparative method**. This approach involves testing predictions about the evolution of an interesting trait by looking at animals other than the species whose characteristics are under investigation [225]. We used the comparative method informally earlier when considering the evolution of infanticide (see p. 18), and again when dealing with the responses of moths and other insects to bat ultrasonic calls (see p. 111). The idea was that if a given trait, such as infanticide, is adaptive for one species, such as Hanuman langurs, then it should also evolve in other species, such as lions, that are subject to the same selective pressures as langurs. Here we shall use the comparative method to test the adaptationist explanation for mobbing by black-headed gulls. This technique, as applied to this case, yields the following prediction: if mobbing by ground-nesting black-headed gulls is an evolved response to predation on gull eggs and chicks, then other gull species whose eggs and young are at low risk of predation should not exhibit mobbing behavior.

The rationale behind this prediction is as follows: The costs of mobbing can be outweighed only by the benefits derived from distracting predators. If predators were not a major problem for a species, then the benefits of mobbing would be reduced, increasing the odds that the overall benefit-to-cost ratio of mobbing would be negative, leading to its eventual loss from a species whose ancestors possessed the trait.

There is good reason to believe that the ancestral gull was a ground-nesting species with a range of nest-hunting predators against which the mobbing defense would have been helpful. If we look at the 50 or so species of gulls living today, we find that the large majority nest on the ground and exhibit communal mobbing behavior against enemies that hunt for their eggs and chicks [1198]. These similarities among gulls, which also have many other features in common, are believed to exist in part because all gulls are the descendants of a relatively recent common ancestor, from which they all inherited the genetic package that predisposes all gulls to develop a similar set of traits. However, a few of today's gull species nest on cliff ledges, rather than on the ground. Perhaps these gull species are the descendants of a more recent cliff-nesting gull that evolved from a ground-nesting ancestor. The alternative possibility, that the original gull was a cliff nester, requires the cliff-nesting trait to have been lost and then regained, which produces an evolutionary scenario that requires more changes than the competing one (Figure 6.5). Many evolutionary biologists, although not all, believe that simpler scenarios involving fewer transitions are more probable than more complicated alternatives. In this, the majority accepts a commonly held philosophical principle, known as Occam's razor or the principle of parsimony, which holds that simpler explanations are more likely to be correct than complex ones. (The contrary view is that, because of the many chance events that influence the course of evolution, historical pathways may often be quite convoluted, with traits lost, then regained, then lost again, violating expectations based on Occam's razor [1007].)

In any event, cliff-nesting gulls currently have relatively few nest predators because it is hard for small mammalian predators to scale cliffs in search of prey, while predatory birds have a difficult time maneuvering near cliffs in turbulent

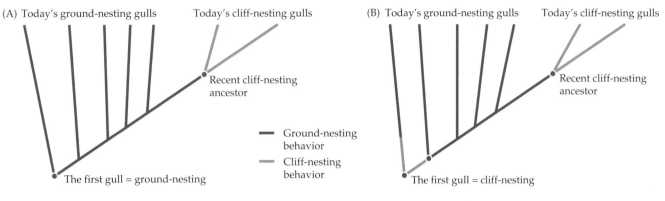

(A) Today's ground-nesting gulls   Today's cliff-nesting gulls

Recent cliff-nesting ancestor

— Ground-nesting behavior
— Cliff-nesting behavior

The first gull = ground-nesting

(B) Today's ground-nesting gulls   Today's cliff-nesting gulls

Recent cliff-nesting ancestor

The first gull = cliff-nesting

Figure 6.5   Gull phylogeny and two scenarios for the origin of cliff-nesting behavior. Hypothesis A requires just one behavioral change, from ground nesting to cliff nesting. Hypothesis B requires two behavioral changes, one from the ancestral cliff nester to ground nesting, and then another change back to cliff nesting.

coastal winds. Thus, a change in nesting environment surely meant a change in predation pressure, which ought to have altered selection on these gulls. The evolutionary result should have been a shift away from the ancestral mobbing behavior pattern. By finding such cases of **divergent evolution** and identifying the selective reason for the change, one can, in principle, establish why the ancestral trait has been retained in some species but has been modified or lost in other species.

The kittiwake nests on nearly vertical coastal cliffs (Figure 6.6), where its eggs are relatively safe from predators. These small, delicate gulls have clawed feet,

Figure 6.6   Not all gulls nest on the ground. (Left) Steep cliffs are utilized for nesting by kittiwake gulls, which appear in the lower half of this photograph. (Right) Kittiwakes are able to nest on extremely narrow ledges; their youngsters crouch in the nest, facing away from the cliff edge. Left, photograph by Bruce Lyon; right, photograph by the author.

and they can land and nest on tiny ledges. As a result, predation pressure on kittiwake eggs and young has been greatly reduced, although not eliminated altogether [793], compared with that affecting their ground-nesting relatives. The relatively small size of these gulls may also make the adults more vulnerable to personal attack by nest predators, making the benefit-to-cost ratio for mobbing still less favorable. As predicted, groups of nesting adult kittiwakes do not mob their predators, despite sharing many other structural and behavioral features with black-headed gulls and other ground-nesting species. The kittiwake's behavior has become less like that of its close relatives, providing a case of divergent evolution in support of the hypothesis that mass mobbing by black-headed gulls evolved in response to predation pressure on the eggs and chicks of nesting adults [262].

The other side of the comparative coin is that species from different evolutionary lineages that live in similar environments, and therefore experience similar selection pressures, can be predicted to evolve similar traits, resulting in cases of **convergent evolution**. Such species evolve the same adaptive solution to a particular environmental obstacle to reproductive success, despite the fact that their different ancestral species had very different genes and attributes (Figure 6.7). We can predict, for example, that if mobbing by colonial, ground-nesting gulls is an adaptation to distract predators from vulnerable offspring, similar behavior should have evolved in other, quite unrelated animals that nest or breed in groups and are often visited by predators intent on eating their offspring.

As predicted, mobbing behavior has evolved convergently in other bird species, including the bank swallow. This species also nests in colonies, which are visited by predators, including snakes and blue jays, that enjoy eating swallow eggs and nestlings [561]. The common ancestor of bank swallows and gulls occurred long, long ago, with the result that the lineages of the two groups have evolved separately for eons—a fact recognized by the placement of gulls and swallows in two different families, the Laridae and the Hirundidae. Despite their evolutionary and genetic differences from gulls, bank swallows behave

Figure 6.7   The logic of the comparative method. Members of the same evolutionary lineage (e.g., gull species of the family Laridae) share a common ancestry, and therefore share many of the same genes, and thus tend to have similar traits, such as mobbing behavior. But the effects of shared ancestry can be overridden by a novel selection pressure. A reduction in predation pressure has led to divergent evolution by the cliff-nesting kittiwake, which no longer mobs potential enemies. The various ground-nesting gulls, including the black-headed gull, mob nest predators, as do some colonial swallows, including the bank swallow, even though gulls and swallows are not related, having come from different ancestors long ago. These colonial swallows and gulls have converged on a similar antipredator behavior in response to shared selection pressure from predators that have fairly easy access to their nesting colonies. What kind of evolution is responsible for the difference between bank swallows, which mob their enemies, and rough-winged swallows, which do not? This difference constitutes evidence in support of what evolutionary hypothesis about mobbing?

Figure 6.8   Colonial California ground squirrels have evolved mobbing behavior. One squirrel kicks sand at a rattlesnake, while others give a variety of alarm calls. Courtesy of R. G. Coss and D. F. Hennessy.

like gulls when they are nesting. As they swirl around and dive at their predators, they sometimes distract hunting jays or snakes that would otherwise destroy their offspring.

Even some colonially nesting mammals have evolved mobbing behavior [914]. Adult California ground squirrels, which live in groups and dig burrows in the ground, react to a hunting rattlesnake by gathering around it and kicking sand in its face, thereby preventing the snake from exploring nest burrows for prey (Figure 6.8). Although adult squirrels have a partial antivenin that reduces the costs of being bitten by a rattlesnake, squirrels that are struck do not get off scot-free. Given the costs of a rattlesnake bite, one team of researchers led by Ronald Swaisgood predicted that squirrels would be able to adjust their behavior in relation to the level of risk of snakebite, which they could assess by listening to the rattling buzzes made by irritated rattlesnakes. These sounds vary according to the size and body temperature of the snakes. Larger snakes present a greater risk to squirrels, as do warmer snakes, which can move more quickly. Squirrels were, as predicted, less eager to approach speakers playing the rattles of larger and warmer snakes. In nature, this ability to assess the risk of snakebite would enable the squirrels to reduce the costs of their mobbing behavior. In addition, as their offspring matured and became better able to avoid snakes on their own, mother squirrels were less likely to mix it up with rattlers, as measured by the reduced time they spent close to a tethered snake (Figure 6.9). In other words, as the benefits of mobbing behavior fell, the benefit-to-cost ratio of the activity declined, and as one would predict, the willingness of adult squirrels to engage in the behavior also decreased [1173].

Because mobbing behavior has evolved independently in several unrelated species whose adults can sometimes protect their vulnerable offspring by distracting predators, we can conclude tentatively that mobbing is an antipreda-

Figure 6.9 **Evidence for a cost of parental mobbing behavior** in the California ground squirrel. As their pups mature, and so become better able to evade rattlesnakes on their own, mother squirrels tend to spend less time near tethered snakes that experimenters have placed near their offspring. After Swaisgood et al. [1173].

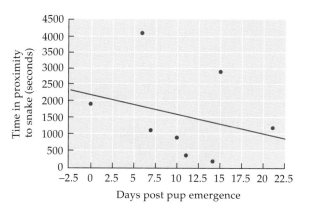

tor adaptation. But what if I have presented only supportive examples while ignoring other colonial species in which parents do not mob the consumers of their youngsters (and do not have alternative means of protecting their offspring)? If for every group-living species in which mobbing occurred under the expected conditions there were two in which the behavior was absent, you would be skeptical of the predator distraction hypothesis, and rightly so. For this and other reasons, researchers increasingly require that the comparative method be used in a statistically rigorous fashion [518]. For our purposes, however, the point is that one can, in principle, test adaptationist hypotheses by predicting that particular cases of divergent or convergent evolution will have occurred—a prediction about the past that can be checked by making disciplined comparisons among species living today.

Discussion Questions

6.5   The ability to hear ultrasound in one species of noctuid moth is considered an antipredator adaptation because it apparently enables individuals to hear and avoid nocturnal, ultrasound-using bats. Imagine that you wished to test this hypothesis via the comparative method. Identify the utility of each of the following lines of evidence on the hearing abilities of other insect species. Specify whether these cases involve convergent evolution, divergent evolution, or neither.

1. Almost all other species of noctuid moths also have ears that respond to ultrasound.
2. Almost all the species in the evolutionary lineage that includes the noctuid moths and many other moths belonging to several other superfamilies also have ears that respond to ultrasound [1346].
3. Some diurnal noctuid moths have ears, but are largely or totally incapable of hearing ultrasound [422].
4. Almost all butterflies, which belong to the same large evolutionary grouping as the noctuids but are usually active during the day, lack ears and so cannot hear ultrasound [420].
5. Six species of noctuid moths found only on the Pacific Ocean islands of Tahiti and Moorea have ears and can hear ultrasound, but do not react to this stimulus with anti-bat responses [424].
6. Members of one small group of nocturnal butterflies have ears on their wings and can hear ultrasound; they respond to ultrasonic stimulation by engaging in unpredictable dives, loops, and spirals [1346].
7. Lacewings and praying mantises fly at night and have ears that detect ultrasound and lead to anti-bat defensive behavior (see p. 111).

6.6   Some persons would say that the fact that most noctuid moths have ultrasound-sensitive ears is "simply" a reflection of their shared ancestry, a holdover from the past, and therefore that ultrasound sensitivity is not an adaptation in these species [150]. Others disagree, arguing that it makes no sense to limit adaptations to just those traits that have diverged from the ancestral pattern [1007]. Who is right?

The Cost–Benefit Approach to Antipredator Behavior

We have used mobbing behavior to illustrate how adaptationists can employ natural selection theory to analyze the possible adaptive value of behavioral traits. All behaviors have fitness costs and fitness benefits, but only those whose benefits exceed their costs can be directly selected for. A demonstration that a trait has substantial reproductive benefits, as mobbing does by virtue of helping parents protect their offspring, constitutes evidence in support of the trait in question being the evolutionary product of natural selection.

Many investigators have tried to test whether a putative antipredator adaptation really does confer significant benefits on individuals that employ the behavior of interest. Some persons, for example, have been interested in why butterflies aggregate in large, densely packed groups around mud puddles on tropical riverbanks, where they suck up fluid containing valuable mineral nutrients from the soil (Figure 6.10A). While they are "mud-puddling," the butterflies run the risk of being attacked by various birds. Perhaps the butterflies form large, compact groups simply to dilute the risk of being attacked and captured. This dilution hypothesis is certainly plausible. Imagine that five insect-eating birds visit the riverbank, and that each bird kills two prey there, each day. Under these conditions, the risk of death for a member of a group of 1000 butterflies is 1 percent per day, whereas it is ten times higher for members of a group of 100 butterflies. This benefit associated with being part of a large group applies to the mud-puddling butterflies studied by Joanna Burger and Michael Gochfeld (Figure 6.10B). The observations of these researchers in the field show that any butterfly puddling with fewer than 40 other individuals would decrease its risk of predation by moving to a larger group [171]. Thus, the dilution effect provides a benefit for individual butterflies that tend to mud-puddle with many others, which means that the benefit-to-cost ratio for any tendency to aggregate could well be positive, facilitating the evolution of the behavior.

Could the dilution effect also contribute to the tendency of some mayflies to synchronize their metamorphosis from aquatic nymphs to flying adults so that most individuals emerge from the water during just a few hours on a few days each year [1175]? If so, then the higher the density of emerging individuals, the lower should be the risk to any one mayfly of being inhaled by a trout as it makes the risky transition to adulthood. To check this prediction, Bernard

Figure 6.10   The dilution effect in butterfly groups. (A) Masses of butterflies on a Brazilian riverbank. (B) Individual butterflies that "mud-puddle" in large groups experience a lower risk of predation than those that suck up fluids from the ground by themselves or in small groups. A, photograph by Joanna Burger; B, after Burger and Gochfeld [171].

(A)

(B)

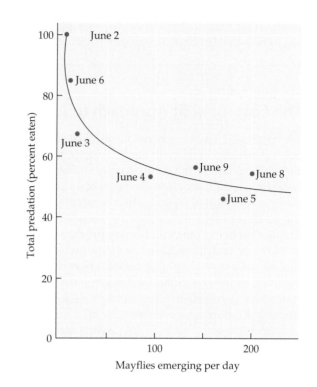

Mayfly

Figure 6.11   The dilution effect in mayflies. The more female mayflies that emerged together on a June evening, the less likely any individual mayfly was to be eaten by a predator. After Sweeney and Vannote [1175].

Sweeney and Robin Vannote placed nets in streams to catch the cast-off skins of mayflies, which molt on the water's surface as they change into adults, leaving the discarded cuticle to drift downstream on the current. Counts of the molted cuticles revealed how many adults emerged on a particular evening from a particular segment of stream. The nets also caught the bodies of females that had laid their eggs and then died a natural death; a female's life ends immediately after she drops her clutch of eggs into the water, provided a nighthawk or a whirligig beetle does not consume her first. Sweeney and Vannote measured the difference between the number of molted cuticles of emerging females and the number of intact corpses of spent adult females that washed into their nets on different days. The greater the number of females emerging together on a given day, the stronger the dilution effect, and the better the chance each mayfly had to live long enough to lay her eggs before expiring (Figure 6.11) [1175].

Of course, there are other possible benefits of getting together with others, including the potential for group attack on a shared enemy (Figure 6.12A), which could prevent it from doing damage. The social insects, including termites, ants, wasps, and bees, are famous for the ability to go after a predator together, using their stingers or jaws to injure or even kill intruders (Figure 6.12B). The race of the honey bee that lives in Africa, where bee colonies have been relentlessly exploited over the millennia for their honey and larvae by a host of mammalian predators, including human beings, is renowned for the ferocity of its response to creatures that threaten to rob its nest. In fact, more than 80 times as many African honey bees will pursue beekeepers that have disturbed their hive for 25 to 50 meters compared with Europeanized honey bees, which have been artificially selected for docility [490]. The willingness of the African race of honey bees to sting potential predators en masse is reflected in the fact that more than a thousand people have died from its attacks as the bee has spread through the Americas following its release by Brazilian beekeepers [139].

Figure 6.12   Fighting back by terns and wasps. (A) A group of royal terns confronts a gull interested in stealing a tern egg. (B, C) A colony of *Polybia* wasps (B) just before and (C) just after a nest was tapped by an observer. These wasps will leave their nest to assault any predator foolish enough to persist in disturbing them. A, photograph by Bruce Lyon; B–C, photographs by Bob Jeanne.

Among the insects, organized defensive attack is not limited to stinging bees and biting termites. In southwestern Australia, one frequently comes across sawfly larvae clustered in balls of ten or so individuals. These caterpillar-like insects feed on eucalyptus leaves, which contain highly resinous, toxic oils that deter most herbivores—but not sawflies. The sawfly grubs not only ingest the eucalyptus oils safely, but also store them in special sacs, from which they can be regurgitated onto attacking ants or birds [853]. Disturb a group of sawfly larvae resting in a defensive circle during the day, and they all spew out large, pale, sticky droplets of resinous fluid (Figure 6.13), which they are prepared to smear communally on any approaching enemy, the better to send it on its way. In this case, and in the others mentioned above, because researchers have been able to identify the antipredator benefits that individuals derive from their behaviors, we now have the minimum evidence needed to label these behaviors evolved responses to predation.

Figure 6.13 Communal defense by sawfly larvae. These larvae form clusters that rest with their heads facing outward during the day. When threatened, they raise and wave their abdomens above their heads in warning while regurgitating droplets of repellent eucalyptus oils, which they hold in their mouths to apply to an enemy. Photograph by the author.

6.7 In my front yard, I sometimes find several hundred male native bees clustering in the evening on a few bare plant stems (Figure 6.14). An assassin bug sometimes approaches the cluster and kills some bees as they are settling down for the night. Devise at least three alternative hypotheses on the possible anti-assassin bug value of these sleeping clusters, and list the predictions that follow from each hypothesis.

The Costs and Benefits of Camouflage

One can also test hypotheses on the possible antipredator benefits of behaviors used by solitary animals. For example, many persons have thought it likely that certain apparently camouflaged animals have evolved the ability to select the kind of resting background where they would be hard to see (Figure 6.15). A classic attempt to test whether preferred resting places really do enhance the camouflage of a prey species involved the peppered moth, *Biston betularia* (Figure 6.16). In some parts of Great Britain and the United States, the melanic (black) form of this moth, once extremely rare, almost completely replaced the once abundant whitish salt-and-pepper form in the period from about 1850 to 1950 [467]. Most biology undergraduates have heard the standard explanation for the initial spread of the melanic form (and the special allele associated with its mutant color pattern): As industrial soot darkened the color of forest tree trunks in urban regions, the whitish moths living in these places became more conspicuous to insectivorous birds, which ate the whitish form and thereby removed the genetic basis for this color pattern. Despite some recent claims to the contrary [249], this story remains largely valid [466], especially with respect to the significance of H. B. D. Kettlewell's famous experiments [637]. Kettlewell placed the two forms of the peppered moth on dark tree

Figure 6.14 A group of sleeping bees. In this species, males spend the night clustered together. Photograph by the author.

Figure 6.15   **Cryptic coloration depends on background selection.**   The
Australian thorny devil possesses remarkable camouflage, but it works only when the
lizard is motionless in areas littered with bark and other varicolored debris, not on roads.
Photographs by the author.

Dark moth

Light moth

Figure 6.16   **The camouflaged moth,** *Biston betularia.*
One typical (whitish) individual and one melanic (black) indi-
vidual are shown in each photograph. Left, photographs by
Michael Tweedie; right, photograph by Bruce Grant.

Figure 6.17 Predation risk and background selection by moths. Specimens of typical and melanic forms of the peppered moth were pinned to tree trunks or to the undersides of limb joints. Moths of both types were less likely to be found and removed by birds on limb joints than on trunks, but overall, melanic forms were less often discovered by birds in polluted (darkened) woods, while typical forms "survived" better in unpolluted woods. After Howlett and Majerus [568].

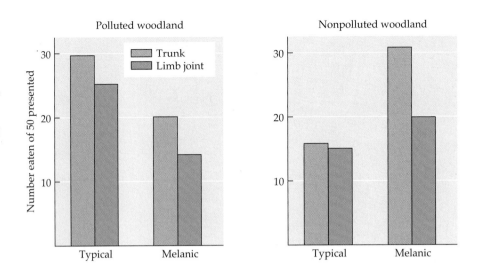

trunks and on pale tree trunks, finding that whichever form was more conspicuous to humans was also taken by birds much more quickly than the other form. Thus, paler individuals were at special risk of attack when they perched on dark backgrounds.

In nature, *Biston betularia* apparently rarely sits on tree trunks; instead, it tends to settle within the shaded patches just below the junctions of branches with the trunk. If the moth's actual perch site selection is an adaptation, then we can predict that individuals resting underneath tree limbs should be better protected against predators than if they were to choose an alternative location. R. J. Howlett and M. E. N. Majerus glued samples of frozen moths to open trunk areas and to the undersides of limb joints. Their data showed that birds were particularly likely to overlook moths on shaded limb joints (Figure 6.17) [568]. Here is another demonstration that a putative adaptation almost certainly does provide a survival benefit.

Discussion Question

6.8 "In the years since 1950, pollution controls have reduced the amount of soot deposited on tree trunks and the melanic form of *Biston betularia* has correspondingly become increasingly scarce in Europe [135, 237] and North America [468], where the species also occurs." Put this statement in the context of a scientific investigation into whether the typical salt-and-pepper coloration of some members of this species constitutes an adaptation. Begin with a research question and proceed through hypothesis, prediction, test, and conclusion.

Moths other than *Biston betularia* make decisions about where to perch during the day. For example, the whitish moth *Catocala relicta* usually perches head-up, with its whitish forewings over its body, on white birch and other light-barked trees (Figure 6.18). When given a choice of resting sites, the moth selects birch bark over darker backgrounds [1061]. If this behavior is truly adaptive, then birds should overlook moths more often when the insects perch on their favorite substrate. To evaluate this prediction, Alexandra Pietrewicz and Alan Kamil used captive blue jays, photographs of moths on different backgrounds, and operant conditioning techniques (Figure 6.19) [952]. They trained the blue jays to respond to slides of cryptically colored moths positioned on an appropriate background. When a slide flashed on a screen, the jay had only a short time to react. If the jay detected a moth, it pecked at a key, received a food reward, and was quickly shown a new slide. But if the bird pecked incorrectly

Figure 6.18 Cryptic coloration and body orientation. The orientation of a resting *Catocala* moth determines whether the dark lines in its wing pattern match up with the dark lines in birch bark. Photograph by H. J. Vermes, courtesy of Ted Sargent, from Sargent [1061].

Figure 6.19   Does cryptic behavior work? Images of moths on different backgrounds and in different resting positions are shown to a captive blue jay, which is rewarded for detecting a moth. Photograph by Alan Kamil.

when shown a slide of a scene without a moth, it not only failed to secure a food reward, but had to wait a minute for the next chance to evaluate a slide and get some food. The caged jays' responses showed that they saw the moth 10 to 20 percent less often when *C. relicta* was pinned to pale birch bark than when it was placed on darker bark. Moreover, the birds were especially likely to overlook a moth when it was oriented head-up on birch bark. Thus, the moth's preference for white birch resting places and its typical perching orientation appear to be anti-detection adaptations that thwart visually hunting predators such as blue jays.

Some insects apparently can conceal themselves even while remaining active. For example, an African assassin bug covers its body with sand and dust and then piles on a backpack of odds and ends, especially the dried corpses of small insects that the bug has killed and drained of their fluids. The bug can then move about hidden beneath the material on its back. This concealment works. When visually hunting army ants were presented with an opportunity to attack bugs with or without their dust coats and backpacks, they assaulted those bugs that the experimenters had disrobed, but ignored those that still had their camouflage in place, much to the bugs' benefit (Figure 6.20) [136].

Visual cues are not, however, the only ones available to predators, many of which track prey down by their odor. This may be why skipper butterfly lar-

Figure 6.20   The value of a backpack. (A) Assassin bugs cover themselves with dust and pick up debris to build a "backpack." (B) Bugs that retain the dust and debris that they apply to themselves are ignored by predatory army ants, whereas those that have had a portion or all of their camouflage removed by an experimenter are subject to predation. A, photograph by Miriam Brandt, courtesy of Dieter Mahsberg; B, after Brandt and Mahsberg [136].

(A)

(B)

(A)

(B)

Figure 6.21   Personal hygiene by a skipper butterfly larva may be an antipredator adaptation. (A) A larva of a skipper butterfly inside its partly opened leaf shelter. Butterfly larvae expel pellets of waste from their shelters. (B) When waste pellets are added experimentally to shelters from which the larvae have been removed, wasps focus on those shelters in preference to shelters containing an equivalent quantity of odorless glass beads. Shown are the percentage of initial visits by ten wasps to the two kinds of shelters and the percentage of time the wasps spent investigating the two kinds of shelters. A, photograph by the author; B, after Weiss [1266].

vae that live within rolled-up leaves (Figure 6.21A) fire their fecal pellets away from their leaf shelters [1266]. These caterpillars employ a special device that holds a firm waste pellet in the anus until a rapid change in blood pressure expands the terminal segment of the abdomen, causing the little fecal ball to be discharged explosively into space like a miniature pea from a peashooter. Some skipper caterpillars regularly blast their waste pellets 20 body lengths or more from their shelters.

If this form of waste management is an antipredator adaptation, then these larvae should be hunted by enemies that could use the odor cues associated with the fecal pellets to locate their prey if those cues were available. Martha Weiss tested this prediction by removing skipper larvae from a set of leaf shelters and adding fresh waste pellets to some of the shelters, while the remainder received the same number of odorless glass beads about the same size and color as the pellets. The experimental and control shelters were then placed in a cage with a colony of paper wasps, known predators of butterfly larvae. The wasps were far more likely to inspect leaf shelters with caterpillar waste than those with glass beads (Figure 6.21B). Moreover, when Weiss offered a foraging paper wasp a choice between two shelters, each containing a hidden caterpillar but one surrounded with 25 waste pellets and the other accompanied by 25 glass beads, the wasp first found and killed the caterpillar associated with fecal matter on 14 of 17 trials. It would appear that getting rid of one's waste products has survival benefits for these butterfly larvae.

Discussion Question

6.9   Weiss also collected information on the growth rates of caterpillars forced to inhabit shelters that she contaminated with their feces and of others she allowed to mature in clean shelters. She found no difference in the weight of the pupae that experienced these two different conditions as larvae; moreover, the days required for the larvae to pupate also did not differ between individuals growing up with and without waste pellets in their shelters. Why did Weiss gather these data?

Some Darwinian Puzzles

Although many insects possess beneficial attributes that reduce their risk of being spotted or smelled, other species seem to go out of their way to make themselves obvious to their predators (Figure 6.22). This is the kind of thing that attracts the attention of adaptationists because of the obvious fitness costs

(A)

(C)

(B)

Figure 6.22   Warning coloration and toxins. Animals that are chemically defended typically behave in a conspicuous manner. (A) The monarch butterfly's body and wings may contain potentially lethal cardiac glycosides, which it sequesters from its milkweed diet. (B) Blister beetles, which have blood laced with cantharidin, a highly noxious chemical, may mate for hours right out in the open on flowering plants. (C) This conspicuous moth exudes a toxic froth from thoracic glands when disturbed. A–B, photographs by the author; C, photograph by Tom Eisner.

incurred by a prey animal that is conspicuous to its predators. Traits whose costs seem likely to exceed their benefits qualify as **Darwinian puzzles** because they would seem unlikely to evolve by natural selection. Take the monarch butterfly, for example, a species whose orange and black wing pattern makes it easy to spot. Bright coloration of this sort is correlated with greater risk of attack in some cases [1159]. How can it be adaptive for monarchs to flaunt themselves in front of butterfly-eating birds in this way? In order to overcome its costs, such conspicuous coloration ought to have some very substantial benefits, and in the case of the monarch, it does. These benefits appear to be related to the ability of monarch larvae to feed on poisonous milkweeds, from which they sequester an extremely potent plant poison in their tissues [154]. In dealing with the monarch butterfly, you would be wise to ignore the recommendation of the lepidopterist E. B. Ford, who wrote, "I personally have made a habit, which I recommend to other naturalists, of eating specimens of every species which I study" [397]. Any bird that makes the mistake of accepting the Ford challenge with respect to monarchs usually finds the experience most unpleasant, albeit highly educational (Figure 6.23). After vomiting up a noxious monarch just once, a surviving blue jay will avoid this species religiously thereafter [152, 154].

If you have absorbed the essence of the cost–benefit approach, you will now be asking how an eaten monarch could gain fitness by inducing vomiting, given that regurgitated monarchs rarely fly off into the sunset. Is this case an exception to the rule that dead animals cannot pass on their genes? Probably not, although the possibility exists that toxicity has evolved via indirect selection (see p. 451–452) if dead individuals educate the potential predators of their close relatives, thereby helping these genetically similar monarchs to pass on shared genes to the next generation. Another hypothesis, however, is that monarchs

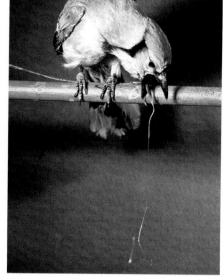

Figure 6.23   Effect of monarch butterfly toxins. A blue jay that eats a toxic monarch vomits a short while later. Photographs by Lincoln P. Brower.

recycle poisons from their food plants to make themselves so bad-tasting that most birds will release any monarch immediately after grabbing it by the wing [154]. In fact, when caged birds are offered a frozen but thawed monarch, many pick it up by the wing and then drop it at once, evidence that in nature monarchs could benefit personally from being highly unpalatable, a fact advertised by their color pattern.

Bright coloration is not the only means by which a prey species can make itself conspicuous. Take the tephritid fly that habitually waves its banded wings as if trying to catch the attention of its predators. This puzzling behavior attracted two teams of researchers, who noticed that the wing markings of the fly resemble the legs of jumping spiders, an important fly predator. The biologists believed that when the fly waves its wings, it creates a visual effect similar to the aggressive leg-waving displays that the spiders themselves use (Figure 6.24) [475, 797]. Thus, the fly could be a code breaker (see Chapter 4) whose appearance and behavior releases escape behavior on the part of the predator.

In order to test the deception hypothesis, one group of researchers became expert at fly surgery. Armed with scissors, Elmer's glue, and steady hands, the fly surgeons exchanged wings between clear-winged houseflies and pattern-winged tephritid flies. After the operation, the tephritid flies behaved normally, waving their now plain wings and even flying about their enclosures. But these modified tephritids with their housefly wings were soon eaten by the jumping spiders in their cages. In contrast, tephritids whose own wings had been removed and then glued back on repelled their enemies in 16 of 20 cases. Houseflies with tephritid wings gained no protection from the spiders, showing that it is the combination of leglike color pattern *and* wing movement that enables the tephritid fly to deceive its predators into treating it as a dangerous opponent rather than a meal [475]. By comparing the proportions of survivors

Figure 6.24   Why behave conspicuously? This tephritid fly (top) habitually waves its banded wings, which give it the appearance of a leg-waving jumping spider (bottom). When the spiders wave their legs, they do so to threaten one another. The fly mimics this signal to deter attack by the spider. Photographs by Bernie Roitberg; from Mather and Roitberg [797].

Figure 6.25 **An advertisement of unprofitability to deter pursuit?** Stotting behavior by a springbok, a small antelope that leaps into the air when threatened by a predator, just as Thomson's gazelles do. Photograph by Jen and Des Bartlett.

among those endowed with a presumptive adaptation and those with an alternative, the researchers showed that the tephritid fly's wings and behavior work better than other options.

Some vertebrates also behave in ways that paradoxically make them easier to spot. Take the Thomson's gazelle. When pursued by a cheetah or lion, this antelope may leap several feet into the air while flaring its white rump patch (Figure 6.25). Any number of possible explanations exist for this behavior, which is called *stotting*. Perhaps a stotting gazelle sacrifices speed in escaping from one detected predator in order to scan ahead for other as yet unseen enemies lying in ambush (as lions often do) [958]. The anti-ambush hypothesis predicts that stotting will *not* occur on short-grass savanna, but will instead be reserved for tall-grass or mixed grass and shrub habitats, where predator detection could be improved by jumping into the air. But gazelles feeding in short-grass habitats do stot regularly, so we can reject the anti-ambush hypothesis and turn to some others [191, 192]:

- *Alarm signal hypothesis:* Stotting might warn conspecifics, particularly offspring, that a predator is dangerously near. This signal could increase the survival of the signaler's offspring and relatives, thereby improving the stotter's fitness (see p. 451).

- *Social cohesion hypothesis:* Stotting might enable gazelles to form groups and flee in a coordinated manner, making it harder for a predator to cut any one of them out of the herd.

- *Confusion effect hypothesis:* By stotting, individuals in a fleeing herd might confuse and distract a following predator, keeping it from focusing on one animal.

- *Pursuit deterrence hypothesis:* Stotting might announce to a pursuing predator that the individual was in excellent condition and therefore unlikely to be captured, which, if true, would favor predators that stopped chasing that gazelle.

Table 6.2 lists some predictions that are consistent with these hypotheses. Because the same prediction sometimes follows from two different hypotheses, we must consider multiple predictions from each one if we are to discriminate among them. This study nicely illustrates the value of coming up with as many

TABLE 6.2 *Predictions derived from four alternative hypotheses on the adaptive value of stotting by Thomson's gazelle*

	Alternative hypotheses			
Prediction	Alarm signal	Social cohesion	Confusion effect	Signal of unprofitability
Solitary gazelle stots	No	Yes	No	Yes
Grouped gazelles stot	Yes	No	Yes	No
Stotters show rump to predator	No	No	Yes	Yes
Stotters show rump to gazelles	Yes	Yes	No	No

plausible hypotheses as possible and then thinking through the predictions derived from each one.

Tim Caro learned that a single gazelle will sometimes stot when a cheetah approaches, an observation that helps eliminate the alarm signal hypothesis (if the idea is to communicate with other gazelles, then lone gazelles should not stot) and the confusion effect hypothesis (because the confusion effect can occur only when a group of animals can flee together). We cannot rule out the social cohesion hypothesis on the grounds that solitary gazelles stot, because there is the possibility that solitary individuals stot in order to attract distant gazelles to join them. But if the goal of stotting is to communicate with fellow gazelles, then stotting individuals, solitary or grouped, should direct their conspicuous white rump patch toward other gazelles. Stotting gazelles, however, orient their rumps toward the predator. Only one hypothesis is still standing: gazelles stot to announce to a predator that they will be hard to capture. Cheetahs get the message, since they are more likely to abandon hunts when the gazelle stots than when the potential victim does not perform the display (Figure 6.26) [191].

The presumption is that stotting gazelles must be honestly advertising their uncatchability to cheetahs. If not, cheetahs would gain by ignoring, or even targeting, gazelles that took the time to stot. The prediction of honesty in signaling has not been tested for gazelles, but has been for an *Anolis* lizard that performs pushup displays when it spots an approaching lizard-eating snake. Snakes tend to break off their hunt in response to seeing the lizard bobbing up and down. Because the number of pushups performed by individual lizards varies, Manuel Leal realized that he had an opportunity to test the prediction that these displays did indeed convey accurate information about a lizard's capacity for escape.

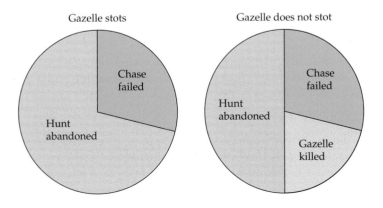

Figure 6.26 Cheetahs abandon hunts more often when gazelles stot than when they do not, supporting the hypothesis that these predators treat stotting as a signal that the gazelle will be hard to capture. After Caro [191].

(A)

(B)

Figure 6.27   **Are pushup displays an honest signal** of a lizard's physiological condition? (A) The lizard *Anolis cristatellius* performs a pushup display when it spots an approaching snake. (B) The time a lizard spent running until exhaustion was positively correlated with the number of pushups performed by that individual under perceived threat from a model snake. A, photograph by Manuel Leal; B, after Leal [703].

In order to determine whether the number of pushups was correlated with the lizard's running endurance, Leal and an assistant first counted the number of pushups individuals performed in the laboratory when exposed to a model snake. They then took the lizard to a circular runway, where they induced it to run by lightly tapping it on the tail. The time a lizard spent running without stopping was proportional to the number of pushups it performed in response to the model of its natural predator (Figure 6.27) [703]. Thus, as the attack deterrence hypothesis requires, predators could derive accurate information about the physiological state of a potential prey by watching its pushup performance. Since this anole sometimes does escape when attacked, it could well pay predatory snakes to make foraging decisions based on the signaling behavior of their would-be victims.

Discussion Question

6.10   The Bonaire whiptail lizard runs a short distance from potential predators and then raises one foreleg, which it waves about ostentatiously (Figure 6.28) [240]. This arm-waving behavior might be another example of a pursuit deterrence signal. What predictions follow from this hypothesis with respect to when the arm-waving behavior should be performed in response to the approach of a human being (a predator substitute)? That is, should arm waving occur more often when a person approaches slowly or rapidly? In response to a direct or a tangential approach? And which arm should be waved when the lizard is not directly facing the human?

One final Darwinian puzzle associated with conspicuous prey behavior comes from the observation that some animals produce extraordinarily loud

Figure 6.28   **The lizard *Cnemidophorus murinus*** often waves a forearm at humans that disturb it. Why? Photograph by Laurie Vitt.

and piercing screams while in the clutches of a predator. Rabbits, for example, respond to being captured in this fashion, despite the fact that they are normally the quietest of animals. What benefit could come from making so much noise? Perhaps there is no benefit of all from screaming, which is simply a nonadaptive product of the prey's pain system. On the other hand, perhaps a loud scream might so startle a predator that it would release its captured prey [546]. Goran Högstedt noted, however, that although fear screams are noisy, as required by the startle hypothesis, the calls persist as long as the victim remains alive, which should reduce their potential to startle (but see [235]).

In his paper, "Adaptation unto Death," Högstedt looked at whether screaming might warn conspecifics of danger from the predator. If so, then the screamer should be easy to locate so that listeners would know where the predator is. The acoustical features of fear screams do make it easy for other animals to pinpoint the source of the noise, but as Högstedt pointed out, other members of the injured prey's species largely ignore its cries. This reaction makes a certain amount of sense, since a predator with a captive will be occupied with that victim for some time, which means that uncaptured specimens nearby have little to fear for the moment.

But perhaps fear screams are given by young animals to enlist the aid of their parents. This hypothesis produces the prediction that the calls should be given only by dependent young, and this is not the case. Adult songbirds captured in mist nets are just as likely to scream as juveniles when handled by a human being (Figure 6.29).

The evidence actually points to a fourth hypothesis: namely, that the captured animal's screams attract other predators to the scene—predators that may turn the tables on the prey's captor, or at least interfere with it, sometimes enabling the prey to escape in the resulting confusion. This hypothesis requires that predators be attracted to fear screams, and they are, as Högstedt showed by broadcasting the taped screams of a captured starling, which brought hawks, foxes, and cats to the recorder. (On the other hand, in another study, when coyotes attracted to starling distress calls came toward a fellow coyote with a captured starling, the attacker intensified its effort to dispatch the bird, which was not good news for the starling [1329].)

Finally, the attract-competing-predators hypothesis produces the prediction that birds living in dense cover should be more prone to give fear screams than birds of open habitats. In areas where vision is blocked, a captured bird cannot rely on other nearby predators to see it struggling with its captor. Therefore, as a final effort to avoid death, it may attempt to call competing predators to the scene. Mist-netted birds of species that live in dense cover are in fact more

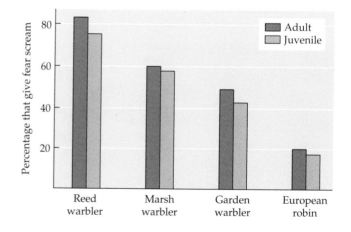

Figure 6.29   Fear screams and age in four European songbirds. In many species, juveniles are no more likely to give fear screams when handled by a human than are adults. This observation suggests that the function of the call cannot be to enlist parental assistance. After Högstedt [546].

likely to give fear screams when handled than are species that occupy open habitats [1329]. Those that survive thanks to their calls to other predators obviously derive a benefit from their behavior, which could make it adaptive to scream at death's door.

Optimality Theory and Antipredator Behavior

We have illustrated how thinking about fitness benefits (*B*) and fitness costs (*C*) has drawn the attention of behavioral researchers to a variety of evolutionary puzzles and motivated a search for the reproductive advantages that might have driven the evolution of behavioral traits that initially appear to make individuals more vulnerable to predators, not less. In this concluding section of the chapter, I shall introduce two cost–benefit approaches, both derived from natural selection theory, that can potentially yield precise quantitative predictions, rather than the more general qualitative ones that we have been discussing up to this point. Both theories attempt to construct hypotheses that take into account the net benefit (*B* – *C*) of a trait, rather than merely focusing on whether a trait provides a benefit or not. Both theories reflect the reality that adaptations have to do more than merely confer a benefit if they are to increase in frequency in a population. An adaptation, by definition, is *better* than the alternatives, and "better" means that the net benefit associated with a true adaptation is greater than that associated with the nonadaptive alternatives that have been replaced, or are being replaced, by natural selection.

We can illustrate our first approach, which is based on **optimality theory**, by looking at the costs and benefits of four alternative hereditary behavioral phenotypes in a hypothetical species (Figure 6.30). Of these four phenotypes (W–Z), only one (phenotype Z) generates a net loss in fitness (*C* > *B*) and is therefore obviously inferior to the others. The other three phenotypes all are linked to a net fitness gain, but just one, phenotype X, is an adaptation, because it produces the greatest net benefit of the four phenotypes. Thus, the alleles for phenotype X will spread at the expense of the alternative alleles underlying the development of the other forms of this behavioral trait in this population. Phenotype X can be considered the *optimal* trait here because it is an adaptation, because the difference between *B* and *C* is greatest for this trait, and because this trait will spread while all others are declining in frequency (as long as the relationship between costs and benefits remains the same).

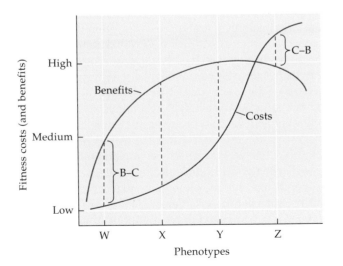

Figure 6.30   An optimality model. If one can measure the fitness costs and benefits associated with four alternative behavioral phenotypes in a population, then one can determine which trait confers the greatest net benefit on individuals in that population. This trait would be considered an adaptation—an optimal trait that would replace the alternatives given sufficient evolutionary time.

If it were possible to measure *B* and *C* for a set of alternatives that might have existed within a population, one could predict that the phenotype with the greatest net benefit would be the one observed in nature. Unfortunately, it is often difficult to secure precise measures of *B* and *C* in the same fitness units. But if this can be done, then optimality theory makes it possible to produce a hypothesis founded on the premise that traits currently observed in a population will be optimal, and from this hypothesis can come quantitative predictions. Optimality theory is most strongly associated with the analysis of foraging behavior (as we will see in the next chapter) because in this context, it is sometimes possible to measure both benefits and costs in the same currency: calories gained from collected foods (the benefits) and calories expended in collecting those foods (the costs). But optimality theory has been applied to some antipredator behaviors as well.

For example, northern bobwhite quail spend the winter months in small groups, called coveys, which range in size from 2 to 22 individuals, but with a very strong peak around 11 individuals, in the midwestern United States. These coveys almost certainly form for antipredator benefits. For one thing, members of larger coveys are safer from attack, judging from the fact that overall group vigilance (percentage of time that at least one member of the covey has its head up and is scanning for danger) increases with increasing group size, and then levels off around a group size of 10. Moreover, in aviary experiments, members of larger groups react more quickly than smaller ones when exposed to a silhouette of a predatory hawk. The antipredator benefits of being in a large group are, however, almost certainly offset to some degree by the increased competition for food that occurs in larger groups. This assumption is supported by evidence that relatively large groups move more each day than coveys of 11. (Small groups also move more than groups of average size, probably because these birds are searching for other groups with which to amalgamate.) The mix of benefits and costs associated with coveys containing different numbers of individuals suggests that birds in intermediate-sized coveys have the best of all possible worlds, and in fact, daily survival rates are highest during the winter in coveys of this size (Figure 6.31). This work demonstrates not just that the quail form groups to detect their predators effectively, deriving a benefit from their social behavior, but that they attempt to form groups of the optimal size, deriving the greatest possible net benefit in terms of survival from their preference.

Game Theory Applied to Social Defenses

In addition to the optimality theory approach, we can look at behavioral evolution through the lens of **game theory**. Here too, both the costs and the benefits of behavioral decisions are considered, under the assumption that individuals are attempting unconsciously to maximize their reproductive success. But game theoreticians focus on cases in which individuals are competing with one another in such a way that the fitness consequences of a given behavioral option depend on the actions of the other competitors. Under this approach, decision making is treated as a game, just as it is by the economists who invented game theory in order to understand the choices made by people as they compete with one another for consumer goods or wealth. Game theoretical economists have shown that a strategy that works well in one situation may fail when matched against another way of making economic decisions. Figuring out what strategy will win often depends on what the other game players are doing.

The fact that all organisms, not just humans interested in how to spend their money or get ahead in business, are engaged in competitions of various sorts means that the game theoretical approach is a natural match for what goes on

(A)

(B)

(C)

Northern bobwhite

Figure 6.31   Optimal covey size for northern bobwhite quail is a function of the costs and benefits of belonging to groups of different sizes. (A) The probability that an individual would survive on any given day, its survival rate, a benefit, is highest for birds in coveys of about 11 birds. (B) Distance traveled, a cost, is lowest for individuals in coveys of about 11 birds. (C) Most bobwhite are found in coveys of about 11 birds. After Williams et al. [1306].

in the living world. The fundamental competition of life revolves around getting more of one's genes into the next generation than one's fellows. Winning this game almost always depends on what those other individuals are up to, which is why taking this factor into account makes a great deal of sense to evolutionary biologists.

One of the premier evolutionary biologists of all time, W. D. Hamilton, was a pioneer in thinking about evolution as a game between competing phenotypes. Hamilton argued that under some conditions, a behavioral strategy that caused individuals to be social could, over time, spread through a population in which the other individuals lived alone. The end result, according to Hamilton, could be a **selfish herd** [503] in which all the individuals were trying to hide behind others to reduce the probability of being selected by a predator. Imagine, for example, a population of antelope grazing on an African plain in which all individuals stay well apart, thereby reducing their conspicuousness to their predators. Now imagine that a mutant individual arises in this species, one that approaches another animal and positions itself so as to use its companion as a living shield for protection against attacking predators. The social mutant that employs this tactic would incur some costs; for example, two animals may be more conspicuous to predators than one, and so attract more attacks than scattered individuals, as has been demonstrated in some cases [1255]. But if these costs were consistently outweighed by the survival benefit gained by social individuals, the social mutation could spread through the population. If so, then eventually all the members of the species would be aggregated, with individuals jockeying for the safest position within their groups, actively attempting to improve their odds at someone else's expense. The result would be a selfish herd, whose members would actually be safer if they all could agree to spread out and not try to take advantage of one another. But since populations of individuals employing the solitary strategy would be vulnerable to invasion by an exploitative social mutant that uses a hide-behind-another strategy to take fitness from its companions, the exploitative tactic could spread through the species—a clear illustration of why we define adaptations in terms of their contribution to the fitness of individuals relative to that of other individuals with alternative traits.

Game theory, like its cousin optimality theory, can be used to generate mathematical models (hypotheses) from which can be drawn precise, quantitative predictions. But for the moment, let's apply the theory's general approach to a case involving Adélie penguins. These birds often wait on the ice near open water for some time until a group assembles, and only then do they all jump into the water more or less together to swim out to their foraging grounds. The potential value of this social behavior becomes clearer when one realizes that a leopard seal may be lurking in the water near the jumping-off point (Figure 6.32) [245]. The seal can capture and kill only a certain small number of penguins in a short time. By swimming in a group through the danger zone, many penguins will escape while the seal is engaged in dispatching one or two of their unfortunate fellows. If you had to run a leopard seal gauntlet, you would probably do your best to be neither the first nor the last into the water. If penguins behave as we would, then the groups that form at the water's edge qualify as selfish herds, whose members are engaged in a game whose winners are better than others at judging when to plunge into the water.

The selfish herd hypothesis generates testable predictions that can be applied to Adélie penguins or to any other prey species that forms groups. A key prediction is that individuals in these groups should compete with one another for the safer positions. Bluegill sunfish breed in colonies (see p. 442), with each male defending a nesting depression in the lake bottom against rivals [485]. As predicted from a selfish herd model, males compete intensely for nesting ter-

Figure 6.32   Selfish herds may evolve in prey species. Adélie penguins have formidable predators, such as this leopard seal. While the seal is dispatching one penguin, others can more safely enter the water and escape to the open ocean. Photograph by Gerald Kooyman/Hedgehog House.

ritories at the center of the colony, and larger, generally older, males are more likely to win. Thus, it is a younger, smaller, peripheral male who first confronts a foraging bullhead or cannibalistic bluegill, and as a result, the eggs of central males are only half as likely to be taken as the eggs of males in peripheral nests.

Game theory has many more applications, and we will deal with some of them in the chapters ahead. For now, it is enough to realize that although optimality theory and game theory are both derivatives of natural selection theory, they provide somewhat different tools to help us identify why animals behave they way they do.

Discussion Question

6.11   Consider Figure 6.33, a game theory diagram based on the concept of a selfish herd (with thanks to Jack Bradbury). In a population of prey animals, most individuals are solitary and stay well apart from others. But some mutant types arise that search out others and use them as living shields against predators. The mutants take fitness from the would-be solitary types by making them more conspicuous to their predators. We will set the fitness payoff for solitary living in a population composed only of solitary individuals at P. But when a solitary individual is found and used by a social type, the solitary animal loses some fitness (B) to the social type. There is a cost (C) to social individuals in terms of the time required to find another individual to hide behind, and a cost arising from the increased conspicuousness to predators of groups composed of two individuals rather than one. When two social types interact, we will say that they each have one chance in two of being the one that happens to hide behind the other when a predator attacks. If B is greater than C, what behavioral type will come to predominate in the population over time? Now compare the average payoff for individuals in populations composed entirely of solitary versus social types. If the average fitness of individuals in a population of social types is less than that of individuals in a population composed of solitary types, can hiding behind others be an adaptation?

Figure 6.33   A game theoretical model in which the fitness gained by a solitary or social focal animal is dependent on the behavior of its opponent, which may be either a solitary or a social individual.

Summary

1. An adaptation is the product of natural selection, which means that it is a hereditary trait that does a better job at promoting individual reproductive or genetic success than any available alternative form of that trait. Another way to say the same thing is that an adaptation has a better ratio of fitness benefits to fitness costs than any alternative characteristic that has happened to appear in the history of a species.

2. The adaptationist approach is the research procedure followed by persons who wish to test hypotheses on the possible adaptive value of attributes of interest to them. Traits that particularly intrigue adaptationists are those that have substantial fitness costs, which must therefore generate major fitness benefits if they are to spread and persist in populations under natural selection. Traits whose costs seem to exceed their benefits constitute Darwinian puzzles; the solutions to these puzzles are prized by adaptationists.

3. To test adaptationist hypotheses, a scientist must check the validity of predictions derived from these potential explanations. Relevant evidence on the fitness costs and benefits of traits can be gathered through field observations, controlled manipulative experiments, or natural experiments involving comparisons among living species. The comparative method of testing adaptationist hypotheses is based on two key assumptions: (1) that related species will exhibit differences in their attributes if they face different selection pressures, despite their having a common ancestor and thus a similar genetic heritage, and (2) that unrelated species that share similar selection pressures will converge on the same adaptive response, despite their having a different genetic heritage.

4. Two additional tools for the development and testing of adaptationist hypotheses are optimality theory and game theory. Optimality theory is most useful in cases in which it is possible to measure both benefits and costs of alternative traits in the same fitness currency. If this can be done, one can check whether individuals actually behave in an optimal manner; that is to say, in such a way as to maximize their net benefit. Game theory comes into play when the benefit of a behavioral option to one individual depends on what the other members of its population are doing. This theory views evolution as a game in which the players are armed with different strategies that are in competition with one another, with the winners creating a population over time that cannot be invaded by a player with an alternative strategy.

Suggested Reading

Two books, one by Wolfgang Wickler [1291] and the other by Rod and Ken Preston-Mafham [976], contain many amazing examples of animal coloration and behavioral defenses. John Endler has provided a modern review of the interrelationships between predator adaptations and prey counteradaptations [370]. The cost–benefit approach to antipredator behavior is described by Steven Lima and Lawrence Dill [724]. Tim Caro's papers on gazelle stotting [191, 192] illustrate the adaptationist approach very well, as does his review of how the pursuit deterrence hypothesis has been tested [193]. S. J. Gould and R. C. Lewontin's attack on adaptationism [463] is worth reading—critically. Bernie Crespi succinctly reviews the various definitions of adaptation and the several reasons for the occurrence of maladaptations in nature [252]. *The Selfish Gene* (second edition) by Richard Dawkins is still the premier book for explaining behavioral game theory to a general audience [304].

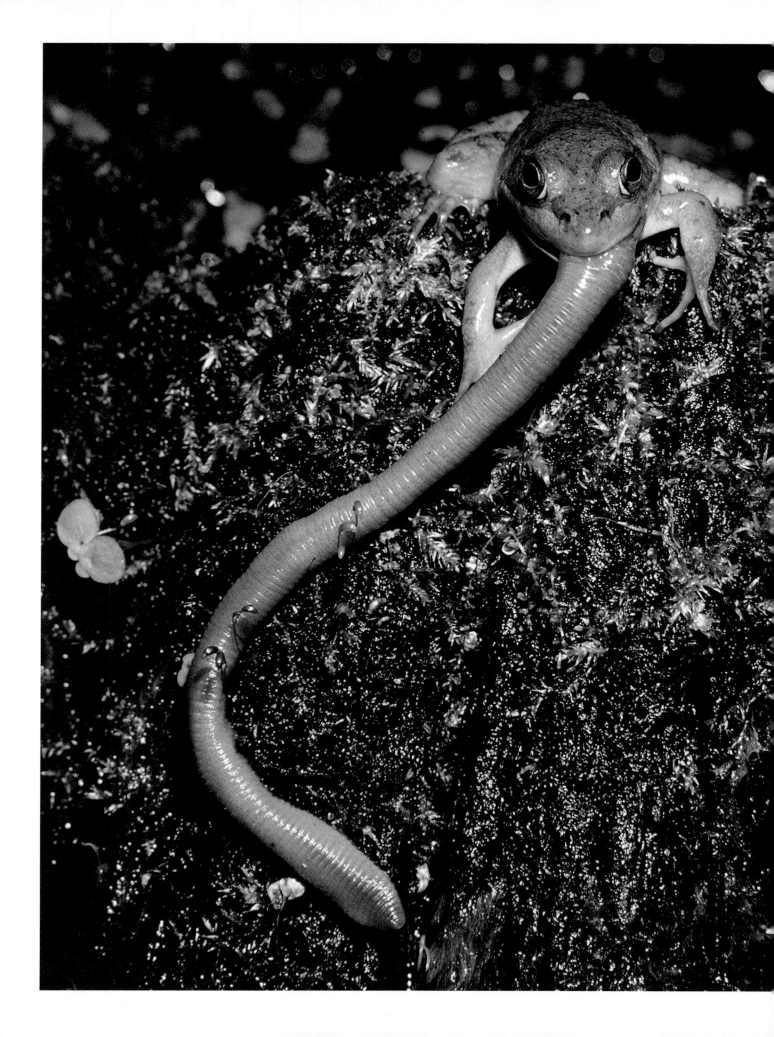

7 | *The Evolution of Feeding Behavior*

*I*n our examination of antipredator behavior, we showed how a cost–benefit approach can be used to establish whether a behavioral trait has an adaptive survival-promoting function. Prey species have evolved any number of adaptations of this sort. Given that most animals are so good at postponing their demise, you may now possess a certain sympathy for foraging predators, which have to overcome a whole series of obstacles if they are to get something to eat. On the other hand, if you have ever observed a cactus wren or a blue jay plucking one insect after another from their hiding places in dense vegetation—a katydid here, a caterpillar there, then another caterpillar—you might conclude that predators can take care of themselves. In a way, the hunting skills these animals exhibit are the product of their prey's defenses, which have favored the evolution of predatory counteradaptations in an arms race that continues to this day. As a result, animals hunting for food often do get enough to eat.

This chapter examines hypotheses on the adaptive value of various elements of animal feeding behavior, with the goal of illustrating how behavioral ecologists use the theoretical tools at their disposal, including optimality theory and game theory, two explanatory ideas introduced in the preceding chapter. In addition to looking at how feeding behavior currently contributes to individual reproductive success, we shall consider some

◀ *A **bullfrog** has many dietary choices to make. Does its selection of nightcrawlers maximize the calories it gains? Photograph by Kenneth H. Thomas* attempts to outline the historical sequence of events leading to modern feeding adaptations—the other component of an ultimate analysis of behavior.

Optimal Foraging Behavior

A foraging crow has many decisions to make. Where should it search for food? At what time of day? For what prey? How long should it spend trying to process the prey that it has found? Users of optimality theory could potentially analyze each of these foraging decisions in terms of their contribution to the crow's fitness by testing whether the bird is behaving in an optimal (fitness-maximizing) manner. Let's use this approach to examine the choices crows make about what kinds of shellfish to open.

When a beachcombing northwestern crow spots a clam, snail, or whelk, it sometimes, but not always, picks it up, flies into the air, and then drops its victim onto a hard surface. If the mollusk's shell shatters on the rocks below, the bird flies down to the prey and plucks out the exposed flesh. The adaptive significance of the bird's behavior seems straightforward: It cannot use its beak to crack open the extremely hard shells of certain mollusks. Therefore, it breaks open its prey by dropping them on rocks. This seems adaptive. Case closed. But we can be much more ambitious in our analysis of the crow's foraging decisions. When a hungry crow is searching for food, it has to decide which mollusk to select, how high to fly before dropping the prey, and how many times to keep trying if the mollusk does not break on the first try.

Reto Zach made several observations while watching foraging crows:

1. The crows picked up only large whelks about 3.5–4.4 centimeters long.

2. They flew up about 5 meters to drop their chosen whelks.

3. They kept trying until the whelk broke, even if many flights were required.

Zach sought to explain the crows' behavior by determining whether the birds' decisions were optimal in terms of maximizing whelk flesh available for consumption per unit of time spent foraging [1351]. The optimality hypothesis yielded the following predictions:

1. Large whelks should be more likely than small ones to shatter after a drop of 5 meters.

2. Drops of less than 5 meters should yield a reduced breakage rate, whereas drops of much more than 5 meters should not greatly improve the chances of opening a whelk.

3. The probability that a whelk will break should be independent of the number of times it has already been dropped.

Zach tested each of these predictions in the following manner: He erected a 15-meter pole on a rocky beach and outfitted it with a platform whose height could be adjusted and from which whelks of various sizes could be pushed off to drop onto the rocks below. He collected samples of small, medium, and large whelks and dropped them from different heights (Figure 7.1). He found, first, that large whelks required significantly fewer 5-meter drops before they broke than either medium-sized or small whelks. Second, the probability that a large whelk would break improved sharply as the height of the drop increased—

Figure 7.1 Optimal foraging decisions by northwestern crows when feeding on whelks. The curves show the number of drops at different heights needed to break whelks of different sizes. Northwestern crows pick up only large whelks, increasing the calories available to them, and they drop their whelks from a height of about 5 meters, thereby minimizing the energy they expend in opening their prey. After Zach [1351].

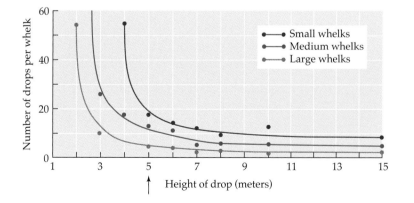

up to about 5 meters—but going higher did little to improve the breakage rate. Third, the chance that a large whelk would break was not affected by the number of previous drops and was instead always about one chance in four on any given drop. Therefore, a crow that abandoned an unbroken whelk after a series of unsuccessful drops would not have a better chance of breaking a replacement whelk of the same size on its next attempt. Moreover, finding a new prey would take time and energy.

Zach went one step further by calculating the average number of calories required to open a large whelk (0.5 kilocalories), a figure he subtracted from the food energy present in a large whelk (2.0 kilocalories), for a net gain of 1.5 kilocalories. In contrast, medium-sized whelks, which require many more drops, would yield a net loss of 0.3 kilocalories; trying to open small whelks would have been even more disastrous. Thus, the crows' rejection of all but large whelks was adaptive, assuming that fitness is a function of energy gained per unit of time. The birds' selection of dropping height and their persistence in the face of failure also enabled them to reap the greatest possible energy return for their foraging activities [1351].

Discussion Questions

7.1 In some places, American crows open walnuts by dropping them on hard surfaces. Unlike northwestern crows opening whelks, American crows reduce the height from which they drop walnuts from about 3 meters on the first drop to about 1.5 meters on the fifth drop. If this tendency is adaptive, what prediction follows about a difference between whelks and walnuts in the likelihood of breaking on successive drops? In addition, American crows tend to drop walnuts from lower heights when other crows are present. If this trait is an adaptation, what prediction must be true? Check your answers against data in [260].

7.2 A badger living in Oklahoma could hunt for either scorpions or ground squirrels (or both). Scorpions provide only 10 calories each, but require only 2 minutes to find, on average, with an additional 3 minutes to remove the stinger; ground squirrels offer 1000 calories, but take an average of 3 hours to find and an additional 90 minutes to capture, kill, and consume. If the badger's ultimate goal is to maximize its rate of caloric gain, should it forage for squirrels, scorpions, or both? Show your math. (Thanks to Doug Mock for this question.)

As just noted, a central assumption of the crow–whelk optimality study was that crows would achieve maximum reproductive success by maximizing the number of calories ingested per unit of time. This is an assumption that deserves checking. The relationship between whelk opening efficiency and fitness has not been established for crows, but in an experiment with captive zebra finches, which were given the same kind of food but under different feeding regimes so that some birds had higher foraging costs than others, the individuals with the highest daily net caloric gains survived best and reproduced the most [703].

In a "natural" experiment of a somewhat similar sort, food intake has been shown to be related to the survival of the red knot (a sandpiper). In the spring, migrating red knots stop off on the shores of Delaware Bay in the eastern United States to feed on horseshoe crab eggs before completing their journey to northern Canada, where they breed. Horseshoe crab populations have crashed in recent years due to intense overharvesting by fishermen who use these unfortunate animals as bait. As a result, the crabs no longer spawn in the huge numbers that they once did, and the knots find it much harder to gain the 60 to 100 grams of body weight that they need to complete their journey and breed in Canada. If there is a direct link between food and survival, as assumed in optimal foraging models, then we would expect a sharp drop in the population of red knots that uses the migratory flyway with the Delaware Bay stopover. This

Oystercatcher

Figure 7.2   Available prey versus prey selected. Foraging oystercatchers choose mussels that are larger than the average available mussel, but they do not concentrate on the very largest mussels. After Meire and Ervynck [816].

population did indeed decline from 51,000 in the year 2000 to 27,000 just 2 years later. In addition, researchers captured, weighed, and marked red knots in Delaware Bay, then recaptured some of the same birds in subsequent years. The weights of birds that survived to return a second year were greater upon their initial capture than those of knots that were not recaptured or resighted a second time, a further indication that acquiring energy efficiently is directly associated with fitness [51].

How to Choose an Optimal Mussel

The Eurasian oystercatcher is a shorebird whose foraging decisions can also be matched against predictions taken from optimality models. Two Belgian researchers, P. M. Meire and A. Ervynck, developed a calorie maximization hypothesis to apply to oystercatchers feeding on mussels [816]. Like Reto Zach, they calculated the profitability of different-sized prey, based on the calories contained in the mussels (a fitness benefit) and the time required to open them (a fitness cost). Even though mussels over 50 millimeters long require more time to hammer or stab open, they provide more calories per minute of opening time than smaller mussels. Therefore, the model predicts that oystercatchers should focus primarily on the largest mussels. But in real life, the birds do not prefer the really large ones (Figure 7.2). Why not?

Hypothesis 1: The profitability of very large mussels is reduced because some cannot be opened at all, reducing the average return from handling these prey.

In their initial calculations of prey profitability, the researchers had considered only those prey that the oystercatchers actually opened (Figure 7.3, model A). As it turns out, oystercatchers select some large mussels that they find impossible to open, despite their best efforts. The handling time wasted on these large impregnable mussels reduces the average payoff for dealing with this size class

Figure 7.3   Two optimal foraging models yield different predictions because they calculate prey profitability differently. Model A calculates the profitability of a mussel based solely on the energy available in opened mussels of different sizes divided by the time required to open these prey. Model B calculates profitability with one added consideration; namely, that some very large mussels must be abandoned after being attacked because they are impossible to open. After Meire and Ervynck [816].

of mussels. When this factor is taken into account, a new optimality model results, yielding the new prediction that the oystercatchers should concentrate on mussels 50 millimeters in length, rather than the very largest size classes (Figure 7.3, model B). The oystercatchers, however, actually prefer mussels in the 30- to 45-millimeter range. Therefore, time wasted in handling large, invulnerable mussels fails to explain the oystercatchers' food selection behavior.

Hypothesis 2: Many large mussels are not even worth attacking because they are covered with barnacles, which makes them impossible to open.

This additional explanation for the apparent reluctance of oystercatchers to feast on large, calorie-rich mussels is supported by the observation that oystercatchers never touch barnacle-encrusted mussels. The larger the mussel, the more likely it is to have acquired an impenetrable coat of barnacles, which eliminates these prey from consideration. According to a mathematical model that factors in (1) prey opening time, (2) time wasted in trying but failing to open a mussel, and (3) the actual size range of realistically available prey, the birds should focus on 30- to 45-millimeter mussels—and they do. Note that these researchers used optimality theory to produce an initial hypothesis, which they rejected on the basis of evidence collected. They then modified their model and subjected to it a new test, with the result that they gained an improved understanding of oystercatcher feeding behavior.

Discussion Questions

7.3 The pike cichlid is a predatory fish that feeds on guppies, a smaller fish. In the laboratory, at least, and probably in the rivers of Trinidad as well, this predator tends to attack and consume relatively large guppies. A Swedish research team measured the time it took for pike cichlids to detect, approach, stalk, and attack guppies of four size classes (10, 20, 30, and 40 mm long). They also recorded the capture success rate for each size class, as well as noting the time it took to handle and consume those prey that were definitively captured. With these data, they constructed two models of prey value as measured by weight of prey consumed per unit of time. Model A considered only the time to attack and the capture rate, whereas model B incorporated these two factors plus the post-capture handling time in calculating the weight of food consumed per unit of time (Figure 7.4). Which of the two models strikes you as the most realistic optimal foraging model, and why? In light of the two models, what evolutionary issues are raised by the actual preference of pike cichlids for 40 mm guppies?

7.4 Red knots often feed on small shelled mollusks buried in mudflats. These prey are distributed in a patchy fashion. The sandpipers enter a patch, and probe for prey; if they find some, they have to pause to digest their crunchy catch before resuming their search. At some point, the birds leave to go to another patch, where they engage in another round of hunting (Figure 7.5) [1228]. How would you go about determining whether the birds are foraging optimally or not? Should you calculate their intake rate only when foraging, not when digesting? Would you expect the birds to take into account the average prey density in the patches in which they have already hunted? Might they gain by paying attention to their discovery rate of mollusks in their current foraging patch? When should a bird leave a patch?

Criticisms of Optimal Foraging Theory

By developing and testing optimality models, researchers have concluded that northwestern crows and European oystercatchers choose prey that provide the maximum caloric benefit in relation to time spent foraging. But some persons have criticized the use of optimality theory on the grounds that animals do not always behave as efficiently as possible. As we have seen, however, optimality models are not constructed to make statements about perfection in

Pike cichlid

Figure 7.4 Two optimal foraging models of the value of hunting guppies of different sizes by a predatory fish, the pike cichlid. After Johansson et al. [601].

Figure 7.5   A foraging bout by the red knot. After a period of successful searching, the bird pauses to digest the prey it has found and eaten. When it resumes foraging, it must then decide how long to persist in a given patch before moving to a new one. After van Gils et al. [1228].

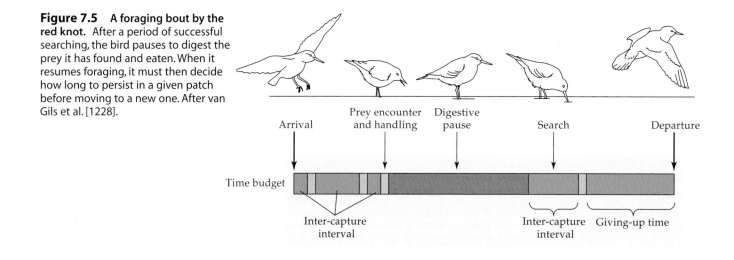

evolution, but rather to make it possible to test whether one has correctly identified the variables that might have shaped the evolution of an animal's behavior. As we have also seen, the factors included in an optimality model have a large effect on the predictions that follow. If an oystercatcher is assumed to consider every mussel in a tidal flat as a potential prey item, then it is predicted to make different foraging decisions than if the modeler assumes that oystercatchers simply ignore all barnacle-covered mussels. If the predictions of an optimality model fail to match observations, researchers can make progress by rejecting that model and going on to develop and test alternative hypotheses that take other factors into consideration.

If ecological factors other than caloric intake affect oystercatcher foraging behavior, for example, then a caloric maximization model will fail its test, as it should. And for most foragers, foraging behavior does indeed have consequences above and beyond the acquisition of calories. If you suspected, for example, that predators had shaped the evolution of an animal's foraging behavior, then the kind of optimality model you might choose to construct and test would not focus solely on calories gained versus calories expended. If foraging exposes an animal to the risk of sudden death, then when that risk is high, we would expect foragers to sacrifice short-term caloric gain for long-term survival [342, 1249].

Such a sacrifice has been demonstrated experimentally for garden skinks. These lizards were reared in enclosures divided into two microhabitats: shaded areas containing wood blocks and open sunlit areas. Crickets were placed in the enclosures as food for the skinks; the insects were free to move between habitat types. In some of the enclosures, the open areas were daubed with the scent of a lizard-hunting snake. The skinks in these enclosures were far more likely to spend time in the shaded, cover-containing portion of their living space than were skinks kept in enclosures without snake scent; the control animals moved about more often, especially in the open microhabitats, with the result that they captured more crickets and grew more rapidly during the first 6 months of life (Figure 7.6) [342]. An optimality model that failed to consider the trade-offs beween foraging success and predation risk would fail to predict the skinks' behavior accurately.

Leaf-cutter ants (*Atta cephalotes*) collect plant material that they use to grow fungus gardens in their underground nests. The ants feed on the fruiting bodies (mushrooms) of the fungus, which depends on the ants for its propagation. Indeed, the ants combat another undesired species of fungus by regurgitating a bacterial solution on their gardens. The antibiotic by-products produced by

(A)

(B)

Figure 7.6   Young garden skinks lower their foraging success in order to reduce the risk of predation. Lizards growing up in experimental cages with predatory snake scent (A) moved about less and (B) avoided open areas more often than individuals in control cages lacking these snake odors. After Downes [342].

the bacteria block the growth of the unwanted fungus, thereby increasing the production of the favored fungus in the ants' gardens [268]. So far, so good for the assumption that the ants are working to maximize their food intake.

However, although the ants could also increase the food available to them by collecting leaves night and day, their workers are primarily nocturnal. During the day, foraging occurs only sporadically, and the foragers at work then are small and relatively inefficient at cutting up and transporting sections of leaves back to the nest [903]. Why does the colony forage this way? Because *large* foragers that work *during the day* are hard-hit by a parasitic fly, which lays its eggs on them; when the fly larvae hatch, they burrow into the head of their victim with lethal effect. In contrast, nocturnal foragers of all sizes are safe from the fly. As a result, the colony trades off energy gain during the daylight hours for a longer total foraging life for its valuable large workers (Figure 7.7). A simple optimality model that predicted that worker leaf-cutters of all sizes should forage around the clock would be wrong, but helpful, because testing that model would lead eventually to a better understanding of leaf-cutter foraging behavior.

Figure 7.7   Foraging efficiency is compromised when the risk of predation is high. During the day, but not at night, foraging leaf-cutter ants tend to be smaller than the optimal size for efficient food collecting. But during the day, large foraging worker ants are at risk from parasitoid phorid flies, which ignore ants with head widths of 1.8 millimeters or less. After Orr [903] and Wilson [1317, 1318].

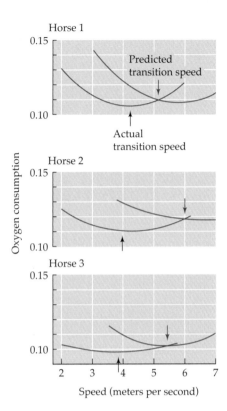

Oxygen consumption

Horse 1

0.15

Predicted
transition speed

0.10

Actual
transition speed

Horse 2

0.15

0.10

Horse 3

0.15

0.10

2   3   4   5   6   7

Speed (meters per second)

Figure 7.8   Records of energy consumption by three horses in relation to two locomotory modes: trotting (red line) versus galloping (green line). After Farley and Taylor [381].

7.5   Various persons have proposed that when a horse switches from slower trotting to faster galloping, it changes gaits to minimize the energetic expense of locomotion, under the assumption that animals able to minimize the energetic costs of getting from point A to point B will enjoy greater reproductive success than individuals that squander their energy reserves in inefficient locomotion. The energy minimization hypothesis was tested by Claire Farley and Richard Taylor with the help of three cooperative horses willing to run on a treadmill while outfitted to provide data on their oxygen consumption, a factor directly related to energy use [381]. What scientific conclusion is justified on the basis of Figure 7.8? Does this study support those who claim that optimality theory is not useful because the assumptions underlying particular hypotheses are often oversimplified and incorrect?

7.6   The golden-winged sunbird feeds exclusively on nectar from certain flowers during the winter in South Africa. Some birds are territorial at some patches of flowers, but may abandon these defended areas at other times. Frank Gill and Larry Wolf devised a way to measure the rate of nectar production per bloom at a given site. They also secured previously published information on the caloric costs of the territorial chase flights of birds defending a feeding territory (2000 calories per hour), as well as the costs of foraging for nectar (1000 calories per hour) and resting (400 calories per hour). Table 7.1 shows the calories saved by birds by holding territories with different nectar production rates in relation to foraging in other undefended flower patches with different nectar production rates. Gill and Wolf assumed that the nonbreeding birds' goal was to collect sufficient nectar to meet their daily survival needs. (Why was this assumption reasonable?) They found that some sunbirds were territorial only when nectar production rates were higher in defended territories than in undefended areas. They also found that territory size contracted as the intrusion rate by nonterritorial birds increased; when the number of intruders was very high, the birds simply stopped defending their foraging area [452, 453]. Based on the data in Table 7.1, how many minutes of territorial defense would be worthwhile if a bird had access to a 2 microliters per bloom per day site while other flower patches had half that rate of nectar production? How many minutes of territorial defense would be worthwhile if a bird's choices were to defend a 3 microliters per bloom per day site versus foraging nonaggressively at another site where the blooms each produced 2 microliters of nectar per day.

Game Theory and Feeding Behavior

The preceding examples show how optimality theory can be used to develop mathematical models of foraging behavior that specify exactly what variables are believed to be important in shaping the behavior under investigation. Some-

TABLE 7.1   *Benefits of territoriality to golden-winged sunbirds under different conditions*

Nectar production (microliters/blossom/day)		1	2	3	4
Foraging time (hours) required to meet caloric need per day		8	4	2.7	2

Nectar production in			
Territory	**Undefended site**	**Hours of rest gained**	**Calories saved**[a]
2	1	8 − 4 = 4	2400
3	2	4 − 2.7 = 1.3	780
4	4	2 − 2 = 0	0

Source: Gill and Wolf [452]

[a]For each hour spent resting instead of foraging, a sunbird expends 400 calories instead of 1000, saving 600 calories. Total calories saved = 600 × hours of rest gained by not having to forage for nectar.

times models of this sort permit the production of quantitative predictions that can be checked against reality with real precision, enabling researchers to reach firm conclusions about the validity of a hypothesis.

Another similar tool for exploring the possible adaptive value of feeding behavior is game theory, which, as we outlined in the preceding chapter, is especially useful in cases in which individuals are in competition with one another for a valuable resource. This approach can be applied, for example, to cases in which two or more foraging techniques exist within the same species. Some examples of this phenomenon, including the slug-eating garter snakes and their tadpole- and fish-eating cousins, as well as the rovers and sitters found in populations of larval fruit flies [41, 308], were described in Chapter 3. Many other cases are known in which individuals of the same species specialize in different foraging techniques [1130]. Some roseate terns, for example, get their food exclusively by stealing prey from their fellow terns, which capture fish the honorable way by diving into the ocean after their prey [1089]. Whenever two or more distinctive foraging phenotypes are found within a species, the obvious question is, why hasn't the type associated with higher fitness replaced its rival over evolutionary time? The odds, for example, that thieves and hunters will secure exactly the same amount of food, on average, per unit of time spent foraging seem vanishingly small. If one type of tern did even slightly better on average than the other, the genes associated with that trait would spread and replace the alternative alleles linked to the less effective food-acquiring behavior.

With respect to the fruit fly and garter snake examples, it is possible—even likely—that both hereditary phenotypes are still present in their species primarily because of the ecological diversity present over the large geographic ranges of these animals. In some places, fruit fly larvae occur at high density, which favors the rover form; in other locations, larvae are scarce, which selects for the sitter phenotype. Likewise, the slug-eating snake phenotype and its alleles can prevail in areas where slugs are present, but not in places where slugs are absent. A much greater explanatory challenge is posed, however, when two hereditary phenotypes coexist within a species at exactly the same place and time. But the challenge can be met, as illustrated by the following example.

If thieving and hunting terns differ in their foraging behavior because of genetic differences between them (although this point has not yet been established), the two kinds of birds would be said to employ two different **strategies** [484]. In the lingo of game theory, evolved strategies are not consciously adopted game plans of the sort humans often employ, but rather behavioral traits that differ among individuals because of their hereditary makeup (like the rover and sitter phenotypes in fruit fly larvae). As we have seen, if one strategy confers higher fitness than the other, eventually only the superior strategy will persist. But under some special circumstances, two strategies can coexist indefinitely, thanks to the effects of what is called **frequency-dependent selection** (Figure 7.9). This kind of selection occurs when the fitness of one phenotype is a function of its frequency relative to the other phenotype. When the fitness of one type increases

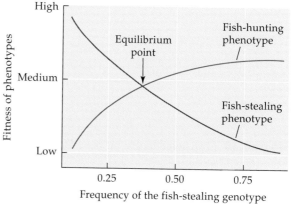

Figure 7.9 **Frequency-dependent selection** provides an answer to the question, how can two hereditary phenotypes coexist in the same population? When the fitness of each phenotype is dependent on the frequency of the other in the manner shown here, the two phenotypes will have equal fitness at some equilibrium point. In this hypothetical case, when the fish-stealing phenotype makes up about 35 percent of the population, both phenotypes have the same fitness on average. Change in their relative frequencies will result in frequency-dependent selection that will drive the two phenotypes back to the equilibrium point.

as that type becomes rarer, then that phenotype will eventually increase in frequency in the population—but only until such time as it has the same fitness as individuals playing the other strategy. Frequency-dependent selection will then act against either type if it becomes still more common, pushing the proportion of that form back toward the equilibrium point at which both types have equal fitness. When this happens, the two phenotype frequencies will be stable from generation to generation.

Discussion Question

7.7   Imagine a population of 1000 individuals in which there are two hereditarily distinct foraging phenotypes, thief and hunter. Imagine that there are 195 thieves and 805 hunters. Let's say that both thieves and hunters have 1.2 surviving offspring on average. What were the frequencies of the two phenotypes in the parental generation? What will be the frequencies in the generation composed of their offspring? What would happen if thieves have 1.1 surviving offspring on average, whereas hunters have 0.9? What's the point of this question?

Although no one has actually studied the roseate tern case from this perspective, it is known that the frequency of thieves in this species is very low relative to that of hunters, which makes sense, given that the thieving terns are totally dependent on their hunting counterparts to produce fish for them to steal. Another example of a case involving frequency-dependent selection, which has been studied with the aid of game theory, involves an African cichlid fish, *Perissodus microlepis*. This cichlid comes in two forms, one with the jaw twisted to the right and the other with the jaw twisted to the left. The fish makes a living, believe it or not, by darting in to snatch scales from the bodies of other fish in Lake Tanganyika. Individuals with the jaw twisted to the right always attack the prey's left flank, while the other phenotype goes in on the right (Figure 7.10). Right-jawed parents usually produce offspring that inherit their jaw shape and

Figure 7.10   Two hereditary forms of an African cichlid fish. Right-jawed and left-jawed cichlids have asymmetrical mouths that they employ to snatch scales from the left and right side, respectively, of the fish they prey on.

"Right-jawed" *Perissodus* attack prey from the left rear side

"Left-jawed" *Perissodus* attack prey from the right rear side behavior; ditto for left-jawed parents, indicating that the difference between the two forms is hereditary.

So why do both phenotypes occur in this species? Michio Hori proposed that the fishes this predator attacks could learn to expect a raid on their scales from the left if most attacks were directed at the prey's left flank. Thus, in a population in which right-jawed predators predominated, a lefty would have an advantage because its victims would be less vigilant with respect to scale-snatchers darting in on the right flank. This advantage would translate into higher reproductive success for the rarer phenotype and an increase in its frequency until left-jawed fish made up half the population. With a 50:50 split, the equilibrium point would have been reached.

If Hori's hypothesis was correct, he knew that he should find the frequency of any one phenotype oscillating around the equilibrium point. Checking his prediction required measurements of the relative frequencies of the two types over a decade (Figure 7.11). His data supported the frequency-dependent hypothesis for the maintenance of two foraging phenotypes in this species.

But not every example of multiple foraging phenotypes resembles the *Perissodus* case. Take the ruddy turnstone. This small sandpiper has many different ways of finding prey items on beaches, ranging from pushing strands of seaweed to one side, to turning stones over, to probing in mud and sand for little mollusks. Some individuals specialize in one tactic, while others prefer a different technique, but individual turnstones are rarely committed to just one foraging method. This observation suggests that the differences between them are not caused by genetic differences, but instead reflect an environmental difference of some sort [1283]. Philip Whitfield wondered if that environmental factor might be the dominance status of the foragers. Turnstones often hunt for food in small flocks, and the individuals in these groups form a pecking order. The birds at the top can displace subordinates merely by approaching them, thereby keeping them from exploiting the richer portions of their foraging areas. Dominants use their status to monopolize patches of seaweed on beaches, which they push about and flip over; subordinates keep their distance, and are often forced to probe into the sand or mud instead of feasting on the invertebrates contained within the seaweed litter.

The turnstones exhibit flexibility in their foraging behavior, as individuals are apparently capable of adopting feeding tactics in keeping with their ability to control different sources of food on the beach. The capacity to be flexible is provided by what game theoreticians have labeled a **conditional strategy**, which

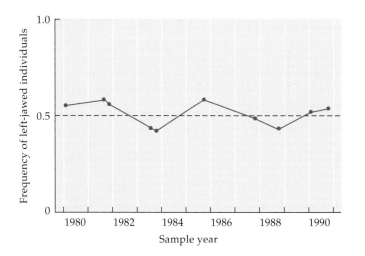

Figure 7.11   The results of frequency-dependent selection in *Perissodus microlepis*. The proportion of the left-jawed form in the population oscillates from slightly above to slightly below 0.5 because whenever it is more common than the alternative phenotype, it is selected against (and becomes less numerous); whenever it is rarer than the alternative phenotype, it is selected for (and becomes more numerous). After Hori [563].

is an inherited mechanism that gives the individual the ability to alter its behavior adaptively in light of the conditions it confronts (such as having high or low status in its social environment). Unlike the left- and right-jawed cichlids, which are locked into one particular phenotype, turnstones can almost certainly switch from one feeding phenotype to another. But subordinate turnstones tend to stick with the probing technique because if they were to challenge stronger rivals for seaweed patches, the low-ranking birds would probably lose, in which case they would have wasted their time and energy for nothing while also running the risk of being injured by an irate dominant. Instead, by knowing their place, these subordinates make the best of a bad situation and presumably secure more food than they would if they tried without success to explore seaweed patches being searched by their superiors. It can be adaptive to concede the better foraging spots to others if you are a turnstone with more powerful rivals.

We will have much more to say about conditional strategies in the chapters ahead, but here I stress again that either genetic or environmental differences can provide a proximate explanation for why alternative behavioral phenotypes occur within a population (see Chapter 3). Ultimate explanations based on game theory have explained this kind of behavioral variation as the product of frequency-dependent selection or as the result of selection for an adaptive flexibility that enables individuals to adjust their behavioral phenotype in response to variable environments.

More Darwinian Puzzles in Feeding Behavior

Both optimality theory and game theory employ a cost–benefit perspective, which we will now use again to solve some other puzzling features of animal feeding behavior. For example, a great many animals form dense nesting colonies or nighttime roosts that bring large numbers of individuals into close contact. If we think about these social aggregations in terms of fitness costs and benefits, it is often easier to see how being in groups could be more costly than beneficial, especially in terms of the rapid depletion of local food resources.

One possible benefit that might overcome the costs of being clumped together with others would be the ability of some individuals to take advantage of the food sources known to others. So, for example, the members of a colony of nesting seabirds might monitor the foraging success of returning neighbors and follow the successful hunters back to where they had been hunting. In this way, an animal that had been unsuccessful in its own attempts to detect food could find good foraging areas.

Discussion Question

7.8   Imagine that someone were to propose that nesting colonies of seabirds have formed to permit rapid information transfer about ephemeral food sources among their members, which in turn results in efficient collecting of food and maximization of the colonies' reproductive output. What theoretical objections would critical colleagues have about this person's hypothesis? What is the difference between this hypothesis and another based on game theory [67], which emphasizes the use of two tactics by the members of a colony: one a "producer" tactic, in which individuals seek out food patches, such as schools of fish, and the other a "scrounger" tactic, in which individuals exploit the searching successes of others by observing producers in action?

The idea that nesting colonies serve as information centers [1250] generates several predictions. One is that birds waiting in the colony should tend to follow successful hunters back out to a productive foraging area. But in an experimental study of a black-headed gull colony in which food platforms were set out at sea, the birds that found these sites were not followed by others on subsequent trips. Even when gulls landed by their nests with fish held conspicuously in their beaks, neighboring birds nesting nearby not only did not fly after the fish finder when it left to go on another foraging trip, but when these neighbors did head out to sea, they did not often fly in the same direction as their successful colonymates [32].

Although most attempts to test the information center hypothesis have yielded negative results [1017], Erick Greene's work on ospreys is an exception. These fish-eating hawks form loose nesting colonies in some coastal areas, and the birds do watch their colonymates to learn where to find fish species that appear here and there in large schools [473]. Ospreys are not only more likely to begin foraging shortly after a fellow osprey returns to the colony with a schooling alewife or smelt, but also tend to fly out in the direction taken by the successful forager. In addition, birds that see a successful hunter with prey are able to capture the same prey much more quickly than are ospreys that hunt without benefit of observing a successful forager (Figure 7.12).

Still more support for the information center hypothesis comes from John Marzluff and his colleagues, who watched young, nonterritorial ravens in Maine leaving the groves of trees where nonbreeding birds had spent the night. On morning after morning in the winter, dozens of ravens left one communal roost in the space of a few minutes, with most individuals headed in the same direction. To find out whether naive followers could gain information about the location of food, such as a deer or moose carcass, from other birds, the raven research

Figure 7.12   Do osprey nesting colonies serve as information centers? (A) When ospreys leave their nests at a time when no colonymate has returned with prey in the preceding 10 minutes, they fly off in all directions. But if another osprey has come back with an alewife (a schooling fish), the departing birds tend to fly off in the same direction (represented by an arrow) as the successful hunter. (B) Naive ospreys (those that have not seen a prey-carrying colonymate prior to departure) take much longer to find alewife schools than informed ospreys (those that have just observed a returning neighbor carrying an alewife in its talons). After Greene [473].

Raven

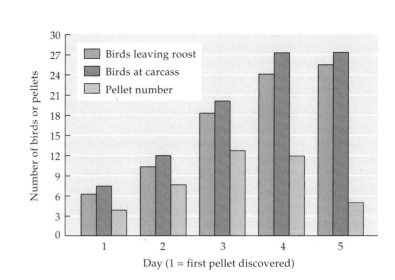

Day (1 = first pellet discovered)

- Birds leaving roost
- Birds at carcass
- Pellet number

Number of birds or pellets

Figure 7.13 **Roosting ravens follow informed foragers to a food bonanza.** After the discovery of a new food source (a sheep carcass), the number of regurgitated pellets marked with beads from the carcass increases at the roost. Likewise, the number of ravens leaving together from the marked-pellet portion of the roost increases, as does the number of ravens feeding on the dead sheep. After Wright et al. [1341].

team captured some ravens and held them for a number of days in an aviary. They then released some of the birds near roosts of ravens that had just found a dead moose (which had been generously donated by the researchers while the captured ravens were cooling their heels in the aviary). All 14 birds released at roosts in the evening followed their roostmates out the next morning to the bait. In contrast, only 4 of 15 captured ravens that were released away from roost sites found the food on their own [790].

Likewise, in an English raven roost, flocks sometimes flew off together in the morning in the direction of dead sheep baits that the experimenters had deposited in the countryside. These baits had been salted with marked beads, which the birds regurgitated later while ridding themselves of pellets of indigestible material. These pellets were collected, broken apart, and any beads tallied. As the number of birds in groups headed toward a carcass increased from day to day, so too did the number of beads recovered from pellets at the roost (Figure 7.13). These findings support the contention that English ravens, like North American ones, follow successful foragers to new resources as these are found by a lucky or skillful food finder [1341]. The osprey and raven studies suggest that there is life still in the information center hypothesis, even though the idea may have only limited application.

Why Do Some Spiders Make Their Webs Conspicuous?

Another Darwinian puzzle is provided by orb-weaving spiders that incorporate striking zigzag lines of white ultraviolet silk into their webs (Figure 7.14), which would seem to make the trap more obvious to the prey that the spider needs to catch. Given that many insects can see ultraviolet light, one wonders how orb weavers could gain by alerting potential victims to the danger posed by their webs. But perhaps these web

Figure 7.14 **Web ornament of an orb-weaving spider.** The large female spider has added three thick, conspicuous zigzagging lines of ultraviolet-reflecting silk to her web, which radiate out from her central resting point. Photograph by Marie Herberstein.

additions actually attract ultraviolet-detecting bees that specialize in taking nectar from ultraviolet-reflecting flowers. The fact that bees constitute the large majority of prey of some spiders with ornamented webs supports this hypothesis, as does the finding that in those webs in which only a single line of decorative silk has been laid down by the owner, insects are more likely to become trapped in the half of the web where the decorative silk resides than in the other half (Figure 7.15) [250]. Similarly, webs with more decorations capture more prey per hour than webs with fewer decorations [538], and isolated decorations cut from webs also attract prey [1252]. These lines of evidence suggest that what looks like a costly feature of the web actually yields a food benefit.

On the other hand, it could be that spiders that have been catching many prey, simply because their webs are in a good spot, are the spiders that tend to add decorations to their webs. If this is true, then the correlation between web decorations and higher rates of prey capture need not mean that the decorations *cause* the increase in prey capture. Instead, perhaps high feeding rates provide the extra energy that the spiders need to spin their web decorations, which they do for some purpose other than prey attraction. This hypothesis can be evaluated by testing the prediction that spiders given an abundance of food should invest more in web decorations than spiders deprived of food. Indeed, in two species of orb-weaving spiders, females that received more food made significantly larger decorations, and in one species, well-fed females were more likely to add decorations to their webs than were food-deprived individuals [108].

If one places an orb-weaving spider in a wooden frame and lets it build a web there, so that the web can be moved to a site chosen by the experimenter, then one can put two such webs side by side in a field. In this way, one can compare the rate of prey capture in an ornamented web with that in another web in the same spot whose decorations have been removed—something that can be easily accomplished by excising the two web lines that support the decoration (two web lines are also cut from the other web, but not those that hold the decoration). Such an experiment controls for the effects of site productivity. Under these conditions, the decorated web catches about a third less prey than the undecorated one. Thus, a foraging cost, not a benefit, applies to web decorations under some conditions [109].

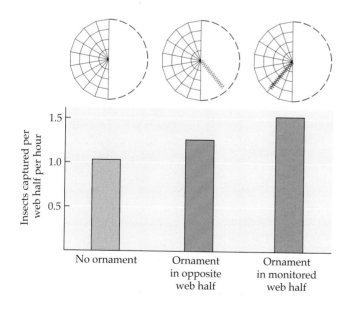

Figure 7.15 Do web ornaments lure prey? Garden spider webs without ultraviolet-reflecting ornaments capture fewer prey per hour than those containing ornaments. Furthermore, in webs with only one ornament, more flying insects are trapped in the half of the web containing the ornament than in the half lacking these structures. After Craig and Bernard [250].

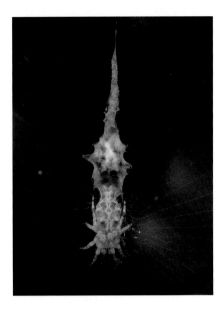

Figure 7.16   Some spiders appear to hide on the large silken egg sac in their webs. Because these spiders will replace experimentally removed egg sacs with a silky web ornament on which they perch, some ornaments seem to help conceal the web-building spider. Photograph by Bill Eberhard.

If it is true that decorating one's web carries a cost in lost calories, what counterbalancing benefit exists for this behavior? One possibility is that when passing birds see the bright decorations, they swerve to avoid colliding with the web, the better to keep from getting covered in sticky silk. If this is true, then in the experiment above, webs with decorations should have been less often damaged by birds than webs without any "keep away" signals. And in fact, the actual results from the experiment matched the predicted ones, with decorated webs damaged by birds 45 percent less often than undecorated ones [107].

Another benefit of web decorations might come from having the added silk camouflage the body of the web builder, making the spider less vulnerable to attack from its predators. Evidence in support of this explanation comes from a set of spiders whose females regularly incorporate a silky egg sac into their webs (Figure 7.16). The spider perches on this moderately conspicuous vertical tube of silk, eggs, and debris in ways that make her much less visible to human observers, especially because the spider's color pattern matches that of the egg sac so closely [356]. Moreover, when the egg sac is experimentally removed, the spider replaces it with a vertical strip composed entirely of silk, on which she perches as if it were an egg sac. Given that the egg sac seems to be used as a concealment aid, its replacement can be assumed to serve the same function.

More data on this point come from a study in which orb-weaving spiders that had built webs with or without extra silk decorations in an enclosure were exposed to wasps that hunt these animals to feed to their own offspring. In one experiment, only 32 percent of the spiders with web decorations were captured, as opposed to 68 percent of those without [110].

However, in yet another study, spiders with decorated webs were *more* likely to disappear than those with undecorated webs, suggesting that at least some predators may actually target the occupants of more conspicuous webs [538]. Praying mantises appear to do so. When given a choice of going down either of two arms of a Y-shaped maze, one with a decorated web and the other with an undecorated web at the end, the predators headed more often toward the web with ornaments (Figure 7.17) [165].

The work on web decorations offers a case history of a not uncommon phenomenon: a trait whose adaptive function is uncertain. Some findings seem to indicate that web decorations (in at least some spider webs) help attract prey,

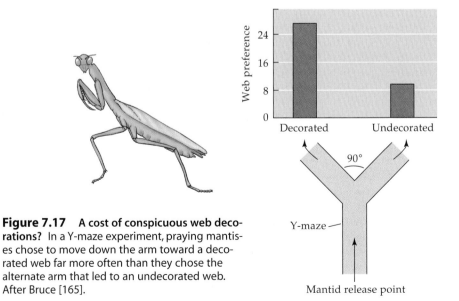

Figure 7.17   A cost of conspicuous web decorations? In a Y-maze experiment, praying mantises chose to move down the arm toward a decorated web far more often than they chose the alternate arm that led to an undecorated web. After Bruce [165].

but also attract predators. Other researchers have concluded that the decorations repel prey, but help protect the spider against predators. The lack of agreement may stem from the use of different spider species, different experimental procedures, or different spider prey and predators [1147]. The situation cries out for additional studies, and you can be sure that more papers on spider web decorations are on the way.

Why Do Humans Consume Alcohol, Spices, and Dirt?

One more Darwinian puzzle involves the apparently harmful or irrational food choices made by some people. For example, a great many persons in modern societies are addicted to alcohol, often at great fitness cost to themselves. The overconsumption of alcohol is clearly maladaptive. But is the trait an inexplicable artifact of human culture? Not according to Robert Dudley, who presents a historical hypothesis on the origins of alcoholism [347]. Dudley points out that since our closest relatives, the chimpanzees, derive most of their calories and nutrients from ripe fruits, it is likely that the ancestral species that gave rise eventually to modern chimpanzees and humans was also a frugivore. Fruit-eating chimpanzees, and primates in general, prefer ripe fruits because these fruits have the highest concentrations of sugars (Figure 7.18). Ripe fruits also happen to contain a certain amount of ethanol, which is volatile and so provides a potential olfactory cue to the location of high-payoff foods, all the more so because ethanol itself is a calorie-rich food. Thus, it seems likely that an ancestral fondness for ethanol would have been adaptive. However, the proximate mechanisms that facilitate alcohol consumption evolved in a world without six-packs of beer, cheap wine, and gin and tonics. Today these same mechanisms can be employed maladaptively in ways that would have been impossible even a few thousand years ago.

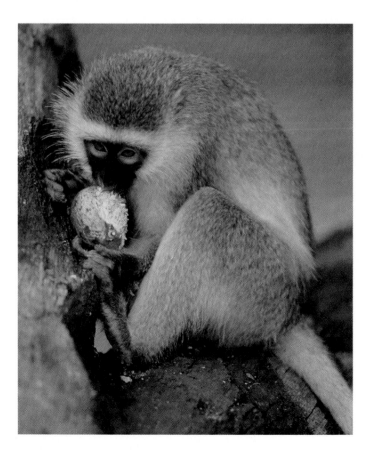

Figure 7.18 **A primate eating ripe fruit.** Photograph by Mary Beth Angelo.

226   Chapter 7

Or take the addition of spices to our food, a subject explored by Jennifer Billing and Paul Sherman [100]. These researchers were attracted to the topic by some apparent disadvantageous aspects of spice use, which include the high prices of many of the spices that humans add to their food. The Countess of Leicester, who lived in England in the 13th century, was willing to pay as much for a pound of cloves as she would for a cow [986]. Most of the major voyages of the early European explorers, Christopher Columbus included, were motivated by the desire for spices, as all North American schoolchildren learn. And yet the caloric and nutritive value of most spices is small, given that they are often used sparingly in any one dish. Perhaps, then, spice use is simply an arbitrary product of cultural invention—a plausible argument, since culinary traditions vary so greatly from culture to culture.

Billing and Sherman, however, proposed and tested the adaptationist hypothesis that spices served (and may continue to serve) a fitness-enhancing function by virtue of their antimicrobial properties. This hypothesis, needless to say, requires that spices kill the dangerous bacteria that sometimes contaminate our food, especially meats held under imperfect refrigeration. The microbiological literature provides evidence pertinent to this prediction (Figure 7.19). Moreover, the antimicrobial hypothesis also predicts that the extent to which spices are used should be a function not of their local cultivation, but of the risk of dangerous microbial contamination, which is related to local climate and the nature of the food being prepared for consumption. As expected, traditional recipes in hot, tropical countries call for more bactericidal spices than do recipes in, say, Norway or Sweden [100]. Moreover, in a sample of cookbooks from 36 countries, recipes devised for meat-containing dishes, which are especially vulnerable to microbial contamination, call for more antimicrobial spices than vegetable dishes, which are less likely to support dangerous bacteria and other pathogens [1097]. Thus, we can conclude that spices have been used adaptively

Figure 7.19 The antimicrobial properties of the major spices, most of which inhibit the growth of one-half or more of the bacterial types against which they have been tested. After Billing and Sherman [100].

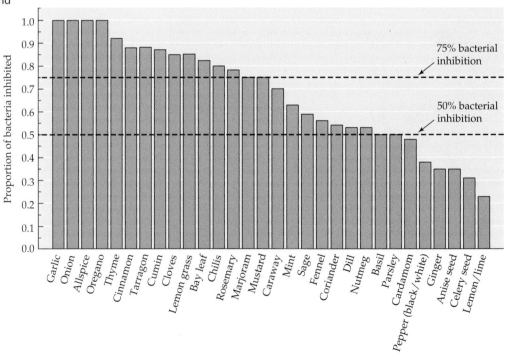

by humans even during the relatively short period of our evolution when these substances have been made widely available.

Finally, consider another oddity: the eating of dirt, a trait that is widespread among primates, including humans [325]. No one has ever claimed that the clayey soil consumed by chimpanzees or humans has any caloric value whatsoever. Among the hypotheses advanced to explain this habit in our species is the possibility that clay eating is a pathological, aberrant behavior of no functional significance. In contrast, an alternative adaptationist hypothesis proposes that clays are consumed to detoxify certain kinds of foods, thereby improving their nutritional value [602].

The pathology hypothesis predicts that only a relatively few, possibly somewhat deranged, individuals will eat clay. This prediction does not withstand scrutiny, since clay eating is, or was, common practice in a diverse set of cultures, including the Aymara of Andean Bolivia, the Hopi of the American Southwest, and the natives of Sardinia, an island in the Mediterranean Sea. What the people in these cultures have in common is a dietary reliance on bitter tannin-rich acorns or bitter alkaloid-laden varieties of potatoes. When potatoes are dipped in a clay slurry and then baked, or when acorns are baked into a clay-containing bread, the tannins and alkaloids in these foods are either bound to the clays or otherwise altered chemically, rendering the foods more palatable and less toxic [602].

As a comparative test of the detoxification hypothesis, we can predict that animals other than our closest relatives whose diets feature foods high in tannins or alkaloids will also seek out and eat certain clays. Rhesus monkeys and some other primates not especially closely related to humans do eat clay when feeding on tannin-rich vegetation [669]. In addition, two species of lemurs that live together in the same Malagasy rainforest, but differ in their reliance on alkaloid-containing seeds, also differ in their consumption of dirt in the predicted manner [974]. Finally, even nonprimates, such as African elephants, consume dirt when feeding on potentially toxic vegetation [567].

And it is not just mammals that ingest dirt from time to time. Gorgeous red-and-green macaws and other parrots regularly visit certain South American riverbanks to gnaw off and eat chunks of clay. The diet of these birds includes certain seeds, unripe fruits, and leaves that are high in alkaloid content (Figure 7.20) [862]. But perhaps the parrots feed on soils to secure gravel for their gizzards, or, alternatively, they could be eating dirt to acquire essential minerals absent in their vegetarian diets. Neither of these two hypotheses passes muster, given that the clays consumed by parrots at one famous source in the Peruvian jungle are composed of extremely fine particles, eliminating the gizzard grit hypothesis; moreover, the clays also contain very little in the way of minerals useful as dietary supplements [325]. What the special clays selected by the parrots do have is negatively charged cation exchange sites that bind with positively charged alkaloids and other toxic chemicals found in unripe fruit and certain seeds. The clays line the birds' digestive tract for hours, inactivating plant alkaloids and thereby protecting the cells of the gastrointestinal lining. When captive parrots were given a slug of an alkaloid with or without a dose of the clay they prefer, the subjects with clay-protected guts had 60 to 70 percent lower levels of the toxin in their blood when tested 3 hours later [451]. The fact that humans, lemurs, and macaws have converged on similar dietary solutions to similar ecological problems shows again how the right comparisons among species can provide useful evidence on adaptationist hypotheses.

Figure 7.20 **Clay eating has evolved in several species of parrots** that feed on foods rich in tannins or toxins, including these macaws, which regularly gather at this Amazonian river bank to collect and consume a particular kind of clay. Photograph by Frans Lanting.

Discussion Question

7.9 I suspect that most readers of this book consider eating dirt an unattractive proposition, but that you would be even less thrilled at the idea of cannibalizing a fellow human being. Nevertheless, try to be dispassionate about the matter, and develop a cost–benefit analysis of human cannibalism as if you were an adaptationist. You should be able to make predictions about the circumstances under which an evolutionary biologist would expect the behavior to occur. Then read Jared Diamond's article on the subject [326] and reconstruct his argument: identify the question he seeks to answer, and produce the alternative hypotheses, the predictions, the tests, and the conclusion.

The Adaptive Value and History of a Complex Behavior

The focus of this chapter up to this point has been on the challenge of developing and testing adaptive explanations for behavioral attributes whose function is not obvious. Here we will shift our focus to include the other component of ultimate analyses of behavior: tracing the evolutionary steps between the origin of a trait and its modern manifestation. A classic challenge in this regard is figuring out how the dances of honey bees evolved over time. These dances are

Figure 7.21 Round dance of honey bees. The dancer (the uppermost bee) is followed by three other workers, who may acquire information that a profitable food source is located within 50 meters of the hive. After von Frisch [1240].

performed by worker bees when they return to their hive after having found a good source of pollen or nectar [1240]. As the dancers move in circuits on the vertical surface of the honeycomb in the complete darkness of the hive, they attract other bees, which follow them around as they move through their dance.

Researchers watching dancing bees in special observation hives have learned that their dances contain a surprising amount of information about the location of a food source (such as a patch of flowers). If the bee executes a *round dance* (Figure 7.21), she has found food fairly close to the hive—say, within 50 meters of it. If, however, the worker performs a *waggle dance* (Figure 7.22), she has found a nectar or pollen source more than 50 meters away from the hive. By measuring either the number of times the bee runs through a complete waggle-dance circuit in 15 seconds or the

Figure 7.22 Waggle dance of honey bees. As a worker performs the waggle-run portion of the dance, she shakes her abdomen from side to side. The duration and the orientation of the waggle runs contain information about the distance and direction to a food source. In this illustration, workers attending to the dancer learn that food may be found by flying 20 degrees to the right of the sun when they leave the hive. (A) The directional component of the dance is most obvious when it is performed outside the hive on a horizontal surface in the sunlight, in which case the bee uses the sun's position in the sky to orient its waggle runs directly toward the food source. (B) On the comb, inside the dark hive, dances occur on vertical combs so that they are oriented with respect to gravity; the deviation of the waggle run from the vertical equals the deviation of the direction to the food source from a line between the hive and the sun.

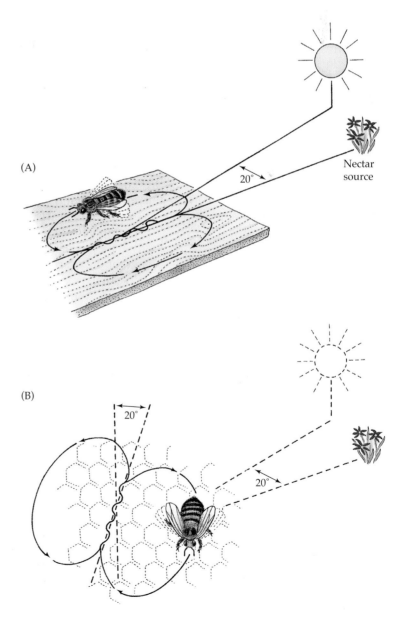

duration of the waggle-run portion of the circuit, a human observer can tell approximately how far away the food source is. The longer the waggle-run portion lasts, the more distant the food.

Moreover, by measuring the angle of the waggle run with respect to the vertical, an observer can also tell the *direction* to the food source. Apparently, a foraging bee on the way home from a distant but rewarding flower patch notes the angle between the flowers, hive, and sun. The bee transposes this angle onto the vertical surface of the comb when she performs the waggle-run portion of the waggle dance. If the bee walks directly up the comb while waggling, the flowers will be found by flying directly toward the sun. If the bee waggles straight down the comb, the flower patch is located directly away from the sun. A patch of flowers positioned 20° to the right of a line between the hive and the sun is advertised with waggle runs pointing 20° to the right of the vertical on the comb. In other words, when outside the hive, the bees' directional reference is the sun, whereas inside the hive, their reference is gravity.

The conclusion that the dances of bees contain information about the distance and direction to good foraging sites was reached by Karl von Frisch after several years of careful experimental work [1240]. His basic research protocol involved training bees (which he daubed with dots of paint for identification) to visit feeding stations, which he stocked with concentrated sugar solutions. By watching the dances of these trained bees, he saw that their behavior changed in highly predictable ways depending on the distance and direction to a feeder. More importantly, his dancing bees were able to direct other bees to a feeder they had found (Figure 7.23), leading him to believe that bees use the information in the dances of their hivemates to find good foraging sites.

Over the years, von Frisch's conclusions have been debated, with some scientists arguing that the workers recruited to a food source rely exclusively on the flower odors present on the dancing bee's body as a guide for their search, rather than deriving information from the dance movements per se [1270]. However, when foragers from a hive were trained to two feeders that were given the same scent, but were located in opposite directions from the hive, and only one feeder was advertised by waggle dances (because only that feeder contained a concentrated sugar solution), the bees recruited from the hive arrived primarily at the feeder advertised by the dancers. This kind of experiment offers support for the proposition that workers gain useful information from the movements of dancing bees [1080].

Figure 7.23   Testing directional and distance communication by honey bees. (A) A "fan" test to determine whether foragers can convey information about the direction to a food source they have found. After training scout bees to come to a feeding station at F, von Frisch collected all newcomers that arrived at seven feeding stations with equally attractive sugar water. Most new bees arrived at the feeder in line with F. (B) A test for distance communication. After training scouts to come to a feeding station 750 meters from the hive, von Frisch collected all newcomers arriving at feeders placed at various distances from the hive. In this experiment, 47 newcomers were captured at the two feeders closest to 750 meters, far more than were caught at any two other feeders. After von Frisch [1239].

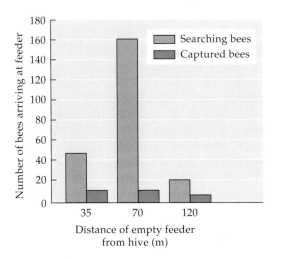

Figure 7.24 Honey bee recruits really do "read" the symbolic information in dances, as shown by their readiness to fly to empty feeders if misdirected by scouts that had been deceived into encoding the wrong information in their dances. The scouts had actually secured food from a site only 11 meters from the hive, but had been fooled into dancing as if the spot was much farther away. When workers attended to scouts whose dances announced (falsely) that food could be found 70 meters from the hive, more recruited bees (orange bars) showed up near an empty feeder 70 meters from the hive than at two feeders at other locations. Any recruit that actually landed on the feeder was collected (blue bars) to remove any individuals that might have been recruited to the site by foragers other than the experimental scout bees. After Esch et al. [371].

Additional evidence that workers attend to the dances of their hivemates comes from an experiment in which recruiter bees were trained to collect food at the end of an 8-meter-long tunnel set 3 meters away from the hive. When the recruiters advertised their find, they performed a waggle dance, not a round dance—which is surprising, given that the food source was only 11 meters distant, well within the range that would normally trigger a round dance. The reason for the recruiters' error has to do with the proximate mechanism by which bees determine the distance they have flown, which is based not on a direct calculation of the absolute distance flown, but instead on a function of the total amount of image motion their visual system has recorded as they fly to a foraging site. Because the researchers forced foragers to travel through a narrow tunnel, the amount of image motion recorded by the bees' retinas was much greater than if the scouts had been traveling in the open air with objects far away from them. Thus, when they came back to announce their discovery, they danced as if they had found food about 70 meters away, judging from the mean duration of the waggle component of their dances, which was about 350 milliseconds. When the experimenters put out three empty feeders at 35, 70, and 140 meters from the hive, recruited workers showed up primarily at the 70-meter site, even though it contained neither food nor any special floral scent (Figure 7.24) [371]. The fact that the recruits came to this site shows that they could "read" the dances of their hivemates that had been tricked into giving incorrect information about the location of a nectar source.

Discussion Question

7.10 Wolfgang Kirchner and Andreas Grasser evaluated the performance of honey bee recruits from a special hive that could be turned on its side or held upright in the standard position [642]. They found that when the hive was on its side, bees continued to dance in the darkness, but on a horizontal surface, not a vertical one. Under these conditions, recruitment at distant feeders (more than 100 meters from the hive) that had been visited by dancing bees was very poor. When, however, the hive was returned to an upright position and the comb surface on which recruiters danced was vertical (as it would be in natural hives), most recruits appeared at the feeders that the scouts had visited. How do you interpret these results? What bearing do they have on the argument about whether recruits derive information from the dances of their colonymates? What prediction can you make about relative rates of recruitment to sites less than 50 meters from the hive when it is turned on its side as opposed to when it is upright?

The Adaptive Value of Honey Bee Dances

Tom Seeley and Kirk Visscher have examined how the time and energy costs of dancing might be offset by fitness benefits to the queen whose daughters perform this behavior [1082]. Since worker honey bees are sterile females, their activities cannot promote their own reproductive success, but their behavior could help their reproductively capable relatives, especially their mother (see pp. 460–462). For example, if a hive's dancers could contribute to the rapid recruitment of a large forager workforce to a food patch, then the colony could harvest more resources there before other bee colonies and competitors arrived to deplete the food. If dancing has this effect, then the buildup of bees at a flower patch after a scout discovers and begins advertising it should be faster than if each bee had to discover the patch without guidance from colonymates.

To test this prediction, Seeley and Visscher moved a colony of bees to a barren island off the coast of Maine, along with a number of pots of flowering plants. They then shifted these portable flower patches from place to place on the island and measured the time it took scout bees to locate each one. Scouts came by at the rate of about one bee every 1 to 3 hours. But once a scout returned to her hive and danced there, recruited bees quickly flooded the food source. Even though any one recruit took an average of 2 hours to find a newly advertised food source, some of the many bees following the scout's dances arrived at the advertised site quickly, resulting in a fairly rapid buildup of recruits at a given site (Figure 7.25).

Nevertheless, this work does not establish definitively that it was the symbolic dances per se, and not the flower scent information provided by the dancers, that was responsible for the recruitment of workers to profitable food patches. A more direct way to examine the fitness consequences of dancing would be to weigh colonies during periods when they had access to dance recruitment information and when they did not. If dancing is adaptive, then colonies should gain more weight (through improved pollen and nectar collection) when accurate dance information is available. To test this prediction, Gavin Sherman and Kirk Visscher devised an experiment in which they used four hives outfitted with horizontal dance platforms only. Half the hives had their dance platforms illuminated by diffuse light, while the other half were outfitted with a unidirectional light source. Under the diffuse light conditions, scouts continued to dance, but their movements were disoriented, since they had no one reference point against which to orient their dances. Under the unidirec-

Time since flowers were set out (minutes)

Figure 7.25 **Rapid buildup of recruited foragers at experimental flower patches** after their discovery by scout bees. Researchers placed pots of borage flowers, treated with different scents, at three locations on an island. At each of the three locations, considerable time passed before a scout found the patch (arrow), but shortly thereafter, many additional bees arrived at the flowers. The green bars indicate the presence of the scout bee at the food patch. After Seeley and Visscher [1082].

(A)

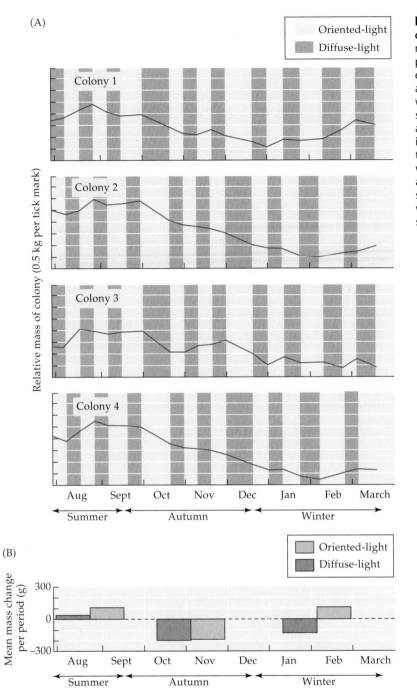

Figure 7.26   The adaptive value of the dance communication system. (A) Records of colony mass (the line graphs) plotted over alternating periods when the bees in each colony were exposed to diffuse light, and so did not have access to effective dance communication, and when they were exposed to a unidirectional light source, and so could communicate the distance and direction to a food source accurately by dancing. (B) A comparison of mean mass changes for the four colonies over the summer, fall, and winter when the bees did and did not have access to accurate dance information. The colonies made significant weight gains during the winter only when they could use their dance communication system. After Sherman and Visscher [1093].

tional light conditions, dancing recruiters used the light bulb as a substitute for the sun, so that their dances were oriented with respect to this cue, enabling recruits to derive useful information from their colonymates.

By alternating periods when the colonies were under diffuse-light versus unidirectional-light conditions, Sherman and Visscher could add up the gains or losses in mass during the two experimental treatments. The results (Figure 7.26) indicate that during the summer and fall, the occurrence of oriented dances did not have a statistically significant effect on colony weight. During the winter season, however, colonies gained mass during the oriented dancing periods and lost mass during the disoriented dancing periods [1093]. The ability of recruits to secure directional information about food location from others in the hive apparently has no effect during some times of year, but a positive, fitness-enhancing effect at other times. As Sherman and Visscher point out, if honey bee colonies are locked into a boom-or-bust economy, then the ability to take full advantage (via dancing-directed foraging) of brief periods of resource abundance may be highly advantageous.

An Evolutionary Scenario for Honey Bee Dances

Having described honey bee dances and shown that they are almost certainly adaptive, we can now try to figure out how such complex behavior could have originated and changed over time. Martin Lindauer was the first to speculate on the history of the dances [728]. He began by looking at three other members of the genus *Apis*, which he found to perform dance displays identical to those of the familiar honey bee (*Apis mellifera*), except that in one species, *A. florea*, the bees dance on the horizontal surface of a comb built in the open over a tree branch (Figure 7.27). To indicate the direction to a food source, a worker simply orients her waggle run directly at it. Because this is a less sophisticated maneuver than the transposed pointing done in the dark on vertical surfaces by *A. mellifera*, it may resemble a form of dance communication that preceded the dances of *A. mellifera*.

Lindauer then looked to stingless tropical bees not belonging to the genus *Apis* for recruitment behaviors that might provide hints about the steps preceding the first waggle dancing. Although debate exists on just how closely related stingless bees are to honey bees [878], different stingless bees employ different communication systems, which Lindauer organized into the following evolutionary hypothesis.

The possible first stage: Workers of some species of stingless bees in the genus *Trigona* run about excitedly, producing a high-pitched buzzing sound with their wings, when they return to the nest from flowers rich in nectar or pollen. This behavior arouses their hivemates, which detect the odor of the flowers on the dancer's body. With this information, they leave the nest and

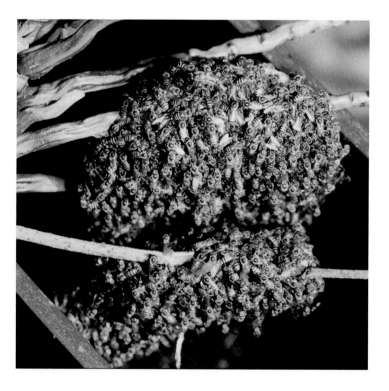

Figure 7.27   The nest of an Asian honey bee, *Apis florea*, is built out in the open around a branch. Dancing workers on the flat upper surface of the nest (two nests are shown here) can run directly toward the food source when performing the waggle dance. Photograph by Steve Buchmann.

Figure 7.28   Communication by scent marking in a stingless bee. In this species, workers that found food on the side of a pond opposite from their hive could not recruit new foragers to the site until Martin Lindauer strung a rope across the pond. Then the scouts placed scent marks on the vegetation hanging from the rope and quickly led others to their find. Photograph by Martin Lindauer.

search for similar odors. The actions of the dancing bee do not provide specific signals indicative of the direction or distance to the desirable food. The same kind of behavior also occurs in the bumblebees, which form small colonies with "dancing" scouts that do not provide signals correlated with the distance or direction to a food source [336].

A possible intermediate stage: Workers of other species of *Trigona* do convey information about the location of a food source. In these species, a worker that makes a substantial find marks the area with a pheromone produced by her mandibular glands. As the bee returns to the hive, she deposits pheromone on tufts of grass and rocks every few yards. Inside the hive entrance, other bees wait to be recruited. The successful forager crawls into the hive and produces buzzing sounds that stimulate her companions to leave the hive and follow the scent trail that she has made (Figure 7.28).

A still more complex pattern: A number of stingless bees in the genus *Melipona* convey separate distance and directional information. A dancing forager communicates information about the distance to a food source by producing pulses of sound; the longer the pulses, the farther away the food is. In order to transmit directional information, she leaves the nest with a number of followers and performs a short zigzag flight that is oriented toward the food source. The scout returns and repeats the flight a number of times before flying straight off to the food source, with the recruited bees in close pursuit.

Please realize that when, for example, we speak of a *Trigona* bee exhibiting an "intermediate stage" in the evolution of communication about food location, we are not implying that this species has somehow failed to move along toward a more complex, more adaptive "final stage." For that *Trigona* bee in its current environment, trail marking may well be superior to any other option. The existence of trail marking in this modern species simply provides a clue about

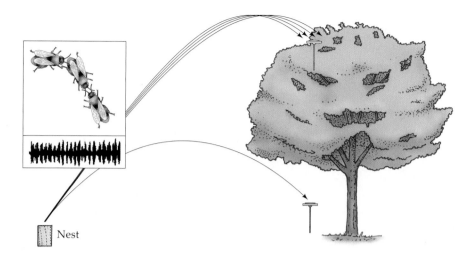

Figure 7.29 Acoustical communication of the height of a food source by the bee *Melipona panamica*. When a scout that has been trained to collect food from a feeder high in a tree interacts with her nestmates (left panel), she produces a specific kind of acoustical signal (shown as a black sonogram in the panel beneath the bees) while unloading the food. After listening to these sounds, recruits are much more likely to locate the training feeder than a control feeder placed at the base of the tree. After Nieh [875].

Nest

the possible behavior of a now extinct bee whose trail marking behavior has been modified in some lines derived from this ancestral species.

That all modern bee species exhibit communication systems well suited for their particular environment is a message supported by studies of tropical bees done since Lindauer's pioneering work. For some of these bees, the ability to communicate whether a food source is high in the canopy of the forest or lower down is ecologically highly relevant, and some species have evolved the requisite signals. Thus, the bee *Melipona panamica* produces one kind of buzzy acoustical signal when unloading food in a funnel-like entryway to the nest and then generates another kind of signal by dancing in the entryway. The sounds made while unloading food inform recruits where the resource is in the vertical plane (Figure 7.29) [875]. The dance conveys information about the *distance* to the food source advertised by the dancer. Finally, in order to convey information about the *direction* to the food source, the scout leaves the nest to guide recruits in the right direction. Only recruits that have access to all three sources of information about the location of a flower patch will get close enough to detect the scent marks deposited at the site by the scout before she returned to the hive to buzz, dance, and lead [874].

Although, judging from *M. panamica*, some stingless bees have communication systems every bit as complicated as that of the honey bee, not all do. The diversity of communication systems among stingless bees suggests that communication about *distance* to a food source by an ancestor of the honey bee probably at first involved only agitated movements by a food-laden worker [876]. Other workers that were stimulated by the activity of the returning forager would then have left the hive to find food, perhaps aided by memory of the odors associated with the food. In some species, selection may have subsequently favored standardization of the sounds and movements made by successful foragers, as in *Melipona*. These actions may have set the stage for further changes that have been incorporated into the round and waggle dances of *Apis* bees, which contain symbolic information about how far food is from the hive [728, 1314].

In contrast, communication about the *direction* to a food source appears to have originated with personal leading, with a worker guiding a group of recruits directly to a nectar-rich area. Here the evolutionary sequence has involved less and less complete performance of the guiding movements as generations of queens produced workers with a greater and greater tendency to perform incomplete leading. At first this may have taken the form of partial leading (as

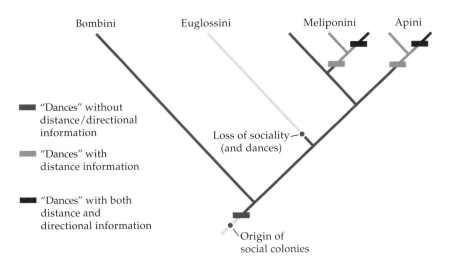

Bombini   Euglossini   Meliponini   Apini

■ "Dances" without distance/directional information

▨ "Dances" with distance information

■ "Dances" with both distance and directional information

Loss of sociality (and dances)

Origin of social colonies

Figure 7.30 Evolutionary history of the honey bee dance communication system. Because bees in three of four closely related groups are social, the ancestor of all four groups may well have been social. If this ancestor evolved a simple form of "dancing," the trait may have been lost in the Euglossini along with the capacity for sociality, since euglossine bees do not form colonies with queens and workers. The retention of the dances in the Meliponini and Apini set the stage for evolutionarily more recent elaborations of this behavior, enabling dancers in some species to convey information about the distance to a food source and then both distance and directional information. The honey bee (*Apis mellifera*) possesses yet another modification to its dances, which encode symbolic, gravity-based information about the direction to a food source.

in some *Melipona*) and later involved simply pointing in the proper direction with a waggle run on a horizontal surface (as in *A. florea*). From antecedents like this came the transposed pointing of *A. mellifera*, in which the direction of flight relative to the sun is converted into a signal (the waggle run) oriented relative to gravity. This evolutionary sequence of events can be plotted on a phylogeny of four closely related groups of bees, which include the honey bees (Figure 7.30).

Discussion Question

7.11 Phylogenetic diagrams are widely used to develop historical scenarios for behaviors of interest. Use the information contained in Figure 7.31 to produce a hypothesis on the possible evolutionary history of foraging modes (visual ambushing versus olfactory searching) of lizards. What is the minimum number of times that olfactory foraging has evolved among these lizards?

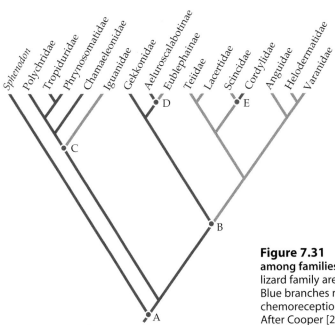

Figure 7.31 A phylogenetic diagram of the evolutionary relationships among families of lizards. Orange branches indicate that the members of this lizard family are olfactory foragers, which flick their tongues to detect prey odors. Blue branches represent ambush predators, which employ vision rather than chemoreception to find their meals. Extinct ancestral species are labeled A to E. After Cooper [239].

Summary

1. Optimality theory and game theory have both played a role in the study of the possible adaptive value of feeding behavior. Optimality approaches often focus on the net caloric gains associated with a feeding decision. Game theoretical approaches consider the payoffs for individuals competing with others for valuable resources.

2. Optimality theory has attracted some controversy, largely on the grounds that some common assumptions underlying optimality hypotheses are unlikely to always be true, such as the notion that maximizing the rate of caloric intake will also maximize fitness. There is no question, for example, that in many species individuals sacrifice energy maximization in order to reduce the risk of predatory attack. Tests of optimality hypotheses, however, are designed to help researchers identify the factors involved in the evolution of animal behavior. Hypotheses with incorrect assumptions will be rejected if properly tested.

3. In the category of feeding behavior, Darwinian puzzles that have attracted adaptationist researchers include such things as the clumping of competitors for limited food resources in small areas, the addition of highly visible ornamental silk to the orb webs of spiders, and the consumption of dirt and other non-nutritive materials. All these actions would seem to have negative fitness consequences and therefore require a demonstration of special benefits if they are to be considered adaptations.

4. Ultimate studies of feeding behavior are not limited to research into the adaptive properties associated with getting enough to eat. In addition, evolutionary biologists have attempted to outline possible scenarios for the origin and subsequent modification of behaviors that have resulted in the complex attributes of living things, such as the astonishing dances of honey bees. Comparisons among living species can provide important clues in the development of these kinds of historical hypotheses.

Suggested Reading

For a general review of predator decisions, see John Krebs and Alejandro Kacelnik's book chapter [668]. Their review also covers some mathematical models based on optimality theory, an approach that has often been applied to foraging behavior. The mathematics of optimality modeling are presented in detail by Dennis Lendrem [709] and by Marc Mangel and Colin Clark [772]. Reto Zach has written a paper on optimal foraging in crows that is a model of clarity [1351]. For a critique of optimality models, see a paper by C. J. Pierce and J. C. Ollason [950], which can be contrasted with Krebs and Kacelnik's chapter [668]. Bernd Heinrich's *Bumblebee Economics* [530] is a good companion to this chapter, as it deals clearly and simply with optimality theory as applied to bumblebees. For a game theoretical approach to foraging, see [454]. To read more on the amazing lives of honey bees, see Tom Seeley's *The Wisdom of the Hive* [1080]. Philip Starks provides a very useful review of the puzzles posed by conflicting conclusions on the function of web decorations [1147].

Choosing Where to Live

During the years when I was a teenager, my father and I joined some friends in devoting one Saturday in May to bird-watching near our home in southeastern Pennsylvania. We usually chalked up about a hundred species, including yellow warblers singing high in the sycamore trees by White Clay Creek, blue-winged warblers flitting among the little trees in shrubby, overgrown farm fields, and common yellow-throats (another species of warbler) in the marshy spots. The male warblers calling loudly from their perches not only had made the effort to find a very specific habitat, but were also prepared to defend their turf against others of their species, as they announced over and over again (see Chapter 2). In addition to the costly investments each bird had made in selecting a particular habitat and defending a territory, the warblers that my father and I saw had traveled hundreds or even thousands of miles from wintering grounds as far away as South America in order to set up a breeding territory in our neighborhood. In a few months, if they survived, they would head south again on a great migration, which their offspring would undertake as well.

Each of the decisions that these little songbirds make about where to live comes with obvious and substantial costs, which is why these decisions are interesting to evolutionary biologists. Why should a yellow warbler

◀ **Monarch butterflies** *cluster in winter in only a handful of locations in Mexico and California, an example of highly refined habitat selection. Photograph by Frans Lanting* refuse to nest in a cattail marsh? Why should a common yellowthroat spend hours each day singing to keep rival males informed of his willingness to fight for space in the same marsh that yellow warblers avoid? And why should a blue-winged warbler born in Landenberg, Pennsylvania, fly all the way to Honduras, only to turn around in a few months and fly all the way back again? This chapter takes a look at Darwinian puzzles of this sort.

Habitat Selection

The rule that certain species live in particular places applies to all groups of animals, not just warblers, perhaps because in so many cases, the opportunities for successful reproduction are so much better in habitat A than in habitat B for members of species X. The importance of access to appropriate habitat has been dramatically illustrated by the link between habitat destruction and the declining populations of certain animals. Loss of isolated people- and dog-free beaches has, for example, endangered the western snowy plover and the least tern, both of which nest on open beaches and small sandy islands. If the populations of these birds are falling because of habitat loss, then creation of artificial beaches in areas inaccessible to people and their dogs ought to attract breeding plovers and terns, helping local populations rebound. This experiment has been done in Batiquitos Lagoon in southern California and elsewhere. When material dredged off the sea bottom was used to produce artificial beaches and sandbars, both plovers and terns occupied the sites (Figure 8.1) and reproduced successfully [771, 971].

Habitat restoration may help rebuild snowy plover and least tern populations. Habitat *management* is another conservation tool based on the premise that particular species are matched to particular environments. The Florida scrub jay, another endangered species, lives only in sandy, scrubby areas of Florida. Once this habitat burned irregularly but naturally as a result of lightning strikes, creating a mosaic of scrub oak woodland with open sandy patches surrounded by other plant associations. The scrub jays prefer to nest in substantial areas of open oak habitat, and their nesting success is highest there [140]. When natural wildfires are quickly put out, however, the scrub oak thickets eventually become tall and dense, egg-eating blue jays move in, and scrub jays lose out [1338]. Any management scheme devised to increase the Florida scrub jay population will surely have to employ fire as a tool to shape the habitat to match the preferences and needs of scrub jays.

Given the importance of being able to breed in specific habitat types, we would expect animals to have evolved strong preferences for some places over others, even if they are capable of reproducing in a variety of environments. The European great tit, for example, can nest either in mixed woodland or in hedgerows. But the birds prefer mixed woodland to hedgerows, as demonstrated by the shifts made by hedgerow birds into woodland sites upon the experimental removal of breeding pairs from the favored habitat [666].

If habitat preferences are adaptive, then individuals that are able to fulfill their preferences ought to leave more descendants than those unable to acquire prime real estate. This proposition is true for great tits [666]. It is also consistent with the finding that in many species, some individuals occupy *source habitats* (where the population grows), while others are relegated to *sink habitats* (where the population declines). Poor-quality sink habitats are generally utilized by competitors who are unable to insert themselves into superior source habitats, often because they are excluded by older, more accomplished opponents [327] and so must make the best of a bad situation elsewhere.

Figure 8.1   Habitat selection and conservation.
Knowledge of the habitat preferences of nesting least terns has enabled conservation biologists to create suitable environments for this endangered species. (Left) A least tern on its nest in the open, sandy beach habitat the bird requires if it is to breed successfully. (Right) This large sandy island, built with dredged materials, has attracted many pairs of nesting seabirds, including the least tern. Left, photograph by T. Vezo; right, photograph by Troy Mallach.

The predicted relationship between habitat preference and reproductive success does not always hold, however. In the Czech Republic, for example, the blackcap warbler has a choice between stream-edge deciduous forests and mixed coniferous woodlots away from water. The bird prefers the stream-edge habitat, which attracts the initial colonists in spring. Yet the reproductive success of breeding pairs in the two environments is essentially the same [1263]. Why? One answer has been provided by Steve Fretwell and his colleagues, who used game theory to predict what animals should do when faced with a choice between alternative habitats of different quality and different levels of competition. They demonstrated mathematically that as the density of resource consumers in the superior habitat increased, there would come a point at which an individual could gain higher fitness by settling for a lower-ranked habitat that had fewer settlers of the same species and thus less competition for critical resources [411]. In fact, the Czech ornithologist Karel Weidinger found that the density of nesting blackcaps was four times higher in the preferred stream-edge forests than in the second-ranked habitat. Thus, these birds apparently make habitat selection decisions based not just on the nature of the vegetation and other markers of insect productivity, but also on the intensity of the competition from others of their species.

This same ability is possessed by male side-blotched lizards, a species for which the value of a territory is a function of the number of rocks it contains, apparently because an abundance of rocks attracts females that use this resource to control their body temperature by basking or seeking cover. In their experimental study of territoriality, Ryan Calsbeek and Barry Sinervo waited until the male lizards had subdivided the prime habitat into a patchwork of territories; they removed rocks from some territories and added them to others. They predicted that males with rockier territories should draw in more females *and*

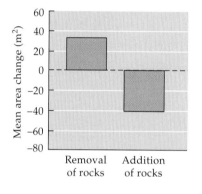

Figure 8.2   Change in territory size in relation to an experimental change in territory quality in the side-blotched lizard. Male lizards expanded their territories when valuable rocks were removed from these sites, whereas males that received the extra rocks contracted their territorial borders. As a result of expanding their territories, males on rock-depleted sites were able to acquire about one mate on average, which meant they did as well as males on improved territories. After Calsbeek and Sinervo [183]; photograph by Paul Hamilton.

more male competitors. In contrast, males that had lost valuable rocks should expand the size of their now less valuable territories in order to compensate for the lower density of rock perches. The males with the more hotly contested territories contracted the area they defended, while the males whose territories had lost value increased the size of their territories (Figure 8.2). The end result was that males in both categories wound up with about one female on average, showing that evenly matched competitors can adjust their territory size in ways that enable them to do as well as their rivals.

The capacity to evaluate habitat quality is also highly developed in the honey bee, a point that becomes evident when a colony splits in two. One half of the colony, which contains the old queen and half her worker force, fly off in a swarm, leaving the old hive and the remaining workers to a daughter queen. The departing swarm settles temporarily in a tree, where the workers hang from a limb in a mass around their queen (Figure 8.3). Over the next few days, scout workers search for chambers in the ground, in cliffs, or in hollow trees. Of the sites within range of the waiting swarm, only those with a volume of 30 to 60 liters cause returning scouts to perform a dance back at the swarm [1079], which communicates information about the location of the potential new home (see pp. 228–230). Other workers around a dancing scout may be sufficiently stimulated to fly out to the spot themselves. If it is attractive to them, they too will dance upon their return and send still more workers to the area.

One highly significant feature of the dancing behavior of bees scouting for a nest site is that they typically make only a few trips out to a potential site, and each time they return, they dance for shorter periods before giving up altogether (Figure 8.4) [1236]. As a result, if any one site is to build up a population of adver-

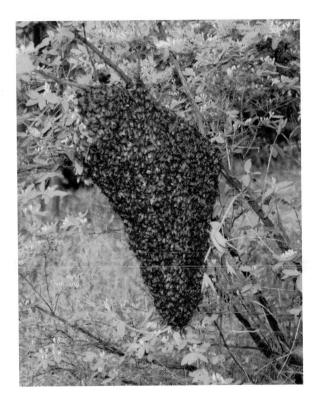

Figure 8.3 **Finding a new home.** A swarm of honey bees waits while its scouts go off to search for a new hive site. Photograph by Kirk Visscher.

tisers, it must be so attractive that the rate of recruitment of new scouts exceeds the dropout rate of old scouts. Sites that are capable of inspiring energetic new recruiters will generate an exponentially increasing population of scouts, while those that are less appealing will produce an ever-decreasing, and eventually nonexistent, population of advertisers [864]. As a result, at some point, many or all of the active recruiters will be advertising the same location, leading to the presence of many scouts investigating this potential new homesite (Figure 8.5) [1081]. When a new scout encounters several dozen other workers at a site, she may return to the swarm and begin "piping," which is to say that she produces a vibrational signal that others in the swarm can sense. If enough scouts switch to the piping mode, all of their sisters in the swarm may get the message, which stimulates them to begin shivering their wing muscles. The muscle contractions raise the bees' internal temperature to the level necessary for flight. When the bees are ready, they take off, and the airborne swarm heads to the selected nest site, utilizing the information they have received from the swarm scouts [1083].

What is especially interesting about this process from an optimality perspective is that when some bees were experimentally offered two identical high-quality nest sites at different distances from the home hive—say, 50 and 200 meters away—the bees chose the more distant of the two [728]. One might think that they would prefer the closer spot, since covering even a few hundred meters can exhaust their queen, a prodigious egg layer but a poor flier. However, because the value of a site is also affected by its proximity to other colonies, a dispersing swarm that moves a considerable distance probably reduces the likelihood that mother and daughter colonies will compete for the same flower patches.

We can test this hypothesis by predicting that a swarm's readiness to fly the extra distance should be correlated with the intensity of the potential competition between mother and daughter hives. As it turns out, northern European

Figure 8.4 **Different patterns of dancing behavior** exhibited by nest-site scouts and foraging workers. Scouts perform dances with many waggle runs after their first few trips home from an attractive, potential new home for their colony, but then their dance activity falls off sharply. In contrast, the more trips a forager makes to and from a good nectar- or pollen-producing patch of flowers, the more waggle runs she is likely to incorporate into her recruitment dances. After Beering [82] and Visscher [1236].

Figure 8.5 Reaching a decision about where to live. Changes occurred over a 3-day period in the number of scouts advertising the locations of potential nest sites for their swarm (whose location is represented with a circle). The width of the arrows radiating from the circles is proportional to the number of different bees dancing for that potential nest site during the period shown at the top of each panel. The swarm left for its future home in the morning of 21 June. After Seeley and Buhrman [1081].

bee colonies are much larger, and thus more likely to compete for food, than those in southern Europe, a difference caused by the importance of having a large number of bees as a living insulation blanket for the queen and her core workers during cold winters [593, 594]. As expected, large honey bee colonies in cold Germany move farther when finding new hive sites than do the much smaller swarms of bees that live in warmer Italy [459].

Discussion Question

8.1 The pattern of dancing by nest-site scouts returning to a swarm differs markedly from the pattern exhibited by foragers returning to a hive (see Figure 8.4). The fact that foragers continue to recruit over many trips and gradually increase, rather than decrease, the number of waggle runs incorporated into their recruitment dances means that instead of having the entire population focusing on one foraging patch, several patches are likely to be exploited simultaneously. Does this fact suggest why the behavior of nest-site scouts and foraging recruiters differs? Someone might say that we do not need evolutionary explanations for the collective decisions made by the foragers in a hive or the bees in a swarm because these colony-wide decisions are the inevitable consequence of mindless behavioral rules used by individual members of the group. In other words, as is sometimes claimed, groups possess self-organizing properties that stem from the simple behaviors of their members, and knowledge of the "emergent properties" of groups negates the need for other explanations of their social activities. How would you respond to this assertion?

Habitat Preferences of a Territorial Aphid

Honey bee colonies do not attempt to defend a foraging area as their exclusive feeding preserve, and therefore dispersing swarms can settle wherever they choose without direct interference from other colonies. For other species, such as territorial yellow warblers and side-blotched lizards, a truly free choice among habitats is not possible because some individuals can exclude others from superior sites. Tom Whitham examined the effects of this kind of aggressive competition on habitat selection in a tiny insect, the poplar aphid [1284–1286].

Each spring in Utah, vast numbers of aphid eggs hatch in crevices in the bark of cottonwood poplar trees. The emerging aphids walk to leaf buds on cottonwood branches. Each female—and there may be tens of thousands per tree—actively selects a leaf, settles by its midrib, almost always near the base, and in some way induces the formation of a hollow ball of tissue—a gall—in which she will live with the offspring she bears parthenogenetically (Figure 8.6). When her daughters are mature, the gall splits, and the aphids disperse to new plants.

Whitham found that newly emerged females quickly occupied all the very large leaves on poplars, but that since there were about 20 aphids for every large leaf on the trees he studied, these females regularly encountered competitors, with whom they fought for as long as 2 days (Figure 8.7), sometimes dying in the attempt to secure a territory [1287]. Given the costly nature of these fights, we would expect that major benefits come from possessing large leaves—and they do (Figure 8.8).

Defeated females and small ones incapable of effective fighting are forced to accept inferior habitats. If the defeated individuals are able to make the best of a bad situation, they will settle on the best of the lower-value habitats available to them. The options for these individuals include finding a smaller, unoccupied leaf or settling with an established territorial foundress on a large leaf. If a leaf already has a resident female, the latecomer will have to form her gall farther out on the midrib, where it will provide fewer nutrients than the gall at the prime location near the leaf petiole. If the other aphid is already enclosed in a gall, the newcomer will not have to fight, but she will have to settle for a reduc-

Figure 8.6   Poplar galls are occupied by poplar aphids. (Left) A poplar aphid gall at the base of a leaf. (Right) A scanning electron micrograph of a female aphid inside her gall. Courtesy of Tom Whitham.

Figure 8.7   Territorial dispute between two poplar aphids. Females may spend hours kicking each other to determine who gets to occupy a preferred leaf or the superior location on a leaf. Courtesy of Tom Whitham.

tion in fitness. However, the second colonist on a medium-sized leaf can do just as well as a single aphid on a small leaf (Table 8.1). As predicted, when aphid females do double up, they choose medium and large leaves, despite their scarcity [1286].

Discussion Question

8.2   In American robins, habitat selection is affected by experience. In particular, birds whose nesting attempts have failed in one year often do not return to the same spot to try again in the subsequent year. Several different hypotheses can account for this behavior. Deduce what some of these hypotheses might be, based on the following data from an experiment in which researchers destroyed the nests of a randomly selected subsample of nesting robins and then compared their return rate in the next year with that of other robins whose uninterrupted nesting attempts had resulted in the production of fledglings: Of the subsample of robins with experimentally induced nest failures, only 18 percent came back, whereas 44 percent of the successful nesters returned.

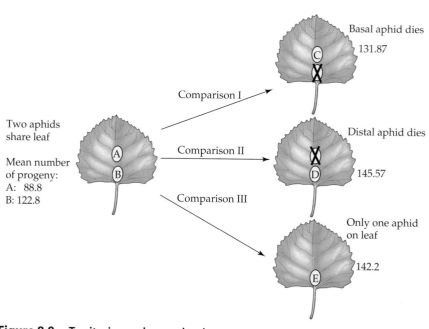

Figure 8.8   Territories and reproductive success. The average number of progeny produced by aphids that succeed in monopolizing a poplar leaf versus those that are forced to share a leaf of the same size with a rival. After Whitham [1286].

TABLE 8.1   *Effect of leaf size and position of gall on the reproductive success of female poplar aphids*

Number of galls per leaf	Mean leaf size (cm)	Mean number of progeny produced		
		Basal female	Second female	Third female
1	10.2	80		
2	12.3	95	74	
3	14.6	138	75	29

Source: Whitham [1286]

The Costs and Benefits of Dispersal

Once a poplar aphid finds a leaf to live on, she will spend the rest of her life there. But for other species, such as the honey bee, dispersal from one home base to another is a regular occurrence. In order to move from point A to point B, animals must expend calories not only while moving, but even before dispersal, when they invest in the development of locomotory muscles. Consider, for example, that if a cricket is to leave a deteriorating environment and move to a new and better place, it will need large flight muscles to fly away. The calories and materials that go into flight muscle development and maintenance presumably have to come out of the general energy budget of the animal. This means that other organ systems, such as a female's ovaries, cannot develop as rapidly as they could otherwise, which imposes a fitness cost on the flight-capable individual.

This trade-off hypothesis has been tested ingeniously in the following way. In some cricket species, two phenotypes occur, one that has the requisite muscular machinery for flying and another form with small wing muscles and low lipid (fuel) reserves that cannot get off the ground. Females that can fly do not start producing eggs as quickly as flightless females, as one would predict from the trade-off argument [1360]. Researchers can also create flightless females in a cricket species that normally has only flight-capable individuals by dabbing young adults with a chemical similar to juvenile hormone (see Chapter 3). If it is true that the ability to fly comes at a reproductive price, then the experimentally produced flightless females should be able to invest more in their ovaries than control females not exposed to the hormone. Indeed, ovarian development occurs much more rapidly in the experimental females, which feed no more or less than the controls, but whose flight muscles deteriorate while resources are directed to their ovaries (Figure 8.9) [1360].

Dispersing individuals not only have to pay energetic developmental and travel costs, but are also more often exposed to predators (e.g., [584])—all of which raises the question, why are animals so often willing to leave home even when this means leaving a familiar, resource-rich location? This question is particularly pertinent for species in which some individuals disperse

Figure 8.9   The trade-off between development of locomotory muscles and reproductive equipment. Female crickets treated with juvenile hormone, which results in flightless individuals with reduced wing muscle mass, followed different developmental curves than control (flight-capable) females. Note especially the difference in the growth rates of the ovaries for the two categories of females. After Zera et al. [1360].

Belding's ground squirrel

Figure 8.10   Distances dispersed by male and female Belding's ground squirrels. Males go much farther on average from their natal burrows than females. After Holekamp [550].

while others do not, or at least do not travel as far. In the fire ant, for example, when a new generation of queens is produced, some fly away from their colony, mate with males at distant sites, and go on to seek out a new location where they will burrow into the ground and try to found a nest on their own. Few females will succeed, but many try, especially those with a particular geno-type that we will label *BB*. In contrast, queens with the genotype *Bb* leave the colony to mate, but apparently often return home, attempting to rejoin their natal colony to become an egg-laying queen there [311].

Thus, one proximate basis for the behavioral difference between the two kinds of queens has to do with their genetic makeup. An ultimate reason for this behavioral difference is linked to what happens to *BB* queens if they try to join a colony where several queens are already present. Many of the workers in a colony will have the *Bb* genotype, and they systematically dismember any new queen that lacks the *b* allele [624]. Under these circumstances, it is not surprising that *BB* queens generally avoid established colonies and do the best they can to found colonies of their own.

Another species in which some individuals disperse farther than others is Belding's ground squirrel. Young male squirrels travel about 150 meters from the safety of their mother's burrow, whereas young females usually settle down only 50 meters or so from the burrow in which they were born (Figure 8.10) [550]. Why should young male Belding's ground squirrels go farther than their sisters?

According to one argument, dispersal by juvenile animals of many species may be an adaptation against **inbreeding depression** [989]. When two closely related individuals mate, the offspring they produce are more likely to carry damaging recessive alleles in double doses than are offspring produced by unre-lated pairs. The risk of associated genetic problems should in theory reduce the average fitness of inbred offspring, and high juvenile mortality does indeed occur in inbred populations of many animals [1000]. Thus, when inbred and non-inbred white-footed mice were experimentally released into a field from which their ancestors had been captured, the inbred mice survived only about half as well as their outbred companions [598]. And even if inbred mice man-age to reach adulthood, they may be less likely to reproduce than outbred indi-viduals (Figure 8.11) [780].

If avoidance of inbreeding is the point of dispersing, however, then one might expect as many female ground squirrels as males to travel 150 meters from their natal burrow. But they do not, perhaps because the costs and benefits of disper-

Oldfield mouse

Figure 8.11   **Inbreeding depression in oldfield mice** may result in the failure of inbred females to reproduce as early as outbred females. After Margulis and Altmann [780].

sal differ for the two sexes. As Paul Greenwood has suggested for mammals generally [477], female Belding's ground squirrels may remain at or near their natal territories because their reproductive success depends on possession of a territory in which to rear their young. Female ground squirrels that remain near their birthplace enjoy assistance from their mothers in the defense of their burrows against rival females. Thus, the benefits of remaining on familiar ground are greater for females than for males, and this difference has probably contributed to the evolution of sex differences in dispersal in this species [549].

There may, however, be another reason why male mammals disperse greater distances than females. The usual rule is that males, not females, fight with one another for access to mates (see p. 332), and therefore loser males may find it advantageous to move away from same-sex rivals that they cannot subdue [847]. Although this hypothesis probably does not apply to Belding's ground squirrels, since young males have not been seen fighting with older ones around the time of dispersal, the idea deserves to be tested in other cases. Lions, as we saw in Chapter 1, live in large groups, or prides, from which young males disperse. In contrast, the daughters of the resident lionesses usually spend their entire lives where they were born (Figure 8.12) [987]. The sedentary females benefit from their familiarity with good hunting grounds and safe breeding dens in their natal territory, among other things.

The departure of many young male lions coincides with the arrival of new mature males, which violently displace the previous pride masters and chase off the subadult males in the pride as well. These observations support the mate competition hypothesis for male dispersal. However, if young males are not evicted after a pride takeover, they often leave anyway, without any coercion from adult males and without ever having attempted to mate with their female relatives. Moreover, mature males that have claimed a pride sometimes disperse again, expanding their range to annex a second pride of females, at the time when their daughters in pride num-

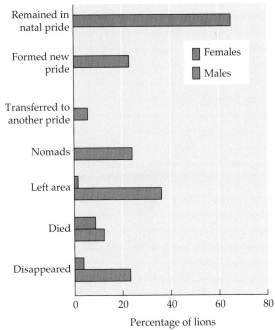

Figure 8.12   **Male and female lions differ in their dispersal patterns.** Lionesses on the Serengeti Plain of Africa tend to remain with their natal pride, whereas males move to other prides or become nomads. After Pusey and Packer [987].

ber one are becoming sexually mature. Proximate inhibitions against inbreeding apparently exist in lions and cause males to leave home. Ultimately, dispersing males may gain by mating with nonrelatives, even though the timing of their departure from their birthplace is not always under their control [509].

Discussion Question

8.3   In Belding's ground squirrels and many other mammals, young males disperse while young females remain on or near their natal territory, but this pattern is exactly reversed in most birds. Why might this be so? In producing your hypotheses, consider the fitness value of a territory to a typical male mammal and a typical male bird [477].

Migration

A familiar, but nonetheless amazing, form of dispersal is **migration**, which involves movement away from and subsequent return to the same location on an annual basis. Many living birds, mammals, fishes, and sea turtles (see Figure 4.45) migrate, and so did some extinct creatures as well [194, 238]. In fact, nearly half of all the breeding birds of North America are migrants that take off in the fall for a trip to Mexico or Central America or South America, only to return in the spring [248]. Tiny ruby-throated hummingbirds, the weight of a penny, fly nonstop 850 kilometers across the Gulf of Mexico twice a year. Arctic terns breeding in Canada may complete a roughly 40,000-kilometer migration each year (the equivalent of seven trips across the continental United States) (Figure 8.13), much of it over the ocean [1269], where the birds must stay airborne for many days and nights in a row. This is a journey that the average human would find utterly exhausting even in a jumbo jet.

Arctic tern

Figure 8.13   The migratory route of arctic terns. These birds fly from high in the Northern Hemisphere to Antarctica and back each year. Some young birds may spend 2 years circling Antarctica before returning to the northern breeding grounds.

(A)

(B)

Figure 8.14   Short-range migration in the three-wattled bellbird. (A) A male three-wattled bellbird calling from a perch in a Costa Rican forest. (B) After breeding in the mountains of north-central Costa Rica, bellbirds first head to the north and east, then go south and west to reach forests on the Pacific coast before returning north to the mountains. A, photograph by Michael and Patricia Fogden; B, after Powell and Bjork [972].

Migration poses a major historical problem: If sedentary species were ancestral to migratory ones, as they probably were, how could the ability to travel thousands of kilometers each year to specific destinations have evolved? Persons interested in this question have noted that many bird species in the tropics engage in short-range "migrations" of dozens to hundreds of miles, with individuals moving up and down mountainsides or from one region to another immediately adjacent one. The three-wattled bellbird, for example, has an annual migratory cycle that takes it from its breeding area in the mid-elevation mountainous forests of north-central Costa Rica to lowland forests on the Atlantic side of Nicaragua, then to coastal forests on the Pacific side of southwestern Costa Rica, from which the bird returns to its mountain breeding area (Figure 8.14) [972]. The distances traveled by migrating bellbirds are substantial (up to 200 kilometers), but not breathtaking.

Douglas Levey and Gary Stiles point out that short-range migrants occur in nine families of songbirds believed to have originated in the tropics. Of these nine families, seven also include long-distance migrants that move thousands of kilometers from tropical to temperate regions. The co-occurrence of short-range and long-distance migrants in these seven families suggests that short-range migration preceded long-distance migration, setting the stage for the further refinements needed for the hugely impressive migratory trips of some species [713]. Thus, long-distance migrants are probably descended from species that moved far less on an annual basis.

C. *minimus* (breeding)
C. *fuscescens* (breeding)
C. *ustulatus* (breeding)
C. *bicknelli* (breeding)
C. *guttatus* (breeding)

North America

C. *guttatus* (wintering)

Mexico

Greater Antilles

C. *frantzii* (resident)
C. *aurantiirostris* (resident)
C. *occidentalis* (resident)
C. *mexicanus* (resident)
C. *guttatus* (wintering)
C. *ustulatus* (wintering)

C. *bicknelli* (wintering)

Central America

C. *aurantiirostris* (resident)
C. *occidentalis* (resident)
C. *gracilirostris* (resident)
C. *fuscater* (resident)
C. *dryas* (resident)
C. *ustulatus* (wintering)

South America

C. *fuscater* (resident)
C. *aurantiirostris* (resident)
C. *dryas* (resident)
C. *ustulatus* (wintering)
C. *fuscescens* (wintering)
C. *minimus* (wintering)

Migrant breeding ranges

C. *guttatus* wintering range

Resident ranges

Migrant wintering ranges

Figure 8.15 **The geographic distribution of *Catharus* thrushes,** a genus that contains both resident and migratory species. Note that the resident species are concentrated in and around Central America. After Outlaw et al. [910]; photograph of a Swainson's thrush (*C. ustulatus*) by R. Curtis.

One genus of birds, the *Catharus* thrushes, may shed some light on this theory of avian migration. The genus contains twelve species, seven of which are resident in areas from Mexico to South America; the other five are migratory species that travel between breeding areas in northern North America and wintering zones to the south, especially in South America (Figure 8.15). These observations suggest that the ancestors of the migratory species lived in Mexico or Central America. Moreover, the most parsimonious interpretation of a phylogeny of this genus is that migratory behavior has evolved three times, with subtropical or tropical resident species giving rise to migratory lineages each time (Figure 8.16). Thus, the history of this genus supports the hypothesis that migratory species evolved from tropical nonmigratory ancestors.

The Costs of Migration

If the long-distance migratory abilities of today's birds evolved by natural selection in stages from the traits of a sedentary resident ancestor, then each increase in migratory capacity must have been adaptive, with the benefits of the modified trait outweighing the costs associated with the new ability. The problem with this argument is that the costs of migration are not trivial. For birds, they include the extra weight that the migrant has to gain in order to build up energy reserves for the trip. Some migrant songbirds nearly double their body weight

Figure 8.16   The long-distance migratory trait overlain on the phylogeny of *Catharus* thrushes. The phylogenetic tree was constructed on the basis of similarities among the species with respect to two genes. Although this tree is a current best estimate, there still remains some uncertainty about the relationships shown here. After Outlaw et al. [910].

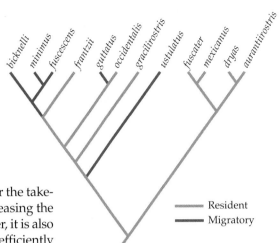

prior to their long flights. Even moderate fuel loads dramatically alter the take-off angles of individuals startled by a predator, almost certainly increasing the chance that the predator will catch the fleeing songbird [727]. However, it is also true that fully loaded red knots, once under way, actually fly more efficiently in terms of turning fuel into wing power than when they are at their nonmigratory body weight [687]. But flight, no matter how efficient, still costs calories, and there is always the chance that a migrant will run out of gas before reaching its destination—not a fitness-enhancing event. (Lawrence Swan once saw a migrating hoopoe, a bird with the exquisite scientific name *Upupa epops*, forced by exhaustion to hop up a Himalayan pass at 20,000 feet [1174].)

An optimality approach to migration generates the prediction that migrants will evolve tactics that reduce the costs of the trip, which obviously include the energy expended in flight. Many persons have wondered, for example, whether the V-formation adopted by many large birds when migrating is an energy-saving adaptation. A team led by Henri Weimerskirch was able to test this proposition by enlisting the assistance of a group of great white pelicans that had been imprinted (see Chapter 3) on an ultralight aircraft [1264]. The pelicans' heart rates were monitored with electronic heart-rate loggers attached to their backs, and their wingbeat rates were derived from film taken with a digital camera. The data on the pelicans' heart rates and wingbeat frequencies revealed that birds flying alone had to work harder than those flying in V-formation (Figure 8.17). The overall energy savings for pelicans that can take advantage of the updrafts created by the wingbeats of their companions in this way is about 11–14 percent, not a trivial sum for birds traveling long distances.

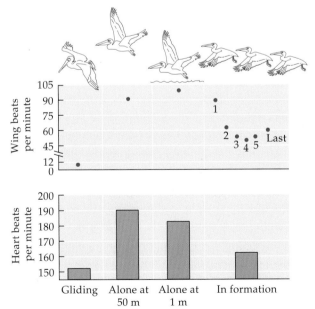

Figure 8.17   Flying in V-formation is an energy saver. Data on wingbeat frequency and heart rate are presented for various flight options available to the great white pelican. After Weimerskirch et al. [1264].

Discussion Question

8.4   If an adaptationist were to consider the wingbeat frequencies of pelicans in formation flight, that person would have some questions to ask. What are they?

If saving energy were the overriding goal of migrating birds, however, then we would not expect to find so many European songbirds traveling from all across the continent in the fall to cross the Mediterranean at the narrow point between southern Spain and northern Africa (see Figure 3.13) [96]. This route greatly lengthens the total journey for birds headed to central Africa, but reduces the overwater component of the trip, perhaps decreasing the risk of drowning at sea, a survival advantage that trumps the extra calories expended.

If this hypothesis is true, then we would expect other songbirds to have evolved an unconscious sensitivity to the mortality risks of their migratory decisions. Red-eyed vireos migrating in the fall from the eastern United States to the Amazon basin of South America must either cross a large body of water, the Gulf of Mexico, or stay close to land, moving in a southwesterly direction along the coast of Texas to Mexico and then south. The trans-Gulf flight is shorter, but vireos that cannot make it all the way to Venezuela are dead ducks, so to speak.

In light of this danger, Ronald Sandberg and Frank Moore predicted that red-eyed vireos with low fat reserves would be less likely than those with considerable body fat to risk the long journey due south across the Gulf of Mexico. They captured migrating vireos in the fall on the coast of Alabama, classified each individual as lean or fat, and placed the birds in orientation cages similar to those shown in Figure 3.14. Vireos with less than about 5 grams of body fat showed a mean orientation at sunset toward the west–northwest, whereas vireos that had been classified as having more fat tended to head due south, just as Sandberg and Moore had predicted (Figure 8.18) [1058].

Some songbirds even smaller than the red-eyed vireo attempt a still more impressive journey across water in the fall, one that requires a nonstop flight from Canada to South America over 3000 kilometers of ocean (Figure 8.19) [1310]. At first glance, a blackpoll warbler that selects this migratory route would seem to have a death wish. Surely the birds should take the safer passage along the coast of the United States and down through Mexico and Central America. But migratory blackpolls commonly appear on islands in the Atlantic and the

Red-eyed vireo

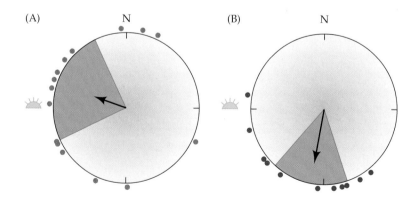

Figure 8.18   Body condition affects the migratory route chosen by red-eyed vireos. (A) Birds with low fat reserves do not head south toward the Gulf of Mexico, but instead head west (sunset symbol) as if to begin an overland flight toward Mexico. (B) Birds with ample energy reserves orient due south. The central arrow shows the mean orientation for the birds tested in each group. After Sandberg and Moore [1058].

Blackpoll warbler

Figure 8.19   Transatlantic migratory path of blackpoll warblers (arrows) from southeastern Canada and New England to their South American wintering grounds. Courtesy of Janet Williams.

Caribbean, and many are in good condition when they arrive, demonstrating their capacity to make long nonstop ocean crossings [700, 812].

The blindly courageous blackpoll warbler that manages this overwater trip has substantially reduced some of the costs of getting to South America. First, the sea route from Nova Scotia to Venezuela is about half as long as a land-based trek, although admittedly it requires an estimated 50 to 90 hours of continuous flight. Second, there are very few predators lying in wait in mid-ocean or on the islands of the Greater Antilles that the blackpolls may reach. Third, the birds leave the Canadian coast only when a west-to-east-traveling cold front can push them out over the Atlantic Ocean for the first leg of their journey, after which the birds use the westerly breezes typical of the southern Atlantic to help them reach an island landfall.

Discussion Questions

8.5   When blackpolls return to Canada from South America in the spring, they do not retrace their fall migratory path, but instead travel mostly over land. Why?

8.6   Swainson's thrushes (*C. ustulatus*), one of the species in the genus *Catharus* mentioned earlier, breed in a large band right across North America. Those birds that live in the northwestern part of North America do not all follow the same migratory route. Some birds go right down the Pacific coast and winter in Central America. But others travel all the way to the eastern part of North America before flying south to winter in South America (Figure 8.20) [1041]. One hypothesis to account for the behavior of the thrushes taking the long way south is that these birds are descendants of those that expanded the species' range from the East Coast far out to the west and north after the retreat of the glaciers about 10,000 years ago. What kind of evolutionary hypothesis is this? Is the behavior maladaptive? How can you account for the persistence of the trait?

Figure 8.20   The two migratory routes of Swainson's thrush. Although some birds travel more or less directly from the Pacific Northwest to Central America, others fly across North America before heading to South America to spend the winter. After Ruegg and Smith [1041].

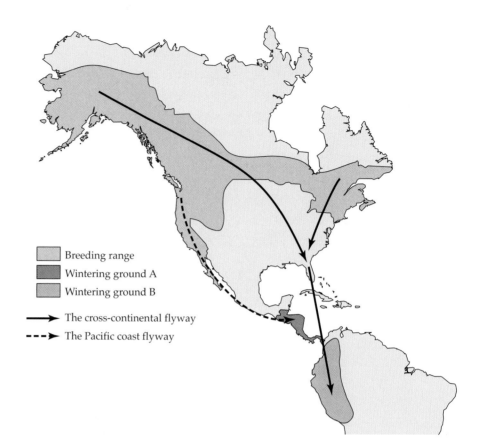

Breeding range
Wintering ground A
Wintering ground B

⟶   The cross-continental flyway
╌╌►   The Pacific coast flyway

The Benefits of Migration

For all their navigational and meteorological skills, migrating blackpolls and other birds cannot eliminate the costs of their travels altogether. What ecological conditions might elevate the benefits of migration enough to outweigh these costs, leading to the spread of migratory abilities by natural selection? The answer for many migratory songbirds in the Americas probably lies in the immense populations of protein-rich insects that appear in the northern United States and Canada in the summer, when long days fuel the growth of plants on which herbivorous insects feed [157]. Moreover, the many hours of summer daylight mean that migrant songbirds can search for food longer each day than tropical bird species, which have only about 12 hours each day to harvest prey for their offspring. But a summertime food bonanza is not the only factor favoring migration, given that many migrants abandon areas where food is still plentiful in order to winter elsewhere [86].

Discussion Question

8.7   For some whale species that migrate from Arctic or Antarctic oceans to give birth in warmer water nearer the equator, food cannot provide an ultimate benefit, since the adults do not feed on the calving grounds. Other hypotheses for whale migration have been advanced, including the idea that whale calves can gain weight more quickly in subtropical waters, where they need to invest less energy in keeping warm. Alternatively, some persons have suggested that infant whales in these waters are less likely to be attacked by predators, especially killer whales [215, 242]. How would you test these hypotheses, given the practical difficulties of directly measuring the metabolic costs of thermoregulation by whale calves or of actually observing killer whale attacks on other whales in any environment?

Resources other than food can also vary in availability seasonally, making migration adaptive. In the Serengeti National Park of Tanzania, over a million wildebeest, zebra, and gazelles move from south to north and back again each year. The move north appears to be triggered by the dry season, while the onset of the rains sends the herds south again. It might be that the herds are tracking grass production, which is dependent on rainfall. Eric Wolanski and his colleagues have established, however, that the most important factor sending the animals north is actually a decline in water supplies and an increase in the saltiness of the water in drying rivers and shrinking waterholes. If one knows the salinity of the water available to the great herds, one can predict when they will leave on their march north [1332].

Food definitely is not the primary factor causing the monarch butterfly to migrate from the eastern half of North America to central Mexico [1040, 1222] (see Figure 4.42). Many migrating monarchs wind up in a very few high mountain patches of Oyamel fir forest in central Mexico, where they form spectacularly dense nonfeeding aggregations [153] (Figure 8.21). Why should some adult monarchs fly as many as 3600 kilometers to spend the winter clinging to a conifer in the cold high mountains of Mexico? Surely there are more suitable places much closer to the milkweed-producing areas that monarchs use as their spring and summer breeding grounds.

But perhaps not, since killing freezes occur regularly at night throughout eastern North America during winter. In contrast, freezes are very rare in the Mexican mountain refugia used by the monarchs. In these forests, at about 3000 meters elevation, temperatures rarely drop below 4°C, even during the coldest winter months. Occasionally, however, snowstorms do strike the mountains, and when this happens, as many as 2 million monarchs can die in a single night of subfreezing temperatures. The risk of freezing to death could be completely avoided in many lower-elevation locations in Mexico. But William Calvert and Lincoln Brower note that in warmer and drier areas, the monarchs would quickly use up their water and energy reserves. By remaining moist and cool—

Figure 8.21   Monarch butterflies spend the winter resting in huge clusters on fir trees in a few Mexican mountain sites. Photograph by Lincoln P. Brower.

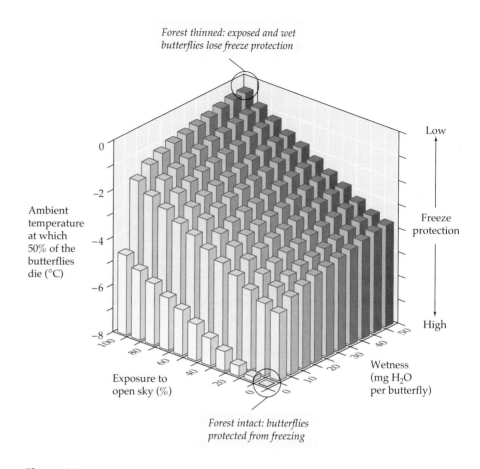

Figure 8.22   **Habitat quality and survival of monarchs overwintering in central Mexico.** Protection from freezing depends on a dense tree canopy that reduces wetting of the butterflies by rain or snow and their exposure to open sky, factors that greatly increase monarch mortality. After Anderson and Brower [26].

without freezing to death—the butterflies conserve vital resources, which will come in handy when they start back north after their 3 months in the mountains [184].

The hypothesis that the stands of Oyamel fir used by the monarchs provide a uniquely favorable microclimate that promotes winter survival is being tested in an unfortunate manner. Even in supposedly protected reserves, an alarming amount of wood cutting and logging has occurred [1001]. Brower and his associates believe that timber removal causes butterfly mortality, even when some roosting trees are left in place. Opening up the forest canopy increases the chances that the butterflies will become wet and exposed, which increases the risk that they will freeze (Figure 8.22). Thus, even partial forest cutting may destroy the conditions needed for the survival of monarch aggregations. If the loss of a relatively small number of Oyamel firs causes the local extinction of overwintering monarch populations, it will be a powerful but sad demonstration of the value of a specific habitat for this migratory species [26].

Migration as a Conditional Tactic

Both migratory and nonmigratory individuals occur in some species, just as winged and wingless forms coexist in some cricket species. The European blackbird is an example, since in some portions of its breeding range, some individuals migrate in the fall, leaving the area to others that overwinter on the breeding grounds [1073]. As discussed earlier, if these two behaviors represent different hereditary strategies, then they would have to provide equal fitness in order to persist together within a population (see p. 217). Therefore, a two-strategies hypothesis for blackbird behavior generates two predictions: (1) that the lifetime fitness of the two types should be the same, on average, and (2) that the differences between migratory and nonmigratory individuals are caused by differences in their genetic constitution.

Data on the fitness consequences of being migratory versus being a resident are not available, but individual blackbirds regularly change their behavior from one year to another (Figure 8.23). The fact that an individual bird can switch from the migrant strategy to the resident strategy indicates that the differences between the two behavior patterns are not hereditary. Therefore, because the two-strategies hypothesis seems unlikely, we shall consider a conditional strategy alternative (see pp. 219–220). All the blackbirds in a population could share a conditional strategy that enables them to choose between migrating or staying put, depending on the social conditions they encounter. If so, then we would expect individual birds to adopt whichever tactic yields higher fitness for them, given their social status. Socially dominant individuals should be in a position to select the better of the two tactics, forcing subordinates to make the best of a bad situation by adopting the option with a lower reproductive payoff (but one that is greater than they could get from futile attempts to behave like dominants) [301]. For example, perhaps an area can support only a few resident blackbirds during the winter. Under such circumstances, subordinates faced with a cadre of more powerful residents might do better to migrate away from the competition, returning in the spring to occupy territories made vacant by the deaths of some rivals over the winter.

Given the logic of this hypothesis, we can make several predictions: (1) blackbirds should have the ability to switch between tactics, rather than being locked into a single behavioral response, (2) socially dominant birds should adopt the superior tactic, and (3) when choosing *freely* between tactics, individuals should choose the option with the higher reproductive payoff (as poplar aphids do when choosing which leaf to colonize). In light of these predictions, it is significant that migratory European blackbirds head off in the fall at times when dominance contests are increasing in frequency [750]. Moreover, blackbirds do switch fairly often from the migratory option to staying put, typically when they are older and presumably more socially dominant [751].

European blackbird

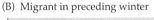

Figure 8.23   Is migratory behavior the less preferred tactic for European blackbirds? (A) Birds that were residents in the preceding winter tended to be nonmigratory the next winter as well. (B) In contrast, birds that were migrants in the preceding winter often switched to the resident option the following winter. After Schwabl [1073].

Discussion Question

8.8   Here's another way of thinking about the blackbird case, courtesy of Steve Shuster [1107]: Instead of most birds having one particular kind of genetic constitution that enables them to switch between migratory and resident behavior, perhaps any given population contains a host of different genotypes that affect just how individuals decide to stay put or to migrate. The acquisition and retention within a population of multiple genotypes with these effects could be the result of several phenomena. For example, although there might be just one currently favored, relatively common genotype in a given part of the species' range, the genotype favored in one place might not be the same as in another. As a consequence of gene flow from one population to the next, genetic variation would be maintained within all the populations. Additional genetic variation could come from mutations within a population. Thus, there might be all sorts of genetically different behavioral phenotypes within the species, with different individuals making different assessments about when to do what in response to various environmental circumstances. According to this view, conditional strategy theory is fatally flawed by its assumption that there is genetic uniformity in a population with respect to the genes that influence the kinds of behavioral flexibility that we are talking about. What does such an approach predict about the degree of genetic variation underlying the "stay put" and "migrate" phenotypes in European blackbirds? Does Chapter 3 contain any helpful information on how much fitness-relevant genetic variation exists within animal populations? (You may want to reexamine the section on artificial selection.) How does the genetic diversity approach account for the existence of individuals whose behavior yields lower fitness than that of other individuals with different behavioral phenotypes? How might an advocate of conditional strategy theory respond to this counter-view?

The migratory component of a species can also be composed of subgroups that vary in where they breed and overwinter, as in the case of the blackcap warbler (Chapter 3) [95] and Swainson's thrush [1041]. As more studies are done using stable isotope technology [1040], more and more cases of subpopulations that utilize different portions of their species' geographic ranges are emerging. Stable isotope technology is based on the fact that when animals feed on different foods in different parts of the world, the ratios of different forms of key elements they ingest may vary, which in turn affects the chemical composition of body parts into which these elements are incorporated. So, for example, the feathers of black-throated blue warblers that winter on islands in the western Caribbean are chemically distinct from the feathers of birds of the same species that winter in the eastern Caribbean. These differences can be traced back to where the feathers were produced in the United States. Thus, it was discovered that black-throated blue warblers that breed in the mountains of the southern United States go to the eastern Caribbean for the winter, carrying their chemically distinctive feathers with them, whereas their fellow black-throated blues that reproduce in the northeastern United States migrate to the western Caribbean [1039].

The occurrence of migratory variation of this sort poses the same kind of evolutionary question that we discussed in the context of the coexisting resident and migrant strategies of blackbirds: Why do members of the same species vary in their choice of breeding and wintering sites? This problem can be illustrated by examining where six different populations of the fox sparrow breed and spend the winter in western North America (Figure 8.24) [85]. The members of each population are sufficiently different in appearance to warrant placing them in six different subspecies. One subspecies (A) in the northwestern United States and southwestern Canada is essentially sedentary, but members of the other five populations migrate. They do not, however, select the same destination. Instead, the birds that breed in southern Alaska (subspecies D, E and F) travel great distances (probably across the Pacific Ocean) to winter in south-

Fox sparrow

Figure 8.24   **The leapfrog migratory pattern** of western North American populations of the fox sparrow. The populations that breed farthest to the north migrate farthest to the south in the fall. (Population A is composed of birds that live year-round in the same area.) After Bell [85].

ern California. Other populations (B and C) breeding closer to the United States travel many fewer kilometers to wintering grounds in central to northern California. How can we account for this pattern?

One hypothesis is that individuals in the different populations achieve equivalent fitness by adopting the optimal migration pattern for their particular breeding site. Thus, the birds that form the southern Alaskan populations (D, E and F) benefit by moving to southern California to exploit the surge of food available there in the mid- to late spring, which enables these individuals to put on the large fat reserves needed for their long flight north. If sparrows from a breeding population in southern Canada (B or C) also went to southern California for the winter, they would have to compete for food with the Alaskan birds that were there, and they would be late in getting back to Canada, where the breeding season begins sooner than in Alaska. Late arrival in southern Canada would damage their chances in territorial competition with other fox sparrows that had migrated earlier over shorter distances. Thus, members of the southern Canadian breeding population winter closer to Canada than Alaskan fox sparrows, and they start their migration back earlier in the year as well. According to this view, members of each subspecies balance travel costs and arrival times in selecting the optimal sites in which to overwinter and breed [85].

Discussion Question

8.9   As an exercise, apply conditional strategy theory to the western fox sparrow case in the same way as we did with European blackbirds. Use your hypothesis to make predictions about what decisions individuals of different competitive abilities would make about remaining resident in an area year-round versus migrating various distances. For example, if your hypothesis were correct, what should happen if a bird improved its condition from one year to the next? What information presented above permits you to evaluate your predictions? Is it useful to know that in the white-crowned sparrow, resident and migrant birds have almost identical estimated annual survival rates, or that those sparrow species that migrate relatively long distances do not have lower survival rates than other species that travel shorter distances [1059]?

Territoriality

Fox sparrows everywhere defend the areas in which they breed, which is to say that they are **territorial**. Migrant males arrive on the breeding grounds before females and stake out their claims, singing their special song lustily and charging after any other male fox sparrow that dares come close to their real estate. When females come north, they choose among the territorial males, settling with a partner on his patch of brush and small trees, often next to a stream. After a bond has formed between male and female, she is prepared to help him chase away other fox sparrows, especially if they happen to be females. Although such aggressive behavior in defense of space is very widespread, occurring in creatures as different as poplar aphids and Belding's ground squirrels, many other animals, including honey bees and monarch butterflies, ignore or tolerate their fellows. So once again we have an evolutionary issue to resolve: Why spend the time and energy to be territorial?

A cost–benefit approach to territoriality requires that we consider the disadvantages of territorial defense, one of the most obvious of which is the time cost of the behavior. A territorial surgeonfish, for example, chases rivals away from its algae-rich turf on a Samoan reef an average of 1900 times each day [251]. The wear and tear of these territorial pursuits, to say nothing of the out-and-out fighting that occurs in some territorial species, can lead to a shortened life, as has been documented for a damselfly species with both territorial and nonterritorial individuals [1218].

In addition to the risks of injury or exhaustion, other damaging effects can arise indirectly from the underlying mechanisms of territorial aggression. For example, in species in which testosterone promotes territorial defense, the effects of the hormone may exact a toll via a reduction in parental care or loss of immune function (see p. 168) [1323]. Moreover, the hormone may so increase the activity level of males, even if they are not actually fighting, that they suffer as a result. Catherine Marler and Michael Moore demonstrated this point experimentally with Yarrow's spiny lizards [782, 783]. They inserted small capsules containing testosterone beneath the skin of some male lizards they captured in June and July, a time of year when the lizards are only weakly territorial. These experimental animals were then released back into a rockpile high on Mt. Graham in southern Arizona. The testosterone-implanted males patrolled more, performed more push-up threat displays, and expended almost a third more energy than controls (lizards that had been captured at the same time and given a chemically inert implant). As a result, the hyper-territorial males expended their energy reserves and died sooner than males with normal concentrations of testosterone (Figure 8.25). (The tendency of testosterone-boosted males to give more displays, be more active, and lose weight has also been observed in male Yarrow's spiny lizards kept in an experimental enclosure [647, 648].) Likewise, unusually aggressive territorial males of the side-blotched lizard (*Uta stansburiana*) die at a faster rate than males that invest less in territorial behavior (see Figure 5.31).

Because territoriality is exceedingly costly, we can predict that peaceful coexistence on an undefended home range should evolve when the benefits of owning a valuable resource do not outweigh the costs associated with its monopolization. In Yarrow's spiny lizard, for example, males do live together in relative harmony in June and July, when females are not sexually receptive, but when fall comes and females are ready to mate, the males become highly aggressive, a pattern that applies to many other animals. This capacity to adjust the level of aggression up or down is also seen in the tropical pseudoscorpion *Cordylochernes scorpioides*. The life cycle of this arthropod involves periods of colonization of recently dead or dying trees, followed in a few generations by dis-

(A)

Yarrow's spiny lizard

(B)

Figure 8.25   Energetic costs of territoriality.
(A) Males of Yarrow's spiny lizard that received an experimental implant of testosterone spent much more time moving about than did control males. (B) Testosterone-implanted males that did not receive a food supplement disappeared at a faster rate than did control males. Testosterone-implanted males that received a food supplement (mealworms) survived as well as or better than controls; thus the high mortality experienced by the unfed group probably stemmed from the high energetic costs of their induced territorial behavior. After Marler and Moore [782, 783].

persal from the now older dead trees to fresh sites. Dispersing pseudoscorpions make the trip under the wing covers of a huge harlequin beetle (Figure 8.26), disembarking when the beetle touches down on an appropriate tree. The pseudoscorpions mate both on trees and on the backs of beetles, but male pseudoscorpions attempt to control real estate only when they occupy the small, and therefore economically defensible, patches on a beetle's back [1353]. Under these circumstances, the benefits of territoriality, in terms of number of mates monopolized, can exceed the costs of defense.

If individuals vary in their territory-holding abilities, then those with superior competitive ability should secure reproductive advantages over others excluded from favored habitat. A case in point is the American redstart. These warblers compete for territories during the nonbreeding season, when they are on their tropical wintering grounds in Central America and the Caribbean. In Jamaica, males tend to occupy black mangrove forests along the coast, while females are more often found in second-growth scrub inland. Field observations reveal that older, heavier males in mangrove habitat attack intruding females and younger males, apparently forcing them into second-rate habitats [787]. If these older, dominant males benefit from their investment in territorial aggression, then there should be some reproductive advantages for individuals that succeed in occupying the favored habitat. In fact, birds living in mangroves retained their weight over the winter, whereas redstarts in the apparently inferior scrub habitat generally lost weight [787].

Figure 8.26 **A beetle and its pseudoscorpions.** (Top) Two huge male harlequin beetles battle for position on a tree trunk. (Bottom) Groups of the pseudoscorpion *Cordylochernes scorpioides* hitch a ride under the beetle's wing covers as it flies from one tree to another. Photographs by David and Jeanne Zeh.

Probably because they have more energy reserves, territory holders in mangroves leave their wintering grounds sooner than birds living in second-growth scrub (Figure 8.27). In some migrant songbird species, early male arrivals secure a reproductive advantage by claiming the best territories and gaining quick access to females when they arrive (e.g., [520]). Female redstarts also have something to gain by reaching the Canadian breeding grounds early (and in good condition) because the short Canadian summer gives them only a few months to raise their chicks; late starters run out of time to get the resources needed for their young to fledge successfully. In other words, early birds get the fledglings in this species. Because the arrival time of a redstart in Canada is a function of where in Jamaica it spent the winter, overwintering habitat is a key predictor of reproductive success in this species (Figure 8.28) [885].

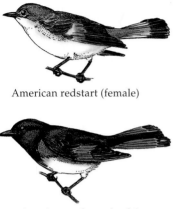

American redstart (female)

American redstart (male)

Figure 8.27   Habitat quality and the date of departure from Jamaican wintering grounds by American redstarts. Birds occupying the preferred black mangrove habitat are able to leave sooner for the northern breeding grounds than redstarts forced into second-rate, second-growth scrub. After Marra et al. [787].

Discussion Question

8.10   In some songbirds, a nonterritorial pair sets up housekeeping unobtrusively within the territory defended by another pair of the same species. However, nonterritorial pairs either do not breed at all [1129] or fledge fewer offspring on average than the owners of the territories they live in [323]. Develop a game theoretical hypothesis to analyze why there are any nonterritorial pairs of this sort. Come up with a two-strategies hypothesis and a conditional strategy hypothesis. What predictions follow from your two hypotheses?

Territorial Contests

Studies of territoriality have found that winners in the competition for territories gain substantial indirect and direct reproductive benefits. Given this evidence, it seems paradoxical that when a territory holder is challenged by a rival, the owner almost always wins the contest—usually within a matter of seconds.

Why do the intruders give up so quickly? Game theory has supplied one possible answer with algebraic demonstrations that an arbitrary rule for resolv-

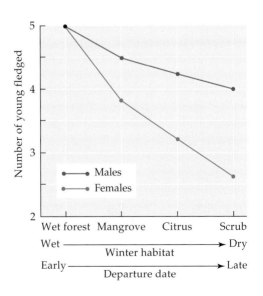

Figure 8.28   Redstart habitat selection on the wintering grounds affects reproductive success. The estimated number of fledglings produced by redstarts in relation to where they spent the winter. Birds that wintered in the most productive habitat, wet forest, are believed to enjoy the highest reproductive success. After Norris et al. [885].

ing conflicts between residents and intruders could be an **evolutionarily stable strategy**—that is, one that cannot be replaced by an alternative strategy. One simple *arbitrary* rule might be, "The resident always wins." If all competitors for territories were to adopt the "resident always wins" rule, so that intruders always gave up and residents never did, a mutant with a different behavioral strategy could not spread its special allele. Therefore, the "resident always wins" strategy would persist indefinitely. (Similar algebraic maneuvers demonstrate that the contrary convention, "The intruder always wins," is also a possible evolutionarily stable strategy. Therefore, according to this hypothesis, any given territorial species should be as likely to have evolved an "intruder wins" rule as a "resident wins" rule.)

Attempts to locate species that employ an arbitrary "resident always wins" rule have generally failed, although at one time the speckled wood butterfly was believed to employ this strategy. Males of this insect defend small patches of sunlight on the forest floor, where they may occasionally encounter receptive females. Nick Davies found that territorial males always defeated intruders, which invariably departed rather than engaging in intense territorial conflicts [288]. The butterflies were capable of such fights, as Davies showed by capturing and holding territorial males in an insect net until a new male had arrived and claimed the seemingly empty sunspot territory. When the prior resident was released, he returned to "his" territory, only to find it taken by the new male. This male, having claimed the site, reacted as if he were the resident, with the result that the two males engaged in what passes for a fight in the butterfly world. The combatants circled one another, flying upward, occasionally clashing wings, before diving back to the territory, sometimes repeating this maneuver. Eventually the previous resident gave up and flew away, leaving the site under the control of the replacement resident (Figure 8.29).

Although Davies's results were consistent with the hypothesis that male butterflies use an arbitrary rule to decide winners of territorial contests, Darrell Kemp and Christer Wiklund decided to repeat the experiment, but without subjecting the captured initial resident to the potentially traumatic effects of being held in an insect net before release. Instead, they put their captives in a cooler,

Figure 8.29   Does the resident always win? An experimental test of the hypothesis that territorial residents always win conflicts with intruders was performed with males of the speckled wood butterfly. When one male ("White") is the resident, he always defeats intruders (1, 2). But when "White" is temporarily removed (3), permitting a new male ("Black") to settle on his sunspot territory (4), then "Black" always defeats "White" upon his return after release from captivity (5, 6). After Davies [288].

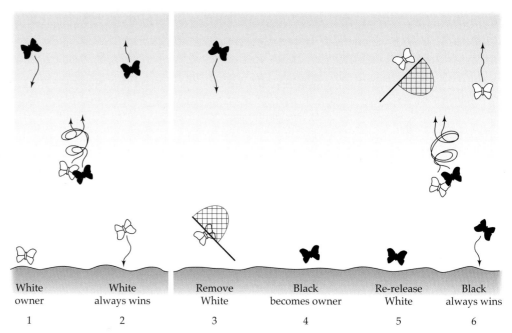

White owner	White always wins	Remove White	Black becomes owner	Re-release White	Black always wins
1	2	3	4	5	6

retrieving them after a replacement had taken up residence in their sunspot territories for about 15 minutes. After placing the original resident on the ground near his old territory, they tossed a wood chip over him. Butterflies pursue visual stimuli of this sort, so the male could be steered toward his rival in this way. When the perched replacement male saw the incoming initial resident, he reacted in the manner of a territorial speckled wood butterfly by engaging his opponent in a spiral flight. But these flights lasted much longer than the ones that Davies had observed, and their outcome was strikingly different, with the original sunspot holders winning 50 of 52 contests with new residents [626]. So the current resident does *not* always win, a result that eliminates the arbitrary rule hypothesis. (Kemp and Wiklund believe that in Davies's experiment, the original resident males that were held in nets until their release just wanted to escape rather than fight for their old territories, and thus the relatively short interactions and the new resident "victories.")

If the arbitrary "resident always wins" rule does not explain why residents usually repel intruders, what hypothesis will do the job? Perhaps males that succeed in acquiring territories have some sort of nonarbitrary advantage over others that translates into superior **resource-holding power**. One such advantage might be large body size, which would provide a physical edge for the territory holder. In fact, in species ranging from rhinoceroses [999] to fiddler crabs [595] to wasps [893], territorial individuals are relatively large individuals, which gives them the strength to drive smaller rivals away.

Nevertheless, being larger and so presumably stronger is not the key to territorial success in every species. In the red-shouldered widowbird, an African species that looks remarkably like the North American red-winged blackbird, males with larger and redder shoulder patches are more likely to hold territories than are rivals with smaller, duller epaulets. The less gaudy males become nonterritorial "floaters," hanging around the territories of other males, conceding defeat whenever challenged, but ready to assume control of vacant territories should a resident disappear. Territorial males and floaters do not differ in size, but even so, when males in the two categories were captured and paired off in a cage unfamiliar to either individual, the ex-resident almost always dominated the floater, even when the resident's red shoulders had been painted black (Figure 8.30) [983]. These results suggest that some intrinsic feature of certain males other than body weight, which is advertised by the size and color of their epaulets but can be expressed in the absence of this signal, enables these males to win fights with others.

Just as is true for red-shouldered widowbirds, males of some damselflies can be divided into territory holders and floaters. Although territorial male damselflies are not necessarily larger or more muscular, those that win the lengthy back-and-forth aerial contests that occur over streamside mating territories almost always have a higher fat content than the males they defeat (Figure 8.31) [778, 964]. In these insects, contests do not involve physical contact, but instead consist of what has been called a war of attrition, with the winner outlasting the other, and the ability to continue displaying is related to the individual's energy reserves.

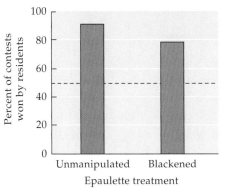

Figure 8.30   Territorial male red-shouldered widowbirds have higher resource-holding power than nonterritorial floaters. When residents compete with floaters for food in captivity, residents usually win, even though they have been removed from their breeding territories for the experiment, and even if their red epaulets have been painted black to eliminate this signal of dominance. The dotted line shows the result that would occur if residents and floaters were equally likely to win dominance contests. After Pryke and Andersson [983]; photograph by P. Craig–Cooper.

Black-winged damselfly

Figure 8.31   Fat reserves determine the winner of territorial conflicts in black-winged damselflies. In this species, males may engage in prolonged aerial chases that determine ownership of mating territories on streams. (A) Larger males, as measured by dry thorax weight, do not enjoy a consistent advantage in territorial takeovers. (B) Males with greater fat content, however, almost always win. After Marden and Waage [778].

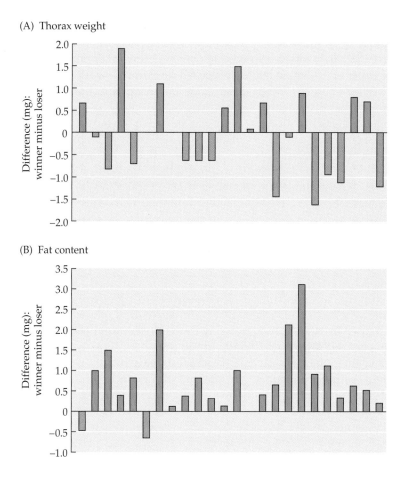

Differences in resource-holding power, however, whether based on size or on energy reserves, cannot account for every case in which territory owners seem to win at will. For example, older female Mediterranean fruit flies, who are almost certainly not as physically fit as younger flies, can keep younger females away from the fruits in which the insects lay their eggs [925]. By the same token, older males of the song sparrow respond more intensely to simulated intrusion of their territories via song playback than younger ones [581], and eggfly butterfly males are more willing to persist in their aerial war-of-attrition contests as they age than when they were younger (Figure 8.32) [625]. Findings of this sort do not make sense in light of the hypothesis that territorial winners are in better condition or more physically imposing than losers.

Therefore, let us consider a third hypothesis for resident success in territorial defense: namely, that when the fitness payoff for holding a territory increases over time, owners have more to fight for than intruders. On the one hand, as territory holders age, their opportunities for future reproductive success fall, and as a result, the costs of engaging in energetically expensive or risky contests also decrease. This fact of life history could in some cases boost the net gain from territorial defense by elderly residents.

Not only do the costs of territorial behavior decline for aged territory holders, but the benefits of territorial possession may grow over time because of the nature of interactions among territorial neighbors. An aggressive newcomer initially has to spend a lot of time dealing with boundary disputes with neighbors, but once territorial borders have been agreed upon, everyone calms down, producing what has been called the "dear enemy" effect [390]. For example, the mean distance at which territorial males of an African lizard will charge at a

Figure 8.32   Older males fight harder in the eggfly butterfly. (A) An old male missing large chunks of his hind-wings perched on a leaf in his territory in Queensland, Australia. (B) A researcher staged a series of encounters in the labo-ratory between a hand-reared male of known age and a fresh, young rival placed in his cage. In the ensuing aerial clashes, the resident persisted longer as he grew older. Each color represents a different individual resident. A, photo-graph by Darrell Kemp; B, after Kemp [625].

familiar neighbor is less than one-fifth that for strangers. Moreover, when chas-ing a neighbor, the resident pursues him for only a few centimeters, whereas he keeps after an unfamiliar intruder for a meter and a half on average [1288].

Thus, once a territory owner and his neighbors have learned who is whom, they no longer need to expend time and energy in lengthy chases. If a resident is ousted, however, the new resident will have to fight intensely for a time with his neighbors in order to get the borders of their territories right. The original resident therefore has more to gain by holding onto his present territory than the new intruder can secure by acquiring it, since any new boy on the block will still have to deal with his unfamiliar neighbors even after he has ousted the res-ident.

A logical extension of the dear enemy hypothesis is that neighbors may find it advantageous to combine forces to repel an intruder that might other-wise displace one of them, which would result in the remaining residents being left to cope with a rambunctious newcomer. This factor is believed to be behind the formation of occasional coalitions between two neighboring males in a ter-ritorial fiddler crab [48]. Every so often a male of *Uca mjoebergi* will leave his burrow in a mudflat and join a neighbor in fending off a wandering male that has challenged that neighbor. Typically the helper intervenes when the encroach-ing intruder is larger than his "dear enemy" neighbor but smaller than the helper. The two-on-one game plan usually works in that when a crab has a helper, the two are able to drive off the intruder 88 percent of the time, whereas defenders that have to cope alone with a takeover threat win only 71 percent of the time. The benefit to the helper appears to be the time and energy savings that come from having a familiar neighbor as opposed to an aggressive new neighbor intent on establishing himself in the neighborhood. Males that know one another almost never fight, but newcomers are far more combative, at least initially [48].

The dear enemy effect could contribute mightily to an asymmetry between what an established territory holder has to lose in a territorial contest and what a challenger has to gain. Given this payoff asymmetry, we can predict that when a newcomer is permitted to claim a territory from which the original resident has been temporarily removed, the likelihood of the replacement resident win-ning a fight against the original resident will be a function of how long the replacement has occupied the site. This experiment has been done with birds, such as the red-winged blackbird, as well as with some insects and fishes. In redwings, when captive ex-territory holders are released from an aviary and allowed to return to their home marsh to compete again for their old territories, they are more likely to fail if the new males have been on the experimentally vacated territories for some time [84].

The payoff asymmetry hypothesis also predicts that contests between an ex-resident and his replacement will become more intense as the tenure of the replacement increases because longer tenure boosts the value of a site to the current holder and thus his motivation to defend it. This prediction has been supported in animals as different as tarantula hawk wasps [13] and songbirds [667]. If, for example, one removes a male tarantula hawk wasp from the peaktop shrub or small tree that he is defending and pops him into a cooler, his vacant territory will often be claimed within a few minutes. If the ex-territory holder is quickly released, he usually returns promptly to his old site and chases the newcomer away in less than 3 minutes on average. But if the ex-territory holder is left in the cooler for an hour, then when he is released and hurries back to his territory, a battle royal ensues. The newcomer resists eviction, and the two males engage in a long series of ascending flights, in which they climb rapidly up into the sky side by side for many meters before diving back down to the territory, only to repeat the activity again and again until finally one male—usually the replacement wasp—gives up and flies away. The mean duration of these contests is about 25 minutes, and some go on for nearly an hour [13].

Although cases of this sort support the payoff asymmetry hypothesis for why residents usually win, it is possible that lengthier contests occur after replacements have been on territory for some time because the removed resident has lost some resource-holding power while in captivity. This possibility was checked in a study of European robins by taking the first replacement away and permitting a second replacement to become established for a short time before releasing the resident. In this way, Joe Tobias was able to match 1-day replacements against residents that had been held captive for 10 days. Under these circumstances, the ex-territory holders always won, despite their prolonged captivity [1202]. In contrast, when ex-residents that had been caged for 10 days went up against replacements that had been on the territory for 10 days, the original territory holders always lost. Therefore, contests between replacements and ex-residents were decided by how long the replacement had been on the territory, not by how long the ex-resident had been in captivity. The payoff asymmetry hypothesis appears to explain why the resident almost always wins in European robins. Perhaps it applies to many other species as well.

Summary

1. In choosing where to live, many animals actively select certain places over others. If living in a certain kind of habitat enhances fitness, then individuals able to occupy preferred habitats should have higher fitness than their competitors, unless they are forced to share the more desirable places with a higher density of rivals.

2. The selection of living space often occurs in the context of leaving one spot for another, as when juvenile animals leave the place where they were born to go in search of new homes. Cost–benefit analyses have helped to explain why young male mammals typically disperse farther than females of their species—probably because the costs of dispersal are greater for females than for males.

3. Migration is a special form of dispersal in which the migrant eventually returns to the place it left. This phenomenon raises interesting questions about its evolutionary origin and subsequent modification as well as its adaptive value. The ability to migrate very long distances probably evolved in populations that had acquired short-range migratory capacity. Studies of the adaptive value of the behavior have focused on learning how migrants lower the costs of travel while maximizing the benefits gained.

4. The coexistence of resident versus migratory individuals (or territorial versus nonterritorial members of a species) has often been explained with conditional strategy theory. Possession of a conditional strategy confers behavioral flexibility on individuals, which are therefore able to choose adaptively among several options or tactics.

5. Territorial behavior is costly, which is why it is far from universal. Defense of living space evolves only when territorial individuals can gain substantial benefits, such as special access to mates or food. Territorial contests are usually won quickly by owners over intruders. The competitive edge held by territorial residents may stem from superior physical strength or energy reserves (which gives them high resource-holding power), or it may exist because residents have more to lose than intruders can gain (a payoff asymmetry) thanks to the dear enemy effect (in which familiar neighbors no longer fight intensely with one another over territorial boundaries).

Suggested Reading

The concept of an evolutionary strategy has been made clear by Richard Dawkins [300, 303] and Mart Gross [484]. The related concept of polyphenism, which reflects the adaptive ability of organisms to develop or behave in different ways depending on their environment, has been explored in depth by Mary Jane West-Eberhard [1273]. This concept is closely linked to conditional strategy theory, an approach that Steve Shuster and Michael Wade criticize in their book [1107]. Tom Whitham [1284, 1286] shows how to use evolutionary theory when studying habitat selection. Dispersal in Belding's ground squirrels is analyzed at the proximate and ultimate levels by Kay Holekamp and Paul Sherman [550]. The cost–benefit approach to territoriality was first used by Jerry Brown and Gordon Orians [159], and the approach has been applied subsequently with special skill by Nick Davies [296]. The use of stable isotope technology, an important new tool in the study of migration, is clearly explained by D. Rubenstein and K. Hobson [1040].

9 The Evolution of Communication

*I*f you were on an African safari and had a chance to watch a clan of spotted hyenas for some time, you might see one hyena present its large, engorged penis to a companion, who would inspect the organ closely while perhaps also offering the other hyena a chance to nuzzle, sniff, or lick its own erect penis (Figure 9.1). This rather indelicate operation catches the eye of most people who see it, but what really gets their attention is the news that they may very well have been watching two *females* in action, not two males. In the spotted hyena, both sexes are endowed with a "penis" that can be erected, and both use the appendage in ways that appear to communicate something to their fellow hyenas in what has been called the greeting ceremony.

Although we discussed some evolutionary aspects of communication in Chapter 2, this chapter will tackle a variety of topics not covered in detail previously, especially issues associated with the origin and modification of communication systems over evolutionary time. We begin with the spotted hyena, a particularly good case with which to illustrate two key points: first, the difference between questions about the evolutionary origin of a behavioral trait versus questions about the adaptive value of the behavior, and second, the ability of scientists to make progress by continually evaluating and re-evaluating one another's hypotheses.

◀ *When a bull elk bugles, other males listen. What do they learn from their companion's communication signals? Photograph by John Cancalosi*

Figure 9.1   The pseudopenis of the female spotted hyena can be erected (top), in which condition it is often presented to a companion (bottom) in the greeting ceremony of this species. Top, photograph courtesy of Steve Glickman; bottom, photograph by Heribert Hofer.

The Origin and Modification of a Signal

At some point, a female ancestor of today's spotted hyenas must have been the first female ever to greet a companion by offering her a chance to inspect her "penis." How did this female ever come to behave in such an utterly bizarre way? To answer this question, we first turn to comparisons among the relatives of spotted hyenas, which include a large number of species in the mammalian order Carnivora. In many carnivores, including some other species of hyenas, males and females inspect the anogenital regions of their companions [678, 911] (a trait that should be familiar to you if you own a dog). Because of the widespread distribution of anogenital sniffing among the mammalian carnivores, we can conclude that the ancestor of the spotted hyena probably engaged in this behavior [678]. At some point, females of this pre-spotted hyena shifted their inspections from the anal glands of other individuals to the erect penises of their fellow hyenas, male and female alike. But in order for this to happen, the female's "penis" had to evolve. Where did this organ come from?

Progress on this front came in 1939, when L. Harrison Matthews published a paper on reproduction by spotted hyenas in which he speculated that development of the pseudopenis might be the result of exposure of the female to high levels of male sex hormones [798]. Stephen Jay Gould brought this proximate hypothesis to a large audience in 1981 [460] in one of his popular essays in the magazine *Natural History*. Gould noted that in mammals, the male penis and the female clitoris develop from exactly the same embryonic tissues. The general mammalian rule seems to be that if these tissues are exposed early on to male sex hormones (testosterone and other androgens), as they almost always are in male embryos, a penis is the end result. If the same target cells do not interact with androgens, as is the case for the typical embryonic female mammal, then a clitoris develops. Indeed, when a female embryo of most mammals comes in contact with testosterone, either in an experiment or because of an accident of some sort, her clitoris becomes enlarged and looks like a penis [376]. This effect has been observed in our own species in the unlucky daughters of pregnant women who received medical treatment that inadvertently exposed their offspring to extra testosterone [843], as well as in females whose adrenal glands produce more testosterone than normal [963].

Given the general pattern of development of mammalian external genitalia, Gould thought it likely that female embryos in the spotted hyena must be exposed to unusual concentrations of male androgens. In support of this hypothesis, he pointed to a paper written in 1979 by P. A. Racey and J. D. Skinner [998], who reported that female spotted hyenas had circulating levels of testosterone equal to those of males, unlike other hyenas and unlike mammals generally. This discovery appeared to confirm the hypothesis that an unusual hormonal environment led to the origin of the spotted hyena's false penis, setting the stage for the origin of the dual-sex penis-sniffing greeting ceremony.

Discussion Question

9.1   When Gould wrote his article on the spotted hyena [460], he concluded that asking the question, What is the pseudopenis for? was unnecessary and indeed unwise in that it distracted attention from what he considered to be the more important question, How is it built? What do you think?

Work on the extra androgen hypothesis continued with some researchers checking on androgen levels in wild female spotted hyenas (not an easy task). These workers found to their surprise that testosterone levels in free-living females are actually *lower* than in reproductively active males [464]. Nonetheless, *pregnant* females hyenas do have higher testosterone levels than lactating females (Figure 9.2). This latter finding leaves the door open to some extent for the extra androgen hypothesis [344], as does the discovery that males and females have similar levels of a precursor of testosterone [464] called androstenedione (a substance made famous by baseball slugger Mark McGwire, who used

Figure 9.2   Concentrations of testosterone in male and female spotted hyenas. Testosterone was measured in the blood and in the feces of four categories of free-living hyenas: dominant (reproductively active) males, subordinate (reproductively inactive) males, lactating females, and pregnant females. After Dloniak et al. [333].

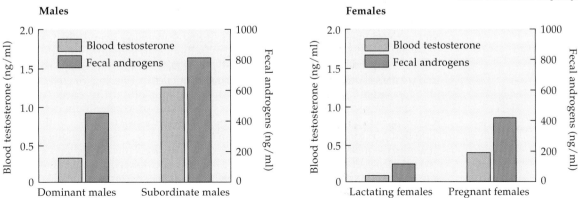

this hormone regularly during 1998 when he hit 70 home runs). Because the placenta of a pregnant female converts androstenedione to testosterone, her female embryos could conceivably be exposed to masculinizing levels of testosterone [455].

But if the extra androgen hypothesis is correct, then the experimental administration of anti-androgenic chemicals to pregnant adult females should abolish the pseudopenis in their subsequent daughters while also feminizing the external genitalia of their newborn sons. When anti-androgens are administered to pregnant female hyenas, they do cause male hormone levels to fall, but the sons of these females are feminized only slightly in the sense of having a penis that resembles the female pseudopenis, which is subtly different from the real thing [963]. Moreover, the daughters of the treated females retain their elaborate pseudopenis in only slightly altered form. Both results are at odds with the view that early exposure to male sex hormones is sufficient to cause female embryos of the spotted hyena to develop their unusual genital equipment [345]. Some persons have concluded, therefore, that the extra androgen hypothesis is effectively dead [858], although others still believe that androgens are involved—along with other "androgen-independent" mechanisms of genital development [333]. Clearly our understanding of the proximate basis for the pseudopenis is less certain than it seemed a few decades ago.

If the hyena pseudopenis did originate as a result of a mutation that exposed the female fetus to high concentrations of androgens, and if the development of the pseudopenis is now somewhat (or totally) independent of fetal androgen exposure, then some genetic changes must have occurred during the spotted hyena's evolution. These changes have evidently resulted in the partial or complete decoupling of the development of the pseudopenis from its original hormonal context, just as the activation of sexual behavior in red-sided garter snakes and white-crowned sparrows has been liberated from sex hormone control (see Chapter 5). In the spotted hyena, we can speculate that certain mutations may have spread, one after the other, in part because they lowered male-like androgen concentrations present in ancestral females, or modified their effects, thereby reducing the damaging developmental by-products of exposure to androgens for females (see below). Perhaps now an entire battery of altered genes contributes to the development of the secondary sexual characteristics of today's spotted hyena females. As a result, these creatures currently possess a less costly developmental mechanism for the production of an elaborate pseudopenis, which they apparently use to communicate with other hyenas in ways that may advance their fitness.

Discussion Question

9.2   In studying the courtship behavior of the empid flies, E. L. Kessel was amazed to find a species, *Hilara sartor*, in which males gather together to hover in swarms, carrying empty silken balloons [635]. In this species, females fly to the swarm, approach a male, and receive a balloon, which they hold while mating occurs. In the overwhelming majority of fly species, including some other empid flies, courtship does not involve the transfer of any object from male to female. But in addition to (1) the empty balloon gifts of *H. sartor* and (2) the lack of courtship gifts in some empids, males of other species in the group court females by offering (3) gifts of an edible food item (a freshly killed dead insect), (4) gifts of a dried insect fragment wrapped in a silken covering, or (5) gifts of an edible prey insect wrapped in silk. Construct a behavioral phylogeny that yields a hypothesis, consistent with the notion that evolutionary history involves a layering of change upon change, on the historical sequence leading to empty balloon gift-giving. How would you test your hypothesis? (A fairly recent comparative analysis of empid courtship behavior appears in [264]; for a detailed study of a balloon-making species, see [1055].)

The Adaptive Value of a Signal

Understanding the history of a trait with its origin and subsequent changes is one goal of evolutionary biologists, but the other main aim of these scientists is the discovery of why some changes have spread and persisted, while others have disappeared over time. These goals come together in questions such as, why would the original mutant female hyena whose daughters had masculinized external genitalia have had more descendants than the other females of her era? Surely her masculinized daughters would have been harmed by the disruption of long-tested patterns of sexual development. In our species, females exposed to higher than normal amounts of androgens as embryos not only develop an enlarged clitoris, but they may also be sterile as adults, illustrating just how reproductively damaging the developmental effects of male sex hormones can be. And even in modern populations of spotted hyenas, 10 to 20 percent of all females die the first time they attempt to give birth because their pups have to pass through the clitoris (Figure 9.3), which is too small for this purpose; moreover, about 60 percent of all first-born cubs die as well [344]. Given the huge reproductive costs of masculinized genitalia in the spotted hyena, some biologists have argued that there must have been very substantial benefits associated either with the pseudopenis itself or with the underlying mechanism that results in the development of this phenotype [410, 858].

A focus on underlying developmental mechanisms has led to a number of "by-product" explanations for the pseudopenis in which selection for some other truly adaptive attribute has as its by-product the development of an enlarged clitoris. For example, perhaps a mutant gene that boosted androgen levels in females spread because it helped to make adult females larger and more aggressive, not because it helped the females acquire a pseudopenis. By all accounts, spotted hyena females are indeed unusually aggressive, at least by comparison with males, which are always subordinate and deferential to the opposite sex. Females not only keep males in their place, but they also compete intensely with one another for status, with the winner or alpha female gaining priority of access for herself and her offspring to the wildebeest and zebra that

Figure 9.3   A cost of the pseudopenis for female spotted hyenas. The birth canal of this species extends through the pseudopenis, which greatly constricts the canal and often leads to fatal complications for the mother and her first-born pup during birth. Drawing courtesy of Christine Drea, from Frank et al. [410].

Figure 9.4 Competition for food is fierce among spotted hyenas, which may favor highly aggressive individuals. A hyena clan can consume an entire giraffe in minutes. Photograph by Andrew Parkinson.

her clan kills or steals from lions (Figure 9.4) [408, 409]. With high social status comes dramatic gains in reproductive success. Top females are always able to hunt within the clan's territory, while subordinate females under some circumstances must travel outside their home ground to get enough to eat, a risky endeavor. The young of alpha females grow faster, are more likely to survive, and are more likely to become dominant themselves than are the offspring of subordinate hyenas (Figure 9.5) [541]. All these reproductive benefits of the hormonal mechanisms that promote aggression in adult females could possibly support some costly side effects, such as the pseudopenis.

But there are problems here. For one thing, if having a pseudopenis was strictly a liability, selection surely would have favored mutant alleles that happened to reduce the harmful effects that androgens (or other developmentally potent chemicals) have on female embryos. In addition, the extra androgen–extra aggression hypothesis is at odds with the finding that adult male spotted hyenas are less aggressive than females despite having the same or higher androgen levels than females (see Figure 9.2). If exposure to extra androgen in pregnant females really did lead to hyperaggressivity in their

Figure 9.5 Dominance greatly advances female reproductive success in the spotted hyena. A mother's social status is directly linked to cub survival to age 2 years and the dominance rank of her daughters as determined by observation of interactions between pairs of hyenas. After Hofer and East [541].

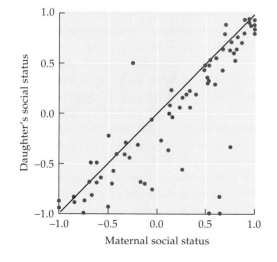

daughters, why wouldn't the hormone have the same effect on male embryos?

Issues of this sort have led some to wonder whether the pseudopenis might not have adaptive value in and of itself, despite the obvious costs of the thing. The first adaptationist hypothesis of this sort was offered by Hans Kruuk (the same biologist who studied mobbing in black-headed gulls; see p.177). In 1972 he suggested that the erect female genitalia were used to reduce tensions among the highly aggressive female members of a clan, promoting social bonds that kept the group together to work for the mutual benefit of all [678]. But Kruuk did not specify why an erect pseudopenis was required for bonding among females in this species when other hyenas have perfectly good greeting ceremonies based on inspection of their companions' anal glands. In these other species, females give birth via the vagina, with none of the complications that spotted hyenas regularly endure as a result of having to give birth through the clitoris.

Because of the deficiencies in the social bonding hypothesis, Martin Muller and Richard Wrangham have more recently offered another reason why the pseudopenis might be adaptive in and of itself. They argue that the structural mimicry of males by females provides a "male-like" camouflage that diverts aggression away from them, presumably because females are tricked into treating individuals with a pseudopenis as if they were males [858]. As Muller and Wrangham themselves realize, however, the evidence in favor of their idea is weak. For example, the hypothesis leads to the implausible prediction that aggressive females primarily use the visual cues associated with the external genitalia to identify the sex of a companion and so should have difficulty telling the difference between males and females. But because hyenas in particular, and mammals in general, rely heavily on olfactory, as well as acoustical, cues to discriminate between males and females [858], adult females almost certainly can quickly tell whether a fellow adult is a male or a female, particularly when inspecting the other individual at very close range in a greeting ceremony.

All of this brings us back to this ceremony in which the pseudopenis currently plays a role of some sort. Could this role be so important as to outweigh the reproductive handicaps imposed by the female's enlarged clitoris?

An Adaptationist Hypothesis

I can suggest a hypothesis on why female spotted hyenas might benefit from having a mimetic penis to use in their greeting ceremony, an idea that unites elements of Kruuk's "social harmony" explanation with aspects of Muller and Wrangham's "sexual mimicry" hypothesis. My proposal requires that we consider the evolutionary context in which the pseudopenis may have originated (note the complementarity of the different levels of analysis offered by hypotheses on the origin of a trait and those on its adaptive value). I suspect that the female-dominant, male-subordinate relationship evolved *before* the pseudopenis appeared. Furthermore, I suggest that one of the cues that sexually motivated, subordinate male hyenas supplied to dominant females they were courting was an erect penis. Presentation of this organ to the female would clearly signal the suitor's male identity and thus his subordinate, nonthreatening status. Females may have come over time to respond to this signal with acceptance of the male's presence rather reacting with aggression.

Subsequent to the evolution of this kind of male–female communication system, mutant females with a pseudopenis would have been able to tap into an already established system for signaling subordination, a useful thing to be able to do in a highly competitive society with a strict dominance hierarchy among females. The benefit to a subordinate of being able to signal acceptance of her low status in such a hierarchy would come from being allowed to stay with the clan, rather than being tossed out to fend for herself in an environment in which lone hyenas do not live for long, let alone reproduce. Hyenas that can live within a clan have a chance to reproduce eventually; hyenas that die young do not.

Evidence in support of this hypothesis comes from the observation that subordinate females and youngsters are far more likely than dominant animals to initiate interactions involving pseudopenile display, indicating that they gain by transferring information of some sort to dominant individuals [353]. Cooperation between subordinate and dominant hyenas might be advanced if the dominant female could accurately assess the physiological (e.g., hormonal) state of a subordinate by inspecting its erect, blood-engorged penis or clitoris. Perhaps truly submissive subordinates are signaling that they lack the hormones (or some other detectable chemicals) needed to initiate a serious challenge to the dominant individual, who can then afford to tolerate these others because they do not pose an immediate threat to her status. Note that this hypothesis resembles Kruuk's to the extent that "social harmony" follows from the subordinate's acceptance of its status relative to a more dominant individual. Moreover, male mimicry is also involved to the extent that subordinate females behave like males to demonstrate their rank relative to other females (but not to deceive their companions about their actual sex).

Discussion Question

9.3   How might you use comparative data to test the hypothesis that the use of a pseudopenis in the spotted hyena's greeting ceremony has adaptive value for the subordinate signaler as a means of demonstrating its willingness to submit to a dominant female? Take advantage of the fact that information exists on the behavior of two *nonsocial* species of hyenas [830, 912]. In addition, another highly social mammal, the naked mole-rat (see Figure 13.34), a member of the order Rodentia, is somewhat similar to spotted hyenas in that reproduction in any one group is monopolized by a single female; the "queen" is dominant to all of the many other females in her group with whom she interacts [781]. Although naked mole-rat females do not have masculinized genitalia, an enlarged clitoris is characteristic of female lemurs in Madagascar [909, 1238], and the trait appears in a number of other mammals as well [963].

The History of a Signal-Receiving Mechanism

The spotted hyena story illustrates, among other things, that we definitely do not know everything there is to know in behavioral biology. Having gone as far as we can go for the moment with the possible origin of the greeting ceremony and the possible adaptive value of the behavior to a signal giver, let us now focus on the evolution of the signal-receiving side of the equation. Our focal species will be the whistling moth, *Hecatesia exultans*, the males of which produce loud ultrasonic pulses of sound by beating their knobbed wings together (Figure 9.6, right). Although we cannot hear these sounds, other whistling moths can, as they demonstrate when they fly to a speaker playing a recording of whistling moth signals in the appropriate habitat (Figure 9.6, left) in an attempt to find and interact with the signaling male [12]. Thus, the sounds made by males may play a role in defense of a calling territory as well as in attracting receptive females to the caller.

Both male and female whistling moths possess mechanisms for hearing their species' high-pitched calls [1166]. The sound-receiving system consists in part of an ear on either side of the thorax, near the point where the hindwing attaches to the thorax. The whistling moth is a noctuid moth, a member of the same family as the bat-detecting moths we met in Chapter 4. Like the ears of those moths,

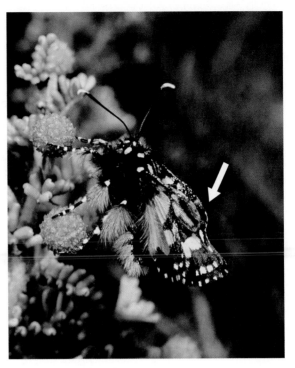

Figure 9.6 Ultrasonic communication. (Left) A male whistling moth has been attracted by the recorded calls of another male played from a small speaker in his territory, confirming that this species uses these sounds for communication. (Right) Male whistling moths produce ultrasounds by striking the knobby "castanets" (arrow) on their forewings together. Photographs by the author.

its outer ear is composed of a small ellipse of thin cuticle, called the tympanic membrane, which covers an air sac. When the membrane vibrates in response to ultrasound, the air sac also moves, supplying mechanical energy to associated sensory receptors. These mechanoreceptors then send signals to other parts of the nervous system. How might such a sophisticated device have evolved for detecting ultrasonic signals from other whistling moths?

Researchers interested in this question have operated with the same ground rules as those persons interested in the origin of the pseudopenis in spotted hyenas. They have assumed that in changing from the presumed ancestral state, which would have been no ears at all (even today, most moths lack ears), the whistling moth's ear probably did not arise via one or two mutations. The system is much too complex, and composed of too many interrelated parts, to have developed from a change in just one or two genes. Instead, the ear is much more likely to have been assembled gradually, with one small adaptive change layered on another, eventually producing the complex acoustical mechanism in its present form. But where do we go to test this hypothesis?

James Fullard and Jane Yack [425] decided to look for clues about the past in the bodies of today's moths. They examined some saturniid moths, members of a group that lacks ears and therefore cannot hear anything. They paid special attention to those parts of the saturniid thorax where one can find ears in noctuid moths (Figure 9.7). Fullard and Yack found mechanoreceptor cells attached to the saturniid thoracic cuticle that are remarkably similar in structure to the mechanoreceptor cells in the ears of noctuid moths, suggesting that the ears of noctuids were not built from scratch, but rather involved modifications of existing sensory cells that performed some other function for the earless moths.

What do these "nonacoustical mechanoreceptor cells" do for saturniids? They are attached to a part of the thorax that moves rhythmically as the wings move, particularly when the moth is vibrating its wings rapidly to generate metabolic heat prior to takeoff. These mechanoreceptors translate mechanical energy from the motion of the cuticle into messages that are relayed elsewhere in the moth's

Figure 9.7   Evolution of a sensory system. Sphingid moths, like saturniid moths, cannot hear, but noctuid moths can. The arrows in the upper panels point to the same location on the thorax of the two moths. At this site, sphingids do not have an ear with a tympanic cavity, but noctuids do. The lower panel shows the relationship between the thoracic plates (the two plain sheets at the top of the drawings), certain thoracic muscles (stippled bands), and a branched sensory nerve. Note the anatomical similarities between the two moths, especially with respect to the sensory nerve. The branch of the nerve labeled b1 carries information about the position of the hindwing in the sphingid; in the noctuid, this same branch relays acoustical information from the tympanic membrane to the moth's central nervous system. After Yack [1344] and Yack and Fullard [1345]; drawing of adult sphingid moth by Diane Scott.

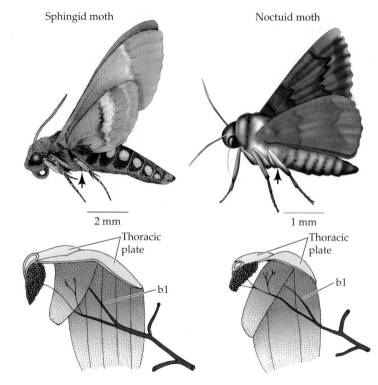

nervous system, where information about the alignment of various body parts is used to adjust the moth's position in space. A similar kind of mechanoreceptor in an ancestral moth in the noctuid lineage might have conferred a weak ability to hear certain loud sounds because loud sounds can make the insect exoskeleton vibrate [425]. Interestingly, an ancestral stretch receptor for monitoring body position has also evolved into an auditory receptor for detecting airborne sounds in other insects unrelated to moths [1230], demonstrating that the function of receptors can indeed change over time.

Almost certainly, the first hearing noctuid moth had nowhere near the sophisticated sensory skills of the current descendants of that moth. But with the addition of other small changes, such as a slight thinning of the cuticle over the "acoustical" mechanoreceptors, an enlargement of the respiratory chamber behind this part of the thorax, and a mechanoreceptor design that was somewhat more sensitive to sounds of particular frequencies, these mutant moths could have heard bat cries more reliably. Receivers with superior ultrasonic hearing might have survived better, reproduced more, and spread the genetic basis for these variations, thus setting the stage for the next slight improvement to spread through the species via natural selection.

The History of Insect Wings

Darwin recognized long ago that shifts in adaptive function occurred commonly over evolutionary time. In a wonderful book on the evolution of orchids, he wrote, "The regular course of events seems to be, that a part which originally served for one purpose, becomes adapted by slow changes for widely different purposes" [287]. This rule appears to apply to whistling moth hearing mechanisms, which are derived from systems that once monitored wing vibrations rather than airborne sound vibrations. Moreover, whistling moth ears now provide individuals with information about the acoustical signals made by other whistling moths, whereas the ancestors of whistling moths almost certainly used their hearing ability for a different purpose: the detection of ultrasound produced by bats.

The point that structures take on different functions over evolutionary time can be reinforced by examining where insect wings came from and identifying what the antecedent structures did for their owners. In modern whistling moths, wings are used both for flight and for sound production. The design of these wings and their control systems are extremely complex; they cannot have arisen in a single evolutionary step. Some persons believe that the precursors of insect wings were outgrowths of the respiratory apparatus—the gills—of an ancestral aquatic arthropod in which the gills were attached to the legs [45]. This ancestor endowed its modern descendants with a set of shared genes that now play a key role in the development of the body plan of crustaceans, insects, millipedes, horseshoe crabs, and spiders. Two of these genes, *pdm/nubbin* and *apterous*, are active in the development of the gills of modern crustaceans, which are also appendages attached to the legs of these animals. In horseshoe crabs, the same genes are turned on during the development of this animal's underwater breathing apparatus, the book gills. In two related groups of animals that left the water long ago, the spiders and the insects, these genes have been retained, but now contribute to the development of book lungs in spiders and wings in insects (Figure 9.8) [283].

An intermediate stage in the evolution of insect wings probably involved insects with immature aquatic life stages that had respiratory gill plates on the thorax that could also be moved as an aid to underwater locomotion (Figure 9.9) [681]. If some of these creatures retained their movable plates when they metamorphosed into terrestrial adults, then the structures could have been used as sails to send the newly formed adults skimming across the water to land. A speedy trip may have been advantageous if there were fish predators eager to

Figure 9.8   Arthropod gills have evolved into many different structures with different functions. In an aquatic ancestral arthropod, the gill plates served a respiratory function; they were appended to the legs of this animal. In the modern descendants of this ancient creature, the gills have retained their underwater respiratory role in crustaceans and have been modified somewhat as book gills in horseshoe crabs. In the three terrestrial lineages derived from the ancestral arthropod, the gill plates have been lost altogether from the myriapods, retained but modified as book lungs and spinnerets in spiders, and retained in highly modified form as wings in insects. After Damen et al. [283].

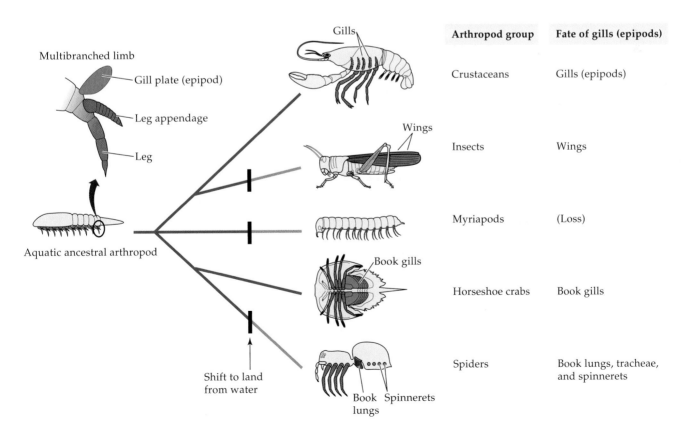

Arthropod group	Fate of gills (epipods)
Crustaceans	Gills (epipods)
Insects	Wings
Myriapods	(Loss)
Horseshoe crabs	Book gills
Spiders	Book lungs, tracheae, and spinnerets

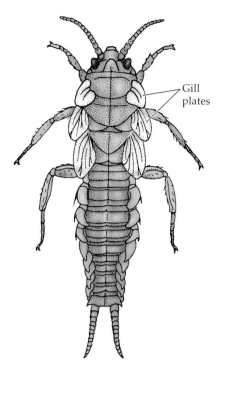

Gill
plates

Figure 9.9   Evolutionary precursors of insect wings? The movable gill plates of this extinct stonefly-like insect may have been used to propel the insect underwater. Structures of this sort may then have been used as sails by adult aquatic insects as they skimmed the water surface. After Kukalová-Peck [681].

capture the new adults as they floated on the water. Even today, some flightless species of stoneflies—an ancient group of aquatic insects—stand on the water when they emerge as adults and use their stumpy wings as sails to catch the wind (Figure 9.10) [773]. So does at least one mayfly species, a member of another ancient group of aquatic insects [1042]. (The mayfly is almost certainly descended from ancestors that flew, but it has lost this ability because it lives in waters where fish are extremely scarce; as the risk of predation has been reduced for this insect, the advantages of full flight have fallen correspondingly.) When Jim Marden and Melissa Kramer (an undergraduate at the time) clipped the wings of flying stonefly species so that the insects could not take off, the wing-clipped individuals could still sail along faster than they could swim as immatures. So here we have another demonstration that the winglets of ancestral flightless insects might have had survival value, and so could have spread through populations in the past, even though they did not confer the ability to fly [774, 775].

Once sailing wings had become standard for some ancient insect species, mutant individuals able to move these appendages in a rowing motion might have been able to scull across the water, something that a living Chilean stonefly can do [777]. Subsequently, other more powerful wing movements driven by larger muscles might have carried a rower or a skimmer faster across the water. These simple flappers would have set the stage for yet another modification, an increase in flight muscle capacity that could lift a powered skimmer off the water. Some modern stoneflies demonstrate how the transition between water skimming and full-fledged aerial flight might have come about. Adult stoneflies in the genus *Leuctra* beat their wings to achieve sufficient lift so that only the two hind legs remain in contact with the water. This position stabilizes the insect while permitting it to move its wings over a much wider arc than is possible for those stoneflies that keep all six feet on the surface of the water. Six-

Figure 9.10   A surface-skimming stonefly (left) standing on the water. (Right) The stonefly has raised its short, stubby wings, which it uses strictly to catch the wind as it skitters over the water while keeping all six feet on the water's surface. Photographs courtesy of Jim Marden.

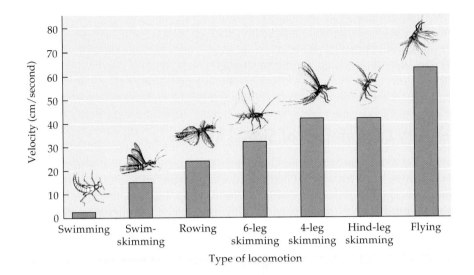

Figure 9.11   A possible evolutionary pathway from swimming to full flight in the stoneflies. Shown here are data on the mean speed of travel across or over water by living representatives of seven different kinds of stoneflies. The groups are organized by travel velocity from slowest to fastest. If evolution proceeded in the sequence shown here, each modified stonefly would have been able to move more rapidly across and then over water. After Marden and Thomas [777].

leg skimmers travel about 40 percent slower than hind-leg skimmers (Figure 9.11), showing how mutant individuals of some ancestral insect species that happened to use the hind-leg technique might have enjoyed a selective advantage [665]. From hind-leg skimming it is a short additional step to full flight, which would have carried the first true fliers away from insect-eating fish into the relative safety of the air. In fact, modern flying stoneflies employ almost exactly the same set of leg movements as they go from sitting on the water surface to jumping into the air as do the modern hind-leg skimmers, which demonstrates the plausibility of the scenario in which a hind-leg-skimming ancestral species gave rise to a fully flighted descendant [776].

Discussion Question

9.4   The diversity of behavior among living stoneflies suggests a historical scenario that goes from floating on the water surface to jumping from the water surface and engaging in free flight (see Figure 9.11). This historical hypothesis can be tested by plotting the behavior of stonefly species on a phylogeny of this group (see [1187] or http://www.bio.psu.edu/People/Faculty/Marden/skim.html). Given the available evidence, what is your assessment of the validity of this hypothesis?

By the time certain aquatic insects had evolved into terrestrial species, what were once gills in their ancestors had been fully modified to serve as functional wings capable of powered flight. With the advent of powered flight, mechanoreceptors in the thorax that once provided sensory information about the position of sailing wings relative to the thorax now could have helped their owners control their wings by monitoring wing-flapping movements. When the ancestor of moths appeared on the scene, this species probably already possessed wing position sensors. Over time, these mechanoreceptors gradually evolved into ultrasound detectors with a new antipredator function (see p. 108). However, ears that evolved because of their anti-bat advantages can also detect ultrasound coming from other sources, including members of one's own species. The whistling moth is active during the day, so it does not have to contend with bat predators. Thus this moth's ear was free to take on yet another function—the detection of ultrasonic signals produced by wing-beating male conspecifics [1166].

Sensory Exploitation of Signal Receivers

According to the scenario outlined above, male whistling moths use ultrasound to communicate with other whistling moths because an ancestor of *Hecatesia exultans* was subjected to predation by bats, which made perception of ultrasound reproductively advantageous in the past. Once the ultrasonic detectors were in place, their existence affected the reproductive payoff to mutant individuals that happened to generate signals in the ultrasonic range, which receivers, because of their evolutionary baggage, could hear more easily than sounds of lower frequencies.

Likewise, I suggested earlier that we could better understand the female spotted hyena's use of her pseudopenis in greeting displays if this species already had evolved a male–female communication system that enabled males to use their penis to signal their subordinate status to females. By the same token, in one group of bowerbirds, males use a harsh and raspy "skrraa" call when courting females (a subject explored in more detail in the next chapter), almost certainly because this call has taken on a new role after originally having evolved as a male–male aggressive signal (Figure 9.12) [123]. In cases of this sort, biologists talk of a trait having been *co-opted* for a new function, as in the case of bird wing feathers, which took on their flight function after modifications to feathers that had evolved originally because of their value as thermoregulatory or courtship display devices.

When signal givers tap into preexisting perceptual mechanisms in a signal receiver, we have a case of **sensory exploitation** [69, 1046]. In one of the best-

Figure 9.12 **An ancestral signal has been co-opted** in some bowerbirds. In most or all bowerbirds, males use the "skrraa" call to threaten rivals. In one cluster of closely related species, this signal now also serves a courtship role. After Borgia and Coleman [123].

(A)

(B)

Figure 9.13   Sensory exploitation and the evolution of a courtship signal in the water mite *Neumania papillator*. (A) The female (on the left) is in her prey-catching position (the net stance). The male approaches and waves a trembling foreleg in front of her, setting up water vibrations similar to those a copepod might make. The female may respond by grabbing him, but releases him unharmed. (B) The male then deposits spermatophores on the aquatic vegetation in front of the female before waving his legs over them. After Proctor [979].

studied cases of this sort, Heather Proctor hypothesized that modern courtship by male water mites began when males happened to "exploit" the predatory behavior of females waiting to ambush small aquatic invertebrates called copepods [979]. While the predatory female is in her attack position, called the "net stance," the male vibrates a foreleg in front of her. She, in turn, may grab him, using the same response that she uses to capture passing copepods. However, she then releases the male unharmed. He turns around and places spermatophores (packets of sperm) near the female, which she picks up in her genital opening if she is receptive (Figure 9.13).

The apparent use of a predatory grab by the female in response to the male's trembling display suggested to Proctor that males were mimicking the stimuli produced by copepod prey. Perhaps the female's reaction identified her as a potential mate and showed the male where to position his spermatophores to best effect (water mites cannot see). If males trigger the prey detection response of females, then unfed, hungry female water mites held in captivity should be more responsive to male signals than well-fed females. They are, providing support for the contention that once the first ancestral male happened to use a trembling signal, the male's behavior and its hereditary basis spread because it effectively activated a preexisting prey detection mechanism in females [979].

Among the water mites, trembling courtship occurs in only a few species; namely, those in which females adopt the net stance. By measuring a large number of characteristics in several species of water mites, Proctor produced a phylogeny for these animals. The phylogeny indicates that the net stance of females originated in an ancestral water mite that eventually gave rise to eight descendant species. Within this lineage, male courtship trembling subsequently appeared twice, once in the ancestor of the genus *Unionicola* and once in the ancestor that split into two species of *Neumania* (Figure 9.14). If this phylogeny is correct, it would mean that male courtship trembling might have originated well after females had adopted a copepod-ambushing lifestyle and had evolved a sensitivity to underwater vibrations associated with those prey [980]. Courting males that first happened to mimic those vibrations improved their chances

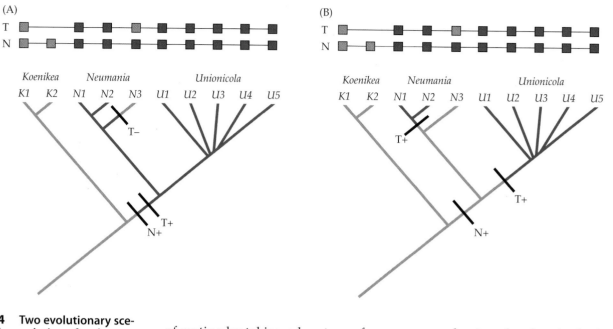

Figure 9.14 Two evolutionary scenarios for the evolution of male courtship trembling in water mites of the genera *Koenikea* (two species), *Neumania* (three species) and *Unionicola* (five species). (A) Male courtship trembling (T) may have evolved at about the same time that predatory females adopted the net stance (N) in the ancestor of *Neumania* and *Unionicola* (N+, T+), with trembling being lost in the line leading to one species of *Neumania* (T–). (B) Another equally parsimonious scenario calls for the evolution of the net stance in the ancestor of the two genera *Neumania* and *Unionicola*, but instead of one origin for trembling, there might have been two origins, one in the ancestor of *Unionicola* and another in the ancestor of *Neumania papillator* (N1) and its close relative (N2). After Proctor [980].

of mating by taking advantage of a sensory mechanism that females had evolved for another reason.

Discussion Question

9.5 Females of an African cichlid fish lay their eggs in depressions in the lake bottom made by males [1291]. The female picks up her large orange eggs in her mouth almost as quickly as she lays them. (She will protect the fertilized eggs and the fry, when they hatch, by holding them in her mouth.) As the female lays her eggs, the male who made the "nest" may move in front of her and spread his anal fin (Figure 9.15), which is decorated with a line of large orange spots. The female moves toward him and attempts to pick up the objects on the fin. As she does, the male releases his sperm, which are taken up in the female's mouth, where they fertilize her eggs. Use sensory exploitation theory to explain the evolutionary origins of the male's behavior. Was the first male to use this signal taking advantage of his mate?

The argument that has been developed for water mite courtship has also been applied to the signals and responses of male and female guppies. Females in some populations of this small fish prefer to mate with males that have bright

Figure 9.15 **A female cichlid fish** (left) is attracted to the anal fin of a male by the orange spots on the fin.

(A)

(B)

Figure 9.16   **Food, carotenoids, and female mate preferences in the guppy.** (A) Males have to acquire orange pigments from the foods they eat, like this *Clusia* fruit that has fallen into a stream where guppies live. Males that secure sufficient amounts of carotenoids incorporate the chemicals into ornamental color patches on their bodies. (B) Females (the larger fish on the right) find males with large orange patches more sexually appealing than males without them. Photographs by Greg Grether.

orange spots on their skin (Figure 9.16) [478]. Male guppies cannot synthesize the orange pigments that go into their body coloration, but instead have to acquire these chemicals, labeled carotenoids, from the plants they eat. Those that get enough carotenoids from their food become attractive to females, but why?

One hypothesis on the origin of the female preference for orange-spotted males suggests that when this mating preference first appeared, it was a by-product of a sensory preference that had evolved in another context. This hypothesis has received support from observations of female guppies feeding avidly on rare but nutritionally valuable orange fruits that occasionally fall into the Trinidadian streams where guppies live [1026]. Thus, it is possible that females evolved visual sensitivity to orange stimuli because of the foraging benefits associated with this ability, not because of any fitness benefits from selective mate choice. If so, then female guppies are predicted to respond more strongly to orange stimuli than to other colors when feeding. In fact, females guppies do approach and try to bite orange discs far more often than discs of any other color. The research team that studied this phenomenon was able to take advantage of the fact that female guppies living in different streams differ in how strongly they prefer mates with orange spots. The strength of the mate choice preference was matched by the relative rates at which females pecked at orange discs presented to them underwater (Figure 9.17).

Sensory Preferences May Precede the Evolution of a Signal

If sensory exploitation is a major factor in the origin of effective signals, then it should be possible to create novel experimental signals that generate responses from animals that have never encountered

Figure 9.17   **Sexual preferences for orange spots** match foraging preferences for orange foods by female guppies. The strength of the female response to orange spots on males varies from population to population and is proportional to the strength of the foraging response to an orange disc. Each point represents a different guppy population living in a different stream. After Rodd et al. [1026].

(A)

(B)

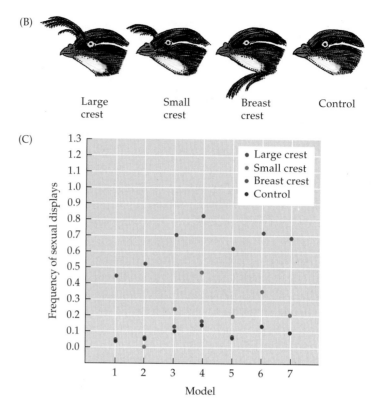

Large   Small   Breast   Control
crest   crest   crest

(C)

Figure 9.18   The response of least auklets to three novel artificial signals. (A) A stuffed male least auklet with an artificial crest of the sort used in an experiment on female sexual preferences in this species. (B) A diagram of some of the heads of models used in the experiment. (From left to right) The control (which lacks a crest as do males of the least auklet), a small crest model, a large crest model, and a breast crest model. (C) The models with large crests elicited the highest frequency of sexual displays by the females during a standard presentation period. A, photograph by Ian Jones; B–C, after Jones and Hunter [610].

those stimuli before. To test this prediction, researchers have played sounds to frogs that contain acoustical elements not present in the species' natural calls [1047, 1048], attached strips of yellow plastic to the tails of male platyfish [69], added feather crests to the heads of least auklets [610], glued plumes to the tails of house finches [1203], and supplied male sticklebacks with bright red spangles to add to their nests [907]. In all of these cases, except for the house finch study, researchers found that the artificial attributes elicited stronger reactions from females than the natural alternatives (Figure 9.18).

Although these data provide support for the sensory exploitation hypothesis, there is an alternative explanation for the response of the test species to these elaborate artificial courtship signals: perhaps the test species evolved from populations that used similar signals in the past. If so, then today's descendants of these ancestral populations might still retain the old sensory preferences even though they no longer exhibit the complex signals themselves [1051]. This conjecture is plausible, given the numerous cases in which elaborate male traits used in courtship and aggression have been lost after having once evolved, as indicated by the fact that all the close relatives of the ornament-less species possess the ornaments in question (which suggests that their mutual common ancestor did as well) [1297]. An example is provided by the lizard *Sceloporus virgatus*, which lacks the large blue abdominal patches of many other members of its genus. These other species use their blue patches in the male threat posture, in which a lizard elevates and compresses its body laterally, making the abdomen visible. But although the signal has apparently been lost in *S. virgatus*, the behavioral response to the signal has not, as shown when some experimental subjects had blue patches painted on them. Lizards that observed a rival giving a threat display enhanced by the blue paint were far more likely to back off than were lizards that saw a displaying opponent that did not have blue patches painted on its body (Figure 9.19) [997].

But might some signal receivers have a preference for signals that neither they nor their ancestors ever possessed or responded to? One way to test this

Figure 9.19   Receivers can respond to an ancestral signal not present in their species. Lizards of a species whose relatives have blue patches on the abdomen are more likely to abandon a conflict when confronted by a conspecific rival with blue patches painted on its abdomen than when seeing the threat display of an unmanipulated control individual or one that has had white patches or black dots painted on the abdomen. After Quinn and Hews [997]; photograph by Paul Hamilton.

Sceloporus virgatus

hypothesis is to provide subjects with a novel test trait not exhibited by *any* closely related species. Under these conditions, the test species' ancestors are unlikely to have exhibited a **sensory bias** for the test trait (based on specialized attributes of their perceptual mechanisms), and so they cannot have passed a preference for the trait in question on to their descendant species. For example, the modern fish species *Priapella olmecae* does not have a colorful, elongate caudal fin (a "swordtail"), nor is it likely that its ancestor had one (Figure 9.20). And yet when Alexandra Basolo provided males of this fish with an artificial yellow sword, females found males with this novel trait highly attractive;

Figure 9.20   Sensory exploitation and swordtail phylogeny. The genus *Xiphophorus* includes the swordtails, which have elongated caudal fins, and the platyfishes, a group without tail ornaments. Because the closest relatives of the platyfishes and swordtails belong to a genus (*Priapella*) in which males lack long tails, the ancestor of *Xiphophorus* probably also lacked this trait. The long tail apparently originated in the evolutionary lineage that diverged from the platyfish line. Even so, females of the platyfish *X. maculatus* find males of their species with experimentally lengthened tails more attractive, suggesting that they possess a sensory bias in favor of these kinds of tails. After Basolo [69].

furthermore, the longer the sword, the greater the female's desire to stay close to the amended male. Basolo concluded that a female preference for long tails evolved in an ancestor of *Priapella* before a long tail ever evolved in this lineage [70].

Nancy Burley and Richard Symanski reached a similar conclusion from their experiments, which involved outfitting zebra finches and long-tailed finches with rather ridiculous-looking white feathers glued to their heads. These Australian finches do not have crests, nor do any of their close relatives, so presumably, neither did their ancestors. Yet females of both species associated more with the novel white-crested males than with normal-looking males (Figure 9.21) [174]. A result of this sort suggests that the origins of the naturally occurring exaggerated crests in some birds could lie in a series of mutations that resulted in longer-than-average crest feathers, for which females of these species had a previously unexpressed aesthetic preference. In addition, the fact that female zebra finches favor males with a red feather glued to their forehead if they were reared by fathers similarly ornamented shows that the preexisting capacity for sexual imprinting could promote the spread of a novel trait, once it originated by chance in a male [1330].

But why should female zebra finches find males that look like dancers in a Las Vegas revue irresistible, while females of *Priapella olmecae* prefer males with an absurdly long artificial tail? In the case of the zebra finch, one suggestion is that females are predisposed to favor white feathers on their partners because they line their nests with white feathers [174]. This hypothesis is consistent with the failure of females to favor males with headdresses of red or green feathers, but it has not yet been tested. Likewise, for the female fish of short-tailed species

(A)

(B)

Figure 9.21   Mate preferences for a novel ornament. (A) A male long-tailed finch (left) and a male zebra finch (right) have been outfitted with bizarre white plumes. (B) The addition of white plumes made male zebra finches more attractive to females than control males without plumes or those given headdresses of red or green feathers. A, photograph by Kerry Clayman, courtesy of Nancy Burley; B, after Burley and Symanski [174].

that respond to males with artificial long tails, one could argue that the preference may not be for long tails per se, but rather for any stimulus associated with large and healthy males [1007]. However, when Basolo examined the relative effects of large body size versus a long tail on female preferences, she found that body size was the less important factor, suggesting that, in this species at least, females are indeed focused on the tail ornament itself, for reasons that are still mysterious [71].

A key point about sensory exploitation is that when mutant males did appear with traits that activated hidden, preexisting aesthetic preferences in females, those males were probably "exploiting" the females only in the proximate sense of activating sensory equipment that had evolved for some other purpose. At the ultimate level, females that responded positively to males with the novel attribute could have gained fitness for any of several reasons. Perhaps their sons could inherit the sexually attractive trait and produce many offspring in turn, or perhaps males capable of producing exaggerated courtship signals were physiologically capable of helping to rear a choosy female's offspring. (We will have more to say about the adaptive value of female choice in the next two chapters.) If it had been reproductively harmful for females to respond positively to males with novel courtship signals, those females in the species that happened to have sensory equipment resistant to exploitation would presumably have had higher reproductive success, leading eventually to the replacement of females predisposed to be taken in by their sexual partners.

The sensory exploitation hypothesis represents a special case of the historical argument that what has already evolved influences what kinds of additional changes are possible and which are not. If natural selection had been given the job of coming up with a jet plane, it would have had to start with what was already available, a propeller-driven plane, changing this "ancestral" form piece by piece, with each modification flying better than the preceding one, until a jet was built [302]. In nature, evolutionary transitions must occur this way because selection cannot start from scratch. As a result, the products of evolution often appear to be jury-rigged devices that might have been designed by Rube Goldberg, a cartoonist famous for his ridiculously overcomplicated inventions designed for everyday tasks.(Goldberg died in 1970 but his wacky cartoons can still be seen at http://www.rube-goldberg.com/). So, for example, the panda's thumb is not a "real" finger at all, but instead a highly modified wrist bone (the panda's hand has the same five digits that most vertebrates have *plus* a small thumb-like projection from the radial sesamoid bone, which is present, although as a much smaller bone, in the feet of bears, dogs, raccoons and the like)[461]. Why do pandas have their own special thumb? According to Stephen Jay Gould, pandas evolved from carnivorous ancestors whose first digit had become an integral part of a foot used in running. As a result, when pandas evolved into herbivorous bamboo eaters, the first digit was not available to be employed as a thumb in stripping leaves from bamboo shoots. Instead, selection acted on variation in the panda's radial sesamoid bone, which now can be used in a thumb-like manner by the bamboo-eating panda.

The "panda principle" can be seen in dozens of other cases. Consider the persistence of sexual behavior in parthenogenetic whiptail lizards. These species are composed entirely of females, and yet if a female is courted and mounted by another female (and females do engage in such pseudomale sexual behavior, for reasons that are not fully understood), she is much more likely to produce a clutch of eggs than if she does not receive this sexual stimulation from a partner (Figure 9.22) [259]. The relationship between courtship and female fecundity in unisexual lizards obviously exists because these reptiles had sexual ancestors. The parthenogenetic females retain characteristics, such as a need for courtship, that their nonparthenogenetic ancestors possessed—characteris-

Figure 9.22   The panda principle is evident in the sexual behavior of a parthenogenetic whiptail lizard. In the left column, a male of a sexual species engages in courtship and copulatory behavior with a female. In the right column, two females of a closely related parthenogenetic species engage in very similar behavior. After Crews [259].

tics that a biological engineer would surely eliminate if he or she could play God by designing an all-female species on paper and then creating it in one go. Natural selection, however, cannot play God, because it is a blind process with no goal in mind and no means to get to a predetermined endpoint.

Discussion Question

9.6   Although natural selection is blind, the products of this process are often amazingly complex. In order to explain how a blind process dependent on random events (mutations) can generate such complexity, Richard Dawkins has provided us with an analogy [303]. He invites us to imagine that a complex current trait is like an English sentence; for example, a line from Shakespeare's *Hamlet:* METHINKS IT IS LIKE A WEASEL. The odds that a monkey would produce this line by tapping at a typewriter are vanishingly small, one in 10,000 million million million million million million. These are not good odds. But if instead of trying to get a monkey or a computer to get the "right" sentence in one go, let's change the rules so that we start with a randomly generated letter set such as SWAJS MEIURNZMMVASJDNA YPQZK. Now we get a computer to copy this "sentence" over and over, but with a small error rate. From time to time, we ask the computer to scan the list and pick the sequence that is closest to METHINKS IT IS LIKE A WEASEL. Whatever "sentence" is closest is used for the next generation of copying, again with a few errors thrown in. The sentence in this group that is most similar to METHINKS … is selected to be copied, and so on. Dawkins performed the experiment and found that it took 40 to 70 generations to reach the target sentence—a few seconds of computer time, not years. What was Dawkins's main point in illustrating what he called "cumulative selection?" In what sense might you say that the sentence analogy more closely resembles artificial selection than natural selection?

Adaptationist Questions about Communication

Sensory exploitation theory has directed our attention to the importance of past changes in shaping the evolution of a new communication signal. Information on the origins of a trait does not, however, make it unnecessary to consider why a signal, once having appeared, was maintained in a species over time. For example, knowing that female guppies probably were sensitive to the color orange because of its association with a favored food does not eliminate the possibility that using this preference in a sexual context raised the reproductive success of the sensory-biased females. If female mate choice based on orange food-mimicking spots on male bodies increased female fitness, then this adaptive effect would have contributed to the selective maintenance of the color bias and its use in sexual encounters as well as when feeding. The fact that male guppies appear to boost their immune systems as well as their attractiveness by consuming carotenoids suggests that by advertising their carotenoid intake via orange spots, males may be signaling their healthiness, hardly an irrelevant factor for female reproductive success [479].

More generally, an interest in the possible adaptive value of various elements of communication systems has resulted in the solution of a class of puzzles that differ from those having to do with the origin and subsequent modification of signals. Darwinian puzzles in communication come to the surface when observers see signal givers or signal receivers apparently disadvantaged by their actions. If giving a signal, or paying attention to a signal, lowered the fitness of the signalers or receivers relative to others, then their behaviors should eventually disappear.

Because he had this selectionist logic in mind, Bernd Heinrich knew he had something worth studying when he heard a mob of ravens yelling loudly while feeding together on a dead moose that a poacher had left in a Maine forest (Figure 9.23) [532, 533]. Ravens are uncommon birds in Maine, and yet somehow fifteen of them had assembled at the hidden carcass, almost certainly because some of the first birds to find the moose had yelled loudly to the others. The behavior did not make sense to Heinrich. Why attract competitors to a food bonanza? Why not stay silent and eat moose meat all winter long by yourself instead of sharing the bounty with other ravenous ravens? In other words, why hadn't selection eliminated any hereditary tendency of ravens to yell at carcasses?

The puzzle might have been solved if the aggregated birds were one big family, with parents yelling to attract distant offspring to the feast. Heinrich felt that this was unlikely, since a pair of ravens produces a maximum of six offspring per breeding season. Subsequently, DNA fingerprinting studies showed unequivocally that these feeding flocks were indeed composed of unrelated individuals [533].

Figure 9.23   A group of ravens feeding on a carcass to which they were attracted by a yelling companion. Photograph by Bernd Heinrich.

Then perhaps ravens yell because the signals get the attention of a bear or coyote, which can open up a tough-skinned moose, eventually providing the caller with access to meat that it could not get in any other way. To test this hypothesis, Heinrich hauled a dead 150-pound goat through the Maine woods, taking it out to various sites during the day and storing the none-too-sweet-smelling carcass in his cabin at night to prevent its loss to a nocturnal scavenging bear. Ravens occasionally approached the decaying goat, though only after making Heinrich wait for hours in cramped and bitterly cold blinds. But contrary to his prediction, the birds never yelled when they found the bait. Moreover, Heinrich sometimes observed ravens yelling at carcasses that had already been ripped open. These findings forced him to abandon the "attract-a-carcass-opener" hypothesis.

Instead, he switched his attention to an alternative explanation, an idea stimulated by seeing how cautiously some ravens approached carcasses when they first found them. Perhaps, he argued, a carcass discoverer yells to draw in other ravens so that if a predator lurks nearby, the other birds will provide possible targets, reducing the yeller's risk of being taken by a hiding coyote or fox. The incoming birds would be attracted because they would gain a wealth of food in return for taking a small risk of being the unlucky victim of a predator. However, this hypothesis generates the prediction that once a group has assembled and feeding has begun, the birds should shut up to avoid attracting more ravens, which would be unwanted and unnecessary for safety purposes. The observation that yelling continued at baits that had already acquired a retinue of actively feeding birds convinced Heinrich to discard the "diluted-risk-of-predation" hypothesis.

As Heinrich continued stoically lugging dead goats into the Maine woods, he came to realize that whenever he saw a single bird or a pair at a feeding site, those ravens were quiet. Yelling occurred only when three or more ravens were present, and it was then, and only then, that large numbers of other ravens

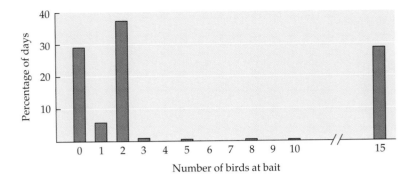

Figure 9.24   Yelling is a recruitment signal. The graph shows the percentage of days on which carcass baits were visited by various numbers of ravens. Carcasses were exploited either by nonyelling territorial singletons and pairs or by large groups of ravens, many of them yelling, most of them young, nonterritorial birds. After Heinrich [532].

came to the area. Heinrich knew that older adult ravens form pairs that defend a territory year-round. Unmated young birds usually travel solo over great distances in search of food. If a singleton attempts to feed in a resident pair's territory, the pair attacks it. Heinrich wondered if yelling could be a signal given by nonterritorial intruders, a signal that attracts other unmated wanderers to a food bonanza that they can exploit if they can overwhelm the defenses of the resident pair.

This "gang-up-on-the-territorial-residents" hypothesis leads to a number of predictions: (1) resident territory owners should never yell, (2) *non*resident ravens should yell, (3) yelling should facilitate a mass assault on a carcass by nonresident ravens, (4) resident pairs should be unable to repel a communal assault on their resources, and (5) a food bonanza should be eaten either by a resident pair alone or by a mob of ravens. Heinrich collected data that supported all of these predictions (Figure 9.24) [532]. He concluded that when a young raven yells, the consequences of providing information ("Food bonanza here") to other nonterritorial birds can include benefits (personal access to food for the yeller) that outweigh the energetic costs of yelling as well as the risk of attack by the resident ravens guarding the carcass [533]. Yelling therefore has the properties of an adaptation because it can yield net fitness benefits to individuals who give the signal under the appropriate circumstances.

Why Do Baby Birds Beg So Loudly?

Heinrich knew how to use the adaptationist approach to clear up a mystery, and other biologists have tried to do the same with other signals that at first glance seem to impose high costs on signalers. For example, you might think it would be suicidal for nestling songbirds to break into noisy begging calls whenever a parent returns to the nest with food. Their loud cheeps and peeps could give a listening hawk or raccoon all the information it needs in order to locate the nest and make off with the baby birds. In fact, when tapes of begging tree swallows were played at an artificial swallow nest containing a quail's egg, the egg in that "noisy" nest was taken or destroyed by predators before the egg in a nearby quiet control nest in 29 of 37 trials [704].

Further evidence for the fitness costs of begging comes from a study of differences in the begging calls of warbler species that nest on the ground versus those that nest in the relative safety of trees [519]. The young of ground-nesting warblers produce begging cheeps of higher frequencies than their tree-nesting relatives (Figure 9.25A). These higher-frequency sounds do not travel as far, and so may better conceal the individuals that produce them, which are especially vulnerable to predators in their low-lying nests. David Haskell created artificial nests with clay eggs and placed them on the ground by a tape recorder that played the begging calls of either tree-nesting or ground-nesting warblers.

(A)

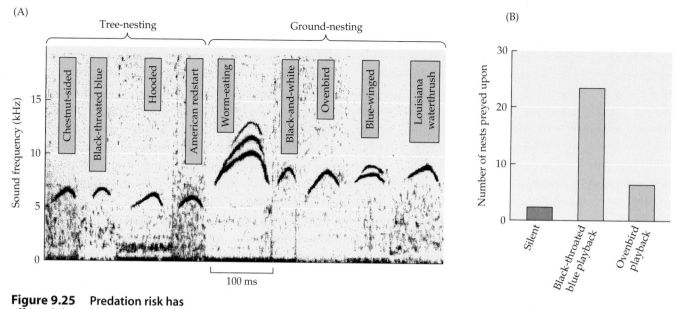

(B)

Figure 9.25   Predation risk has affected the evolution of begging calls in warblers. (A) The sound frequencies of the nestling begging calls of ground-nesting warbler species are higher than those of calls used by young tree-nesting warblers. (B) Experimental playback of the begging calls of the black-throated blue warbler (a tree-nesting species) at artificial nests placed on the ground resulted in higher rates of discovery by predators than did playback of begging calls of the ground-nesting ovenbird. After Haskell [519].

The eggs advertised by the tree-nesters' begging calls were found and bitten significantly more often than the eggs associated with the ground-nesters' calls (Figure 9.25B).

The warbler study suggests that begging calls have evolved properties that reduce their potential for attracting predators. If so, then perhaps baby birds of species that experience high rates of nest predation should also produce softer begging signals of higher frequencies than nestlings of other species less often victimized by nest predators. This prediction was supported by data collected in one survey of 24 species from an Arizona forest [146], more evidence that predation pressure favors the evolution of begging calls that are hard to detect and pinpoint.

Finally, some parent birds give alarm calls when danger threatens the nest, and in at least one species, these calls can induce the baby birds to shut up, which provides another avenue for the reduction of the costs of the nestlings' vocalizations [966]. But all the adjustments we have reviewed merely reduce, rather than eliminate, the risk that a predator will destroy a gaggle of baby birds whose calls seem to say, "Come and eat me." So why make noise at all if the consequence can be lethal?

Currently two hypotheses about begging behavior are under investigation. The "sibling scramble competition" hypothesis focuses on how baby birds might use their vocalizations and other begging responses to compete with nestmates to secure food from their parents in ways that maximize the begging bird's fitness. In contrast, the "honest signaling" hypothesis looks at nestlings as producers of **honest signals** (messages that convey accurate information, in this case about the need for food) that enable their parents to deliver food as effectively as possible (that is, in ways that will maximize parental fitness). According to this idea, any attempt by nestlings to be deceptive or to otherwise disadvantage their parents should generally select for adults that ignore the manipulative "information" the offspring provide. Perhaps surprisingly, the two hypotheses yield similar predictions, such as the prediction that parents should provide food in relation to the intensity of begging by their offspring. Unfortunately, this overlap in predictions makes it difficult to discriminate between the two hypotheses, although the scramble competition hypothesis assumes that the power is more or less in the beaks of the offspring, while the honest signaling hypothesis assumes that the parents are in control [1038].

Figure 9.26 **Testosterone affects begging rate and feeding rate in black-headed gull chicks.** (A) Adult black-headed gull with a chick. (B) Researchers took pairs of first-laid eggs produced on the same day and gave one egg a supplemental injection of testosterone (T) while the other (the control) received an injection of oil. The two eggs were then given to a pair of nesting adults whose own clutch had been removed. When the chicks hatched, those that had received extra T were more likely to approach the adults first, were more likely to "pump" first ("pumping" is an up-and-down motion of the head that is a part of the begging display), and wound up eating more than their oil-treated foster siblings. A, photograph by Corine Eising; B, after Eising and Groothuis [359].

(A)

(B)

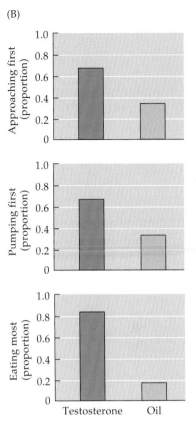

Discussion Question

9.7   In the black-headed gull, the female lays three eggs, but begins incubating before all of the eggs are laid. As a result, the first egg laid gets a head start and produces a senior chick that typically is larger than its siblings. It therefore begs more effectively and usually gets more food than its brothers and sisters. But mother gulls put more androgens into eggs that will hatch later, and the extra androgens enable junior gull chicks to beg more vigorously and claim more food than they would otherwise (Figure 9.26) [359]. How does this example illustrate the difficulties in establishing who has the upper hand in the signaling interactions between young birds and their parents?

In testing the two possible foundations for begging behavior, some researchers have been able to determine how to discriminate between them, and not just in the case of baby birds [1122]. In general, if competition among siblings affects their begging behavior, then it follows that youngsters should adjust their signals in relation to those produced by their siblings, since they are all competing for parental attention. When experimentally food-deprived baby robins were placed in a nest with normally fed siblings, the hungry nestlings begged more loudly than usual—but so did their better-fed siblings [1124], providing evidence that scramble competition for food does occur within robin nests.

On the other hand, we can eliminate competition among siblings as a factor influencing begging tactics by studying a bird species in which parents rear only a single chick. In Wilson's storm-petrels, parents leave their one and only chick in a nest burrow and provision it with regurgitated food at 2-day intervals. When Petra Quillfeldt examined begging by these chicks, she found that they produce a specific call that is used only when a parent comes to the burrow to deliver food (Figure 9.27). In order to evaluate the relationship between chick need and begging behavior, Quillfeldt kept track of the age and body weight of her subjects. These data enabled her to estimate the body condition of the young petrels. Some were in poor condition (well below the average weight for a chick of their age), some were in average condition (at about the mean weight for their age group), and some were in good condition (heavier than average). Chicks in poor condition were presumed to be hungrier than average, and these individ-

(A)

(B)

Time (seconds)   Time (seconds)

Figure 9.27   An honest signal of hunger? (A) A Wilson's storm-petrel chick. Although these chicks do not have nestmates with whom to compete, they do possess a special begging call, produced when a parent returns to the nest burrow with food. (B) Shown here are two records with six and five "long" calls, respectively. The more long calls given, the needier the chick. A, photograph courtesy of Petra Quillfeldt; B, after Quillfeldt [995].

uals produced begging calls at a faster rate, and gave more calls in all, than petrel chicks that were in better condition [995].

Thus, in at least some cases, begging appears to provide an honest signal of need (or perhaps chick viability) that parents could use to decide how much to invest in feeding that chick. But the question arises, why doesn't a Wilson's storm-petrel chick beg especially vigorously even when it isn't all that hungry? In so doing, it might be able to secure food at a higher rate, which should result in more rapid growth or a larger size, either one of which could be advantageous. The young of brood parasites manage this trick and benefit greatly. A young European cuckoo, for example, is able to sound like an entire nest full of baby reed warblers (Figure 9.28), with the result that its host parents work as hard for it as they do for a complete brood of four warbler chicks [639]. If you place another nestling in a reed warbler's nest, choosing a species comparable in size to a cuckoo—say, a baby blackbird—the young blackbird begs less intensely than a cuckoo and, accordingly, is fed less rapidly. But play a recording of a cuckoo's begging calls when the warblers come to the nest, and the lucky blackbird receives a roughly 50 percent increase in the number of loads of food carried to it per hour by its super-stimulated host parents (Figure 9.29) [297].

Although vigorous begging must come with an energetic cost, measurements of the metabolic expenses incurred by begging nestlings demonstrate that this cost is probably small relative to the potential caloric gain [47, 710]. Therefore, the disadvantages of exaggerated begging may lie not in energy expended, but in the damage that any successful food hog might do to its siblings in species in which several offspring are cared for by the same two parents. An individual's success in propagating his or her genes can be affected by more than just

(A)

(B) Reed warbler chick

(C) Reed warbler brood

(D) Cuckoo chick

Time (seconds)

Figure 9.28   The European cuckoo chick's begging call matches that of *four* baby reed warblers. (A) A cuckoo chick begging for food from its reed warbler foster parent. The calls shown below the photograph are those of (B) a single reed warbler chick, (C) a brood of four reed warblers, and (D) a single cuckoo chick. A, photograph by Roger Wilmshurst; B–D, after Davies et al. [297] .

his or her own personal reproductive success (see p. 450). Those animals that harm their close relatives may, in effect, be destroying some of their own genes. Therefore, a begging nestling that happened to do very well at securing food from its parents, but at the expense of its siblings, might actually leave fewer copies of its genes overall than others that held back when begging for food.

(A)

(B)

Figure 9.29   The cuckoo's begging calls stimulate more frequent feeding by its host parents. (A) A blackbird chick has been placed in a reed warbler nest. A speaker is visible in the background. (B) If the begging calls of a cuckoo chick are played to the reed warbler host parents when they visit the nest, the parents deliver more food to the blackbird, matching the amount they bring when a recording of an entire brood of reed warblers is played. A, photograph courtesy of Nick Davies; B, after Davies et al. [297].

Discussion Questions

9.8   The young European cuckoo disposes of all potential competitors for food by pushing its hosts' eggs and babies out of the nest. The cowbird, another parasitic species, behaves differently, often coexisting with one or more of its hosts' offspring. Why is the cowbird's behavior puzzling? How might you account for this parasite's tolerance of nestmates in light of the analysis of begging behavior presented above? See [638] after completing your answer.

9.9   Birds are not the only animals in which offspring have special signals that appear to communicate need to a parent. Develop an evolutionary analysis of crying by human infants in which you consider the two theories outlined above for begging by baby birds. Then employ the following data in evaluating your hypotheses: (1) young infants expend considerable energy when crying, (2) the growth rate for typical infants is highest over the first 3 months of life, with smaller and smaller portions of the energy budget thereafter going to growth as opposed to maintenance, (3) consumption of breast milk peaks at 3 to 4 months of age and then declines, (4) crying peaks at about 6 weeks of age and occurs progressively less often after 3 months of age, except when the child is being weaned, (5) babies who are carried everywhere and nursed on demand (as in traditional societies) cry far less than babies in Western societies, and (6) the high-pitched cries of unhealthy babies are considered especially unpleasant by adult listeners [427, 749, 1268].

How to Deal with Illegitimate Receivers

The raccoon that listens in on the communication taking place between a brood of baby tree swallows and their parents is an **illegitimate receiver**, in the sense that it uses information from the signal to the fitness detriment of the legitimate

(A)

(B)

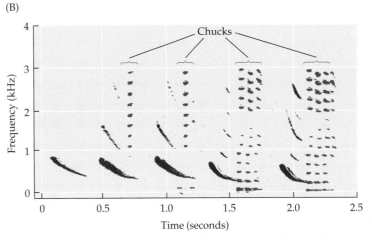

Figure 9.30   Illegitimate receivers can detect the signals of their prey.
(A) A calling male túngara frog may inadvertently attract an illegitimate receiver: the fringe-lipped bat, a deadly predator. (B) The risk of attack is greater if the male's call includes one or more chucks (blue in the sonograms) as well as the introductory whine (purple in the sonograms). A, photograph by Merlin D. Tuttle, Bat Conservation International; B, sonograms courtesy of Mike Ryan.

signalers and receivers. Illegitimate receivers may have powerful effects on the evolution of communication systems. For example, James Nieh and his colleagues have suggested that the remarkable symbolic communication systems of social bees (see p. 236) may have evolved in response to eavesdropping by competitors that flew to food sources discovered and pheromone-marked by other bees [877]. By moving their information-transferring systems inside their own colonies, scout bees may have been able to keep the locations of their finds all in the family rather than losing them to aggressive thieves able to use food signals not intended for them.

Thinking about illegitimate receivers helped Mike Ryan explain why calling males of the túngara frog often give whining calls without chucks, despite the fact that mate-searching females prefer males who add the chuck element to their calls. As it turns out, calling male túngara frogs have another audience: the predatory fringe-lipped bat, which eavesdrops on calling males in order to track them down and haul them away (Figure 9.30). Perhaps this bat makes giving whine-chuck calls especially costly.

This hypothesis generates a number of predictions, which Ryan and his colleagues tested. First, fringe-lipped bats should be attracted to the signals produced by their prey, and indeed, bat predators zero in on speakers broadcasting túngara frog calls while ignoring silent speakers. Second, fringe-lipped bats should be more attracted to whine-chuck calls than to whine calls alone; in fact, these bats were more than twice as likely to inspect, and even land on, a speaker broadcasting a whine-chuck than a whine alone [1045]. Third, the frogs should be less likely to give one-component whine calls when the risk of predation is lower. The chance of becoming a meal for a bat declines for frogs calling in large groups because of the dilution effect (see p. 187). As expected, males calling in large assemblages are much more likely to produce whine-chuck calls than are those in smaller groups [1050].

The risk of exploitation by an illegitimate receiver may also be responsible for the differences between the mobbing call and the "seet" alarm call of the great tit (Figure 9.31) [784]. These small European songbirds sometimes approach a *perched* hawk or owl and give a loud mobbing call whose dominant frequency is about 4.5 kHz. This easily located acoustical signal helps other birds find the mobbers and join in the harassment of their mutual enemy (see p. 177). If, however, a great tit spots a *flying* hawk, it gives a much quieter

(A) Mobbing call

(B) Seet call

Figure 9.31 Great tit alarm calls. Sonograms of (A) the mobbing call and (B) the "seet" alarm call. Note the lower sound frequencies in the mobbing signal. A, courtesy of William Latimer; B, courtesy of Peter Marler.

"seet" alarm call, which appears to warn mates and offspring of possible danger. This signal's dominant frequency lies within 7–8 kHz, so that the sound attenuates (weakens) after traveling a much shorter distance than the mobbing signal. The rapid attenuation of the "seet" call compromises its effectiveness in reaching distant legitimate receivers, but it also lowers the chance that a dangerous predator on the hunt will be able to tell where the caller is. Moreover, the frequencies of the "seet" call lie outside the range that hawks can hear best, while falling within the range of peak sensitivity of the great tit (Figure 9.32). As a result, a great tit can "seet" to a family member 40 meters away, but will not be heard by a sparrow hawk unless the predator is less than 10 meters distant [649].

If the "seet" call of the great tit has evolved properties that reduce the risk of detection by its enemies, then unrelated species should also have evolved alarm signals with similar properties. The remarkable convergence in the "seet" calls

Great tit

Sparrow hawk

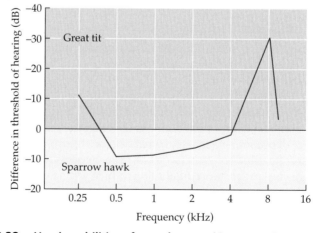

Figure 9.32 Hearing abilities of a predator and its prey. The purple trace shows the difference between the softest sound of a given frequency that great tits and sparrow hawks can hear. A sparrow hawk can hear sounds in the 0.5–4 kHz range that are fainter (5–10 dB lower in intensity) than those that great tits can hear. But great tits can detect an 8 kHz sound (in the range of the "seet" call) that is fully 30 dB fainter than any 8 kHz sound that a sparrow hawk can detect. After Klump et al. [649].

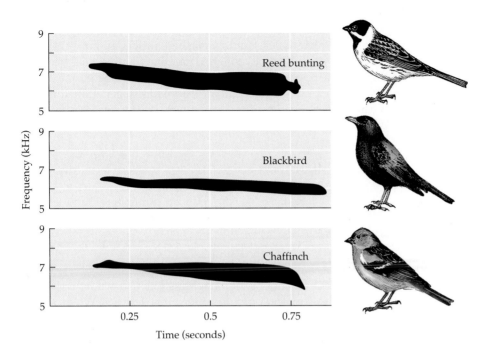

Figure 9.33　Convergent evolution in a signal. The great tit's high-pitched "seet" alarm call (see Figure 9.31B) is very similar to the alarm calls given by other unrelated songbirds when they spot an approaching hawk. After Marler [784].

of many unrelated European songbirds suggests that predation pressure by bird-eating hawks has favored the evolution of alarm calls that are hard for hawks to hear (Figure 9.33) [784].

Discussion Question

9.10　Males of a fish called the northern swordtail attract mates with flashy displays that show off their elaborate tails. But this species has to share its habitat with a lethal predatory fish, the Mexican tetra. A research team established that the bodies of male northern swordtails, especially their tails, reflect a considerable amount of ultraviolet radiation [265]. One of the things they did was to put males and females of this species in tanks with and without ultraviolet filters, finding that females were more attracted to males when the ultraviolet channel was available to them. Given that these researchers were interested in how swordtails might reduce the risk of eavesdropping by Mexican tetras, what other experiment must they have done as well? Outline the science behind this research, starting with the causal question and ending with the possible conclusions the researchers might have reached.

Why Settle Disputes with Harmless Threats?

We have been using an adaptationist approach to examine cases in which signalers appear, at first glance, to be lowering their fitness. Let's now examine some instances in which *receivers* of signals seem to lose fitness by reacting to certain messages from others. First, why do so many animals resolve their disputes with highly ritualized threat displays? Rather than bludgeoning a rival, males of many bird species settle their conflicts over a territory or a mate with much singing and feather fluffing, but without ever touching one another (see Chapter 2). Likewise, caribou stags have magnificent antlers, but they almost never really attempt to harm one another. One team of observers recorded 1308 sparring matches, in which males sedately clashed antlers, but only six real fights during the same period [66]. Even when genuine fighting does occur in the animal kingdom, the opponents often appear to be auditioning for a comic opera. After a body slam or two, a subordinate elephant seal generally lumbers off as fast as it can lumber, inchworming its blubbery body across the beach to the water, while the victor bellows in noisy, but generally harmless, pursuit.

Discussion Question

9.11   One sometimes hears that the reason why so many species resolve their contests via mostly harmless threat signals is to reduce the number of injuries and thereby protect the breeding adults who are needed to produce the next generation of offspring. What's wrong with this hypothesis?

Modern adaptationists have hypothesized that even losers gain fitness when contests are settled quickly without serious fighting [305]. Consider the European toad, *Bufo bufo*, whose males compete for receptive females. When a male finds another male mounted on a female, he may try to pull him from her back. The mounted male croaks as soon as he is touched, and often the other male immediately concedes defeat and goes away, leaving his croaking rival to fertilize the female's eggs. How can it be adaptive for the signal receiver in this case to give up a chance to leave descendants simply on the basis of hearing a croak?

Because European toad males come in different sizes, and because body size influences the pitch of the croak produced by a male, Nick Davies and Tim Halliday proposed that males can judge the size of a rival by his croak. If a small male can tell, just by listening, that he is up against a larger opponent, then the small male ought to give up without getting involved in a fight he cannot win. If this hypothesis is correct, then deep-pitched croaks (made by larger males) should deter attackers more effectively than higher-pitched ones (made by smaller males) [294]. To test this prediction, the two researchers placed mating pairs of toads in tanks with a single male for 30 minutes. The mounted male, which might be large or small, had been silenced by looping a rubber band under his arms and through his mouth. Whenever the single male touched the pair, a tape recorder supplied a 5-second call of either low or high pitch. Small paired males were much less frequently attacked if the interfering male heard a deep-pitched croak (Figure 9.34). Thus, deep croaks do deter rivals to some extent, although tactile cues also play a role in determining the frequency and persistence of an attack.

So why don't small males pretend to be large by giving low-pitched calls? Perhaps they would, if they could, but they can't. A small male toad apparently cannot produce a deep croak, given that body mass and the unbendable rules of physics determine the pitch of the signal that a male toad can generate. Thus, toads have evolved a warning signal that accurately announces their body size. By attending to this honest signal, a male toad can determine something about the size of his rival, and thus his probability of winning an all-out fight with him, just as a parent bird may be able to count on the begging signals of its chicks to indicate just how hungry and healthy they are.

When smaller rivals withdraw upon hearing an honest signal from a larger male toad, both parties gain: small males do not waste time and energy in a battle they are unlikely to win, and large males save time and energy that they would otherwise have to spend fending off annoying smaller toads. Imagine two kinds of aggressive individuals in a population, one that fought with each opponent until physically defeated or victorious, the other that checked out the

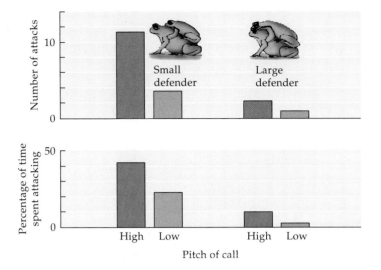

Figure 9.34   Deep croaks deter rivals. When a recording of a low-pitched croak was played to them, male European toads made fewer contacts with, and interacted less with, silenced mating rivals than when a higher-frequency call was played. Tactile cues also play a role, however, as one can see from the higher overall attack rate on smaller toads. After Davies and Halliday [294].

rival's aggressive potential and then withdrew as quickly as possible from superior fighters. The "fight no matter what" types would eventually encounter a superior opponent who would administer a serious thrashing. The "fight only when the odds are good" types would be far less likely to suffer an injurious defeat at the hands of an overwhelmingly superior opponent [800, 1272].

Further, imagine two kinds of superior fighters in a population, one that generated signals that other lesser males could not produce, and another whose threat displays could be mimicked by smaller males. As mimics became more common in the population, natural selection would favor receivers that ignored the easily faked signals, reducing the value of producing them. This, in turn, would lead logically to the spread of the genetic basis for an honest signal that could not be devalued by deceitful signalers.

Discussion Question

9.12   In male cricket frogs, the larger the male, the lower the dominant frequency of his mate-attraction calls. But male cricket frogs will sometimes lower the dominant frequency of their calls when they hear a larger rival calling nearby [1245]. This ability should surprise you in light of our discussion of European toads. Why? William Wagner found that when male cricket frogs lowered their calls in this situation, they were significantly more likely to go on to attack the larger calling neighbor than when they did not do so. Does this finding help you evaluate the hypothesis that males who lower their calls are engaged in deceptive signaling?

If the honest signal theory is correct, we can predict that the male threat displays of many unrelated species should be expensive to produce and therefore difficult for small, weak, or unhealthy males to imitate. This prediction has been tested with the assistance of some side-blotched lizards, a species in which males threaten a nearby opponent by raising and flattening their bodies laterally and then doing a number of pushups. If one runs a male lizard on a treadmill for a time and then gives him a chance to display to his own image in a mirror—something the lizards will readily do—the duration of his threat posture and the number of pushups he performs falls dramatically in this reduced-endurance state compared with his behavior before being forced to run on the treadmill (Figure 9.35). Conversely, males that are induced to perform a series of displays and then placed on the treadmill run for a shorter time than when they have not been displaying. In other words, a male's physiological state is accurately reflected by his display performance [137]

(A)

(B)

Figure 9.35   Threat displays are energetically demanding in (A) the side-blotched lizard. (B) Males that have been run on a treadmill to lower their endurance are not able to maintain their threat posture for as long as they can when they are not tired; they also generate fewer pushup displays. A, photograph by Paul Hamilton; B, after Brandt [137].

Honest signals are also supplied by the bizarre "antlers" and eyestalks of certain insects (Figure 9.36A). Males competing for access to mates confront each other, head to head, permitting rivals to compare the size of their large, and presumably expensive to produce, head projections. Antler span and eye span

(A)

Figure 9.36   Convergent threat displays. (A) Males of this Australian antlered fly in the genus *Phytalmia* confront each other head to head, permitting each fly to assess his own size relative to the other's size. (B) Male red deer clash aggressively, locking their horns and pushing in ways that enable opponents to judge the relative size and strength of their rivals. A, photograph by Gary Dodson; B, photograph by J. H. Robinson.

(B)

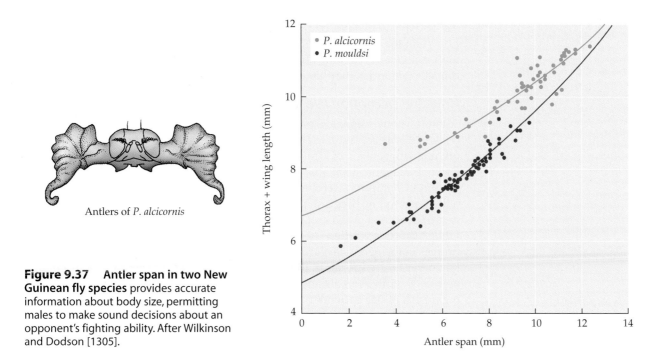

Antlers of *P. alcicornis*

Figure 9.37   Antler span in two New Guinean fly species provides accurate information about body size, permitting males to make sound decisions about an opponent's fighting ability. After Wilkinson and Dodson [1305].

are highly correlated with body size (Figure 9.37), so that contestants can quickly and accurately assess their relative sizes by going head to head [1305]. Smaller males generally abandon the field of battle; the greater the difference between the width of their eye span and their opponent's, the more rapidly they depart, and the more energy they save [924].

Likewise, in the animals called red deer in Europe and elk in North America, antler-clashing and shoving matches permit males to evaluate the size and strength of an opponent (Figure 9.36B), but these animals need not even come into contact in order to make these assessments. In fact, the typical clash between males occurs when the two competitors stand off at a considerable distance and roar loudly at each other (Figure 9.38) [224]. Only males in top condition can sustain roaring at a high rate for many minutes [223] because the activity is costly in terms of energy expended, as is true of many other acousti-

Figure 9.38   An honest signal. Only red deer (elk) stags in top condition can roar for long periods. Photograph courtesy of Tim Clutton-Brock.

cal displays (e.g., [496, 575]; but see [564]). Thus, just as in antlered flies and European toads, red deer opponents can quickly gain honest, accurate information about the fighting ability of a rival, and they use this information to terminate encounters they cannot win.

How Can Deception Evolve?

Although honest signalers provide useful information to one another for mutual benefit, deceptive signalers are not uncommon in nature. Recall the case of the orchids whose decoys lure male wasps into attempting to copulate with the flower (see Figure 3.35). Just as signal givers can lose fitness by providing information to illegitimate receivers, receivers can lose fitness by responding to signals generated by **illegitimate signalers**, which use deception to reduce the fitness of a receiver. A famous example of this phenomenon involves the "femme fatale" fireflies studied by Jim Lloyd [731]. The females of some predatory fireflies in the genus *Photuris* answer the mating signals given by males of certain other firefly species in the genus *Photinus*. Each species of *Photinus* has its own code, with a female answering her male's distinctive flash pattern by giving a flash of her own after a precise time interval. Some *Photuris* females can respond at the correct interval to male signals of three *Photinus* species [732]. If a *Photuris* female succeeds in luring a male *Photinus* close enough, she will grab, kill, and eat him, at which point his reproductive potential drops to zero (Figure 9.39).

Deception of this sort poses a real puzzle for the adaptationist, because the male firefly that pays attention to the wrong "come hither" signal can get himself killed. When an action is clearly disadvantageous to individuals, behavioral biologists have recourse to two main possibilities (see also Table 6.1):

1. Novel environment theory: The maladaptive response is caused by a proximate mechanism that once was adaptive, but is no longer. The current maladaptation occurs because modern conditions are different from those that shaped the mechanism in the past, and because there has not been sufficient time for advantageous mutations to occur that would "fix the problem."

2. Exploitation theory: The maladaptive response is caused by a proximate mechanism that is still adaptive in sum, but can be exploited by some individuals, leading to fitness losses that on average reduce, but do not eliminate, the net fitness gain associated with the mechanism and the behavior it controls.

Novel environments seem unlikely to account for the nature of interactions between *Photinus* males and their predatory relatives. This theory more often applies to cases in which very recent human modifications of the environment appear responsible for eliciting maladaptive behavior (see Figure 4.2). Instead, exploitation theory seems more likely to account for male *Photinus* behavior. The argument here is that, *on average*, the response of a male *Photinus* to certain light flashes increases his fitness, even though one of the costs of responding is the chance that he will be devoured by an exploitative *Photuris* signaler. Males that avoided these deceptive signals might live longer, but they would probably ignore females of their own species as well, and would leave few or no descendants to carry on their cautious behavior.

This hypothesis highlights the definition of adaptation employed by most behavioral biologists. As noted earlier, an adaptation need not be perfect, but it

Figure 9.39   A firefly *femme fatale*. This female *Photuris* firefly is feeding on a male *Photinus* firefly that she lured to his death by imitating the flashes given by females of his species. Photograph by Jim Lloyd.

Figure 9.40   A deceptive signaler. This orange anglerfish has a pale tip on the thin appendage on the front of its head; by moving the lure, the predatory fish attracts prey within striking distance. Photograph by Ann Storrie.

must contribute more to fitness on average than other possible alternative traits. The male firefly that responds to the signals of a predatory female of another species possesses a mechanism of mate location that clearly is not perfect, but is better than those alternatives that improve a male's survival chances at the cost of making him unlikely to reproduce.

   If this adaptationist hypothesis is correct, then deception by an illegitimate signaler should exploit a response that has clear adaptive value under most circumstances [211]. We discussed this idea earlier when explaining why deception works for the orchids that take advantage of the generally adaptive sex drive of male wasps. Likewise, the anglerfish exploits smaller fish that possess a generally adaptive eagerness to attack visual stimuli associated with their prey. The predatory anglerfish provides the stimuli in question by waving a thin rod that projects out of the front of its head. On the end of the rod is a pale tip that looks to smaller fish like something to eat (Figure 9.40), luring them close enough so that the anglerfish can engulf them in its massive mouth [953]. Although the deceived fish pay a heavy price for their interest in the anglerfish's lure, a complete failure to respond to these stimuli would probably doom them to starvation.

Discussion Question

9.13   Develop an honest signal hypothesis and a deception hypothesis to account for the fact that when copulating, males of some butterflies transfer a substance to their mates that makes these females sexually unappealing to other males [28]. What do you need to know to determine which explanation is correct?

Summary

1. A full understanding of the evolution of communication systems requires information about the origins of signals and the pattern of changes that took place in signalers and receivers over time, as well as information about the causal processes, especially those driven by natural selection, that made these changes occur. These two levels of evolutionary analysis complement each other.

2. The origins of signals and of responses to signals generally lie in small changes in behaviors or proximate sensory mechanisms that had different functions in the past. The gradual accumulation of many small adaptive changes over time can result in the formation of traits that differ greatly from the ancestral characteristic from which they are derived.

3. Each evolutionary change must be layered on that which has already evolved. In this sense, the traits already in place constrain or bias the pattern of evolution. One manifestation of this phenomenon is sensory exploitation, which occurs when mutant signalers happen to tap into preexisting sensory biases of receivers that evolved because of their utility in some domain other than communication.

4. Natural selection can cause signal changes to spread through a species only if both signalers and receivers derive net fitness benefits from their participation in the system. Applying the adaptationist approach to communication systems has resulted in solutions to some puzzling cases of signalers and receivers whose behavior seems to reduce, rather than increase, their fitness. In some instances, apparently self-sacrificing communication has been shown in reality to raise an animal's chances of reproducing. In other cases, costly behavioral traits have been shown to occur because certain individuals (illegitimate signalers or illegitimate receivers) are able to exploit an otherwise adaptive communication system for their own benefit.

Suggested Reading

The huge field of animal communication has been reviewed by Jack Bradbury and Sandra Vehrencamp [133]. The way in which science works can be seen in the development of the spotted hyena story; start with Stephen Jay Gould's account [460] and then read the review written by Ned Place and Stephen Glickman [963]. Bernd Heinrich's *Ravens in Winter* illustrates how an adaptationist can use multiple hypotheses to solve an evolutionary puzzle in communication [533]. The same can be said for Jonathan Wells's review of why human babies cry [1268]. For an adaptationist analysis of manipulation and deceit by communicating animals, read [305]; for the classic paper on honest signaling, see [1352].

10 | *The Evolution of Reproductive Behavior*

I was thrilled the first time I saw a male satin bowerbird fly down to his bower, which looked more like something a precocious child might have built than the handiwork of a bird not much larger than a robin (Figure 10.1). The arriving male held a rubber band in his beak, which he dropped among the blue parrot feathers strewn about the bower. Although I did not stay to see a female satin bowerbird visit the bower, others have reported that her arrival sends the male into a wild courtship display that features a bizarre medley of buzzes, screeches, and imitations of other bird songs as well as seemingly aggressive, wing-waving charges at the female. Yet despite the vigor of the male's displays, most courtships end with the abrupt departure of an evidently unreceptive female [121].

In fact, female satin bowerbirds are in no rush to pick a mate [1224]. A female that is ready to reproduce first visits several bowers, scattered widely through the Australian forest, but does not mate with any bower builders on these initial visits. Instead, she takes a break of about a week to construct a nest before returning for another round of inspections, during which time she usually observes the full courtship routine of two to three males, often returning for another look at one or more of her potential partners. (These data come from a study involving color-banded bower-birds and video monitors that recorded what was happening at 34 differ-

Figure 10.1 Bowerbird courtship revolves around the bower. A male satin bowerbird, with a yellow flower in his beak, courts a female that has entered the bower that he has painstakingly constructed and ornamented with blue feathers. Photograph by Bert and Babs Wells.

ent bowers.) During this second round of visits, the female may watch six or so courtship displays in all over a period of several weeks before finally settling on a mate. When she is ready to mate, the female enters the favored male's bower, crouches down, and copulates quickly before flying away. She will have no further contact with her partner, but incubates her eggs and rears the young all by herself. Her mate stays at or near his bower for most of the 8-month breeding season, courting other females that come to inspect his creation and copulating with any that are willing.

Thus, the two sexes of satin bowerbirds not only look different (see Figure 10.1), but their reproductive tactics are totally dissimilar. Although other animals obviously have their own special courtship routines and mating patterns, behavioral differences between the sexes are the rule, whether we are talking about bowerbirds or belugas, aardvarks or zebras. Typically, males do the courting and females do the choosing. This pattern is so widespread that biologists ever since Darwin have tried to provide an evolutionary explanation for it. Thanks to their efforts, we now know a great deal about the history and adaptive value of reproductive behavior—two fascinating topics that this chapter will review.

The Evolution of Differences in Sex Roles

Male satin bowerbirds build bowers; females do not. As always, we can employ two levels of evolutionary analysis in investigating this difference. First, what were the evolutionary origins of bower building by male satin bowerbirds? Second, why has bower building by males of this species been maintained by selection following its origin?

With respect to the first question, the satin bowerbird is one of 20 species in the bowerbird family, 17 of which build bowers [414]. No other bird species builds anything like these elaborate display structures, and so the trait appears to have evolved just once [683, 684]. A single common ancestor gave rise to the cluster of bower-building species, endowing its descendants with the capacity for

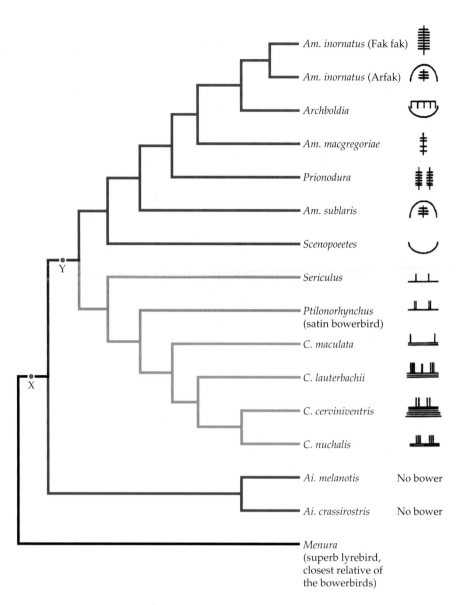

Figure 10.2 Evolutionary relationships among 15 populations of bowerbirds, based on similarities in their mitochondrial cytochrome *b* gene. The icons on the right represent the shape of each species' bower. Thus, *Scenopoeetes* builds a shallow bowl of twigs on the ground, while the satin bowerbird's avenue bower consists of a simple platform with two parallel rows of twigs, and *C. lauterbachii* constructs a more elaborate raised platform with two outer and one inner row of twigs. The really elaborate bowers are of the maypole variety, tall stacks of twigs built around a sapling. Note that two bowerbird species (*Ai. melanotis* and *Ai. crassirostris*) do not build bowers at all, having retained the ancestral trait of extinct species X. Extinct species Y presumably built a simple avenue bower or a cleared display court; this species gave rise to all the current bower-building bowerbirds. After Kusmierski et al. [684].

making bowers (Figure 10.2). Two early species derived from this ancestor subsequently gave rise to the two groups of modern bower-building species, the avenue-bower builders (including the satin bowerbird) and the maypole-bower builders (such as *Amblyornis inornatus*, whose bowers are shown in Figure 10.3). Among the two groups are descendant species whose bower-building behaviors vary greatly, with large differences even among some closely related species, especially the maypole-building bowerbirds. In fact, two geographically separated populations of the same species, *A. inornatus*, build very different maypole bowers (Figure 10.3), even though the birds in the two areas are genetically very similar [1223]. The changes that have occurred have apparently been so rapid and extensive that the bowers of many close relatives lack the kinds of shared features that would help us establish what the first bower might have looked like and what sequence of changes has taken place during the evolution of bower building.

Having provided an admittedly incomplete sketch of the history of bowers, what can we say about the adaptive significance of bower building? In particular, what does a male satin bowerbird gain by spending the better part of 8 months each year constructing his bower, gathering decorations (sometimes

Figure 10.3 Different bowers in different populations of the same bowerbird species. (Top) The huge playhouse bower of one population of the bowerbird *Amblyornis inornatus*. (Bottom) The same species builds a quite different maypole bower in another location. Photographs by Will Betz and Adrian Forsyth.

from other males' bowers), and defending his display site from rival males? One possibility is that the original bower builder conveyed useful information about his quality as a mate to females, and was rewarded when discriminating females copulated with him.

If this hypothesis is true, then we would expect male mating success in modern satin bowerbirds to be correlated with some features of the bower that vary from male to male, such as the skillfulness with which the bower has been constructed and decorated. In fact, even humans can detect differences among the bowers built by different males. Some contain neat rows of twigs, lined up to create a large, tidy, symmetrical bower, while others are obviously messier, less professionally assembled. Bowers also differ markedly in the number of decorative ornaments, such as feathers and rubber bands, that their builders have collected. Female bowerbirds evidently notice these differences too, because they are less likely to exhibit startle responses when visiting well-decorated bowers of high quality. The less jumpy a female, the more likely she is to mate with the bower builder eventually [929, 930], which is why bower quality and the number of bower decorations correlate with male mating success in this [120] and other bowerbird species [767, 768].

If the mate quality advertisement hypothesis is correct, then attractive, well-decorated bowers should be built by males that are superior in some way to those birds that cannot erect a top-flight bower. Stéphanie Doucet and Bob Montgomerie proposed that good bower builders might be healthier birds, unlikely to infect their mates with parasites or disease microbes and more likely to possess sperm with genes for disease resistance that could be passed on to their offspring. In keeping with this proposal, males that build better bowers do have fewer ectoparasitic feather mites than those that make less appealing display structures [338].

Another idea along these same lines is that the bower's quality is in some sense an indicator of the developmental history of the male. Birds that had plenty of food as they matured should have well-constructed brains and should therefore be able to excel at the many demanding manipulative tasks needed

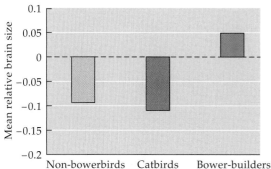

Figure 10.4   Bower building may be an indicator of brain size. Bowerbirds that build bowers have relatively large brains compared with other bowerbirds, the catbirds, that merely construct cleared display courts. The mean brain size (as determined by comparing brain cavity volumes against a measure of body size) of the bower-building bowerbirds also far exceeds that of a sample of other unrelated bird species. After Madden [766].

to assemble a bower. (Note the parallel between this explanation for bower building and the hypothesis presented in Chapter 2 that males of other birds learn a complex song repertoire because in this way they can advertise their developmental past to choosy females of their species.) Joah Madden recognized that if this hypothesis was correct, then the brains of bower-building bowerbirds should be proportionally larger than those of bird species that do not build bowers—a group that includes a few bowerbird species, such as the green catbird, that clear off a display court but build nothing on it. Madden went to the British Museum to X-ray the skulls of a series of stuffed bowerbirds and some of their relatives. Sure enough, bower builders are unusually brainy birds (Figure 10.4) [766].

Discussion Question

10.1   Female satin bowerbirds appear to favor males that give very intense courtship displays, which contain elements of aggression. Perhaps that is why females are often "jumpy" when in the presence of an intensely displaying male. Females that flinch in response to male displays tend to leave the site without mating, whereas females that crouch down in the male's bower tend to stay and may eventually mate with the displaying male. What do these observations suggest about the utility of the bower from the female's perspective? From the male's? What would you predict about a male's response to a model female bowerbird (a mechanical robot enclosed in the skin and plumage of a female) that could be controlled at a distance so as to remain standing throughout the male display, or crouch infrequently, or crouch often? What measurement problems would you have to overcome to test your predictions? See [931].

Because most female bowerbirds mate with just one male—often the same individual that is popular with other females—male reproductive success in any given breeding season is very unequal (Figure 10.5) [120, 1224]. This fact of reproductive life has great significance for a key question that we have not yet addressed: why do male bowerbirds, but not females, build courtship display structures, and why do females evaluate male performance instead of the other way round? Indeed, it is common throughout the animal kingdom for males to try to mate with many females while the objects of their desire are content with one or a few matings, albeit with a male or males they have carefully chosen. This widespread pattern is almost certainly related to a truly fundamental difference between the sexes, which is that males produce sperm and females produce eggs.

In virtually all sexual species, eggs are larger than sperm, which are usually just large enough to contain the male's DNA and enough energy to fuel the journey to an egg. Even in species in which males produce oversized sperm—such as a fruit fly whose males make sperm that are (when uncoiled) ten times

(A)

(B)

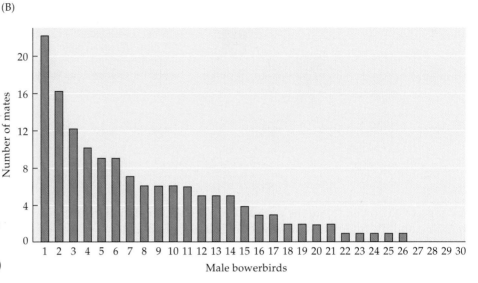

Figure 10.5 **Variance in reproductive success is greater for males than for females** in the satin bowerbird. (A) Very few female bowerbirds have more than two mates per breeding season. (B) Some male bowerbirds, however, mate with more than 20 females in a single season, while others do not mate at all. After Uy et al. [1224].

longer than their bodies [961]—the mass of an egg is still vastly greater than that of a sperm. In contrast, females, are (by definition) the sex that produces eggs, the larger gametes, which in birds and most other animals are enormous relative to sperm (Figure 10.6). A single bird egg may constitute 15–20 percent of the female's body mass, and some go as high as 30 percent [689]. By way of contrast, a male splendid fairy-wren, a very small bird indeed, may have as many as 8 billion sperm in his testes at any given moment, while his mate will lay no more than six eggs per nesting attempt [1219]. The same pattern applies to coho salmon, whose males shower about 100 billion sperm on a typical batch of 3500 eggs, according Bob Montgomerie. Among mammals, the same correlation between gamete size and number of gametes produced holds true. A woman has only a few hundred cells that can ever develop into large mature eggs [276], whereas a single man could theoretically fertilize all the eggs of all the women in the world, given that just one ejaculate contains on the order of 350 million minute sperm.

The critical point is that males almost always have enough gametes to fertilize the eggs of a great many females. Therefore, a male's contribution of genes to the next generation generally depends directly on how many sexual partners he has: the more mates, the more eggs fertilized, the more descendants produced, and the greater the male's fitness relative to less sexually successful individuals. The benefits to males of maximizing the number of different females they inseminate have favored males that can discriminate between familiar mates and unfamiliar females, who may provide new opportunities for males to pass on their genes. Thus, in one *Anolis* lizard species, territorial males direct more than ten times as many courtship displays toward new females that have shown up on their territories for the first time, compared with their rather blasé response toward long-term female companions [904].

Whereas male animals usually try to maximize their sexual encounters, females generally do not, because their reproductive success is typically limited by the number of eggs they can manufacture, not by any shortage of partners. Eggs are costly to produce because they are large, which means that females have to secure the resources to make them. Furthermore, after one batch of eggs has been fertilized, a female may spend still more time and energy caring for the resultant offspring. Thus, during the breeding season of the satin bowerbird, females are likely to spend more of their time foraging, building a nest, or caring for young than looking for mates, while adult male bowerbirds

Figure 10.6   Male and female gametes differ greatly in size. A hamster sperm fertilizing a hamster egg (magnified 4000 times) illustrates the trivial contribution of materials to the zygote by the male. Photograph by David M. Phillips.

are on duty by their bowers every single day. Because the potential rate of reproduction differs so much between the sexes [227], there are many more sexually active males than females at any given time, creating a male-biased **operational sex ratio** (the ratio of sexually receptive males to receptive females) [368]. As a result, each male must compete with many rivals for a limited number of receptive females, each of which donates more material resources to an offspring than her partner.

Expenditures of time and energy and risks taken by a parent to help one offspring are considered **parental investment** in that offspring if they reduce the chance that the parent will reproduce successfully in the future [1212]. Bob Trivers invented this concept to emphasize the trade-offs for parents that make contributions to their progeny. On the plus side, parental investment may increase the probability that an existing offspring will survive to reproduce. But this fitness benefit comes at the cost of the parent's ability to generate additional offspring down the road. In the animal world, the general rule is that females make a larger total parental investment per offspring than males do. As we have seen, the differences start at the level of the gametes. When a female builds an egg, the many nutrients and cell organelles that she donates to it constitute a much larger parental investment than is contained within the single tiny sperm that will fertilize that egg. This difference between the sexes in gametic parental investment is often amplified by other kinds of parental investments, which include the food or protection parents sometimes give to their progeny (Figure 10.7) [1212]. In mammals, for example, females nourish their embryos within their bodies, then care for the babies when they are born and give them milk; the typical male mammal, in contrast, impregnates a female and departs, never to see or interact with his offspring.

But why is it that the sexes differ in the resources they donate to a fertilized egg? Geoffrey Parker and his colleagues have argued that the evolution of the two kinds of gametes, and thus the two sexes, stemmed from divergent selection that favored either (1) individuals whose gametes were good at fertilizing other gametes because of their relatively small size and great mobility (a num-

Figure 10.7 Parental investment takes many forms. (Clockwise from top left) A male frog carries his tadpole offspring on his back. A male katydid gives his mate an edible spermatophore containing orange carotenoid pigments that will be incorporated in her eggs. A female eared grebe protects her young by letting them ride on her back. A female cicada killer wasp drags a cicada she has paralyzed with a sting to a burrow she has dug, where her offspring will feed on the unlucky cicada. Photographs by Roy McDiarmid; Klaus Gerhard-Keller; Bruce Lyon; and the author.

ber-of-offspring strategy) or (2) individuals whose relatively large gametes were good at developing after being fertilized (a parental investment strategy) [927]. Sperm are superbly designed to race toward an egg when released by a male. Bluegill sperm, for example, dart along at up to five sperm lengths per second [175]. In contrast, eggs stay put because they are large, packed with nutrients useful for the development of the zygote following fertilization. No single kind of gamete could be equally good at both tasks, which may have led to the separate evolution of effective fertilization devices (the sperm of males) and effective development devices (the eggs of females).

Testing the Evolutionary Theory of Sex Differences

We have suggested that males usually compete for mates because of inequalities between the sexes in parental investment, which lead to differences in potential reproductive rates and a male-biased operational sex ratio (Figure 10.8). These differences should favor males that compete vigorously with rivals for

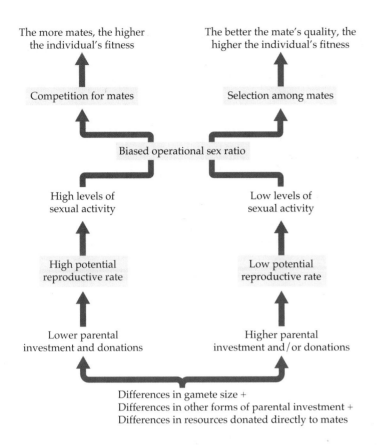

The more mates, the higher the individual's fitness

The better the mate's quality, the higher the individual's fitness

Competition for mates

Selection among mates

Biased operational sex ratio

High levels of sexual activity

Low levels of sexual activity

High potential reproductive rate

Low potential reproductive rate

Lower parental investment and donations

Higher parental investment and/or donations

Differences in gamete size +
Differences in other forms of parental investment +
Differences in resources donated directly to mates

Figure 10.8   Differences between the sexes in sexual behavior may arise from fundamental differences in parental investment that affect the rate at which individuals can produce offspring. The sex that can potentially leave more descendants gains from high levels of sexual activity, whereas the other sex does not. An inequality in the number of receptive individuals of the two sexes leads to competition for mates within one sex, while the opposite sex can afford to be choosy.

mates and pursue females avidly (Figure 10.9). In contrast, female fitness will rarely be advanced by receiving sperm from as many males as possible, so selection should favor instead those females that avoid the costs of additional matings after having chosen partners who have the most to offer in the way of resources or good genes. Satin bowerbirds exemplify this common pattern.

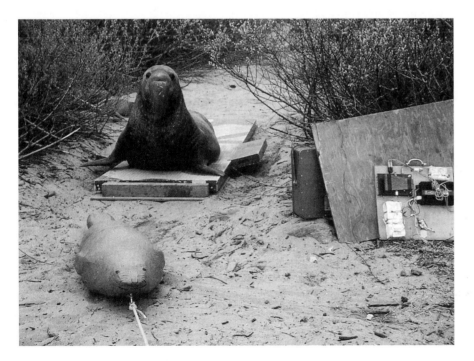

Figure 10.9   Male sex drive is so intense that scientists were able to lure this male elephant seal to this area just by playing a tape recording of the call of a copulating female elephant seal [316]. Once there, the animal saw a moving dummy female seal composed of urethane foam covered with fiberglass. As the male pursued the moving model, he climbed onto a scale, enabling researchers to record his weight without having to sedate and immobilize the animal. Photograph by Chip Deutsch.

Figure 10.10   A sex role reversal in which females, not males, advertise for mates. In the long-tailed dance fly, *Rhamphomyia longicauda*, females fly to swarms where they wait for the arrival of a gift-bearing male. While waiting, the female inflates her abdomen and holds her dark, hairy legs around her body, making herself appear as large as possible in the group of circling females. Photograph by David Funk, from Funk and Tallamy [426].

We can test this theory by finding unusual cases in which males make the larger parental investment or have the lower potential reproductive rate. In species of this sort, males may make contributions other than sperm toward the welfare of their offspring or their mates because if they do not, they may not get a chance to fertilize the eggs of a potential mate. For species in which males make the larger parental investment, the operational sex ratio should be biased toward females, leading to female competition for mates and careful mate choice by males—in other words, a **sex role reversal**. Such a reversal occurs in the mating swarms of certain empid flies, in which the operational sex ratio is heavily female-biased because most males are off hunting for insect prey to bring back to the swarm as a mating inducement [1168] (see also Discussion Question 9.2). When a male enters the swarm bearing his nuptial gift, he may find females advertising themselves with (depending on the species) unusually large and patterned wings or bizarre inflatable sacs on their abdomens (Figure 10.10) [426]. The male gets to choose from among the many ornamented partners available to him.

Likewise, males of some fish species offer their mates something of real value: namely, a brood pouch in which the female can place her eggs. For example, in the pipefish *Syngnathus typhle*, "pregnant" males provide nutrients and oxygen to a clutch of fertilized eggs for several weeks, during which time the average female produces enough eggs to fill the pouches of two males. Since the sex ratio is 1:1, male pouch space is in short supply, and as expected, females compete for the opportunity to donate eggs to parental males. Males with free pouch space actively choose among mates, discriminating against small, plain females in favor of large, ornamented ones, which can provide selective males with larger clutches of eggs to fertilize [92, 1034].

The pipefishes, seahorses, and other closely related species constitute a large family in which the devices that males employ to care for their young range from the relatively simple (a patch on the belly of the male to which females are free to glue their eggs) to the exceedingly complex (the brood pouch of seahorses with its placenta-like features for nourishing the eggs; Figure 10.11 and see also Figure 11.2). If we assume that the more complex the brooding device, the greater the paternal investment in offspring, then we can predict that sex role reversal should be more common at the upper end of the scale. However, this is not the case (Figure 10.11) [1313], showing that sex role reversals are not dependent solely on the degree to which males provide more parental investment than females. Instead, these reversals may depend on the ability of females to produce a sufficient number of eggs in a short enough time to supply sev-

More complex brood pouch ←   → More simple brooding

Hippocampus comes
Hippocampus barbouri
Hippocampus abdominalis
Hippocampus sp.
Hippocampus kudu
Hippocampus erectus
Hippichthys zosterae
Hippichthys penicillus
Syngnathus scovelli
Syngnathus fuscus
Syngnathus louisianae
Syngnathus floridae
Syngnathus taenionotus
Syngnathus typhle
Syngnathus abaster
Syngnathus rostellatus
Syngnathus acus
Syngnathus schlegeli
Vanacampus leptorhynchus
Vanacampus phillipi
Pugnaso curtirostris
Kaupus costatus
Hypselognathus rostratus
Urocampus carinorostris
Solegnathus hardwickii
Phyllopteryx taeniolatus
Corythoichthys intestinalis
Corythoichthys haematopterus
Stigmatopora argus
Stigmatopora nigra
Oostethus brachyurus
Doryrhamphus excisus
Dunckerocampus dactyliophorus
Enteurus aequareus
Nerophis ophidion
Aulichthys japonicus
Pungitius pungitius
Gasterosteus wheatlandii
Gasterosteus aculeatus

— Conventional sex roles

— Reversed sex roles

Hippocampus: brood in pouch

Figure 10.11   The phylogenetic relationship between complex parental care by males and sex role reversal in the fish family Syngnathidae. A very high degree of paternal care is *not* required for the evolution of sex role reversal in this group. After Wilson et al. [1313].

Phyllopteryx: brood on tail

eral males with clutches to brood. In some of the paternal syngnathid species with the most complex brood pouches, females evidently cannot achieve a high enough rate of egg production to warrant mating with more than one male. In these seahorse species, the two sexes have equal reproductive rates, and monogamy is the rule (see Chapter 11). But in some species, such as the sex-role-reversed *Syngnathus scovelli*, some females produce enough eggs to have two, three, or even four mates during a period when no male has more than a single partner, judging from the finding that all the eggs in a given male's brood pouch have genetic fingerprints that can be assigned to a single female [608].

Another example of a species in which the reproductive capacity of females can exceed that of males is the flightless Mormon cricket, which despite its common name has no religious affiliation and is a katydid, not a cricket. When male Mormon crickets mate, they transfer to their partners another kind of nuptial gift, an enormous edible spermatophore (Figure 10.12) [493]. Given that the spermatophore constitutes 25 percent of the male's body mass, most male Mormon crickets probably cannot mate more than once. In contrast, females may be able to produce several clutches of eggs, provided they can persuade several males to mate with them.

Sometimes bands of Mormon crickets march across the countryside, eating farmers' crops and mating as they go. In these high-density populations, when

Figure 10.12 Mormon cricket males give their mates an edible nuptial gift. Here a mated female carries a large white spermatophore received from her partner during copulation. Photograph by Darryl Gwynne.

a male begins to stridulate from a perch, announcing his readiness to mate, females come running and jostle for the opportunity to mount him, the prelude to insertion of the male's genitalia and transfer of a spermatophore. Males often behave in a discriminating manner, refusing to transfer a spermatophore to a lightweight female in favor of a heavier, more fecund partner. A choosy male that rejects a 3.2-gram female in favor of one weighing in at 3.5 grams fertilizes about 50 percent more eggs as a result [493]. So here too, the sex with the higher reproductive potential competes for access to members of the other sex.

This theory of sex role differences also predicts that if the relative reproductive capacities of the two sexes within a species were to change, their reproductive tactics should change as well. A test of this prediction was provided by a study of a skinny Australian katydid (Figure 10.13) whose food supply varies greatly over the course of the breeding season. When these katydids are limited to pollen-poor kangaroo paw flowers, the male's large spermatophore is both difficult to produce and valuable to females as a nuptial gift. Under these conditions, males are choosy about their mates, and females fight with one another for access to receptive, spermatophore-offering males. But when pollen-rich grass trees start to flower, the maximum reproductive rate for males increases (because they can then produce spermatophores more rapidly), whereas the females' reproductive rate is limited by the speed at which they can turn pollen into eggs. At this time, sex roles switch to the more typical pattern, with males competing for access to females and females rejecting some males. This same pattern of changes has been reproduced in the laboratory by controlling the food supplied to captive katydids (Table 10.1) [495].

TABLE 10.1 *Effect of the relative reproductive rates of males and females on sex roles in* Kawanaphila *katydids*

Treatment[a]	Mean number of		Proportion of interactions with	
	Calling males	Matings per female	Male choice of mates	Female competition for mates
Food scarce: Males produce spermatophores slowly	0.4	1.3	0.4	0.2
Food abundant: Males produce spermatophores quickly	6.6	0.7	< 0.1	0.0

Source: Gwynne and Simmons [495]

[a]In each treatment, four cages containing 24 katydids each were observed for equal periods of time.

Figure 10.13   A katydid that shifts sex roles in relation to the availability of spermatophores. Female *Kawanaphila* katydids compete to mate with males that supply them with nutritious gelatinous spermatophores, but only when food is scarce. In this photograph, the female is eating a spermatophore that she received from her mate while she perches on a pollen-poor kangaroo paw flower. The female's two thin antennae and shorter, thicker ovipositor point downward. Photograph by Darryl Gwynne.

Discussion Question

10.2   In the two-spotted goby, a fish that lives on rocky shorelines in northern Europe, males provide parental care for one or more clutches of eggs. Males initially compete fiercely for territories (nest sites in seaweed or mussel shells) during the short breeding season (May to July). But over the course of the breeding season, males become scarce, and females begin to behave aggressively toward one another and to court males. Do the data in Table 10.2 permit you to test the theory outlined above for why the sexes differ in their reproductive tactics?

Sexual Selection and Competition for Mates

We have presented the theory that sex role differences evolve whenever the two sexes differ in the number of offspring they can potentially produce. In most species, males, with their minute sperm, can potentially have a great many descendants, but if they are to achieve even a fraction of their potential, they must deal both with other males, who are attempting to mate with the same limited pool of receptive females, and with the females themselves, who often have a great deal to say about which male gets to fertilize their eggs. The real-

TABLE 10.2   *Seasonal changes in the reproductive biology of the two-spotted goby*

	May	Early June	Late June	Late July
Territorial males per square meter	0.56	0.32	0.13	0.07
Females ready to spawn	0.15	0.39	0.29	0.33
Nest space (number of additional clutches that each nest could accommodate)	2.98	2.53	1.49	0.99

Source: Forsgren et al. [398]

Figure 10.14 **A developmental cost of a sexually selected trait.** Males of the dung beetle *Onthophagus acuminatus* fight for mates. Beetles with long horns (shown in blue) have an advantage in these contests, but the tissues that go into horn construction are unavailable for the building of eyes (shown in yellow). As a result, males with long horns (left) have smaller eyes than rivals with short horns (right). From Emlen [362].

ization that the members of a species can determine who gets to reproduce and who fails to do so led Charles Darwin to propose that evolutionary change could be driven by **sexual selection**, which he defined as "the advantage which certain individuals have over others of the same sex and species, in exclusive relation to reproduction" [286]. Darwin devised sexual selection theory specifically to account for the evolution of costly, survival-*decreasing* traits, such as the extravagant courtship displays of male bowerbirds and the ornamental plumes of male widowbirds and peacocks. The costs of these traits include not only the time and energy required to perform demanding displays and to produce overblown feathers, but also heightened vulnerability to predators, and some subtle developmental trade-offs as well (Figure 10.14) [363]. Darwin argued that although some mate-acquiring attributes surely shorten a male's life directly or indirectly, they may increase his lifetime reproductive success by enabling the well-ornamented male to secure mates in competition with others (Figure 10.15). Darwin contrasted these reproduction-boosting traits with those that enhance survival, which he explained in terms of conventional natural selection.

Today most evolutionary biologists emphasize the similarities between these two types of selection more than the differences. Both processes require that individuals differ in their reproductive success; both will lead to evolutionary change only if these differences have a hereditary basis. However, although sexual selection is "just" a subcategory of natural selection, the distinction Darwin made between the two processes is useful because it focuses attention specifically on the selective consequences of sexual interactions *within* a species.

Figure 10.15   Sexually selected "ornaments" of males all of which impose survival costs but advance success in the competition for mates. Darwin correctly believed that sexual selection via female choice was responsible for the evolution of elaborate plumage and remarkable displays in male birds such as the quetzal (left) and the sage grouse (top right). Darwin argued mistakenly that the strange horns and snouts of certain beetles (bottom right) also arose via female choice; males actually use these structures primarily as weapons when fighting with other males for mates. Left, photograph by Bruce Lyon; top right, photograph by Marc Dantzker; bottom right, photograph by David McIntyre and Joan Gemme.

Discussion Questions

10.3   Take one possibly survival-enhancing element of bowerbird behavior that could be the product of natural selection and another survival-decreasing trait that might be the evolutionary result of sexual selection. For each, list the *essential* conditions that had to occur in the past for natural selection and sexual selection to occur. Is the factor "differences among individuals in age at death" on your list? Why or why not?

10.4   Male rats, sheep, cattle, rhesus monkeys, and humans that have copulated to satiation with one female are speedily rejuvenated if they gain access to a new female. This phenomenon is called the "Coolidge effect," supposedly because when Mrs. Calvin Coolidge learned that roosters copulate dozens of times each day, she said, "Please tell that to the President." When the President was told, he asked, "Same hen every time?" Upon learning that roosters select a new hen each time, he said, "Please tell that to Mrs. Coolidge." Provide a sexual selectionist hypothesis for the evolution of the Coolidge effect. Use your hypothesis to predict what kinds of male animals should lack the Coolidge effect.

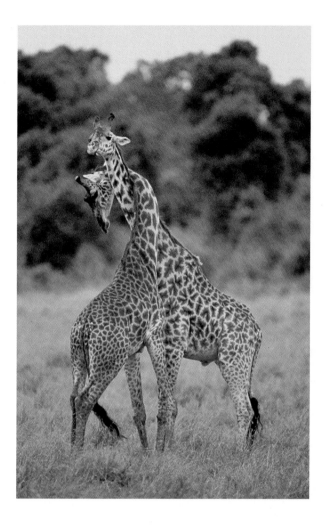

Figure 10.16 **Males of many species fight,** using whatever weapons they have at their disposal. Here two giraffes slam each other with their heavy necks and clubbed heads. Photograph by Gregory Dimijian.

One component of sexual selection (now often called *intrasexual selection*) arises when the members of one sex compete with one another for access to the other sex. Competition is evident, for example, when a male bowerbird has his bower dismantled by another male [981]. Because males with destroyed bowers lose opportunities to copulate, males have been sexually selected to keep a close eye on their display territories and to be willing to fight with rivals. Battles among males are one of the commonest features of life on Earth (Figure 10.16) because winners of male–male competition generally mate more often. In previous chapters, we have made reference to aggressive elk, battling beetles, and fighting flies—and all the fighters were males. Although fights between males are often settled quickly without serious damage to the contestants, in other cases sexual conflict can lead to an unpleasant death. For example, male *Cardiocondyla* ants use "their mandibles to pierce or crush the soft cuticle of young males, to sever their antennae or legs, to decapitate them or to cut their bodies in two" [536]. These ants have evolved large and powerful jaws to eliminate newly adult rivals for the several dozen virgin females that may emerge within their nest [534].

One sexually selected manifestation of conflict among males is the **dominance hierarchy**, in which individuals that live together in groups interact aggressively for a while before sorting themselves out from top dog to bottom mutt. Once everyone knows his place, the alpha male need only move toward a lower-ranking male to have that individual hurry out of the way or otherwise signal submissiveness. If the costly effort to achieve high status in a dominance hierarchy is adaptive, then high-ranking individuals should be rewarded reproductively. In keeping with this prediction, you will remember that in spotted hyenas, a species marked by intense competition among *females* for high social status, the alpha female reproduces far more successfully than her subordinate companions (see Figure 9.5). The same outcome holds for male spotted hyenas [407] and males of many other mammals as well [246].

The relationship between dominance and reproductive success has been studied intensively in groups of savanna baboons, a species marked by intense competition among males for social status. Males are willing to fight to move up the ladder, and as a result, individual males get bitten about once every 6 weeks, an injury rate nearly four times greater than that of females [346]. So does it pay males to risk bites and infections in order to secure a higher ranking? When Glen Hausfater first attempted to test this proposition, he counted copulations in a troop of baboons and found that, contrary to his expectation, males of low and high status were equally likely to copulate [523]. Hausfater subsequently realized, however, that he had made a dubious assumption, which was that any time a male mated, he had an equal chance of fathering an offspring. This assumption would be wrong if matings really only "count" when the females have recently ovulated. When Hausfater reexamined the timing of the copulations, he found that dominant males had indeed monopolized females during the few days when they were fertile. The low-ranking males

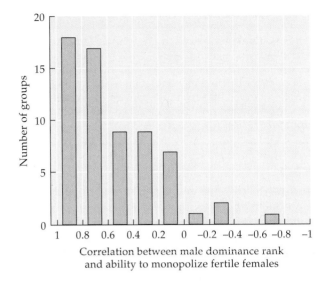

Figure 10.17   **Dominance usually correlates strongly with reproductive success** in savanna baboons. In a Kenyan reserve in which many different troops were followed over many different breeding seasons, the relation between male dominance rank and ability to form consortships with fertile females generally yielded strongly positive correlation coefficients. A correlation coefficient of +1 would indicate a perfect match between the two variables. After Alberts et al. [8].

had their chances to mate, but typically only when females were in the infertile phase of their estrous cycles.

Since Hausfater's pioneering research, others have followed up with many more such studies. A summary of data gathered on troops of savanna baboons observed in a Kenyan reserve over many years reveals that male dominance is almost always positively correlated with male mating success, as indicated by the male's ability to guard a female throughout her fertile phase (Figure 10.17) [8].

A male that copulates with an ovulating female while keeping all other males away from her would seem all but certain to be the father of any offspring that his partner subsequently produces. But we no longer need to guess about paternity in these cases, thanks to recent developments in molecular technology, including DNA fingerprinting and the like, which make it possible to determine with nearly complete certainty whether a given male has indeed fathered a given baby (see Box 11.1). A research team headed by Anja Widdig took advantage of DNA fingerprinting techniques to assign paternity to 94 percent of the 263 infants born in one rhesus monkey troop living on Cayo Santiago over a 6-year period [1293]. (Cayo Santiago is an island off Puerto Rico that is home to the descendants of a population of 400 monkeys transferred there in 1938 from India.) In any given year, about a quarter of the babies were sired by the single most successful male, while almost all of the remaining babies could be traced to between three and six other males. In most years, the other 70 percent or so of the males in the troop did not produce a single offspring.

In the band of rhesus monkeys studied by Widdig and company, reproductive success was tightly correlated with dominance rank. In 1997, for example, when there were 45 adult males competing for mates, the male population could be divided into three groups: top-ranking, mid- ranking, and low-ranking. The top third sired over 70 percent of the offspring produced that year, while just one member of the bottom group reproduced, and his tally was a very modest one baby [1293]. Dominance and reproduction also go hand in hand in savanna baboons.

10.5 Male baboons and rhesus monkeys are much larger than females. Could large body size be a product of sexual selection? If so, what prediction can you make about the *degree* of size differences between the sexes among primates, or any other group of mammalian species for that matter, in relation to the mating opportunities available to males? (Remember that in many groups of animals, some species are monogamous while others are polygynous, as we saw in Chapter 1.) Present your prediction graphically.

Alternative Mating Tactics

Although studies of baboons and many other animals have confirmed that high dominance status yields reproductive benefits for males able to secure the alpha position, the benefits are often not as large as one would expect (Figure 10.18) [8]. As it turns out, socially subordinate baboons can compensate in various ways for their inability to physically dominate others in their group. For one thing, lower-ranking males can and do develop friendships with particular females, relationships that do not depend entirely on physical dominance, but rather on the willingness of a male to protect a given female's offspring [923] (see Figure 12.11). Once a male, even a moderately subordinate one, has demonstrated that he is willing and able to provide protection for a female and her infant, that female may seek him out when she enters estrus again [1158].

Male baboons also form friendships with other males. Through these alliances, they can sometimes collectively confront a stronger rival that has acquired a partner, forcing him to give her up, despite the fact that the high-ranking male could take out any one of his opponents mano a mano. Thus, for example, in one troop of savanna baboons that contained eight adult males, three low-ranking males (fifth through seventh in the hierarchy) regularly formed coalitions to oppose a single higher-ranking male when he was accompanying a fertile female. In 18 of 28 cases, the threatening gang of subordinates forced their lone opponent to relinquish the female [880].

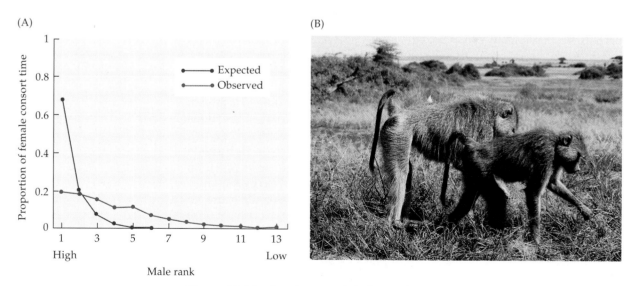

Figure 10.18 Dominant male baboons fail to control fertile females as completely as one would expect if the top-ranked male always had priority of access to estrous females when only one was available in the troop at a particular period. (A) Males of high rank did not remain close to a female for as high a proportion of the time she was in estrous as expected according to their dominance status. (B) A male guarding an estrous female. A, after Alberts et al.; B, photograph by Joan Silk.

Figure 10.19 **Small males of the marine iguana must cope with sexual interference** from larger rivals. Two small males are mounted on the same female; a much larger male is trying to pull them off her back. Photograph by Martin Wikelski, from Wikelski and Baurle [1299].

Baboons are not the only species in which some males use special mating tactics in order to avoid getting thrashed by a stronger or more experienced opponent [31, 484]. While some gray seal males, for example, are fighting with one another to monopolize mating opportunities with females hauled out on beaches with their pups, other males evidently are looking for and finding mates swimming in the water near the colony. These noncombatants father many of the young produced in any one year, judging from tests of the paternity of the babies found in breeding colonies [1312].

The gray seals that intercept females at sea do not have to fight with powerful rivals on land. Another alternative tactic with the same effect has evolved via sexual selection in the famous marine iguana, which lives in dense colonies on the Galápagos Islands. Male iguanas vary greatly in size, and when a small male runs over to mount a female, a larger male may arrive almost simultaneously to remove him unceremoniously from her back (Figure 10.19). Because it takes 3 minutes for a male iguana to ejaculate, one would think that small males whose matings were so quickly interrupted would not be able to inseminate their partners. However, the little iguanas have a solution to this problem, which involves having already ejaculated *prior* to any copulation attempt while retaining their sperm in their bodies. When, in the course of mounting a female, the small male everts his penis from an opening at the base of his tail and inserts it into the female's cloaca, these "old" sperm immediately begin to flow down a penile groove, so that even if he does not have time to achieve another ejaculation, the small iguana can at least pass some viable sperm to his mate during their all too brief time together [1299].

Conditional Mating Strategies

Although the little iguanas' ability to inseminate without ejaculation does help them reproduce, odds are that the big boys, who are able to copulate and inseminate at their leisure, reproduce more successfully. So here we have another example of behavioral variation within a species in which some individuals employ a behavior with a high fitness payoff while others behave differently and appear to be making the best of a bad situation. This pattern is the sort associated with conditional strategies (see Chapter 7).

Applying this approach to the competitively disadvantaged little iguanas, we can argue that these males salvage at least some reproductive chances by inseminating females on the fly, whereas they would have little or no chance of

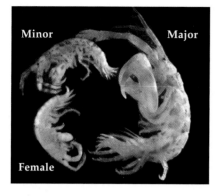

Figure 10.20 Alternative forms of a marine amphipod of the genus *Jassa*. There are two size classes of males, large majors (right) and small minors (top left) that resemble females (bottom left). The differences between majors and minors are caused by differences in their diets, not by heredity. From Kurdziel and Knowles [682].

securing mates via the tactic of fighting with others for access to mates, something that can be done profitably only by large, powerful individuals. Likewise, small males of an amphipod in the genus *Jassa*, which have been stunted by a second-rate diet as they were growing up, try to sneak matings with females, whereas those males that were blessed with a high-protein planktonic diet become large, aggressive individuals able to monopolize some potential mates (Figure 10.20) [682].

In species with a conditional strategy, the ability to adopt the "inferior" tactic can be adaptive (and therefore contributes to the maintenance of the conditional strategy) if small or otherwise handicapped individuals gain more fitness than they would if they were to try to use the tactics of their dominant opponents (Figure 10.21). In the horseshoe crab, for example, some males patrol the water off the beach, finding and grasping females heading toward shore to lay their eggs there. Other males are Johnny-come-latelies that swim onto the beach alone and crowd around paired couples there. As it turns out, an attached male fertilizes at least 10 percent more eggs than any competing satellite male [148, 149]. Although satellite males do not do as well as attached horseshoe crabs, perhaps they do better than they would if they tried to attach themselves to a female at sea only to be pushed aside or displaced by males that are in better physical condition.

To test this proposition, Jane Brockmann removed samples of attached and unattached males from a beach and returned them to the sea with plastic bags

Figure 10.21 Satellite male mating tactics. (Clockwise from top left) A satellite male Great Plains toad crouches by a singing male (note the inflated throat pouch) whose signals he may be able to exploit by intercepting females attracted by the calls. Six subordinate male bighorn sheep trail after a dominant male, who stands between them and the female with whom he will mate at intervals. Two satellite male horseshoe crabs wait by the female on either side of a male that has attached himself to his mate. Photographs by Brian Sullivan; Jack Hogg; Kim Abplanalp, courtesy of Jane Brockmann.

Figure 10.22   A conditional strategy model of the relationship between male condition and reproductive success in the horseshoe crab. This model predicts that when male condition has fallen to 3 on a 5-point scale, males will switch tactics because they will then gain more reproductive success from the satellite option than by trying to attach themselves to females. "Reproductive success" and "Male condition" are given in arbitrary units. After Brockmann [147].

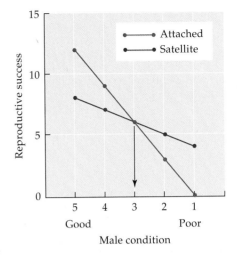

over the claws that they use to attach themselves to females. Before releasing them, she marked her subjects and scored their body condition based on such things as whether the carapace was worn or smooth and whether the animal's eyes were covered by marine organisms or free from obstruction. She found that worn, fouled males in relatively poor condition were more likely to show up again on the beach as unattached individuals. In other words, males in poorer condition were prepared to accept the lower-payoff tactic of coming to shore unattached to search for satellite opportunities, while males in good condition evidently remained at sea, attempting to find a female to grab and defend, even though they were doomed to fail because of the bags over their claws [147]. The decision to be a satellite is an adaptive making-the-best-of-a-bad-job tactic if individuals that select this option gain more than they would by persisting in hopeless attempts to become an attached male (Figure 10.22).

Discussion Questions

10.6   Satellite and attached male horseshoe crabs do not have the same reproductive success. Why hasn't sexual selection eliminated the low-payoff satellite option, if males exercising this option do not leave as many descendants as attached males do?

10.7   Conditional strategies can evolve by sexual selection only if certain conditions are met. What are those conditions, and what kind of conditional strategy among a set of alternatives is likely to be selected for? Make reference to Figure 10.22 in your answer.

Thus, male horseshoe crabs appear to evaluate their body condition, and they use this information to adopt one or another mating tactic, evidence that they possess a conditional strategy. Let's apply the same theory to alternative mating tactics in a *Panorpa* scorpionfly (Figure 10.23) in which (1) some males aggressively defend dead insects, a food resource highly attractive to receptive females, (2) other males secrete saliva on leaves and wait for occasional females to come and consume this nutritional gift, and (3) still others offer females nothing at all, but instead grab them and force them to copulate [1190]. In experiments with caged groups of ten male and ten female *Panorpa*, Randy Thornhill showed that the largest males monopolized the two dead crickets placed in the cage, which gave these males easy access to females and about six copulations on average per trial. Medium-sized males could not outmuscle the largest scorpionflies in the competition for the crickets, so they usually produced salivary gifts to attract females, but gained only about two copulations each. Small males were unable to claim crickets, and they could not generate salivary presents either, so these scorpionflies forced females to mate, but averaged only about one copulation per trial.

Thornhill proposed that in this case, all the males, large and small alike, possessed a conditional strategy that enabled each individual to select one of three options based on his social standing. This hypothesis predicts that the differences between the behavioral phenotypes are environmentally caused, not

Figure 10.23   A male *Panorpa* scorpionfly with its strange scorpion-like abdomen tip, which it can use to grasp females in a prelude to forced copulation, one of three mating tactics available to males of this species. Photograph by Jim Lloyd.

based on hereditary differences among individuals, and that males should switch to a tactic yielding higher reproductive success if the social conditions they experience make the switch possible.

To test these predictions, Thornhill removed the large males that had been defending the dead crickets. When this change occurred, some males promptly abandoned their salivary mounds and claimed the more valuable crickets. Other males that had been relying on forced copulations hurried over to stand by the abandoned secretions of the males that had left them to defend a dead cricket. Thus, male *Panorpa* can adopt whichever of the three tactics gives them the highest possible chance of mating, given their current competitive status. These results clinch the case in favor of a conditional strategy as the explanation for the coexistence of three mating tactics in *Panorpa* scorpionflies [1190].

Distinct Mating Strategies

The conventional wisdom is that almost all cases of alternative mating tactics can be explained via conditional strategy theory ([484], but see [1107]). Exceptions to this general rule exist, however. The ruff, for example, is a sandpiper whose males are either territorial or subordinate satellites on another's territory, a trait that they pass on to their male offspring [698]. The marine isopod *Paracerceis sculpta* is another species with more than one coexisting reproductive strategy. This creature vaguely resembles the more familiar terrestrial sowbugs and pillbugs that live in moist debris in suburban backyards, but resides instead in sponges found in the intertidal zone of the Gulf of California. If you were to open up a sufficient number of sponges, you would find females, which all look more or less alike, and males, which come in three dramatically different sizes: large (alpha), medium (beta), and small (gamma) (Figure 10.24), each with its own behavioral phenotype.

The big alpha males attempt to exclude other males from interior cavities of sponges that have one or more females living in them. If a resident alpha encounters another alpha male in a sponge, a battle ensues that may last hours before one male gives way. Should an alpha male find a tiny gamma male, however, the larger isopod simply grasps the smaller one and throws him out of the sponge. Not surprisingly, gammas avoid alpha males as much as possible while trying to sneak matings from the females living in them [1103]. When an alpha and a medium-sized beta male meet inside a sponge cavity, however, the beta

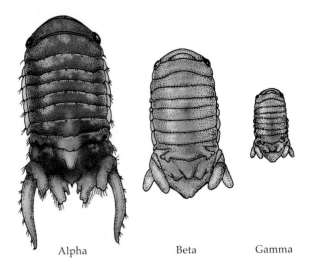

Figure 10.24 **Three different forms of the sponge isopod:** the large alpha male, the female-sized beta male, and the tiny gamma male. Each type not only has a different size and shape, but also uses a different hereditary strategy to acquire mates. After Shuster [1104].

Alpha   Beta   Gamma

behaves like a female, and the male courts his rival ineffectually. Through female mimicry, the female-sized beta males coexist with their much larger and stronger rivals and thereby gain access to the real females that the alpha male would otherwise monopolize.

In this species, therefore, we have three different types of males, and one type has the potential to dominate others in male–male competition. If the three types represent three distinct strategies, then (1) the differences between them should be traceable to genetic differences, and (2) the mean reproductive success of the three types should be equal. If, however, alpha, beta, and gamma males use three different tactics resulting from the same conditional strategy, then (1) the behavioral differences between them should be induced by different environmental conditions, not different genes, and (2) the mean reproductive success of males using the alternative tactics need not be equal.

Steve Shuster and his co-workers collected the information needed to check the predictions derived from these two hypotheses [1105, 1106]. First, they showed that the size and behavioral differences between the three types of male isopods are the hereditary result of differences in a single gene represented by three alleles. Second, they measured the reproductive success of the three types in the laboratory by placing various combinations of males and females in artificial sponges. The males used in this experiment had special genetic markers— distinctive characteristics that could be passed on to their offspring—enabling the researchers to identify which male had fathered which of the baby isopods that each female eventually produced. Shuster found that the reproductive success of a male depended on how many females and rival males lived with him in a sponge. For example, when an alpha male and a beta male lived together with one female, the alpha isopod fathered most of the offspring. But when this male combo occupied a sponge with several females, the alpha male could not control them all, and the beta male outdid his rival, siring 60 percent of the resulting progeny. In still other combinations, gamma males outreproduced the others. For each combination, it was possible to calculate an average value for male reproductive success for alpha, beta, and gamma males.

Shuster and Michael Wade then returned to the Gulf of California to collect a large random sample of sponges, each one of which they opened to count the isopods within [1106]. Knowing how often alphas, betas, and gammas lived in various combinations with competitors and females enabled Shuster and Wade to estimate the average reproductive success of the three types of males, given the laboratory results gathered earlier. When the mathematical dust had settled, they estimated that alpha males in nature had mated with 1.51 females on average, while betas checked in at 1.35 and gammas at 1.37 mates. Since these means were not significantly different, statistically speaking, Shuster and Wade concluded that the three genetically different types of males had essentially equal fitnesses in nature. The requirements for a three-distinct-strategies explanation had been met.

Note that in this example, the reason why three genetically distinct phenotypes can coexist is that *at certain frequencies* all three types have the same fitness. Therefore, any change in the proportion of one type in the population causes a change in the fitnesses of all three types. Imagine, for example, that for one reason or another, beta males became twice as frequent in a population as they were when Shuster and Wade did their collecting. If so, the "extra" betas would have to go somewhere, including sponges that already had one beta male co-residing with an alpha and some females. Then, instead of one beta, there would now be two beta males competing for the standard portion of the harem that alpha males cannot control, which would cut the mean beta fitness in this sponge by half, thereby also lowering the mean fitness of the beta phenotype overall. Relative to the beta phenotype, alphas and gammas would do better, and their *relatively* greater reproductive success would lead to an increase in the proportions of alpha and gammas in the next generation. Eventually, frequency-dependent selection of this sort (see p. 217) would result in a population in which the three types had equal fitnesses on average. Because types that become rarer gain fitness under frequency-dependent selection, no one phenotype is ever eliminated in such a system.

Discussion Questions

10.8 Imagine a species with two kinds of males, fighters and female mimics that pretend to be females to gain access to genuine females within the fighters' territories [400]. Imagine that there are some successful fighters (25 percent of the male population) that mate exclusively with three females each on average; the unsuccessful fighters (50 percent of all males) do not mate with any females because they are chased away by the better fighters. The final quarter of the population practices female mimicry, and these males average one mating each, essentially taking females that would otherwise have mated with the successful fighters. (We shall assume that each female mates just once.) If you were to calculate the fitness of fighters while excluding the unsuccessful fighters, what would you conclude about the proximate control of the two tactics? On the other hand, if you were to calculate the mean fitness of the two behavioral phenotypes while including data from the unsuccessful fighters, would you reach the same conclusion? Given this second method, what would happen in a population in which female mimics made up 50 percent of the population, if the pool of real females available to these males was the same as that available for a population in which only 25 percent of the males were female mimics?

10.9 Figure 10.25 shows three different tactics used by male bluegill sunfish for fertilizing eggs [483]: the territorial nest defense tactic used by large males, the sneaker option used by small males, and the female mimic option. How would you test the competing three-different-strategies and one-conditional-strategy hypotheses to account for the existence of alternative mating tactics in this species? (For test data, see [322].)

Sperm Competition

The alpha male and the beta male sponge isopod differ in how they go about securing copulatory partners. But the reproductive competition among them need not stop there. When females mate with more than one male in a short time, the two males may not divvy up a female's eggs evenly. In the case of

Figure 10.25 Three different egg fertilization behaviors coexist in the bluegill sunfish. (A) A territorial male guards a nest that may attract gravid females. (B) Little sneaker males wait for an opportunity to slip between a spawning pair, releasing their sperm when the territory holder does. (C) A slightly larger satellite male with the body coloration of a female hovers above a nest before slipping between the territorial male and his mate when the female spawns.

bluegill sunfish, for example, older males nesting in the interior of the colony (see Chapter 6) produce ejaculates with more sperm, and their sperm swim faster, suggesting that they would generally have a fertilization advantage over an average bluegill [199]. On the other hand, when a sneaker male bluegill and a guarding territorial male release their sperm more or less at the same time over a mass of eggs (see Figure 10.25), the sneaker male fertilizes a higher proportion of the eggs than the nest-guarding male [417]. What we have here is evidence of competition among males with respect to the fertilization success of their sperm. Such **sperm competition** is a very common phenomenon in the animal kingdom, no matter whether fertilization is external (as in bluegills and many other fishes) or internal (as in most insects, birds, and mammals). If the sperm of some males have a consistent advantage in the race to fertilize eggs, then counting up a male's spawnings or copulatory partners will not measure his fitness accurately [104].

Discussion Question

10.10 More evidence that matings do not necessarily translate into fitness for males comes from a study of a common European frog, *Rana temporaria*. In this species, some males find and grasp egg-laden females and then release their sperm as the female deposits a batch of eggs in a pond. Some other males locate floating egg masses soon after they have been laid. While grasping the clutch as if it were a female, these after-the-fact males release their sperm on the eggs, with the result that more than 80 percent of the clutches in one pond had split paternity [1232]. Some have interpreted this mating system as a means to ensure that the maximum amount of genetic diversity is passed on to the next generation, given that the sex ratio is heavily male-biased. Devise another evolutionary explanation and evaluate the two alternatives.

Sperm wars occur in insects as well as in isopods and fishes (and most other animal groups). Consider the competition for egg fertilization chances in *Calopteryx maculata*, the black-winged damselfly of eastern North America [1241]. Males of this species defend territories containing floating aquatic vegetation, in which females lay their eggs. When a female flies to a stream to lay her eggs, she may visit several males' territories and copulate with each site's owner, laying some eggs at each location. The female's behavior creates competition among her partners to fertilize her eggs [926], and the resulting sexual selection pressure on males has endowed them with an extraordinary penis. To understand how the damselfly penis works, we need to describe the odd manner in which damselflies (and dragonflies) copulate. First, the male catches the female and grasps the front of her thorax with specialized claspers at the tip of his abdomen. A receptive female then swings her abdomen under the male's body and places her genitalia over the male's sperm transfer device, which occupies a place on the underside of his abdomen near the thorax (Figure 10.26). The male damselfly then rhythmically pumps his abdomen up and down, during which time his spiky penis acts as a scrub brush (Figure 10.27), catching and drawing out any sperm already stored

Figure 10.26 Copulation in the black-winged damselfly enables the male to remove a rival's sperm before transferring his own. The male (on the right) has grasped the female with the tip of his abdomen; the female bends her abdomen forward to make contact with her partner's sperm-removing and sperm-transferring penis. Photograph by the author.

Figure 10.27 Sperm competition has shaped the evolution of the black-winged damselfly's penis. (Left) The male's penis has lateral horns and spines that enable him to scrub out a female's sperm storage organ before passing his own sperm to her. (Right) A close-up of a lateral horn reveals rival sperm caught in its spiny hairs. Photomicrographs by Jon Waage, from Waage [1241].

in the female's sperm storage organ. Jon Waage found that a copulating male *C. maculata* removes between 90 and 100 percent of any competing sperm before he releases his own gametes, which he has transferred from his testes on the tip of his abdomen to a temporary storage chamber very near the penis. After emptying the female's sperm storage organ, he lets his own sperm out of storage into his penis and thus into the female's reproductive tract, where they remain for use when she fertilizes her eggs—unless she mates with yet another male before ovipositing, in which case his sperm will be extracted in turn [1241].

If the black-winged damselfly was to be our guide, we might conclude that sperm competition is basically something males do to one another. In reality, females often have an active role in deciding which of their partners' sperm will win the egg fertilization contest [962, 1110]. Even females of the black-winged damselfly could arrange for the removal of some sperm simply by mating with a second male after copulating with an individual that she did not favor. In other cases, females do not have to rely on males for sperm removal, but can expel sperm themselves (Figure 10.28) [289].

Figure 10.28 Sperm competition in the dunnock requires female cooperation. A male pecks at the cloaca of his partner after finding another male near her; in response, she ejects a droplet of sperm-containing ejaculate just received from the other male. After Davies [289].

10.11   The female black-legged kittiwake can eject sperm received from a partner [1244]. If you were told that this bird is monogamous, why would you reject the hypothesis that sperm ejection in this species had evolved by sexual selection? In this species, monogamous pairs begin mating well before egg laying. If sperm ejection is related to an inability to keep sperm alive for long periods within the female's reproductive tract, when should females expel their partner's sperm?

Females of the collared flycatcher, although they cannot physically remove unwanted male gametes as some other birds can, could bias male fertilization chances by copulating in ways that give one male's sperm a numerical advantage over another's. Female birds can store sperm and use it to fertilize eggs that reach the uterus after having been released from the ovaries (Figure 10.29). To determine how many sperm a female flycatcher had in her reproductive tract during the period when her eggs were being fertilized, a research team developed a novel technique [822]. This method was based on the fact that whenever fertilization occurs, masses of sperm that have come out of storage surround the egg before one sperm finally manages to penetrate the egg cell and fertilize it. The remaining "loser" sperm are trapped in what becomes the inner perivitelline layer of the mature egg (the thin, translucent membrane that holds the

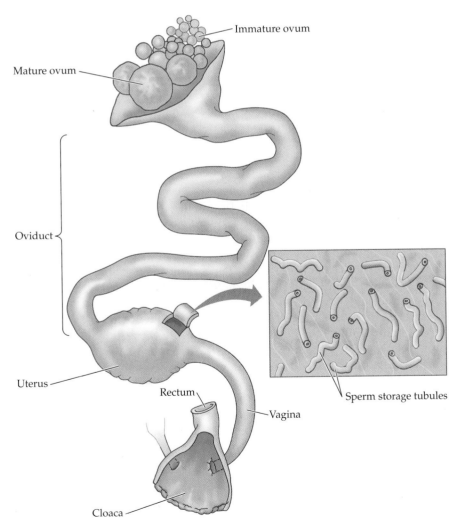

Figure 10.29   The reproductive anatomy of fertilization in birds. When a mature ovum produced in an ovary is released, it travels down the oviduct, where it may encounter sperm and become fertilized. Viable sperm received from males can be stored for lengthy periods in small sacs or tubules, especially in the interior wall of the uterus just above the vagina. The sperm gradually move out of the tubules over time and migrate up the oviduct to meet freshly released eggs. After fertilization, the hard shell is added in the uterus before the egg is laid via the cloaca. After Bakst [54] and Birkhead and Møller [103].

Figure 10.30 **Female collared fly-catchers could bias egg fertilizations** in favor of an extra-pair mate. When females are experimentally prevented from receiving sperm from their social partners, some do not copulate with another male, and the number of sperm trapped in the perivitelline membrane of their eggs falls. But some other females do have an extra-pair mating in the middle or late part of the mating period, and these individuals go on to lay eggs that carry large numbers of nonfertilizing sperm caught in the perivitelline membrane. These "extra" sperm must have come from an extra-pair male. After Michl et al. [822].

Collared flycatcher

yolk of an egg in globular form), where they can be counted by persons who know where to look after an egg has been laid. When researchers did this counting, they found that copulation causes a surge in the number of sperm present in eggs that are laid 2 days later. But if females are not being inseminated, the number of sperm in the eggs laid falls over time, for reasons we need not explore here.

Although we tend to think of birds as monogamous, that is not necessarily the case, as we'll see in the next chapter. Many birds form pair bonds with, and mate with, a social partner, but may also engage in extra-pair matings with other individuals. To keep female flycatchers from receiving sperm from their social partners, researchers glued an anti-insemination ring around the male's cloaca, which did not keep him from mating with the female, but did mean that his sperm could not enter her reproductive tract. Some of these females did not mate with other males thereafter, and, as described above, sperm counts in their eggs declined steadily (Figure 10.30). But some females that were paired with males incapable of insemination subsequently copulated once or twice with another male, boosting sperm levels in their eggs and producing eggs fertilized by this other individual, as revealed by genetic analysis. The number of sperm in the perivitelline layer of eggs laid by females with an extra-pair partner vastly exceeded that in eggs laid by females faithful to a male with an anti-insemination device.

Under natural conditions, therefore, a female could use her control of copulations to manipulate the number of sperm from different males within her reproductive tract. A collared flycatcher that stopped mating with her social partner for a few days and then mated with an attractive neighbor would have five times as many of her extra-pair mate's sperm available for egg fertilizations than she retained from earlier matings with her social partner. This imbalance would give the extra-pair male a big fertilization advantage. In fact, female collared flycatchers that are paired with a male with a small white forehead patch often do secure sperm around the time of egg laying by slipping away for a tryst with a nearby male sporting a larger white patch. Therefore, females of this species can play a major role in determining whose sperm will fertilize their eggs [822], a conclusion that probably applies to most animal species [355].

Mate Guarding

Males must compete for egg fertilization chances on a stage that has been designed to give females partial or complete control of the outcome. Nevertheless, males can sometimes give their sperm an advantage over those of their rivals. One way to do so would be to increase the number of sperm ejaculated into a female that had already mated with another male. In fact, male meadow voles do boost the sperm count of an ejaculate by more than 50 percent when copulating with a female in a place where the odors of another male are present [312].

Although adjustments in sperm donated to a mate can sometimes help males improve the odds that a female will use their gametes to fertilize her eggs, a more powerful measure would be to ensure that she never receives any sperm from anyone else. By standing guard over a mate, a male can sometimes physically prevent his partner from mating again, thereby improving the odds that his sperm will not have to compete with those of other males for her eggs. **Mate guarding** after insemination, a common phenomenon, is achieved in many different ways. In some cases, males keep their mates occupied after mating is finished [19]; in others, mated males deceptively lure new suitors away from the female [388], while in still others, males may seal a partner's genitalia with various secretions [330, 969]. A spectacularly final form of mate guarding is practiced by males of an orb-weaving spider, which die within minutes after having inserted both pedipalps (the sperm-transferring appendages) into a female's paired genital openings. The tips of the pedipalps inflate after insertion, promoting insemination while also blocking entry to the female's sperm-receiving apparatus. After his death, the male and his inserted pedipalps constitute a "whole-body mating plug" that other males cannot easily remove [393]. Although many forms of mate guarding exist, most are less extreme than sacrificing one's life to serve as a chastity belt. In fact, the most common tactic by far is simply to stay with a mate in order to see off any other males should they dare to approach (Figure 10.31) [31].

But is remaining with one partner in order to chase away other suitors really adaptive for the mate-guarding male? The benefits of the behavior, extra egg fertilizations with one female, come with costs, including the loss of opportunities to seek out other females. Janis Dickinson measured this opportunity cost in the blue milkweed beetle, in which the male normally remains mounted on the female's back for some time after copulation. When Dickinson pulled pairs

Figure 10.31 Mate guarding occurs in many animals. (Left) A red male damselfly grasps his mate in the tandem position so that she cannot mate with another male. (Right) The male blueband goby, an Indonesian reef fish, closely accompanies his mate wherever she goes. Left, photograph by the author; right, photograph by Roger Steene.

apart, about 25 percent of the separated males found new mates within 30 minutes. Thus, remaining mounted on a female after inseminating her carries a considerable cost for the guarding male, since he has a good chance of finding a new mate elsewhere if he would just leave his old one. On the other hand, nearly 50 percent of the females whose guarding partners were plucked from their backs acquired a new mate within 30 minutes. Since mounted males cannot easily be displaced from a female's back by rival males, they reduce the probability that an inseminated partner will mate again, giving their sperm a better chance to fertilize her eggs. Dickinson calculated that if the last male to copulate with a female fertilized even 40 percent of her eggs, he would gain fitness by giving up the search for new mates in order to guard his current one [328].

In general, the benefits of mate guarding increase with the probability that *unguarded* females will mate again and use the sperm of later partners to fertilize their eggs. But how do you figure out what unguarded females would do in a species in which all the females are guarded? Dickinson's technique involved the simple removal of a male from a female. Jan Komdeur and his colleagues achieved the same effect in their study of Seychelles warblers by tricking males into ending their guarding prematurely by placing a false egg in their nests about 4 days before their social partner was due to lay an egg. Because female Seychelles warblers produce only one egg per nesting attempt, males have nothing to gain from mate guarding after an egg appears in the nest. In the phony egg experiment, males stopped watching females that were still fertile, and in short order, many of these unguarded females copulated with neighboring males (Figure 10.32) [661]. This result demonstrates that mate guarding provides fitness benefits for male Seychelles warblers.

Figure 10.32 Adaptive mate guarding by the Seychelles warbler. The graphs show the rate of intrusions and extra-pair copulations (EPCs) by males other than the female's social mate in relation to the female's fertile period (shaded area). (A) Control pairs, in which the female's mate was present throughout her fertile period. (B) Pairs in which the female's mate was experimentally induced to leave her unguarded by the placement of a false egg in the nest. After Komdeur et al. [661].

Seychelles warbler

Figure 10.33   Male Seychelles warblers adjust their mate guarding in relation to the risk of losing paternity to rivals. The more male neighbors around a breeding pair, the more time male warblers spend guarding their partner. After Komdeur [659].

Because mate guarding is costly, we can also predict that males will adjust their investment in mate guarding in relation to the risk of cuckoldry, which ought to be affected by how many males live in close proximity to a fertile female and her partner. In the Seychelles warbler, breeding pairs may be surrounded by up to six neighbors. As predicted, the more neighbors, the more time a male spends guarding his fertile partner, and the less time he invests in foraging (Figure 10.33) [659]. In a similar vein, male red-winged blackbirds are more vigilant and aggressive toward neighbors that are unusually sexually attractive, and therefore constitute a greater threat to the guarding male's paternity, than neighbors that are less appealing to their mates [894].

Another way to test the hypothesis that mate guarding is an adaptation that improves the guarding male's chance of paternity is to examine the following prediction: if males can *increase* the time spent in close association with their mates, then they should fertilize a higher proportion of the eggs laid by those females than they would have otherwise. When house sparrows breeding at a zoo in Vienna, Austria, were given millet and other birdseed supplements, females were more likely to forage close to their nests, enabling their partners to more easily monitor their behavior. As a result, the time that pairs spent together almost doubled, and the proportion of chicks per brood that were sired by an extra-pair male fell dramatically (Figure 10.34) [1225].

English sparrow

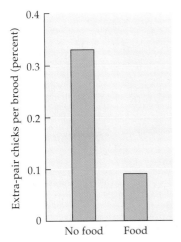

Figure 10.34   Food supplies on a bird's territory affect extra-pair paternity in the house sparrow. When females could forage for food supplements placed close to the nest, and thus could be guarded more easily by their social partners, the proportion of chicks fathered by other males in their broods fell accordingly. After Václav et al. [1225].

Discussion Questions

10.12   Mate guarding should be common in species in which females retain their receptivity after mating and are likely to use the sperm of their last mating partner when fertilizing their eggs. But there are species, including some crab spiders, in which males remain with immature, unreceptive females for long periods and fight with other males that approach these females [335]. How can "guarding" behavior of this sort be adaptive? Produce sexual selectionist hypotheses and allied predictions.

10.13   When male bluethroats (see Figure 4.33) have their blue throats experimentally blackened, they become less attractive to females. Outline the costs and benefits to these males of guarding their mates, once they have secured a partner, and the costs and benefits of attempting to secure extra-pair matings with other females by visiting the territories of other pairs. Use conditional strategy theory to predict the tactics of experimental and control males. Discuss the significance of the finding that experimental males actually fathered fewer of their social partner's offspring on average than did control males.

TABLE 10.3   *Ways in which females and males attempt to control reproductive decisions*

Key reproductive decisions controlled primarily by females

Egg investment: What materials, and how much of them, to place in an egg

Mate choice: Which male or males will be granted the right to be sperm donors

Egg fertilization: Which sperm to use to fertilize each egg

Offspring investment: How much maintenance and care goes to each embryo and offspring

Ways in which males influence female reproductive decisions

Resources transferred to female: May influence egg investment, mate choice, or egg fertilization decisions by female

Elaborate courtship: May influence mate choice or egg fertilization decisions by female

Sexual coercion: May overcome female preferences for other males

Infanticide: May overcome female decisions about offspring investment

Source: Modified from Waage [1242]

Sexual Selection and Mate Choice

Mate guarding is one of the many dramatic evolutionary consequences of sexual selection arising from competition for mates, but as we have also noted, mate choice can also create sexual selection pressure. In most species, mate choice is exercised primarily by females, a fact that has great evolutionary significance for males (Table 10.3) [1242, 1358]. The attributes of males that females favor vary greatly from species to species, but in some cases female preferences are based on highly utilitarian considerations, such as the ability of the male to supply them with a good meal. A classic example comes from Randy Thornhill's study of the black-tipped hangingfly, an insect in which female acceptance of a male depends on the nature of his nuptial gift (Figure 10.35). In this species, a male that tries to persuade a female that an unpalatable ladybird beetle is a good mating present is out of luck. Even males that transfer an edible prey item to their mates will be permitted to copulate only for as long as the meal lasts. Moreover, because a 5-minute meal is a prerequisite for the start of sperm transfer, if a nuptial gift has been polished off in less than 5 minutes, the female will separate from her partner without having accepted a single sperm from him. When, however, the nuptial gift is large enough to keep the copulating female feeding for 20 minutes, she will depart with a full complement of the gift-giver's sperm (Figure 10.36) [1189]. Males of many other animals provide food presents before or during copulation [1226], including male katydids and crickets that give their mates edible spermatophores (see Figure 10.12).

Some researchers have suggested that a special class of nuptial gift givers should be recognized: namely, those male mantids and spiders that wind up being eaten by their mates (see Figure 5.3). Indeed, it could be adaptive, under some special circumstances, for a male to conclude a copulation by becoming a meal for his recent sexual partner [177, 977]. This possibility, however, cannot apply to cases in which males are eaten by females *prior* to the transfer of sperm [854]. And in many cannibalistic species, the risk of becoming a precopulatory meal has favored males that approach females with the utmost caution. But male redback spiders seem to take extraordinary measures to put themselves

Figure 10.35   A potential nuptial gift. A male hangingfly has captured a moth, a material benefit to offer to his copulatory partner. He advertises the availability of his gift by releasing a pheromone from abdominal glands. Photograph by the author.

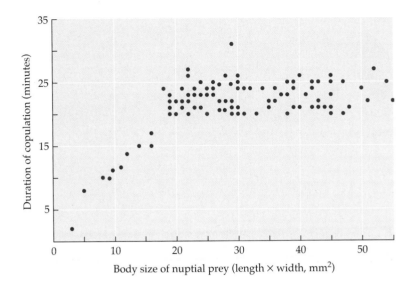

Figure 10.36   Sperm transfer and the size of nuptial gifts. In black-tipped hangingflies, the larger the nuptial gift, the longer the mating, and the more sperm the male is able to pass to the female. After Thornhill [1189].

in harm's way. As a male redback transfers sperm to a female, he performs a somersault that literally throws his body into his partner's jaws (Figure 10.37). In about two-thirds of these cases, the female accepts the invitation and devours her sexual companion [34].

A male redback weighs no more than 2 percent of what a female weighs—barely more than a mouthful for his mate. Nevertheless, hunger may be part of the proximate basis for cannibalism, since females that are deprived of food are more likely to dine on a male [35]. Once eaten, for whatever reason, the deceased male redback does in fact derive substantial benefits from what might seem like a genuinely fitness-reducing experience. Maydianne Andrade showed that eaten males fertilized more of their partner's eggs than uneaten males did, partly because a cannibalistic female spider is less likely to mate again promptly [34]. Moreover, the cost of being cannibalized is very low for male redbacks. Young adult males in search of a mate are usually captured by predatory ants or other spider hunters long before they find a web with an adult female. The intensity of predation on wandering male redbacks is such that fewer than 20 percent manage to locate a first mate, suggesting that the odds that a male could find a second partner if he should survive his initial mating are exceedingly poor [36]. Under these circumstances, the benefit-to-cost ratio of male sexual suicide favors its evolution.

Males of some other animals offer parental care for the female's offspring, rather than a nuptial gift, and in these species we would expect females to exert sexual selection pressure in favor of males that provide more paternal care than average. In keeping with this expectation, females of the fifteen-spined stickleback, a small fish with nest-guarding males, associate more with courting males that shake their bodies relatively frequently. Males capable of more body shaking also perform more nest fanning after courtship is over and egg laying has occurred (see Figure 12.3). Nest fanning sends oxygenated water flowing

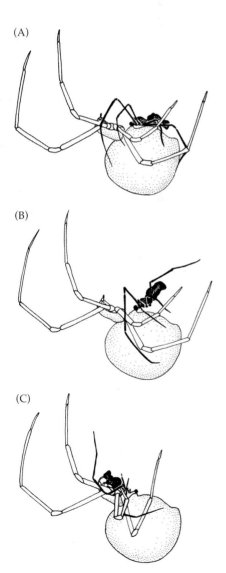

(A)

(B)

(C)

Figure 10.37   Sexual suicide in the redback spider. (A) The male first aligns himself facing forward on the underside of the female's abdomen while he inserts his sperm-transferring organ into her reproductive tract. (B) He then elevates his body and (C) somersaults backward into the jaws of his partner. She may oblige by consuming him while sperm transfer takes place. After Forster [399].

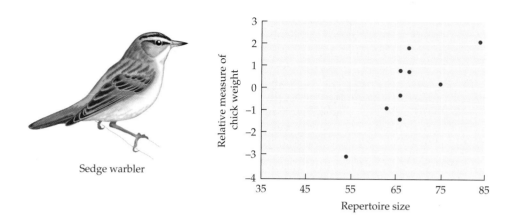

Figure 10.38 **Song repertoire size as an honest signal of male paternal effort.** In the sedge warbler, males with larger repertoires (as measured by the number of different syllable types used to construct the warbler's songs) help produce chicks that weigh more upon fledging. After Buchanan and Catchpole [167].

past the eggs, which increases gas exchange and, eventually, the rate of egg hatching [908].

The female stickleback that evaluates the courtship display of a male sees him engaging in behavior that is directly linked to his parental capacities. Males of many other parental species, however, provide at best only indirect cues that may or may not be related to their capacity to be a helpful father. In Chapter 2, we reviewed some evidence that certain female songbirds prefer males whose songs reveal that they had a privileged pre-adulthood. Males that grow up under better than average conditions are presumably more likely to be in good shape as adults and therefore capable of offering more parental investment to their offspring. In a few species, such as the sedge warbler, the link between song quality and parental ability has been definitively established (Figure 10.38) [167].

Could the quality of a male bird's plumage also be an indicator of his potential to be paternal? Perhaps so, since females of many animal species are especially attentive to differences among males in the red or yellow components of their color patterns [478, 810]. Reds and yellows are often produced by carotenoid pigments derived from the foods the animals consume, as is true of guppies (see Figure 9.16), and so could be an indicator of the foraging ability of an individual, provided that carotenoid pigments are scarce in the preferred foods of the species in question [896].

Another, not mutually exclusive, possibility is that carotenoid-based color patterns reveal something about the health of a male; specifically, the quality of the bird's immune system, which is believed to benefit from access to an adequate supply of carotenoids. Individuals in good condition could therefore afford to advertise this fact by employing some carotenoids in their display ornaments. If so, then experiments in which only some zebra finch males were provided with extra carotenoids should produce two classes of males: males with improved immune systems and brighter beaks (which normally range from dull to bright orange-red) and males with inferior immune systems and duller beaks. As expected, when the experiment was done, the carotenoid-supplemented males had more carotenoids in their blood, brighter beaks, and a stronger immune response [809]. Moreover, female zebra finches find experimentally carotenoid-enhanced males more attractive than those that have been kept on normal diets [111]. In nature, female zebra finches that acquired a more brightly colored mate might benefit by enabling their offspring to acquire "good genes" for foraging ability or by having a healthy mate capable of providing superior care for his offspring.

Much the same thing may be happening in the blue tit, another small songbird with a carotenoid-based ornament, a bright yellow breast. Male blue tits collect and deliver food for their nestlings, usually in the form of carotenoid-containing caterpillars. If the amount or quality of food supplied by a male is related to how bright his yellow feathers are, then the offspring of brilliantly yellow males should be larger when they fledge than the youngsters of less brightly colored males.

But wait a minute. This same prediction would follow if bright yellow males were themselves large at fledging, thanks to their genetic makeup, in which case their offspring would simply inherit their relatively large body size. Because a team of Spanish behavioral ecologists recognized this problem, they correctly controlled for heredity through a cross-fostering experiment. They took complete clutches of eggs and transferred them between nests, moving the offspring of one set of parents to another pair's nest. The foster parents were willing to rear these genetic strangers, and when their adopted chicks had reached the age of independence, the size of the fledglings was a function of the brightness of their *foster* father's yellow plumage, not the color of their genetic father. Bright yellow foster males produced larger fledglings. Because the parental effort of bright yellow males is greater than that of duller individuals, females could benefit by choosing males on the basis of their plumage, although whether blue tit females actually use the quality of the yellow plumage of potential mates to make their choices has yet to be determined [1086].

Discussion Questions

10.14   Males of the barn swallow have long, thin outer tail feathers. When Anders Møller analyzed the effect of tail length on male mating success in the barn swallow in Europe, he did an experiment in which he made some males' tail feathers shorter by cutting them, and made other males' tail feathers longer by gluing feather sections onto their tails [837]. But he also created a group in which he cut off parts of the males' tail feathers and then simply glued the fragments back on to produce a tail of unchanged length. What was the point of this group? And why did he randomly assign his subjects to the shortened, lengthened, and unchanged tail groups? And why did a team of Canadian biologists repeat Møller's experiment on another continent [1125]?

10.15   The manes of African lions vary in color from very dark to quite pale. When two models of male lions, one with a dark mane and the other with a light-colored mane, were put out in lion territories, the males that found the models first approached the light-colored one more often than the dark-maned model. In contrast, lionesses almost always walked close to the dark-maned model first. Peyton West and Craig Packer found that in general, mane color was darker in older males and in those with higher testosterone levels [1275]. Knowing what you know about lion social behavior (see Chapters 1 and 8) and sexual selection theory, how would you explain the differences in the reactions of males and females to the dark-maned models?

Mate Choice without Material Benefits

Many of the examples that we have just reviewed are consistent with **good parent theory**, which explains aspects of male color, ornamentation, and courtship behavior as sexually selected indicators of a male's capacity to provide parental care. Female choice based on these signals makes intuitive sense: the offspring of female sedge warblers or blue tits that pair off with a more-paternal-than-average male are going to be well fed as a rule. Likewise, it is easy to see why females of some species might prefer males able to supply them personally with food or some other present before copulating. For example, female black-tipped hangingflies that receive a large nuptial gift do not have to search for prey (the food they need to live and produce eggs), a task that involves the possibility of flying into a spider's web [1188]. However, males of many species, such as the satin bowerbird, do not provide food or any other material benefit to their mates or their offspring. Even so, as we have seen, female bowerbirds prefer males capable of more intense courtship displays and brighter plumage. The same is true for many other animals [651].

In the canary, another species lacking paternal care, a female's choice of a mate appears to be heavily influenced by his ability to sing a certain portion of the male song, the "A phrase," which is composed of many two-note syllables (Figure 10.39). Females that hear an A phrase trill that packs many syllables into a second of song readily adopt the precopulatory position (if they have been primed with estradiol in advance of the experiment) [1227]. In order to sing the A phrase in a sexy manner, males have to control the syrinx, their vocal apparatus, with great precision. Not only must the movements of the two tubes of the syrinx be integrated perfectly, a neurally demanding exercise, but what goes

Figure 10.39 Mate choice based on male performance of a physiologically challenging task. (A) Male canaries sing songs that contain a special trill that is composed of a series of syllables, each composed of two notes that require the coordinated action of the syrinx and the respiratory system to produce. One property of the trill syllables is their frequency bandwidth, which can be small (ranging from 2 to 4 kHz, for example) or large (ranging from 2 to 6 kHz, for example). Another is the rate at which the syllables are produced. (B) There is an upper limit on how fast males can sing syllables of a given frequency bandwidth. (C) Females prefer trills that are composed of broad-bandwidth syllables sung at a very fast rate. Tapes of three trill types were played to females: F had the narrowest bandwidth and G the greatest. Each trill type was played at two rates, 16 syllables per second and 20 syllables per second. The measure of female preference for the trills was the mean number of copulation solicitation displays given by listening female canaries. Females responded with significantly more displays to A20 and G20 trills. After Drăgăniou et al. [343]; photograph by Cornelia Voigt, courtesy of Stefan Leitner.

on in the syrinx also has to be in synchrony with the bird's breathing, since air is required for sound production [1167]. In other words, being able to sing a very rapid trill might constitute a demonstration of the male's physiological state, an honest signal of the quality of his neuromuscular system and overall condition.

Passing a female canary's song test requires that males not only generate a rapid trill, but also make the individual syllables in the trill cover a relatively wide range of sound frequencies (the bandwidth of the trill). We know this because of the responses of females to tapes of artificial trills, including some that were impossibly exaggerated versions of the A phrase. The most extreme versions of the trill elicited the most copulation solicitation displays from listening female canaries. Because there is an upper limit for male canaries with respect to how rapidly they can sing syllables of a given bandwidth, the female preference in effect favors males able to sing at the outer edge of song capacity. Females may reward these individuals not just by mating with them, but also by adding testosterone to the eggs fertilized by their sperm, a maternal investment that may enhance the chances of optimal development for these offspring [450].

Although these findings seem solid for canaries, it would be preferable to work with a "real" bird rather than a domesticated one that has been subjected to intense artificial selection on its singing behavior [101]. As it happens, at least one wild bird, the swamp sparrow, has been studied in much the same way as the canary. Barbara Ballentine and her colleagues found that in this species, whose males also sing a trilled song, the rate of syllable production and the frequency bandwidth of the notes that make up the trill are what count. For estradiol-primed female swamp sparrows, just as for female canaries, the faster the rate of syllable production and the broader the range of frequencies in the syllables of a male's song, the more stimulated females become as seen in their greater readiness to invite a male to copulate [59].

The way in which a male songbird belts out his courtship song may provide critical cues to his condition for discriminating females, and the same can be said for the visual display components of so many male animals. Consider that peahens prefer peacocks with relatively large numbers of eyespots in their immense tails, which the males display in their famous spread-and-shake advertisements in front of potential mates [944]. The importance of these decorations was demonstrated when Marion Petrie and Tim Halliday captured some adult peacocks and removed 20 of the outermost eyespots from some males' tails. Birds so treated experienced a significant decline in mating success compared with their performance in the previous year. In contrast, control males, which were captured and handled but whose tails were left intact, were no less attractive to females after the treatment [943].

If greater sensory stimulation from courting males does the trick, then experimentally augmenting a male's courtship ornaments should enhance his copulatory success. The relevant experiment was done first with the long-tailed widowbird, a species about the size of a red-winged blackbird that is endowed with a half-meter-long tail (Figure 10.40). The male flies around his grassland territory in Kenya, displaying his magnificent tail to passing females. Malte Andersson took advantage of the wonders of superglue to perform an ingenious experiment. He captured male widowbirds, then shortened the tails of some by removing a segment of tail feathers, only to glue it onto another bird's tail, thereby lengthening that ornament [29]. The tail-lengthened males were much more attractive to females than those that had suffered the loss of a portion of their ornaments. Moreover, the tail-lengthened males also did much better than controls, whose tails were cut and then put back together with applications of superglue.

Figure 10.40   A sexually selected ornament. The extraordinary tail feathers of the long-tailed widowbird are displayed to choosy females while the male flies above his grassland territory. Photograph by Malte Andersson.

The elements of courtship that females may use to assess the quality of potential mates are obviously not the same across the animal kingdom. Sexy songs and displays of extreme ornaments work for many, but not all. In some species, performance during copulation itself might constitute a kind of test imposed on males by their mates, an idea originally championed by Bill Eberhard [355] and now the subject of much discussion [565]. If females evaluate partners on the basis of copulatory performance, then in species whose females mate with several males in a breeding cycle, males might be predicted to possess more elaborate penises, which would be capable of providing greater sensory stimulation during mating. Among the primates, penises come in all shapes and sizes, including some that have an array of rather nasty-looking spines (Figure 10.41). Such a penis could provide information that would permit females to choose among the sperm provided by an array of partners, each competing to offer the most stimulating copulatory experience. However, although knobbed or spiny penises do occur in some primate species in which females live in groups containing several males, giving them a range of potential partners, the relationship is not a tight one. Overall, elaborate penises are no more likely to be found in primate species with a multi-male social organization than in species in which males and females live apart or in bands with a single reproductive male [511].

All of which leaves us wondering why males of some primates would have evolved such elaborate, even dangerous-looking, sperm-transferring devices. Paula Stockley has suggested that the evolution of elaborate, spiny penises may have occurred in the context of reducing the period of female receptivity [1154]. In other words, instead of cryptic female choice among ejaculates driving the evolution of male genitalia, sperm competition might be the main factor. If the benefit of having a spiny penis is to reduce the willingness of females to mate again, then if we were to compare closely related primates that differed in penile spinosity, the species with the spinier penises would be the ones that also had the shorter periods of female receptivity. When Stockley assembled eight such pairs of species, she found the predicted relationship, suggesting that the mates of males with spiny penises might well be less likely to copulate again with another male, given the relatively short period of female sexual receptivity in these species. Stockley speculated that a reduction in length of time females were receptive might stem from the discomfort and even minor damage that mated females might have endured during sexual intercourse with males whose intromittent organs had backward-pointing spines.

Figure 10.41   **Has cryptic female mate choice resulted in the evolution of stimulating male genitalia?**   The more elaborate penises on the left, which probably provide more sensory stimulation, are from primate species in which females have the opportunity to mate with more than one male, and so might use copulatory performance to decide which of several males' sperm to use to fertilize their eggs. The simpler, and presumably less stimulating, penises on the right are from species in which females typically mate with just one male during an estrous cycle. After Dixson [332] and Eberhard [354].

Discussion Question

10.16   Peacocks and long-tailed widowbirds have truly extravagant ornamentation that could be fairly recently evolved (that is, derived) from an ancestral pattern in which male plumage was not nearly so extreme. In the peacock-pheasants, six species in a genus related to that of the peafowl, there is considerable variation in the degree of male plumage ornamentation. In four species, males have highly elaborate plumage featuring large eyespots, but in two species, they do not. Draw two phylogenies, one in which elaborate plumage is the derived trait and another in which a reduction in ornamentation is the derived condition. Then test your hypotheses with data available in [640].

Male Courtship Signals Mate Quality

Although it remains to be firmly established whether females of some species evaluate males on the basis of the stimulation they provide during copulation itself, there is little doubt that heightened sensory stimulation provided by males during courtship often does influence mate selection by females. The fact that elements of a male's song, or his appearance, or his visual courtship display may determine his acceptability as a mate says something about the proximate mechanisms of mate choice, but does not identify the ultimate benefits, if any, of female mate choice, in species in which males have only their sperm to offer. In these species, females cannot be using the physical attributes of males, or their courtship behavior before, during, or after copulation, as indicators of the material benefits they will receive from their mates, because they will receive no such benefits. In these cases, what do females get from being choosy? Let's consider four hypotheses, which are presented in Table 10.4.

Even if males have nothing useful other than gametes to give their mates, they can also give them a variety of sexually transmitted diseases, which ought

TABLE 10.4 *Four theories on why extreme male ornamentation and striking courtship displays have evolved in species in which males provide no material benefits to their mates*

Theory	Females prefer trait that is	Primary adaptive value to choosy females
Healthy mate	Indicative of male health	Females (and offspring) avoid contagious diseases and parasites
Good genes	Indicative of male viability	Offspring may inherit the viability advantages of their father
Runaway selection	Sexually attractive	Sons inherit trait that makes them sexually attractive; daughters inherit the majority mate preference
Chase-away selection	Exploitative of preexisting sensory biases	No benefit received by female

to favor females that are able to avoid infected males. According to **healthy mate theory**, female preferences are focused on male courtship displays and ornaments that serve as indicators of a potential sexual partner's health or parasite load [1014]. Females could use this information to mate with males that are less likely to give them lice, mites, fleas, or bacterial pathogens, any one of which could harm them or their future offspring. As we noted earlier, male bowerbirds with high-quality bowers are less likely to carry ectoparasites in their feathers [338], so females that copulate with good bower builders should be less likely to pick up feather lice as a result of contact with these males.

Of course, females that mate with healthy males may not only avoid parasites and contagious diseases, but also secure genes that will benefit their offspring. The **good genes theory** proposes that male courtship displays and ornaments provide information to females that enables them to choose males with viability-enhancing genes. "Good genes" from a male could, for example, be those that were simply different from the ones possessed by the female, which would enable the choosy female to help her offspring avoid the developmental problems that often occur when individuals have two copies of the same allele at many genetic loci, an outcome generally associated with inbreeding. Perhaps this is why female sedge warblers prefer males with larger song repertoires. Given that repertoire size is correlated with male heterozygosity in this species, female preferences for males with large repertoires should increase the heterozygosity of their offspring, particularly since females somehow ensure that their eggs are fertilized by those sperm that are genetically least like the eggs' genomes [788].

In addition, "good genes" might advance the survival chances of offspring by conferring hereditary resistance to parasitic infection or disease or by promoting adaptive foraging skills or the ability to avoid predators. In the goldfinch, for example, the brightness of the yellow plumage is correlated with the bird's condition when it is molting—undergoing the annual, physiologically demanding process of shedding and replacing the feathers [810]. Moreover, female goldfinches apparently prefer to mate with bright yellow males, judging from the fact that in this species, very yellow females bond with very yellow males, leaving the less yellow individuals to pair off with one another [763]. Because male goldfinches do not help their mates rear young, the female preference for bright yellow males can be beneficial only if the preferred category of males has something to offer in the way of good genes, perhaps those associated with the development of superior foraging ability.

Figure 10.42 **Female choice for indicators of health, which may be related to good genes.** Male starlings vary in how many bouts of singing they can produce in an hour. This variable is correlated with the strength of an immune response to the injection of a foreign chemical into their wing membrane (measured by the degree of swelling). Because females prefer males with high song rates, they could be acquiring mates with strong immune systems. After Duffy and Ball [348].

Females of other bird species base their mating decisions on other criteria besides color. Female starlings, for example, prefer males able to sing frequently; males of this sort have strong immune systems, which they might be able to pass on to their offspring (Figure 10.42) [348]. And in the bowerbird, males appear to possess a whole array of visual and acoustical signals, which conceivably could inform females about a range of potentially heritable traits (Figure 10.43), including the males' parasite loads [339].

Resistance to parasites, if hereditary, would be a very useful attribute to give one's offspring. So it is perhaps not surprising that this factor took center stage in the version of good genes theory developed by W. D. Hamilton and Marlene Zuk, who realized that rapidly evolving parasites would exert continuous selection pressure on hosts to combat their constantly changing enemies [506]. Considerable evidence indicates that parasites do adapt quickly to locally common host genotypes, which maintains selection for host variants better able to escape parasitic infection [730]. Thus, in species subject to parasitic infection, individuals that can signal their relatively parasite-free state should be desirable mates. Hamilton and Zuk predicted that selection for honest signals of noninfection would lead bird species with numerous potential parasites to evolve strikingly colored plumage. They argued that brightly colored feathers are difficult to produce and maintain when a bird is parasitized because parasitic infection causes physiological stress. Hamilton and Zuk found the predicted correlation between plumage brightness and the incidence of blood parasites in a large sample of bird species, supporting the view that males at special risk of parasitic infection engage in a competition that signals their condition to choosy females [506].

Figure 10.43 **Male satin bowerbirds provide females with multiple signals** that are indicators of their health and physiological condition. After Doucet and Montgomerie [338]; photograph by Cyril Webster.

In contrast, **runaway selection theory** [42, 303] proposes that discriminating females acquire sperm with genes whose primary effect is to influence their daughters to prefer the male traits their mother found attractive and to endow their sons with attributes that will be preferred by most females—even if those traits actually reduce the survival chances of individuals that possess them. Because the runaway selection alternative is the least intuitively obvious explanation for male courtship displays and female mate choice, let us briefly examine the argument underlying the mathematical models of Russell Lande [692] and Mark Kirkpatrick [643].

Imagine that a slight majority of the females in an ancestral population had a preference for a certain male characteristic, perhaps *initially* because the preferred trait was indicative of some survival advantage enjoyed by the male. Females that mated with preferred males would have produced offspring that inherited the genes for the mate preference from their mothers *and* the genes for the attractive male character from their fathers. Sons that expressed the preferred trait would have enjoyed higher fitness, in part simply because they possessed the key cues that females found attractive. In addition, daughters that responded positively to those male cues would have gained by producing sexy sons with the trait that many females liked.

Thus, female mate choice genes and genes for the preferred male attribute can be inherited together. This pattern could generate a runaway process in which ever more extreme female preferences and male attributes spread together as new mutations affecting these traits occurred. The runaway process would end only when natural selection against costly or risky male traits balanced sexual selection in favor of traits that are appealing to females. Thus, if peahens originally preferred peacocks with larger than average tails because such males could forage efficiently, they might now favor males with extraordinary tails because this mating preference has taken on a life of its own, resulting in the production of sons that are exceptionally attractive to females and the production of daughters that will choose this kind of male for their own mates.

In fact, the Lande–Kirkpatrick models demonstrate that right from the start of the process, female preferences need not be directed at male traits that are utilitarian in the sense of improving survival, feeding ability, and the like. Any preexisting preference of females for certain kinds of sensory stimulation (see p. 288–289) could conceivably get the process under way. As a result, traits opposed by natural selection because they reduced viability could still spread through the population by runaway selection [643, 692]. Instead of mate choice based on genes that promote the development of useful characteristics in offspring, runaway selection could yield mate choice for *arbitrary* characters that are a burden to individuals in terms of survival, a disadvantage in every sense except that females mate preferentially with males that have them!

Testing the Healthy Mate, Good Genes, and Runaway Selection Theories

Discriminating among these three alternative explanations for elaborate male courtship displays and female choice in species in which males do not provide material benefits has proved very difficult, in part because these three hypotheses are not mutually exclusive. As just mentioned, female preferences and male traits that originated through a good genes process could then be caught up by runaway selection. Note too that males with hereditary resistance to certain parasites (good genes benefits) would also be less likely to infect their partners with those parasites (healthy mate benefits). Moreover, if at the end of a period of runaway selection, males had evolved extreme ornaments and elaborate displays, then only individuals in excellent physiological condition would be able to develop, maintain, and deploy their ornaments in effective displays.

Males in such superb physiological condition would probably have to be highly effective foragers (good genes benefits) as well as parasite-free (healthy mate benefits), in which case females mating with such males would be unlikely to acquire parasites, and their daughters might well get some survival-benefiting genes while their sons received the pure attractiveness genes of their fathers.

Given the difficulty of separating these three possible processes, the best we can do for the moment is to look at predictions derived from one or another theory and test them by applying them to appropriate species. As an example, let's apply the good genes theory to the gray treefrog, a species in which males offer their mates nothing but sperm. Female gray treefrogs are attracted to males that give calls of relatively long duration. If the benefits of female choice are good genes, then if we were to take half of a female's clutch of eggs and fertilize them with sperm from a long-calling male, while dousing the other half with sperm from a short caller, then the tadpoles sired by the long-calling male should develop faster than their half-siblings whose father could manage only short calls [1267]. In three of four trials, the results matched this prediction, providing some support for the good genes argument.

An even more ambitious attempt to test the good genes theory was conducted by Marion Petrie, who applied it to peacocks and derived the following predictions: (1) males should differ genetically in ways related to their survival chances, (2) male behavior and ornamentation should provide accurate information on the survival value of the males' genes, (3) females should use this information to select mates, and (4) the offspring of the chosen males should benefit from their mothers' mate choice. In other words, males should signal their genetic quality in an accurate manner, and females should pay attention to those signals because their offspring will derive hereditary benefits as a result [652, 1352].

Petrie studied a captive but free-ranging population of peacocks in a large forested English park, where she found that males killed by foxes had significantly shorter tails than their surviving companions. Moreover, she observed that most of the males taken by predators had not mated in previous mating seasons, suggesting that females could discriminate between males with high and low survival potential, possibly on the basis of their ornamented tails [941]. The peahens' preferences translate into offspring with enhanced survival chances, as Petrie showed in a controlled breeding experiment. She took a series of males with different degrees of ornamentation from the park and paired each of them a large cage with four females chosen at random from the population. The young of all the males were reared under identical conditions, weighed at intervals, and then eventually released back into the park. The sons and daughters of males with larger eyespots on their ornamented tails weighed more at day 84 of life and were more likely to be alive after 2 years in the park than the progeny of males with fewer eyespots (Figure 10.44) [942].

This combination of results is consistent with the view that peahen mate preferences create sexual selection that is *currently* maintaining or spreading the genetic basis for those preferences because the offspring of the preferred males receive good genes from their fathers. But it is possible that healthy male benefits are involved as well; perhaps the peahens' mate choices lead them to avoid parasitized males and so reduce their risk of acquiring damaging parasites that they would pass on to their offspring, lowering their growth rates and survival chances. Moreover, a demonstration of current selective advantages associated with female preferences and male traits does not rule out the possibility that these attributes *originated* as side effects of runaway selection in the manner described above.

Yet another complication comes from the discovery that females of some species invest more resources in offspring that are sired by preferred males than

Mean area of father's eyespots (mm^2)

Figure 10.44 Do male ornaments signal good genes? Male peacocks with larger eyespots on their tails produced offspring that survived better when released from captivity into an English woodland park. After Petrie [942]; photograph by Brian Lightfoot.

in those fathered by less attractive males. As noted in Chapter 9, female black-headed gulls can adjust the amount of testosterone their eggs receive [359], and hormonal manipulations of this sort are apparently widespread among birds [449, 450, 1074]. Female mallards make larger eggs after copulating with unusually attractive males [267], while black grouse produce and lay more eggs subsequent to mating with a top-ranked male [1019]. All of these effects might easily be attributed to the "good genes" of the male when in reality they arise from the female's manipulation of her own parental investment, not the genetic contribution of her partner.

Sexual Conflict

Brett Holland and Bill Rice have proposed yet another scenario (see Table 10.4) that could result in the evolution of extreme ornaments and displays in males [551]. According to their **chase-away selection theory**, exploitation and conflict between the sexes, rather than cooperation and mutual benefit, could drive the process in the following manner. Imagine that a male happened to have a mutation yielding a novel display trait, one that managed to tap into a preexisting sensory bias of the females of his species. A male that could activate that preexisting preference might induce females to mate with him even though he might not provide the material or genetic benefits offered by other males in the population [1049]. The resulting spread of such exploitative males over time would create selection on females favoring those that had psychological resistance to the purely attractive display trait. As females with a higher threshold for sexual responsiveness to the exploitative trait spread, selection would then favor males able to overcome female resistance, which might be achieved by mutations that further exaggerated the original male signal. A cycle of increasing female resistance to and increasing male exaggeration of key characteristics could ensue, leading gradually to the evolution of costly ornaments of no real value to the female and useful to the male only because without them, he has no chance of stimulating females to mate with him (Figure 10.45).

Chase-away selection theory illustrates dramatically how the evolutionary view of sexual reproduction has changed in recent decades. Once considered a gloriously cooperative exercise between male and female, reproduction is now seen as an arena in which the two sexes battle for maximum genetic advantage.

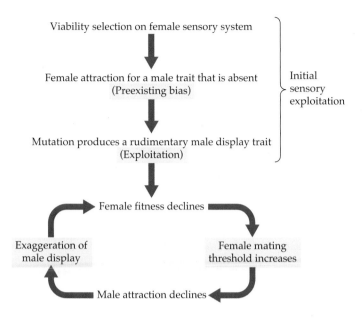

Viability selection on female sensory system

Female attraction for a male trait that is absent
(Preexisting bias)

Mutation produces a rudimentary male display trait
(Exploitation)

Initial
sensory
exploitation

Female fitness declines

Exaggeration of
male display

Female mating
threshold increases

Male attraction declines

Figure 10.45   Chase-away selection theory. The evolution of extreme male ornaments and displays may originate with exploitation of females' preexisting sensory biases. If sensory exploitation by males reduces female fitness, the stage is set for a cycle in which increased female resistance to male displays leads to ever greater exaggeration of those displays. After Holland and Rice [551].

For example, not long ago, termites were believed to be dedicated monogamists who worked together as pairs for years, but careful observation has revealed that males of at least one species regularly abandon relatively lightweight females to search for heavier, more fecund partners, while some females invite male intruders into the nest (probably with pheromonal signals), where they may displace the female's previous mate [1091]. As it turns out, only a minority of individuals in this termite species actually remain with their first partner over the long haul.

Among the termites, conflict between the sexes arises when one or another member of a pair attempts to replace a partner with one that is superior. But even while a mating pair is together, each member may be jockeying for personal fitness gain at the long-term expense of the other. For example, because males of the fruit fly *Drosophila melanogaster* benefit by inducing females to use their sperm rather than those of rivals, their seminal fluid contains chemicals that have damaging side effects on their partners. The prime culprit appears to be the protein ACp62F [752], which boosts male fertilization success (perhaps by damaging rival sperm) at the expense of females, whose lives are shortened and whose fecundity is lowered [960]. Despite the negative long-term effects that toxic protein donors have on their mates, they still gain because they are unlikely to mate with the same female twice. Under these circumstances, a male that fertilizes more of one female's current clutch of eggs in this way can derive fitness benefits even though his chemical donations reduce the lifetime reproductive success of his partner.

If male fruit flies really do harm their mates as a consequence of their success in sperm competition, then placing generations of males in laboratory environments in which no male gets to mate with more than one female should result in selection against donors of damaging seminal fluid. Under these conditions, any male that happened not to poison his mate would reap a fitness benefit by maximizing his single partner's reproductive output. In turn, a reduction in the toxicity of male ejaculates should result in selection for females that lack the chemical counteradaptations to combat the negative effects of the spermicide protein. Indeed, after more than 30 generations of selection in a monogamous environment, females from the monogamous population that were mated once with spermicide-donating "control" males taken from a typical mul-

Figure 10.46 Sexual selection and the evolution of male traits harmful to females. Females from an experimental monogamous lineage of fruit flies have lost much of their biochemical resistance to the damaging chemicals present in the seminal fluids of polygynous male fruit flies. Therefore, monogamous females lay fewer eggs when mated with males from a control polygynous lineage than do control females that evolved with those males. The results of an experiment repeated twice are shown here as replicates 1 and 2. After Holland and Rice [552].

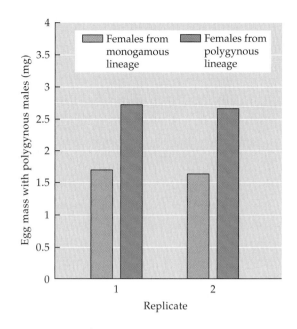

tiple-mating population laid fewer eggs and died earlier than females who had evolved with polygynous, spermicide-donating males (Figure 10.46) [552].

In the light of chase-away selection theory, it is revealing that female fruit flies actually reduce their fitness by preferring to mate with larger males, which they choose either because they find large body size an attractive feature in and of itself or because large body size is correlated with some other attractive feature, such as more persistent courtship. Whatever the reason for their preference, mate choice by females based on this characteristic lowers their longevity (Figure 10.47) and reduces the survival of their offspring [412]. These negative effects on female fitness may arise from the physiological costs of dealing with increased rates of courtship by "preferred" mates and possibly the increased quantities of toxins received from large, "attractive" males [960]. These findings suggest that for the moment, large males are ahead in the chase-away arms race between the sexes in the fruit fly.

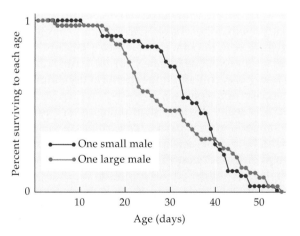

Figure 10.47 Mating with large males reduces female fitness in fruit flies. Even though females prefer to mate with large males, mortality rates are higher for females mated to larger males. After Friberg and Arnqvist [412].

Discussion Question

10.17 Every time a female fruit fly mates, she is exposed to more seminal proteins with potentially negative effects. If one were to perform a selection experiment in which a fruit fly population with high frequencies of mating was created, what counteradaptations should evolve in females over time? Stuart Wigby and Tracey Chapman formed three populations with different sex ratios (female-biased, even sex ratio, and male-biased). Not surprisingly, the frequency with which females mated increased from female-biased to male-biased populations. After 18 and 22 generations of selection, fresh females from the three selected lines were taken from their environments and placed in cages with equal numbers of males. The mortality rate of females from the male-biased line was less than that of females from the even sex ratio line, and much less than that of females from the female-biased line [1298]. What do these results tell us about the evolutionary consequences of sexual conflict between the sexes in this species?

When female fruit flies deal with the rather nasty chemicals they receive during mating, they are engaged in hidden combat with males. Other forms of sexual conflict, however, are far more apparent, as when males commit infanticide (see Chapter 1) [1131] or force themselves on unwilling females [228] (see Figure 14.15). Take traumatic insemination in bedbugs, for example. In this unpopular insect, males use their knifelike intromittent organ to stab a female in her abdomen before injecting their sperm directly into her circulatory system (Figure 10.48) [1161]. This highly unusual method of insemination presumably evolved originally as a result of males injecting their sperm in this way into sexually resistant females that had already acquired sperm from earlier partners via the traditional, less damaging mode of insemination. If so, then traumatic insemination ought to be costly to females, and it is, given that females mating at high frequencies live fewer days and lay fewer eggs [852, 1161].

(A)

(B)

Figure 10.48 A genital product of conflict between the sexes? (A) Male bedbugs have evolved a saberlike penis that they insert directly into the abdomen of their mates prior to injecting them with sperm. The trait may have originated when males benefited by employing traumatic insemination to overcome resistance to mating by unwilling females. (B) The white box shows the site on the female's abdomen that males penetrate with the penis. A, photograph by Andrew Syred; B, photograph by Mike Siva-Jothy.

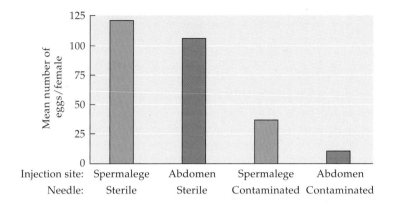

Figure 10.49   A counteradaptation to male sexual exploitation? Female bedbugs have evolved a pad of tissue, the spermalege, in the abdomen that receives sperm inject-ed into them during traumatic insemination. Females that are pricked in the spermalege with a needle produce more eggs than those penetrated elsewhere, although the differ-ence is statistically significant only if the needle has been contaminated with bacteria. After Reinhardt et al. [1011].

But female bedbugs have evidently evolved a counteradaptation that at least reduces the damaging effects of male sexual behavior. Under the abdominal cuticle of female bedbugs, at the site where traumatic insemination typically occurs, lies a special organ called the spermalege. If this pad of tissue has adap-tive value in relation to traumatic insemination, then we can predict that female bugs that are stabbed in the spermalege by needle-wielding human experi-menters will be less damaged than those that are pricked elsewhere in the abdomen. In fact, two groups of researchers reported that egg production was higher in females stabbed in the spermalege, although in one study this bene-fit was observed only if the needle was contaminated with bacteria prior to the experimental piercing (Figure 10.49) [852, 1011]. These results suggest that the spermalege may help females cope with pathogens that males incidentally infect them with during insemination.

In the bedbug, competition among males for egg fertilizations may have led to the evolution of traumatic insemination by males, followed by the evolution of the spermalege in females, which in turn selected for cooperation on the part of males at least to the extent of piercing the female where it would do the least damage. It is not in the male bedbug's interest to do unnecessary or gra-tuitous harm to his partners, given that male fitness depends on females sur-viving long enough to lay at least some eggs fertilized by the inseminator. In some cases, however, sexual conflict results in the injury or even the acciden-tal death of the female. For example, males of our closest mammalian relative, the common chimpanzee, have been known to kill females they were trying to force to mate with them [1131]. The costly nature of sexual coercion and harass-ment has resulted in counteradaptations by female primates, as well as by female bedbugs. Orangutan females have to deal with young males that follow them around, singly or in groups, in an effort to copulate with them against their will. Sexual harassment peaks during the season of the year in Sumatra when fruit abundance is high, which enables young males to reduce the time they spend foraging. During this season, many females form a sexual relation-ship with a huge, older adult male, who accompanies them through the rain-forest and acts as their protector [404]. The much smaller young males know better than to irritate these behemoths.

Even though female orangutans appear to do their best to thwart the eager young males that pursue them through the trees, they occasionally give in and mate with one of their persistent harassers. Why do apparently unreceptive females in this and other species [1036] sometimes let aggressive males mate with them? Do they do so simply because their harassers make the costs of continued resistance so high that it pays to accept defeat? Or is female resistance really a test that reveals whether males have the energy reserves to fuel their persistence, thus demonstrating their good physiological condition and their potential to supply their mates with either good genes or material benefits, such as protection from other males [228, 1277]? Might females even accept a reduction in their lifetime reproductive success by accepting these males because they will produce sons that inherit effective manipulative, even coercive, tactics from their fathers—tactics that may make their sons unusually successful reproducers, thereby endowing their mothers with an exceptional number of grandoffspring [241]? We have much more to learn about the fitness consequences of the apparently hostile interactions that can occur between the sexes.

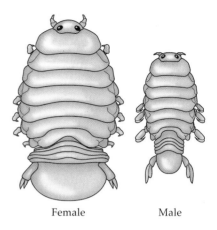

Figure 10.50   A mutually cannibalistic species: the ultimate in sexual conflict. Either the larger female or the smaller male of the freshwater isopod *Ichthyoxenus fushanensis* may kill and consume its partner. Drawing courtesy of C.-F. Dai.

Female   Male

Discussion Question

10.18 In the parasitic isopod *Ichthyoxenus fushanensis* (Figure 10.50), a male lives with a female in a cavity they construct in their victim, a freshwater fish. Sexual cannibalism is not uncommon in this isopod [1216]. Females sometimes eat males early in the April to November breeding season; males sometimes eat females later on. Replacement partners of both sexes are readily available at these times. In addition, this animal is one in which males can transform themselves into females, a phenomenon that occurs only when males have achieved a fairly large size. The larger the female, the more fecund she is. The smaller the difference in size between male and female, the fewer offspring produced by a pair. Why has conflict between the sexes reached such an extreme state in this species? What kinds of males are females expected to eat? What kinds of males are expected to eat their partners? How can you account for the differences in the timing of cannibalism by males and females?

Summary

1. Sexual reproduction creates a social environment of conflict and competition among individuals as each strives to maximize its genetic contribution to subsequent generations. Males usually make huge numbers of very small gametes and often try to fertilize as many eggs as possible, while providing little or no care for their offspring. Because females make fewer, larger gametes and often provide parental care as well, they usually have a lower potential rate of reproduction than males. As a result, receptive females are scarce, and males typically compete for access to them, while females can choose among many potential partners. In a few species, however, these sex roles are reversed, providing an opportunity to test the theory that the differences between the sexes stem from differences in their relative parental investments (or in their potential reproductive rates).

2. Evolution by sexual selection occurs if genetically different individuals differ in their reproductive success because of differences in their ability (1) to compete with others of their own sex for mates or (2) to attract members of the other sex.

3. The competition-for-mates component of sexual selection is behind the evolution of many elements of male reproductive behavior, including competition for social dominance, alternative mating tactics, and mate guarding after copulation. Although in a typical species, males exert selection pressure on one another in the competition for mates, females generally have the last word on reproduction because they control the production and fertilization of eggs.

4. In a typical species, females choose among potential mates, creating the mate choice component of sexual selection. Males of some species seek to win favor with females by offering them material benefits, including nuptial gifts or parental care.

5. Mate choice by females occurs even in some species in which males provide no material benefits of any sort. Choosy females of these species may gain by securing genes that enhance the viability of their offspring (good genes theory). Or female preferences for elaborate ornaments may arise because males with these attributes are healthy and parasite-free (healthy mates theory). On the other hand, extravagant male features could spread through a population in which even arbitrary elements of male appearance or behavior became the basis for female preferences. Exaggerated variants of these elements could be selected strictly because females preferred to mate with individuals that had them (runaway selection theory). A fourth possibility is that the extreme ornaments of males evolve as a result of an upward-spiraling cycle of conflict between the sexes, with males selected for ever-improved ability to exploit female perceptual systems and females selected to resist those males ever more resolutely (chase-away selection theory). The relative importance of these various mechanisms of sexual selection remains to be determined.

6. Interactions between the sexes can be viewed as a mix of cooperation and conflict as males seek to win fertilizations in a game whose rules are set by the reproductive mechanisms of females. Males often attempt to circumvent female mate choice by trying to inseminate unwilling females. Conflict between the sexes also takes other forms, including infanticide, sexual harassment, and the transfer of damaging ejaculates, demonstrating that what is adaptive for one sex may be harmful to the other.

Suggested Reading

To learn more about satin bowerbird sexual behavior, see the papers written by Gerry Borgia [121, 122] and an elegant new book on bowerbirds by Clifford and Dawn Frith [414]. Malte Andersson's book on sexual selection is remarkably comprehensive [31], while alternative mating tactics have been examined by Mart Gross [484] and, quite differently, by Steve Shuster and Michael Wade [1107]. You can learn about sperm competition from the classic paper by Geoff Parker [926], which focuses on insects, as does the book by Leigh Simmons [1110]. Sperm competition in other animal groups is covered by Tim Birkhead and Anders Møller [104], while Bill Eberhard examines how females might control which of several males' sperm stored within them will fertilize their eggs [355]. A book on the house finch by Geoff Hill explores theories on the evolution of colorful ornaments as well as describing how behavioral ecologists actually use sexual selection theory in their research [539]. Finally, although this chapter did not explore the subject of how sexual reproduction evolved in the first place, you can learn about it from G. C. Williams [1308] and Bob Trivers [1214], with an update from Laurence Hurst and Joel Peck [579].

11 | *The Evolution of Mating Systems*

Male satin bowerbirds, as we saw in the previous chapter, are capable of copulating with dozens of females in a single breeding season, although they rarely have the good fortune to do so [120]. In other words, males of this species have the capacity to be polygynous. In contrast, most female satin bowerbirds are monogamous, mating with just one male per nesting attempt. But the satin bowerbird's mating system (males potentially polygynous, females mostly monogamous) is only one of a variety of arrangements found in the animal kingdom. In fact, various mating systems can be found even among the bowerbirds. For example, both males and females of the green catbird, a close relative of the satin bowerbird, pair off one by one before rearing offspring together [413]. In some other birds, such as the spotted sandpiper, polyandry is the order of the day: females copulate with two or three males in a breeding season [902]. A still more extreme version of this pattern is exhibited by the honey bee, whose young queens fly out from their hives into aerial swarms of drones that pursue, capture, and mate with them in midair. The average queen is highly polyandrous, coupling with many males and using the sperm of perhaps a dozen or so during her lifetime of egg laying [1181]. In contrast, drones never mate with more than one queen, because they violently propel their genitalia into their first and

◀ The mating systems of many species, including this long-horned wood-boring beetle of central Arizona, involve defense of food resources by males as a means of encountering mates that are drawn to the resource (the fruit of the saguaro cactus in this case). Photograph by the author

only mate, a suicidal act that ensures that they are both monogamous and, shortly thereafter, dead [1339].

The diversity of mating systems offers a rich banquet of Darwinian puzzles for the evolutionist, two of which are among the most challenging: (1) why are males ever voluntarily monogamous? and (2) why do females of some species practice polyandry? Polygyny is less of a puzzle in this context, but why are there so many different kinds of polygynous mating systems? These questions are the central focus of this chapter.

Is Male Monogamy Adaptive?

Although examples of monogamy by males are not common, they do occur, as we have just indicated. But why should a male restrict himself to a single mate? How can it possibly be reproductively advantageous for a honey bee drone (Figure 11.1), or a male of any other species, for that matter, to fertilize the eggs of just one female? We argued in Chapter 10 that a male's reproductive success increases in direct proportion to the number of females he inseminates. As a result, males typically compete for mates, and the winners are polygynists that are likely to gain higher fitness than their monogamous rivals. For example, male house wrens that attract two females father about nine fledglings per year on average, whereas monogamous male wrens have fewer than six [1143]. Viewed in these terms, males that voluntarily restrict themselves to one mate—and there are species besides the honey bee with this mating system—are something of a mystery.

Many evolutionary explanations have been offered for genuine monogamy, including six hypotheses applied to marine fishes alone [1281]. One possibility is the **mate assistance hypothesis**, which states that males remain with a single female because of ecological factors that make male parental care and protection of offspring especially advantageous [368]. In some environments, the additional offspring that survive because of paternal effort may more than compensate the monogamous male for giving up the chance to reproduce with other females. Males of the seahorse *Hippocampus whitei*, for example, take on the responsibility of "pregnancy," carrying a clutch of eggs in a sealed brood pouch over 3 weeks or so (Figure 11.2). Each male has a durable relationship with one female, who provides him with a series of clutches. Pairs even greet each other each morning before moving apart to forage separately; they will ignore any others of the opposite sex they happen to meet during the day [1234]. Since

Figure 11.1 The monogamous honey bee drone dies after mating. (Left) An intact male. (Right) A queen with the yellow genitalia of a deceased partner attached to the tip of her abdomen. Photographs by Christal Rau, courtesy of Nikolaus Koeniger.

Figure 11.2   Mate assistance monogamy in a seahorse. A pregnant male is giving birth to his single partner's offspring. Several youngsters can be seen emerging, some tail-first, from their father's brood pouch. Photograph by Rudie Kuiter.

a male's brood pouch can accommodate only one clutch of eggs, he gains nothing by courting more than one female at a time. Nor would he have much to gain by switching mates from one pregnancy to the next if his current mate can supply him with a new complement of eggs as soon as his pregnancy is over. Many females apparently can keep their partners pregnant throughout the lengthy breeding season, but cannot produce batches of eggs so quickly that they have some to give to other males [1233]. In other words, the potential reproductive rates (see Chapter 10) of males and females are extremely similar. By pairing off with one female, a male may be able to synchronize his reproductive cycle with that of his partner. When the two individuals are in sync, the male will complete one round of brood care just as his partner has prepared a new clutch. Therefore, he need not spend time waiting or searching for alternative mates, but will be able to secure a new clutch from a familiar mate before returning to his parental duties. In a close relative of *H. whitei*, males that did break off a relationship with a mate appeared to pay a price; they traveled greater distances, and had more time pass between broods, than those males that retained a long-term partner [686].

The **mate guarding hypothesis** offers a different explanation for monogamy, arguing that this mating system arises as a by-product of mate guarding in species whose females would use the sperm of other males to fertilize their eggs if left free to mate with those other individuals [368]. The benefit-to-cost ratio associated with this kind of monogamy should be relatively high if females remain receptive after mating and if they are widely scattered and difficult to locate. These conditions apply to the beautiful clown shrimp, *Hymenocera picta* (Figure 11.3), whose males spend weeks in the company of one female [1292]. As expected, the operational sex ratio (see p. 323) is

Figure 11.3   A monogamous mate-guarding shrimp. When males of the clown shrimp encounter a potential mate, they remain with her, because receptive females are scarce and widely distributed. Here, a couple feeds on the severed arm of a starfish. Photograph by Roger Steene.

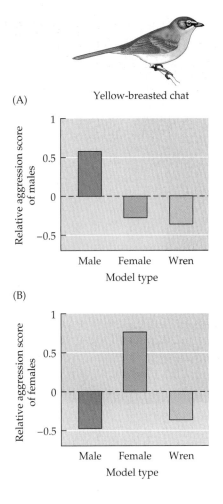

(A)

Yellow-breasted chat

Figure 11.4   Dual mate-enforced monogamy. (A) In the yellow-breasted chat, males are more aggressive toward an apparent male rival (a stuffed model of a male) than are females. (B) The reverse is true for females, which respond with little aggression toward a male model but vigorously attack a female model. Both sexes ignore a control stimulus, a stuffed wren. The higher the relative aggression score, the more aggressive the response of the tested individuals. After Mays and Hopper [803].

highly male-biased in this species because females are receptive for only a short period every 3 weeks or so. Because finding receptive females is so time-consuming, males that encounter a potential mate guard her until she is willing to copulate. Here we have an example of a recurrent theme in this chapter: namely, that ecological factors that determine how females are distributed spatially have major effects on the evolution of male mating tactics.

In other species, although males might gain by acquiring several mates, females may attempt to block their partners' polygynous moves in order to monopolize their parental assistance, leading to **female-enforced monogamy**. This hypothesis is similar to the mate-guarding hypothesis, but with the female as the enforcer of monogamy in this instance. Females of a seabird called the razorbill, for example, often attack their partners when they show sexual interest in a neighbor next to them in their sea-cliff rookeries [1243]. Likewise, female yellow-breasted chats batter stuffed models of females (but not stuffed models of males) that experimenters have placed in the territories that they share with a mate (Figure 11.4) [803]. Paired females of the burying beetle *Nicrophorus defodiens* are also aggressive toward intruders of their own sex. In this species, a mated male and female work together to bury a dead mouse or shrew, which will feed their offspring once they hatch from the eggs the female lays on the carcass. But once the carcass is buried, the male may climb onto an elevated perch and release a sex pheromone to call a second female to the site. If this female also laid her clutch of eggs on the carcass, her larvae would compete for food with the larvae of the first female, reducing their survival or growth rate. Thus, when the paired female smells her mate's pheromone, she hurries over to push him from his perch. These attacks reduce his ability to signal, as Anne-Katrin Eggert and Scott Sakaluk [357] showed by tethering paired females so that they could not suppress their partners' sexual signaling. Freed from a controlling spouse, the experimental males released scent for much longer periods than control males that had to cope with untethered mates (Figure 11.5). Thus, burying beetle males may often be monogamous not because it is in their genetic

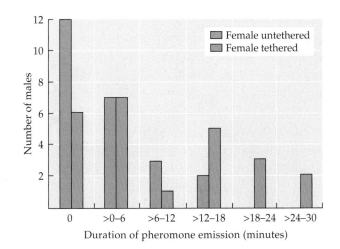

Figure 11.5   Female burying beetles combat polygyny. When a paired female beetle is experimentally tethered so that she cannot interact with her partner, the amount of time he spends releasing sex pheromones rises dramatically. After Eggert and Sakaluk [357].

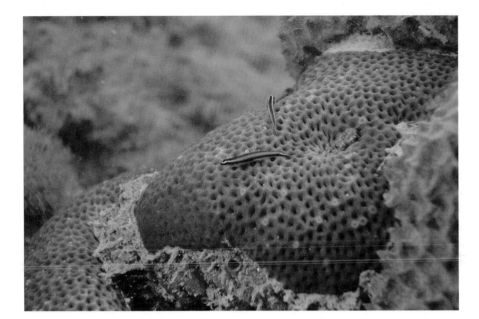

Figure 11.6   A monogamous pair of cleaner wrasses. In this marine fish species, a male and a female cooperate in defense of a territory. Their territory attracts other fish species, from which the wrasses remove parasites and the like. Both individuals attempt to keep same-sex members of their own species away from their partners. Photograph by Liz Whiteman.

interest, but because females make it happen—another example of the kinds of conflicts that occur between the sexes (see Chapter 10).

When males and females both gain by guarding a partner of high quality, as, for example, when a male guards a large, highly fecund female and a female monopolizes a large, highly paternal male, then monogamy provides benefits for both members of the pair, which makes its evolution easier to understand. Elizabeth Whiteman and Isabelle Côté argue that this form of monogamy is relatively common among marine fishes in which pairs defend territories against other duos (Figure 11.6) [1281]. This hypothesis probably also applies to a territorial freshwater fish, the cichlid *Eretmodus cyanostictus*, because when one member of a monogamous pair is experimentally removed, the widowed individual acquires a replacement mate that is smaller, and presumably less valuable, than the original partner [850].

Discussion Questions

11.1   We began this chapter with a mention of male monogamy in the honey bee. Try to explain the male's suicidal mating behavior in light of the alternative hypotheses on male monogamy outlined above. Also include in your list one hypothesis based on group selection theory. What predictions follow from the different explanations you have considered? What data are required to resolve the issue?

11.2   In the starling, some males acquire several mates but do not assist them, whereas other monogamous males work with their only partner to rear their brood together. Sometimes when there are two females nesting on a male's territory, the first female to settle there attacks the clutch of the other female, piercing her companion's eggs with her beak [1010]. Why might she do so, and what kind of monogamy could result from her actions? Under what circumstances would the female's behavior actually qualify as a form of parental investment?

Male Monogamy in Mammals

Although monogamy is not common in any group, this mating system is exceptionally rare in mammals, a lineage notable for the size of female parental investment in offspring. Sexual selection theory suggests that male mammals, which

Figure 11.7   An exceptionally paternal rodent. Male Djungarian hamsters may pull newborns from their mate's birth canal and then clear the infants' airways by cleaning their nostrils, as shown in these photographs (the male is the hamster on the left; arrows point to the pink newborn). After Jones and Wynne-Edwards [611].

cannot become pregnant and do not offer milk to infants (with the exception of one species of bat), should usually try to be polygynous—and they do. However, exceptions to the rule may be useful for testing alternative hypotheses for male monogamy in this group.

For example, if the mate assistance hypothesis for monogamy applies to mammals, then males of the rare mammalian species that exhibit paternal behavior should tend to be monogamous. Although fewer than 10 percent of all mammals exhibit male parental care [1336], some of these exceptions are indeed monogamous, including the Djungarian hamster, whose males actually help deliver their partners' pups (Figure 11.7) [611]. Male parental care contributes to offspring survival in the hamster and also in the monogamous California mouse, pairs of which were consistently able to rear a litter of four pups under laboratory conditions, whereas single females did not do nearly as well [186]. The relationship also held under natural conditions, with the number of young reared by free-living California mice falling when a male was not present to help his mate keep the pups warm (Figure 11.8) [488].

In some mammals, one of the things a paternal male might do for his offspring is to protect them against infanticidal male intruders, which may destroy infants in order to mate with their mothers, as we saw in Chapter 1. In that chapter, we suggested that mate-guarding male prairie voles can and do defend their offspring against infanticidal rivals, so that both mate guarding and mate assis-

Figure 11.8   Male care of offspring affects fitness in the California mouse. The mean number of offspring reared by female mice falls sharply in the absence of a helpful male partner. After Gubernick and Teferi [488].

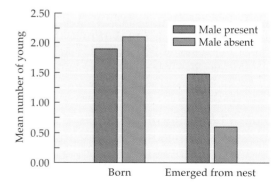

tance contribute to the tendency toward monogamy in this species. If males and females form a social bond to reduce the risk of infanticide, then such monogamous associations should also evolve in primate species in which new mothers tend to travel with their offspring (rather than leaving them in a nest). Mothers with vulnerable young in tow need to have a protective male accompanying them if he is to defend the infant. In one group of primates, the prosimians, a nearly perfect correlation exists between infant carrying and year-round male–female pairings (Figure 11.9) [1229].

On the other hand, in a review of all primates, not just prosimians, Agustin Fuentes found little evidence in favor of the anti-infanticide hypothesis for monogamy. For example, in a number of primates in which a male forms a pair bond with a female, females are larger than or dominant to males and are therefore in little need of help in dealing with aggressive males trying to cause trou-

Figure 11.9   Durable pair bonds between males and females have evolved independently four or five times in prosimian primates in which mothers carry their dependent infants with them. After van Schaik and Kappeler [1229].

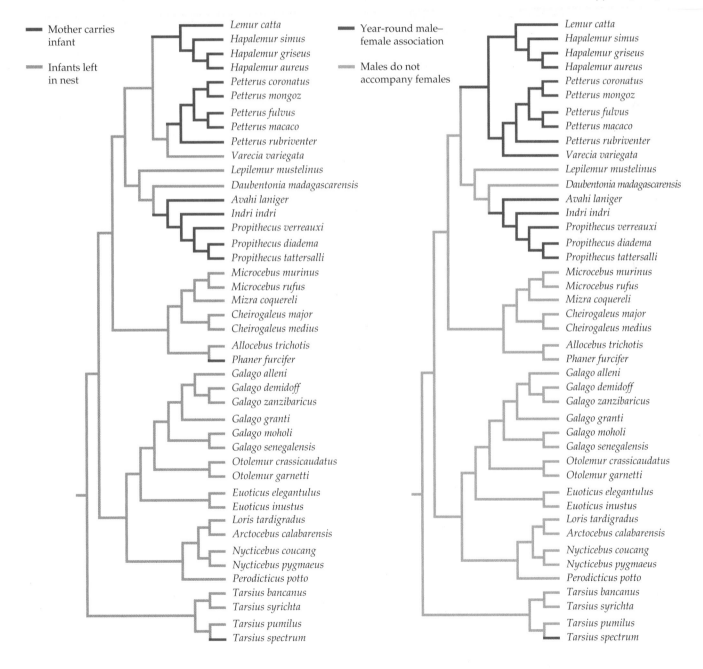

ble [418]. Moreover, a systematic comparative test of the proposition that mammalian monogamy goes hand in hand with male parental care produced completely negative results for primates, rodents, and all other groups [662]. So, for example, although male parental care occurs in half of the 16 primate taxa (species or groups of related species) known to be monogamous, it also characterizes 35 percent of the 20 polygynous taxa. The difference between the two categories is not statistically significant, contrary to the expectation that paternal primate species should be monogamous while nonpaternal species should be polygynous.

In fact, the only mammalian pattern that survives comparative analysis is that males tend to live with females in two-adult units more often when females live well apart from one another in small territories. The monogamous rock-haunting possum of northern Australia is a case in point. A female, her mate, and their young live along the edges of rock outcrops in territories of about 100 meters by 100 meters (Figure 11.10), an area about one-sixth the size of areas occupied by other herbivorous mammals of equivalent weight [1043]. In their small territories, male possums can effectively monitor the activities of one female at relatively modest energetic expense. Any male that attempted to move between the home ranges of several females would run the risk that intruder males would visit his mates after he left. Thus, ecological factors that enable females to live in small, defensible territories tilt the cost–benefit equation toward mate guarding, which then leads to male monogamy.

Figure 11.10   **Mate-guarding monogamy in the rock-haunting possum** is facilitated by the small, discrete home ranges occupied by females of this species, which live along the edges of rock outcrops in northern Australia. After Runcie [1043].

11.3 In a small African antelope called Kirk's dik-dik, most males and females live in monogamous pairs [151]. Evaluate alternative hypotheses for monogamy in this species in light of the following evidence: the presence of males does not affect the survival of their offspring; males conceal the female's estrous condition by scent-marking over all odors deposited by their mates in the pair's territory; males sire their social partner's offspring; females left unaccompanied wander from the pair's territory; some territories contain five times the food resources of others; the few polygynous associations observed do not occupy larger, or richer, territories than monogamous pairs of dik-diks.

11.4 In their classic paper on mating systems, Steve Emlen and Lew Oring [368] suggested that two ecological factors could promote the evolution of monogamy: a high degree of synchrony in reproductive cycling within a population and a highly dispersed distribution of receptive females. Try to reconstruct the logic of these predictions and then make counterarguments to the effect that synchronized breeding could facilitate acquisition of multiple mates while a relatively dense population of receptive females might actually promote monogamy.

Male Monogamy in Birds

In the vast majority of birds, males and females form long-term partnerships for one or more breeding seasons [689]. In keeping with the mate assistance hypothesis, the males in these partnerships often contribute in a big way to the welfare of the offspring produced by their mates [898]. In the yellow-eyed junco, for example, males take care of their mate's first brood of fledglings while the female incubates a second clutch of eggs. The paternal help provided is essential for the survival of these young "bumblebeaks," which are initially highly inept at collecting and processing insect food on their own [1257]. The value of male assistance has also been documented for starlings. In a population in which some males helped their mates incubate their eggs and others did not, the clutches with biparental attention stayed warmer (Figure 11.11) and so could develop under better conditions. As a result, 97 percent of the eggs that had been incubated by both parents hatched, compared with 75 percent of those that had been cared for by their mother only [1010].

Demonstrations of the importance of male parental care include some studies in which females have been experimentally "widowed" and left by themselves to rear their brood. Widowed snow buntings, for example, usually produce three or fewer young, whereas control pairs often fledge four or more [760]. When male parental care is reduced rather than eliminated altogether, this too can have a negative effect on reproductive success. Giving a male spotless starling some extra testosterone lowers his willingness to feed nestlings, whereas males that receive an anti-androgenic chemical, which blocks the effects of naturally circulating testosterone, feed their offspring at increased rates [849].

Figure 11.11 **Paternal male starlings keep their clutches warmer** by helping their mates incubate their eggs. As the graph shows, eggs that were incubated by both parents were kept at about 35°C most of the time, whereas eggs incubated by the female alone were often several degrees cooler. After Reid et al. [1010].

Spotless starling

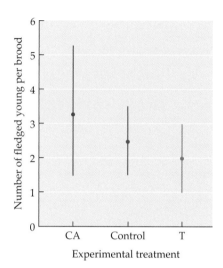

Figure 11.12   Paternal care boosts reproductive success in the monogamous spotless starling. Males whose testosterone levels were reduced by the anti-androgen cyproterone acetate (CA) provided more food for their broods and had the highest fledging rate per brood. Males given extra testosterone (T) provided less food and had the lowest fledging rate. Untreated controls were intermediate with respect to both feeding and fledging rates. After Moreno et al. [849].

The mean number of fledged young per brood was lowest for starlings with extra testosterone and highest for those with the testosterone blocker (Figure 11.12).

Thus, we can safely conclude that paternal, pair-bonded males of at least some bird species really do increase the number of offspring their mates can produce in a breeding attempt. But do monogamous paternal males benefit from their devotion to the offspring of their mate? Because the fitness gained by these males will increase only to the extent that they care exclusively for their own genetic offspring, we can predict that monogamous males will sire all the offspring of their mates. This prediction has been confirmed for some species, such as the common loon: DNA fingerprinting of 58 young from 47 loon families revealed that all were the genetic offspring of the pair that raised them [957]. A similar study of Florida scrub jays (see Figure 13.18) also showed that nestlings were always the offspring of the adults believed to be their parents [996].

But we now know, thanks to advances in molecular genetics (see Box 11.1) that loons and scrub jays are rare exceptions to the rule. In most other birds, social monogamy (the pairing of male and female) does *not* equate with genetic monogamy (which happens when pairs produce and rear only their own genetic offspring). In nearly 90 percent of all bird species, some females engage in **extra-pair copulations** with males other than their social partners and use the sperm they acquire in this manner to fertilize some or all of their eggs [482, 1278]. In other words, socially monogamous males in most birds run the risk of helping offspring other than their own, which clearly reduces the benefits of working with a social partner to rear her offspring.

So why participate in such a system? For one thing, the opportunity for extra-pair copulations enables some males to supplement the fitness gained via their social partner. Such a polygynous male gets the best of both worlds by preventing his primary mate from mating with other males while inseminating other females whose offspring will receive care, but not from him. These preferred males, which often exhibit ornaments and displays thought to be indicative of superior physical condition [482], avoid some or most of the costs of monogamy. Therefore, it falls to the truly monogamous males, which are more likely to be cuckolded by their mates, to accept the disadvantages of monogamy. The question remains, why do these males fail to engage in fitness-boosting extra-pair copulations? The answer is that females have something to say about who gets to mate. If female mate choice revolves around parental contributions from a social partner and honest indicators of quality from extra-pair partners, then males that are unwilling to be paternal and are also unable to offer cues of good condition would find themselves between a rock and a hard place. They could drop out of the reproductive competition altogether, but would obviously fail to leave descendants for as long as they remained on the sidelines. Their only real option is to help a social partner who may fertilize at least some of her eggs with their sperm. When male life expectancy is short, handicapped individuals may leave more descendants by forming monogamous social partnerships than by adopting some other reproductive tactic.

The existence of constraints imposed by competitors and choosy females may also help explain the evolution of mating systems in which most breeding

Box 11.1 | **Genetic fingerprinting and behavioral ecology**

The discovery that social monogamy is not synonymous with genetic monogamy in birds came about primarily through the use of DNA fingerprinting technology, which we have mentioned in earlier chapters. These techniques enabled researchers to establish that mixed paternity of broods is commonplace, revealing that females must have received sperm from males other than their social partner and used this sperm to fertilize some of their eggs. Several different fingerprinting techniques exist, including what is now called multilocus DNA profiling [172] and microsatellite analysis [993].

Because the use of microsatellites has become increasingly popular for paternity assessment, let's briefly outline that technique, which takes advantage of the fact that scattered throughout the chromosomes are stretches of so-called "junk DNA" that consist of repeating units such as

AAT AAT AAT AAT AAT

TTA TTA TTA TTA TTA

Here the base sequence AAT is the microsatellite. The number of times it is repeated at a given location on a chromosome often varies greatly, so that one individual might carry five copies of AAT on one chromosome, whereas the other chromosome of the pair might have eight copies. Still other individuals might have other microsatellite alleles, such as $(AAT)_{10}$ or $(AAT)_{21}$. Microsatellite analysts employ procedures that enable them to identify a given nucleotide sequence, remove it from its chromosome, and copy this region over and over to make sufficient quantities to be detected by the appropriate technique. In the early days of microsatellite technology, a solution containing the amplified DNA was placed at one end of a gel (a thin strip of acrilymide). When an electric field was run through the gel, microsatellite alleles of different sizes distributed themselves in relation to their mass as they moved up the gel. The lighter ones, with fewer repeats, moved farther along than the heavier ones with more repeats. Currently, improved technical means exist to separate the different alleles by their mass.

Imagine that a female with the microsatellite genotype $(AAT)_5(AAT)_8$ mates with two males, one of which has the genotype $(AAT)_{10}(AAT)_{21}$ while the other has the genotype $(AAT)_9(AAT)_{33}$. Each offspring will carry a microsatellite allele from their mother, either $(AAT)_5$ or $(AAT)_8$, and a microsatellite allele present in the DNA of the sperm from their father. In order to determine the paternity of an offspring, one need only establish its microsatellite genotype; if it is, for example, $(AAT)_8(AAT)_{33}$, we know who

A DNA microsatellite gel showing the microsatellite alleles of individuals belonging to two family groups of the splendid fairy-wren, a socially monogamous Australian songbird. Each thick black band (ignore the thin black lines) represents a microsatellite allele. All the birds are heterozygous (have two different alleles of this microsatellite). Family SWG 037 consists of a breeding female (F 111), a male (M 108), and a nestling (N 110), which has inherited one allele from each of its parents. In this instance, social monogamy has yielded genetic monogamy. Family SWG 026 has an adult male (M 79) paired socially with a female (F1 78); another adult female (F2 50) associates with the breeding pair and has helped feed nestlings (N 127, N 128, and Y 54). The helping female and two of the nestlings are the genetic offspring of the socially bonded male and female, from whom they have inherited one microsatellite allele each. On the other hand, N 128 has an allele from female F1 78, but no allele from male M 79. His father is another male (but not M 108) that mated with female F1 78, a genetically polyandrous individual. Photograph by Mike Webster.

its father is (see the accompanying figure). Typically, researchers require a half dozen or so microsatellite alleles to establish paternity with a high degree of confidence. Well over 150 studies of this sort now exist; in about 70 percent of the species studied, extra-pair copulations are known to occur, and these matings result in extra-pair paternity for some philandering males [482].

females form social bonds with several monogamous males, rather than having one social partner and a variable number of extra-pair mates. The mating system of the Galápagos hawk, for example, ranges from monogamy to extreme polyandry [114]. Polyandry appears to be associated with a scarcity of suitable territories, which leads to a highly male-biased operational sex ratio since males outnumber the limited number of territorial, breeding females. The intense competition for these females and the territories on which they live has favored males capable of forming cooperative defense teams to hold an appropriate site. A breeding female may acquire as many as eight mates prepared to pair-bond with her for years, forming a male harem that will help her rear a single youngster per breeding episode [377]. In this and another odd bird, the purple swamphen, all the males that associate with one female appear to have the same chance of fertilizing her eggs [589]. If females do give all their mates equal fertilization opportunities, the mean fitness of each mate is almost certainly higher than that of males that fail to be part of a team.

In some other polyandrous species, the several mates of a female each receive a clutch of eggs from her, which forces them to share the reproductive output of the female with their fellow harem members. In the wattled jacana, for example, aggressive females fight for territories that can accommodate several males. These males mate with the territory owner, who then supplies each of them with one clutch of eggs, which will be cared for exclusively by that male. Males that pair monogamously with a polyandrous female are likely to care for offspring sired by other males, since 75 percent of the clutches laid by polyandrous female jacanas are of mixed paternity [369]. The same sort of thing happens in the red phalarope, a shorebird in which females are both larger and more brightly colored than males (Figure 11.13). Female phalaropes fight for males, which provide all the parental care for the clutch deposited in their nest, whether or not the eggs have been fertilized by other males [273]. Cases of this sort illustrate the disadvantages of being monogamously bonded with a polyandrous female from the male fitness perspective, but male jacanas and phalaropes may have to accept the eggs their partners lay if they are to leave any descendants at all.

However, a monogamous male under the control of a potentially polyandrous female may be able to do some things that at least reduce the probability that he will care for eggs fertilized by other males. For example, males of the red-necked phalarope, like those of its relative, the red phalarope, care for broods

Figure 11.13 Males of the red phalarope may have to share a mate with other males. In this polyandrous species, females defend first one male and then another, donating a clutch of eggs to each male in turn. The more brightly colored bird (on the left) is the female. Photograph by Bruce Lyon.

on their own. Females of this species produce two clutches of eggs in sequence. Polyandrous females almost always draw their second mate from those males that have lost their first clutch to predators. These males, however, favor their original sexual partner over a novel female more than nine times out of ten [1065]. By copulating with this female again and then accepting her eggs, males reduce the risk that the eggs they receive will have been fertilized by another male's sperm.

The same applies to the spotted sandpiper, whose females also behave like males in many ways [902]. In addition to taking the lead in courtship, females are the larger and more combative sex, and they arrive on the breeding grounds first, whereas in most migrating birds, males precede females. Once on the breeding grounds, females fight with one another for territories (Figure 11.14). A female's holdings may attract first one and then later another male. The first male mates with the female and gets a clutch of eggs to incubate and rear on his own in her territory, which she continues to defend while producing a new clutch for a second mate, who may or may not fertilize all of the eggs he broods. As a result, a few females achieve higher reproductive success than the most successful males [900], an atypical result for animals generally (see p. 322).

Lew Oring believes that an understanding of the mating system of spotted sandpipers is advanced by recognizing a key constraint on clutch size [899]. In all sandpiper species, females never lay more than four eggs at a time, presumably because females that produced five-egg clutches would have fewer *surviving* offspring than those that laid four eggs. A reduction in fitness for females laying five-egg clutches could arise from the inability of an incubating adult to accommodate the extra egg. Indeed, egg addition experiments have shown that sandpipers given extra eggs to incubate often damage their clutch inadvertently and lose eggs more often to predators as well [43]. If female spotted sandpipers are "locked into" a four-egg maximum, then they can capitalize on rich food resources only by laying more than one clutch, not by increasing the number of eggs laid in any one clutch. To do so, however, they must acquire more than one mate to care for their sequential clutches, making this a rare case in which female fitness is limited more by access to mates than by production of gametes.

Male spotted sandpipers may be forced into monogamy by the confluence of several unusual ecological features [697]. First, the adult sex ratio is slightly biased toward males. Second, spotted sandpipers nest in areas with immense mayfly hatches, which provide superabundant food for females and for the

Figure 11.14 Female spotted sandpipers fight over males. Two female spotted sandpipers (left) about to fight and (right) in the act of fighting for possession of a territory that may attract several monogamous, paternal males to the winner. Photographs by Stephen Maxson.

young when they hatch. Third, a single parent can care for a clutch about as well as two parents, in part because the young are precocial—that is, they can move about, feed themselves, and thermoregulate shortly after hatching. The combination of excess males, abundant food, and precocial young means that female spotted sandpipers that desert their initial partners can find new ones without harming the survival chances of their first brood. Once a male has been deserted, he is stuck. Were he also to leave the nest, the eggs would fail to develop, and he would have to start all over again. If all females are deserters, then a male that performs the role of sole parent presumably experiences greater reproductive success than he would otherwise, even if his partner acquires another mate to assist her with a second clutch. Furthermore, the *first* partner of a female spotted sandpiper may provide her with sperm that she stores and uses much later to fertilize some or all of her second clutch of eggs (see p. 343). Thus, males that arrive on the breeding grounds relatively early and pair off quickly may not lose much when their mate acquires a second copulatory partner [901]. Once again, the males that get the short end of the stick are the less competitive individuals, the ones slower to arrive on the breeding grounds. They must deal with polyandrous females if they are to have any chance of reproducing, but the female's ability to control the reproductive process puts them at a major disadvantage.

The Puzzle of Polyandry

Having seen that a variety of factors permit females of some species to dictate the mating terms that more or less force some males into accepting monogamy, let's now turn our attention to the surprising aspects of polyandry. Of course, it is *not* surprising that females are polyandrous when they have two or more social partners laboring on their behalf, as is true for Galápagos hawks and spotted sandpipers. Polyandrous female Galápagos hawks turn defense of their rare and vitally important territory over to a band of male protectors, which can surely do a better job than a single male. Polyandrous spotted sandpipers let several devoted partners take over all parental duties. Although these kinds of polyandry make obvious sense from the female perspective, other cases exist that are less clearly adaptive from this viewpoint. Consider those species in which females have one social partner and an indeterminate number of extra-pair mates. Here, the polyandrous female does *not* acquire a stable of helpful mates, inasmuch as she takes sperm and nothing more from her extra-pair partners. Mating with these "extra" males exposes the female to a host of costs. These costs include the time spent searching for and mating with extra-pair males, the risk of losing parental care from a social partner that detects her extra-pair activities, and the chance of picking up a sexually transmitted disease.

If this last potential cost of polyandry is real, then we would expect the immune systems of highly polyandrous species to be stronger than those of species with a greater tendency toward monogamy. One research team led by Charles Nunn tested this hypothesis using comparative data from primates rather than birds. The team took advantage of the fact that there are extremely polyandrous primates, such as the Barbary macaque, whose females mate with as many as ten males in one day; essentially monogamous species, such as the gibbons; and everything in between. Nunn and his colleagues measured immune system strength by looking at data on white blood cell counts from large numbers of adult females of 41 primate species, most of which were held in zoos, where their health is regularly monitored. (Naturally, only healthy specimens were included in the samples.) White blood cell counts were indeed higher in females of the more polyandrous species (Figure 11.15) [892].

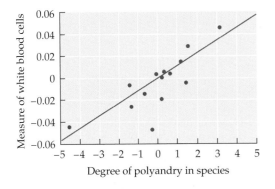

Figure 11.15 **Polyandry has fitness costs.** In primates, the number of mates accepted by females is correlated with the investment made in white blood cells, a component of an animal's immune system. (The higher the measure of white blood cells, the stronger the immune system.) This finding suggests that the more polyandrous the species, the greater the challenge to the female's immune system. Each data point represents a different primate species. After Nunn et al. [892].

Discussion Question

11.5   Why did Nunn and company also look at the relationship between the mean sizes of groups in the different primate species in their study, as well as the extent to which each species was terrestrial as opposed to arboreal?

Given the potential costs of extra-pair copulations for females, it is not too surprising that in at least some species, females attempt to avoid insemination by eager extra-pair males. All birds, waterfowl included, are at risk from sexually transmitted diseases [1090], and perhaps this is why female mallard ducks resist every copulatory attempt by males other than the one partner they choose [266]. The males permitted to mate appear to be individuals of well above average condition, having molted into breeding plumage before their rivals, overcoming the severe energetic demands of the molt before others were able. Males of this sort may be healthier and thus less likely to transmit venereal diseases to their mates.

But although female mallards always say no to unsolicited copulatory attempts, females of other species accept, or even invite, matings from males other than their primary partner [104]. Just what these females might gain from extra-pair copulations has puzzled behavioral biologists, who have nevertheless come up with a list of hypotheses (Table 11.1). Extra-pair fertilizations could, for example, reduce the risk to a female of having an infertile partner as a social mate. This **fertility insurance hypothesis** is supported by the observation that the eggs of polyandrous female red-winged blackbirds are somewhat more likely to hatch than the eggs of monogamous females [470]. Likewise, the rapid loss of stored sperm in the bearded tit may favor females that copulate frequently with their own partner—or someone else's partner—in order to maintain their fertility [1063]. Moreover, in Gunnison's prairie dogs, polyandrous females become pregnant 100 percent of the time, whereas only 92 percent of monogamous females achieve this state [560].

Polyandry and the Acquisition of Good or Compatible Genes

Although fertility insurance may have sometimes contributed to the evolution of polyandry, a more general explanation links mating with several males to certain genetic benefits. In one experiment, female live-bearing Trinidadian guppies that were given the chance to mate with four males produced 80 percent more offspring than those mated to a single male. Not only that, but the gestation period of the multiply mated females was nearly 9 days shorter than that of the monogamous females. In addition, the larger broods of polyandrous mothers behaved more competently than the offspring of monogamous mothers by staying safely in schools longer, rather than wandering off on their own [374]. Similar findings have come from work with the yellow-toothed cavy.

TABLE 11.1 *Why do females mate voluntarily with more than one male?*

Genetic benefits polyandry

1. Fertility insurance hypothesis	Mating with several males reduces the risk that some of the female's eggs will remain unfertilized because any one male may not have sufficient sperm to do the job.
2. Good genes hypothesis	Females mate with more than one male because their social partner is of lower genetic quality than other potential sperm donors, whose genes will improve offspring viability or sexual attractiveness.
3. Genetic compatibility hypothesis	Mating with several males increases the genetic variety of the sperm available to the female, increasing the chance that some sperm will have DNA that is an especially good match with the DNA of her eggs.

Material benefits polyandry

1. More resources hypothesis	More mates mean more resources or parental care received from the sexual partners of a female.
2. Better protection hypothesis	More mates mean more time with protectors who will keep other males from sexually harassing a female.
3. Infanticide reduction hypothesis	More mates mean greater confusion about the paternity of a female's offspring and thus less likelihood of losing offspring to infanticidal males.

Females of this wild guinea pig actively seek out copulations with more than one male when given the opportunity, and they favor relatively heavy males over lightweights [547]. The payoff for polyandry in this species appears to be a reduction in stillbirths and losses of babies before weaning (Figure 11.16) [623].

Polyandrous guppies and cavies that had several partners instead of just one apparently increased their odds of receiving some sperm of exceptional genetic quality. The **good genes hypothesis** for extra-pair copulations suggests that females seek out several males in order to secure superior genes from at least one of these males—genes that will promote faster, more competent development of offspring. In keeping with this hypothesis, when female guppies pair with two males, the male that displays at a higher rate fertilizes more of the female's eggs than expected by chance. Thus, polyandrous females appar-

Figure 11.16 Polyandry has fitness benefits. (A) A male yellow-toothed cavy, showing the very large testes characteristic of this and other species in which intense sperm competition occurs because of a high frequency of multiple mating by females. (B) In this species, females that are experimentally restricted to a single mate have fewer surviving offspring and more offspring that die young than females allowed to mate freely with four males. A, photograph by Matthias Asher, courtesy of Norbert Sachser; B, after Keil and Sachser [623].

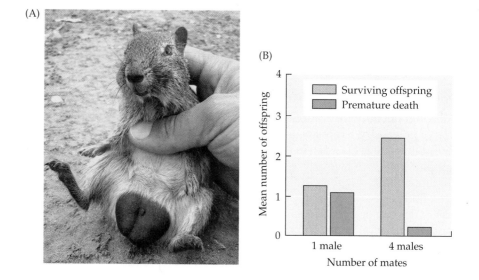

ently permit or encourage sperm competition among males as a means of providing their offspring with genes from vigorous, active males [375].

The superb fairy-wren is another animal for which polyandry and the acquisition of good genes seem to go together. In this gorgeous little Australian wren, a socially bonded pair lives on a territory with a number of subordinate helpers, usually males. When the breeding female is fertile, she regularly leaves her mate and helpers before sunrise and travels to another territory, where she mates with a dominant male before returning home [337]. (Tiny radio transmitters attached to the females revealed what they were up to, and genetic analyses of offspring demonstrated that the females' social partners often lost paternity to distant rivals.) The female wren's apparently careful choice of an extra-pair mate raises questions about what kind of male she prefers and what benefit she gains from her preference. Female fairy-wrens, like mallard ducks, favor males that have managed to molt early, at least a month before the breeding season begins. This preference suggests that females are selecting males that are in such good physical condition that they could meet the energetic demands of molting early, a capacity that these males might be able to pass on to their offspring [351].

The idea that extra-pair copulations enable females to acquire genes that promote offspring survival is compatible with the discovery that older males appear to have greater success in extra-pair matings in some songbirds [329, 605]. Older males have demonstrated an ability to stay alive, and to the extent that their success is linked to their genotype, females could get good genes from them. In the blue tit, the males that females seek out for extra-pair copulations are better than average survivors, and their offspring live longer as well [627, 628].

The conclusion that genetically polyandrous females go out of their way to mate with males likely to possess good genes has received support from studies of birds, fishes, and mammals—and invertebrates as well. The female sierra dome spider routinely mates with more than one male, preferring large and active partners capable of prolonged copulatory stimulation. When Paul Watson experimentally paired females with males of varying size and vigor, he found that the growth rates of the resulting offspring and their eventual size were correlated with their fathers' traits [1254].

Another genetic benefit to females from selective polyandry is the production of "sexy sons" that have inherited the traits that made their fathers attractive. The sons of sexually successful field crickets are indeed more attractive to females than the sons of males that fail to acquire a mate when placed in a cage with a female and a competitor male (Figure 11.17) [1260]. This result suggests that females can gain superior attributes for their sons by choosing some potential

Figure 11.17   A father's mating success can be transmitted to his sons. In experiment 1, two male field crickets were given an opportunity to compete for a female; one male (S) mated successfully, leaving the other male (U) unsuccessful. When the sons of male S were placed in competition for a female with the sons of male U (who had been given a female to mate with after failing to win the initial competition), the sons of S were about twice as likely to mate with the female as the sons of U. In experiment 2, a male that had won a mating competition was later allocated a female at random for breeding, as was a male that had lost the competition. The sons of the two males were then placed in an arena with a female, and as before, the sons of S were much more likely to mate with the female than the sons of U. After Wedell and Treganza [1260].

fathers over others. Remember, however, that in all cases of this sort, it is possible that the improved performance of offspring stems from the willingness of females to invest more resources in the offspring of favored males, not from genetic benefits the offspring have received from their fathers (see pp. 359–360).

Even if some males are capable of offering their mates sperm of exceptional genetic quality, an individual that provided "good genes" for one female might not be the ideal mate for another. Instead, the genetic benefits of mating with a certain male could depend on the compatibility of the sperm and egg genotypes, which could affect the viability of the resulting offspring. According to this **genetic compatibility hypothesis** [1356, 1357], the survival of an offspring might depend (for example) on its heterozygosity, which is usually promoted by outbreeding and decreased by inbreeding. If genetically similar pairs are in danger of producing inbred offspring with genetic defects, then we can predict that any female that has for one reason or another formed a pair bond with a genetically similar individual is a prime candidate to mate with other males.

Some support for the genetic compatibility hypothesis comes from a Norwegian study of the bluethroat in which researchers tested the immune responses of nestlings. In tests of this sort, a foreign substance is injected into the wing; the degree of swelling observed at the injection site is a function of the number of protective T-cells available to disable the foreign chemical. Swelling at the injection site was greater in extra-pair nestlings than in their nestmates that had been produced by the female and her social partner. By mating with a male other than her social partner, the female had evidently acquired genes that helped her produce offspring with more robust immune systems (Figure 11.18). But females that engaged in extra-pair copulations were not acquiring universally good genes, because the paternal half-sibs produced and reared by their extra-pair partners and their social mates had only average levels of immunocompetence [603]. Thus, it must have been the especially good match between the genes of the female and those provided by her extra-pair mate that resulted in an improved immune system for her offspring.

Still more evidence for genotype matching comes from a study of the fertilization success of pairs of male roosters whose sperm were mixed together in equal amounts and then supplied via artificial insemination to a series of hens. The male whose sperm fertilized more eggs in female number 1 was in no way guaranteed to do likewise when his sperm and those of a rival were supplied to female number 2 or female number 10. This result suggests either that chickens have an internal mechanism for favoring some sperm genotypes over others or that some combinations of genes in embryos resulted in their early deaths [102].

Reduction of embryo deaths has been identified as one reason why female pseudoscorpions mate with several males. Polyandrous pseudoscorpions had fewer failed embryos and more offspring surviving to the nymph stage than females that were experimentally paired with a single male (Figure 11.19) [871, 1354]. Yet when Jeanne Zeh compared the numbers of offspring different females produced from matings with the same male, she found no correlation. In other words, as is true for bluethroats and roosters, male pseudoscorpions could not be divided into studs and duds. Instead, the effect of a male's sperm on a female's reproductive success depended on the match between the two gametic genomes, as predicted by the genetic compatibility hypothesis. Because a female's chances of securing genetically compatible sperm increase with the number of males she mates with, we can predict that females of this pseudoscorpion should prefer to mate with a new male rather than with a previous partner. Indeed, when a female was given an opportunity to mate with the same male 90 minutes after an initial copulation, she refused to accept his spermatophore in 85 percent of the trials. But if the partner was new to her, the female took his gametes much more often [1355].

Figure 11.18   Extra-pair matings can boost the immune responses of offspring in the bluethroat. The mean wing-swelling response (an indicator of immune system strength) of youngsters sired by their mother's social partner (within-pair young, or WPY) was less than that of young birds sired by their mother's extra-pair partner (extra-pair young, or EPY) in broods of mixed paternity. After Johnsen et al. [603]; photograph by Bjørn-Aksel Bjerke.

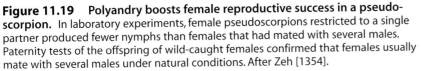

Pseudoscorpion

Figure 11.19 Polyandry boosts female reproductive success in a pseudo-scorpion. In laboratory experiments, female pseudoscorpions restricted to a single partner produced fewer nymphs than females that had mated with several males. Paternity tests of the offspring of wild-caught females confirmed that females usually mate with several males under natural conditions. After Zeh [1354].

Discussion Question

11.6 Genetic incompatibility could explain why inbreeding is deleterious in so many species of animals, including field crickets. If so, what results would you expect from an experiment in which female crickets were paired with two males, one a sibling and the other a nonrelative? (Imagine that you have the ability to determine the paternity of the offspring produced by a given female after her double mating.) In doing this experiment, why would you create two groups: one in which the first male to mate was a sibling and the other in which the first male was a nonrelative? Are the data in Figure 11.20 [144] confirmatory or negative with respect to your expectation? What do these findings reveal about female mate choice even after mating has occurred?

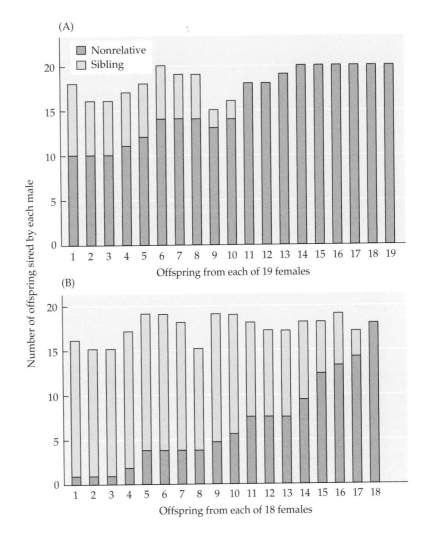

Figure 11.20 Egg fertilizations in female crickets that have mated with a sibling and a nonrelative. (A) The number of offspring sired by the nonrelative and the sibling when the first male to mate was the nonrelative. (B) The number of offspring sired by both males when the first male to mate was the sibling. Sample sizes were 19 and 18 females, respectively. After Bretman et al. [144].

Figure 11.21   Polyandry can yield material benefits. By mating with many males, females of this megachilid bee gain access to pollen and nectar in those males' territories. Photograph by the author.

Polyandry and Material Benefits

Our focus thus far on genetic gains for polyandrous females should not obscure the possibility that females sometimes mate with several males in order to secure certain useful resources, rather than genes, alone from these individuals. As is consistent with this **material benefits hypothesis**, female red-winged blackbirds in some populations are allowed to forage for food on the territories of males with whom they have engaged in extra-pair copulations, whereas truly monogamous females are chased away [471]. Similarly, females of some bees must copulate with territorial males each time they enter a territory if they are to collect pollen and nectar there (Figure 11.21) [14].

Female hangingflies and katydids also have an incentive to mate several times in order to receive material benefits in the form of food presents or nutritious spermatophores from their partners. The spermatophores of highly polyandrous butterfly species contain more protein than the spermatophores of generally monogamous species [105]. Perhaps males of polyandrous butterfly species bribe females to mate with them by providing them with nutrients that can be used to make more eggs (in these species, the male's spermatophore can represent up to 15 percent of his body weight, making him a most generous mating partner). If this hypothesis is true, then the more polyandrous a female, the greater her reproductive output should be. Christer Wiklund and his co-workers used the comparative method to check this prediction. They took advantage of the fact that a number of closely related butterfly species in the same family (the Pieridae) differ substantially in the number of times they copulate over a lifetime. They used the mean number of lifetime copulations for females of each species as a measure of its degree of polyandry. To quantify reproductive output, they weighed the mass of eggs produced by females in the laboratory and divided that figure by the weight of the female to control for differences among the species in body mass. As predicted, across these eight species, the more males a female mated with on average, the greater the average female reproductive output (Figure 11.22) Thus, the reproductive output of these butterflies is correlated with the number of nutritionally valuable spermatophores they receive from mating partners. [1301]

In some butterflies, polyandry means more eggs produced and laid. In some other animals, polyandry enables females to get more parental assistance from their several mates. Thus, in the dunnock, a small European songbird, a female

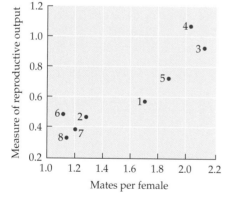

Figure 11.22   Reproductive output is higher in polyandrous pierid butterfly species. The mean number of mates during a female's lifetime and the mean reproductive output of females (measured as the cumulative mass of eggs produced divided by the mass of the female) is shown for eight pierid species. After Wiklund et al. [1301].

Dunnock

Figure 11.23   Adjustment of copulation frequency by polyandrous female dunnocks. Females living in a territory with two males solicit copulations more often from the male that has spent less time with them, whether that is the alpha or the beta male. After Davies [290].

that lives in a territory controlled by one male may actively encourage another, subordinate male to stay around by seeking him out and copulating with him when the alpha male is elsewhere. Female dunnocks are prepared to mate as many as 12 times per hour, and hundreds of times in all, before laying a complete clutch of eggs. The benefit to polyandrous females of distributing their copulations between two males is that both sexual partners will help them rear their brood—provided that they have copulated often enough. Female dunnocks ensure that this paternal threshold is reached by actively soliciting matings from whichever male, alpha or beta—usually the latter—has had less time in their company (Figure 11.23) [290]. Likewise, in the bronze-winged jacana (in which males perform all the egg incubation and chick care duties), females pay close attention to their partners' copulatory requests, which consist of loud calls. The noisiest males sometimes attract neighboring females, who may be able to lure them away from their current mate if she does not attend to the male and repel any intruding females that have responded to his come-hither calls [182].

Material benefits can also come to the polyandrous female in the form of protection against dangerous males. Alternatively, by mating with several males, a female may supply the proximate stimulation that leads all her sexual partners to inhibit their infanticidal tendencies when near her next newborn. In the Hanuman langur, infanticide is common (Chapter 1), but males that have mated with a female generally leave her offspring alone [125]. The fact that females will copulate with more than one male even when they are not ovulating—in fact, even when they are pregnant—suggests that polyandry promotes the female langur's interests by lowering the risk of infanticide [537].

Discussion Questions

11.7   The average proportion of extra-pair offspring in a brood varies among bird species from 0.0 to almost 0.8 [596]. Consider how variation in the benefits and costs of extra-pair paternity might be responsible for variation in the willingness of females to engage in extra-pair copulations. For example, how might low variation in male genetic quality in a species affect the extra-pair paternity figure? What about differences among species in the risk of venereal disease, or in the likelihood that partners will detect and punish the sexual infidelity of a mate? Furthermore, if mating with extra-pair males provides females of a given species primarily with the opportunity to trade up to a genetically or materially superior partner [205], what is the predicted relationship between extra-pair copulations and "divorce" among songbirds?

11.8   In many animals, females mate repeatedly with the *same* male, typically their social partner. Perform a cost–benefit analysis of this kind of multiple mating by females and contrast it with that applied to polyandrous matings.

The Diversity of Polygynous Mating Systems

We have now illustrated what interests evolutionary biologists about male monogamy and female polyandry. There are also some intriguing evolutionary issues associated with polygyny, despite the fact that male attempts to be polygynous are easily understood in terms of sexual selection theory. What is worthy of exploration from an evolutionary perspective is the huge variation from species to species in the tactics employed by males to achieve polygyny. Consider the behavior of bighorn sheep, black-winged damselflies, and satin bowerbirds. Even though some males of all three species acquire many mates by defending territories of one sort or another, they do not all defend the same things. Bighorn rams go where potential mates are (see Figure 10.21), fighting with other males to monopolize females there (**female defense polygyny**). Male black-winged damselflies wait for females to come to them (see Figure 10.26), defending territories that contain the kind of aquatic vegetation in which females prefer to lay their eggs (**resource defense polygyny**). Male satin bowerbirds, on the other hand, defend territories containing only a display bower (see Figure 10.1), not food or other resources that females might use to promote their reproductive success (**lek polygyny**). Moreover, males of many other polygynous species skip territoriality and combat altogether and instead try to outrace their rivals to receptive females (**scramble competition polygyny**).

How can we account for the variety of mating tactics among polygynous species? Steve Emlen and Lew Oring argued that the extent to which a male can monopolize mating opportunities depends on the distribution of females [368]. The degree to which receptive females are spatially clustered in their environment varies as a function of such things as predation pressure and food distribution. Females that live together or aggregate at resource patches can be monopolized economically by males, whereas widely scattered females cannot, because as the size of a territory grows, the cost of defending it also increases.

Female Defense Polygyny

The theory that differences in female distribution patterns underlie the diversity of mating systems generates many predictions, among them the expectation that when receptive females occur in defensible clusters, males will compete directly for those clusters, and female defense polygyny will result. As predicted, social monogamy in mammals never occurs when females live in groups [662]. Instead, in animals as different as bighorn sheep and gorillas, groups of females, which have formed in part for protection against predators, attract males that compete to control sexual access to a group [512, 542]. Likewise, male lions compete to control prides of females, which have formed in part for defense of permanent hunting territories and protection against infanticidal males [919]. Given the existence of these groups, female defense polygyny is the logical tactic. This phenomenon is very widespread (Figure 11.24) [806], even including some birds. Males of the Montezuma oropendola, a tropical blackbird, try to control clusters of nesting females, which group their long, dangling nests in certain trees. The dominant male at a nest tree secures nearly 80 percent of all matings by driving subordinates away [1258]. These dominant males shift from nest tree to nest tree, following females rather than defending a nesting resource, demonstrating that they are employing the female defense tactic.

Figure 11.24   Female defense polygyny in the greater spear-nosed bat. The large male (on the bottom) guards a roosting cluster of smaller females. A successful male may sire as many as 50 offspring with his harem females. Many of the bats have been banded so that they can be recognized as individuals. Photograph by Gary McCracken.

Groups of receptive females also occur in insects and other invertebrates, and female defense polygyny has evolved here as well [1191]. For example, nests of *Cardiocondyla* ants produce large numbers of virgin queens, leading to lethal competition among the males of the colony. The few individuals that survive the mayhem get to copulate with dozens of freshly emerged females [536]. But one does not have to be hyperaggressive in order to practice female defense polygyny. Thus, males of tiny siphonoecetine amphipods construct elaborate cases composed of pebbles and fragments of mollusk shells found in the shallow ocean bays where they live. They move about in these houses and capture females by gluing the females' houses to their own, eventually creating an apartment complex containing up to three potential mates (Figure 11.25) [614].

Female defense polygyny has reproductive consequences for females as well as for males. In yellow-bellied marmots, females often remain near their birth site, so that clusters of mothers, daughters, and aunts end up living together, presumably for improved protection against coyotes and eagles. Males search for locations with one or more resident females, and those males that are able to find and defend a meadow that is home to two or more females practice female defense polygyny. As one might expect, the reproductive success of a male increases as the number of females living in his territory increases. In con-

(A)   (B)

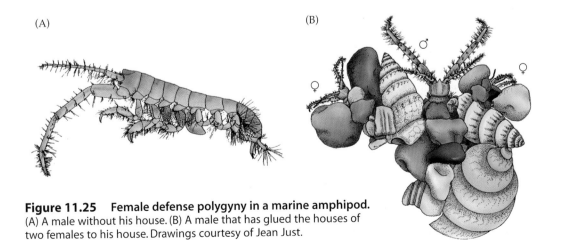

Figure 11.25   Female defense polygyny in a marine amphipod. (A) A male without his house. (B) A male that has glued the houses of two females to his house. Drawings courtesy of Jean Just.

Yellow-bellied marmot

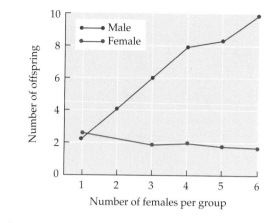

Figure 11.26   Annual reproductive success of male and female marmots. For a male, the number of offspring that survive their first year increases with the size of the group of females he controls, but for females, the number of offspring decreases slightly with group size. After Armitage [40].

trast, females living in the *smallest* marmot groups tend to have the most surviving offspring per year (Figure 11.26) [40].

These observations suggest that female marmots do not benefit from being part of a polygynous male's harem. One hypothesis to account for the failure of females to move from large to small groups is that their *lifetime* fitness is greater when they live with many others. If large groups thwart predators better, then females in large groups should live longer, and reproduce over more years in all, than monogamous females, compensating for their reduced output in any one year. However, contrary to this prediction, no correlation exists between the size of a group of females and female survivorship [360].

Why, then, do females stay together? Just as males have to cope with an assortment of constraints on their reproductive activity, a female's options can be limited [1331]. Perhaps female marmots would indeed enjoy higher lifetime reproductive success if they lived monogamously with a male. But in order to achieve monogamy, they might have to drive off all other females, and this might be costly for a variety of reasons, particularly if those other females would fight tenaciously to stay in high-quality habitat.

Figure 11.27   Resource defense polygyny in an Australian antlered fly. Males compete for possession of egg-laying sites that can be found only on certain species of recently fallen trees. Photograph by Gary Dodson.

Resource Defense Polygyny

In many species, females do not live permanently together, but a male may still become polygynous if he can control a rich patch of resources that females visit on occasion. Some male black-winged damselflies, for example, defend floating vegetation that attracts a series of sexually receptive females, each of which mates with the male and then lays her eggs in the vegetation he controls. Likewise, male antlered flies fight for small rotten spots on fallen logs or branches of certain tropical trees because these locations attract receptive, gravid females, which mate with successful territory defenders before laying their eggs (Figure 11.27) [334].

A safe location for eggs constitutes a defensible resource for a host of animal species; the more of this resource a territorial male holds, the more likely he is to acquire several mates. In an African cichlid fish, *Lamprologus callipterus*, a female deposits a clutch of eggs in an empty snail shell, then pops inside to remain with her eggs and hatchlings until they are ready to leave the nest. Territorial males of this species, which are much larger than their tiny mates, not only defend suitable nest sites, but collect shells from the lake bottom and steal them from the nests of rival males, creating middens of up to 86 shells (Figure 11.28). Because as many as 14 females may nest simultaneously in different shells

Figure 11.28   Resource defense polygyny in an African cichlid fish. (Left) A territorial male bringing a shell to his midden. The more shells, the more nest sites are available for females to use. (Right) The tail of one of the territorial male's very small mates can be seen in a close-up of the shell that serves as her nest. Photographs by Tetsu Sato.

in one male's midden, the owner of a rich territory can enjoy extraordinary reproductive success [1062].

If the distribution of females is controlled by the distribution of key resources, and if male mating tactics are in part dictated by this fact of life, then it ought to be possible to alter the mating system of a species by moving resources around, thereby altering where females are located. This prediction was tested by Nick Davies and Arne Lundberg, who first measured male territories and female foraging ranges in the dunnock. In this drab little songbird, females hunt for widely dispersed food items over such large areas that the ranges of two males may overlap with those of several females, creating a polygynous–polyandrous mating system. However, when Davies and Lundberg gave some females supplemental oats and mealworms for months at a time, female home ranges contracted substantially [298]. Those females with the most reduced ranges had, as predicted, fewer social mates than other females whose ranges had not diminished in size (Figure 11.29). In other words, as a female's home range contracted, the capacity of one male to monitor her activities increased, with the result that female polyandry tended to be replaced by female monogamy. These results support the general theory that males attempt to monopolize females within the constraints imposed on them by the spatial distribution of potential mates, which in turn may be affected by the distribution of food or other resources.

Figure 11.29   A test of the female distribution theory of mating systems. When food supplements reduce the size of a female dunnock's home range, a male can monopolize access to that female and reduce the number of males with whom she interacts. After Davies and Lundberg [298].

Discussion Question

11.9   The alpine accentor is a songbird with a mating system rather like that of its close relative, the dunnock, in that two males, an alpha and a beta, may live in the same area and copulate with the same female(s). However, in contrast to the dunnock, alpha male accentors reduce the amount of parental assistance they give to females as their mating share decreases, whereas the beta male does not make his paternal care donations dependent on his past mating frequency [295]. If female accentors have evolved a mating strategy designed to increase the parental assistance they receive from males, how should their polyandry differ from that of female dunnocks?

What should a female do when the selection of an already mated male means that she must share the resources under his control with other females? According to the **polygyny threshold model**, there is a resource level at which she can gain more by mating with a polygynist on a good territory than by pairing off with a single male on a resource-poor, or predator-vulnerable, territory [898]. If females really are sensitive to this choice threshold, then we can predict that females paired with polygynous and monogamous males should have about the same fitness. Michael Carey and Val Nolan checked this prediction by counting the number of young fledged in a population of indigo buntings, a small songbird whose parental males may attract either one or two mates to their territories in overgrown fields [189]. Monogamous females whose relationship with a male lasted the whole breeding season had only slightly greater reproductive success (1.6 young fledged) than females who participated in a polygynous arrangement (1.3 young fledged). Therefore, females choosing an already occupied territory were not penalized heavily, if at all, in fitness terms.

Stanislav Pribil and William Searcy put the polygyny threshold hypothesis to an experimental test. They took advantage of their knowledge that female red-winged blackbirds in an Ontario population almost always choose unmated males over mated ones. Their usual refusal to mate with a paired male is adaptive, judging from the fact that on those rare occasions when females do happen to select mated males, they produce fewer offspring than females that have their monogamous partner's territory all to themselves. Pribil and Searcy predicted that if they could experimentally boost the quality of territories held by already mated males while lowering the value of territories controlled by unmated males, then mate-searching females should reverse their usual preference. They tested this prediction by manipulating pairs of redwing territories in such a way that one of the two sites contained one nesting female and some extra nesting habitat (added cattails reeds rising from underwater platforms), while the other territory had no nesting female, but the supplemental cattail reed platforms were placed on dry land. Female redwings in nature prefer to nest over water, which offers greater protection against predators. For 14 pairs of territories, the first territory to be settled by an incoming female was the one where she could nest over water, even though this meant becoming part of a polygynous male's harem. Females that made this choice reared almost twice as many young on average as latecomers that had to make do with an onshore nesting platform in a monogamous male's territory [978].

Thus, when given a free choice between a superior territory and an inferior one, it can pay a female to pick the better site even if she has to share it with another female. But why, then, do some female redwings, and females of other birds as well, ever accept a second-rate territory? In the pied flycatcher, second females do not do nearly as well as the first female paired with a male because males devote all their parental care to the brood of the first female, leaving the second female to fend for herself (Figure 11.30) [7]. Polygynous males of this species get their second female by maintaining a second territory well

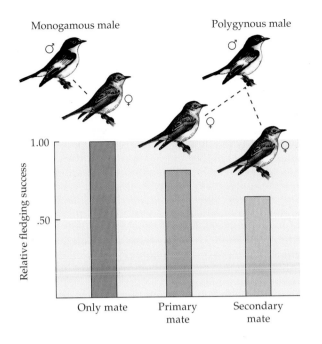

Monogamous male

Polygynous male

Figure 11.30   Does polygyny reduce female fitness? Female pied flycatchers that mate with monogamous males enjoy higher reproductive success than either the primary or secondary mates of polygynous males. After Alatalo and Lundberg [7].

away from the one that is used by their primary mate. Some persons have suggested that the second female may never learn that her partner already has a mate until it is too late.

Svein Dale and Tore Slagsvold have shown, however, that mated male flycatchers spend less total time singing than unmated individuals, thus providing an important potential cue for perceptive females [275]. Because unmated females visit many males and do not rush into a paired relationship [274], a male's mating status should usually become evident before a female makes a final decision. Thus, second females may not be deceived when they accept an already mated male [1182]. Perhaps they choose him anyway because of the high costs of finding or evaluating other potential partners [1151], or because the remaining unmated males have extremely poor territories [1182], in which case the real options for some females may be either to accept secondary status or not to breed at all, just as late-arriving female redwings may have to make do with what is available in their neighborhood, even if it means making the best of a bad job.

Scramble Competition Polygyny

Although female defense and resource defense tactics by competitive males make intuitive sense when females or resources are clumped in small, defensible areas, in many other species receptive females and the resources they need are widely dispersed. Under these conditions, the cost–benefit ratio of mating territoriality usually falls, and males may instead simply try to find scarce receptive females before other males do. Females of a flightless *Photinus* firefly, for example, can appear almost anywhere over wide swaths of Florida woodland. Males of this species make no effort to be territorial; instead, they search, and search, and search some more. When Jim Lloyd tracked flashing males, he walked 10.9 miles in total, following 199 signaling males, and saw exactly two matings. Whenever Lloyd spotted a signaling female, a firefly male also found her in a few minutes [733]. Mating success in this species almost certainly goes to those searchers that are the most persistent, durable, and perceptive, not the most aggressive.

Thirteen-lined ground squirrel

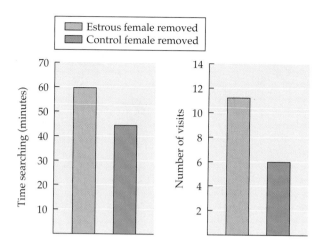

Figure 11.31   Scramble competition polygyny selects for spatial learning ability. Male thirteen-lined ground squirrels remember the locations of females that are about to become sexually receptive. When males returned to an area where such a female had been the previous day, but from which she had been experimentally removed, they spent more time searching for her, and returned to her home range more often, than in the case of removed females that had not been about to enter estrus when the males had visited them the previous day. After Schwagmeyer [1076].

Male thirteen-lined ground squirrels behave like fireflies, searching widely for females, which become receptive for a mere 4 to 5 hours during the breeding season. The first male to find an estrous female and copulate with her will fertilize about 75 percent of her eggs, even if she mates again [396]. Given the widely scattered distribution of females and the first-male fertilization advantage, the ability to keep searching should greatly affect a male's reproductive success. In addition, male fitness may depend on a special kind of intelligence [396]; namely, the ability to remember where potential mates can be located. After visiting a number of females near their widely scattered burrows, searching males often return to those places on the following day. When researchers experimentally removed several females from their home sites, returning males spent more time searching for those missing females that had been on the verge of estrus. Moreover, the males did not simply inspect places that the females had used heavily, such as their burrows, but instead biased their search in favor of the spot where they had actually interacted with an about-to-become-receptive female [1075, 1076]. Individuals with superior spatial memory can probably relocate females more efficiently (Figure 11.31).

Another form of scramble competition polygyny, the **explosive breeding assemblage**, occurs in species with a highly compressed breeding season. One such species is the horseshoe crab, whose females lay their eggs on just a few nights each spring and summer. Males are under the gun to be near the egg-laying beaches at the right times and to accompany a female to shore, where egg laying and fertilization occur [149] (see Figure 10.21). A race to find and pair off with a female also occurs in the wood frog, another species in which the opportunity to acquire a mate is restricted to one or a few nights each year. On that night or nights, most of the adult males in the population are present at the ponds that females visit to mate and lay their eggs. Just as in horseshoe crabs, the high density of rival males raises the cost of repelling them from a defended area. And because females are available only on this one night, a few highly aggressive territorial males cannot monopolize a disproportionate number of them. Therefore, male wood frogs eschew territorial behavior, and instead hurry about trying to encounter one or more egg-laden females before the one-night orgy ends (Figure 11.32) [145].

Lek Polygyny

In some species, males do not search for mates, nor do they defend groups of females or resources that several females come to exploit. Instead, they fight to control a very small area that is used only as a display arena; these mini-ter-

Figure 11.32   An explosive breeding assemblage. A male wood frog grasps a female (upper left) that he has found before rival males, two of which are near the mating pair. Numerous fertilized egg masses float in the water around the frogs. Photograph by Rick Howard.

ritories may be clustered in a traditional display ground, or lek, or they may be somewhat scattered, forming a dispersed lek, as is true for satin bowerbirds [131]. Despite the fact that the males' territories do not contain food, nesting sites, or anything else of practical utility, females come to the lek anyway. When white-bearded manakin females arrive at a lek in a Trinidadian forest, they may find as many as 70 sparrow-sized males in an area only 150 meters square. Each male will have cleared the ground around a little sapling rising from the forest floor. The sapling and cleared court are the props for his display routine, which consists of rapid jumps between perches accompanied by loud sounds produced by snapping his clubbed primary wing feathers together. When a female is around, the male jumps to the ground with a snap and immediately back to the perch with a buzz, and then back and forth "so fast he seems to be bouncing and exploding like a firecracker" [1132]. If the female is receptive and chooses a partner, she flies to his perch for a series of mutual displays, followed by copulation. Afterward, she leaves to begin nesting, and the male remains at the lek to court newcomers. Alan Lill found that in a lek with ten manakins, where he recorded 438 copulations, one male contributed nearly 75 percent of the total; a second male mated 56 times (13 percent), while six other males together accounted for a mere 10 matings [721]. The preferred males tend to occupy sites near the center of the lek, and they engage in more aggressive displays than less successful males [1101].

Huge inequalities in male mating success are standard features of lekking species (see Figure 10.5). Thus, 6 percent of the males in a lek of the bizarre West African hammer-headed bat (Figure 11.33) were responsible for 80 percent of the matings recorded by Jack Bradbury. In this species, males gather in groups along riverbanks, each bat defending a display territory high in a tree, where he produces loud cries that sound like "a glass being rapped hard on a porcelain sink" [130]. Receptive females fly to the lek and visit several males, each of which responds with a paroxysm of wing-flapping displays and strange vocalizations (note the behavioral convergence with manakins and satin bowerbirds).

Figure 11.33   A lek-polygynous mammal: the hammer-headed bat. Photograph by A. R. Devez.

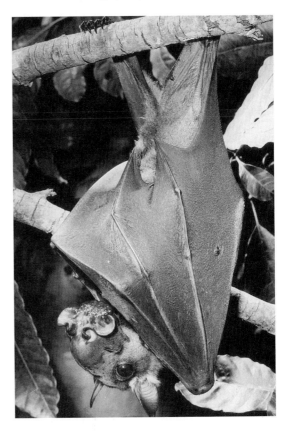

Why do male manakins and hammer-headed bats behave the way they do? Bradbury has argued that lekking evolves only when other mating tactics do not pay off for males, thanks to a wide and even distribution of females [131]. Female manakins and hammer-headed bats do not live in permanent groups, but instead travel over great distances in search of widely scattered sources of food of unpredictable availability, especially figs and other tropical fruits. A male that tried to defend one tree might have a long wait before it began to bear attractive fruit, and when it did, the large amount of food would attract hordes of consumers, which could overwhelm the territorial capacity of a single defender. Thus, the feeding ecology of females of these species makes it hard for males to monopolize them, directly or indirectly. Instead, males display their merits to choosy females that come to leks to inspect them.

But you may recall that the absence of defensible clusters of females was invoked as an explanation for nonterritorial scramble competition polygyny as well. Why some species subject to these conditions simply search for mates while others form elaborate leks is not at all clear. Nor is it known for certain why males of some lekking species congregate in small areas as opposed to displaying by themselves in a dispersed lek *à la* the satin bowerbird, although one possibility is that predation favors adoption of a safety-in-numbers tactic [448]. Any reductions in the cost of displaying go some way toward explaining why lekking males congregate, but we still need to know what benefit males derive from defending a little display territory at a group lek site. Here we shall review three ideas: (1) the **hotspot hypothesis**, according to which males cluster because females tend to travel along certain routes that intersect at particular points, or "hotspots" [134]; (2) the **hotshot hypothesis**, according to which males cluster because subordinate males gather around highly attractive males in order to have a chance to interact with females drawn to these "hotshots" [132]; and (3) the **female preference hypothesis**, according to which males cluster because females prefer sites with large groups of males, where they can more quickly, or more safely, compare the quality of many potential mates [131].

One way of testing the hotspot versus hotshot hypotheses is by temporarily removing males that have been successful in attracting females. If the hotspot hypothesis is correct, then removal of these successful males from their territories will enable others to move into the favored sites. But if the hotshot hypothesis is correct, removal of attractive males will cause the cluster of subordinates to disperse to other popular males or to leave the site altogether.

In a study of the great snipe, a European sandpiper that displays at night, removal of central dominant males caused their neighboring subordinates to leave their territories. In contrast, removal of a subordinate while the alpha snipe was in place resulted in his quick replacement on the vacant territory by another subordinate. At least in this species, the position taken by attractive males, not the real estate per se, determines where clusters of males form [544, 545]. Likewise, in the unrelated black grouse, the exact location of the most successful territory changed slightly from year to year within a long-lasting lek, suggesting that a popular male, rather than a particular spot, most influences the behavior of other males (Figure 11.34A) [1019]. Much the same thing occurs in the lekking marine iguana of the Galápagos [928]. In this species too, the territory at which the most copulations occurred was not the same from one year to the next (Figure 11.34B). In the leks of both black grouse and marine iguanas, the most reproductively successful males are forced to fight most often by their neighbors. In the confusion following an attack when a female is present, a hotshot's neighbor may get a chance to mate.

Although the hotshot hypothesis seems likely in some cases, the hotspot hypothesis has received support in others (Figure 11.35) [447]. For example, a site at which fallow deer bucks gathered to display shifted when logging activ-

Figure 11.34   Hotspots or hotshots? (A) Researchers divided a black grouse lek into 100-square-meter sectors and recorded the total number of copulations in each sector over a 5-year period from 1987 to 1991. The irregular polygons show the location of the top territory for each of the 5 years. The shifts in the preferred territory suggest that male attractiveness, rather than the territory itself, plays the key role in reproductive success in this species, as required by the hotshot hypothesis. (B) Similarly, in the marine iguana of the Galápagos Islands, the territory with the greatest number of recorded matings shifted over the years (in 1987, 1988, 1994, and 1995, four different territories among the 22 occupied sites were ranked highest by females). A, after Rintamäki et al. [1019]; B, after Partecke et al. [928].

Figure 11.35   A test of the hotspot hypothesis. (A) The position of sage grouse leks (numbered red circles) in relation to sagebrush, meadows, forests, and a lake. (B) The distribution of nesting females in relation to the leks where males gather to display. The darker the shading, the more females present. After Gibson [447].

Kafue lechwe

Figure 11.36   Female density is not correlated with lek formation in an African antelope, the Kafue lechwe. The four leks in this region (open circles) are not located in the areas of highest female density, providing evidence against the hotspot hypothesis in this species. After Balmford et al. [60].

ity altered the paths regularly followed by fallow deer does [38]. That the tracks followed by females can sometimes determine where leks form seems likely for a diverse array of tropical birds and insects, given that prominent topographic features, such as mountain ridgelines or forest streams, serve as lekking areas for so many of these animals. In these cases, topography appears to influence where females go, and males follow [10, 1276]. Likewise, female harbor seals appear to use certain routes when traveling underwater between beaches along the California coast. Acoustically signaling male harbor seals station themselves either singly or in groups at points along these corridors, suggesting that hotspot leks can form beneath the surface of the ocean [525].

On the other hand, the hotspot hypothesis has been found wanting for a number of species; in these cases, males do *not* form leks in the places where females are concentrated (Figure 11.36) [60]. Indeed, estrous females of several lekking ungulate species leave their customary ranges to visit groups of males elsewhere, perhaps to compare the performance of many males simultaneously. The possibility that a female preference for quick and easy comparisons pushes male Uganda kob, an African antelope, into large groups has been tested by James Deutsch. He examined the prediction that those leks with a relatively large number of males should attract proportionately more females than leks with fewer males. Contrary to his prediction, however, the operational sex ratio was the same for leks across a spectrum of sizes (Figure 11.37), so that males were no better off in large groups than in small ones [317]. For this species at least, the female preference hypothesis can be rejected. The same is true for the ochre-bellied flycatcher, in which males in leks were not significantly more likely to be visited by females than males that displayed alone [1276]. On the other hand, in an Australian fruit fly, the female encounter rate for males does increase as the number of males in a lek rises to 20, although it then falls off [44], just as it does for the ruff, a lekking sandpiper (Figure 11.38) [1294].

Thus, no one hypothesis on why lekking males form groups holds for every species. Nevertheless, the interactions among males on a lek usually seem to enable individuals of high physiological competence to demonstrate their superior condition to their fellow males and to visiting females [391]. Whether a

Uganda kob

Figure 11.37 **Female Uganda kob do not aggregate disproportionately at leks with large numbers of males.** (A) Female attendance at leks is simply proportional to the number of males displaying there. (B) As a result, the female-to-male ratio does not increase as lek size increases. After Deutsch [317].

lek forms around hotshot males or around centrally located hotspots, males are forced to compete in ways that separate the men from the boys, making it potentially advantageous for females to come to a lek to compare and choose, the better to select a partner of highest quality. Indeed, one of the main themes of evolutionary analyses of mating systems is that in the vast majority of species, female reproductive tactics create the circumstances that determine which competitive and display maneuvers will provide payoffs for males. Thus, male mating systems are an evolved response to female mating systems.

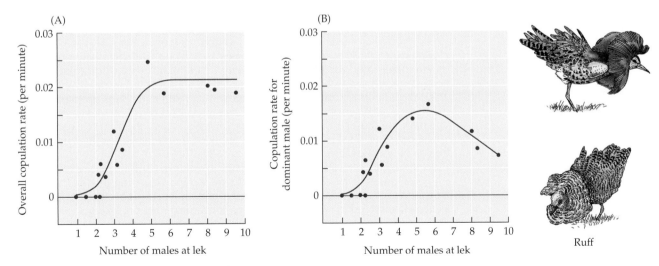

Ruff

Figure 11.38 **Lek size and copulation rate** in the ruff. (A) Up to a point, larger leks attract more receptive females than small leks. (B) At leks of six or more males, however, the average reproductive success of the dominant males falls because the number of arriving females levels off and some lower-ranking males copulate with some of the visiting females. After Widemo and Owens [1294].

Figure 11.39   **The ratios of attracted females to signaling males** of the Mediterranean fruit fly in two "leks" of (A) 6 and 18 males. and (B) 6 and 36 males. Each trial, of which there were 30 in all, generated a single ratio of females to males per "lek." After Shelly [1092].

Mediterranean fruit fly

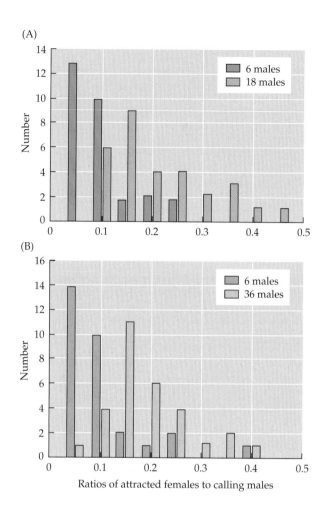

Ratios of attracted females to calling males

Discussion Questions

11.10   Todd Shelly did an experiment with the Mediterranean fruit fly, a lekking species, in which he placed two cups with wire mesh tops containing different numbers of males in a coffee bush [1092]. After the cups were in position, he released 400 females in another coffee bush about 10 meters away. He then counted, at 10-minute intervals, the number of males in each cup releasing pheromones (signaling males evert an abdominal pheromone gland) and the number of females perched on or near each cup. Each trial lasted 80 minutes, and he did 30 trials in all. The data he collected are presented in Figure 11.39. Reconstruct the science underlying this research, working from the graphs in whatever direction makes sense until you have the causal question, hypothesis, prediction, test, and scientific conclusion. You should be able to relate the work to one of the hypotheses described in this section on why males aggregate at leks.

11.11   One of the possible benefits to females of participating in lek mating systems is the selection of mates of exceptional genetic quality. This form of sexual selection should tend to reduce genetic variation among males of lekking species over time. (Why?) What significance would the elimination of genetic variation among males have for the argument that females of modern species go to leks to choose a superior mate? Might the phenomenon of genetic compatibility help us out of this pickle? But is the finding that a very few males monopolize matings at most leks consistent with the suggestion that females visit leks to secure sperm from a genetically compatible partner?

Summary

1. Mating systems can be defined in terms of the number of sexual partners an individual acquires during a breeding season. Males can be monogamous (mating with a single female) or polygynous (mating with several females). Likewise, females can be monogamous (have a single partner) or polyandrous (mating with several).

2. Monogamy by males is a Darwinian puzzle because males that restrict themselves to a single female would seem likely to have fewer descendants than males that successfully mate with many females. Monogamous males can gain fitness, however, if there are large payoffs to parental males or to those that prevent their sole mate from accepting sperm from other males. Alternatively, conflict between females and males may thwart male attempts to be genetically polygynous, even if it would be advantageous from the male perspective.

3. Similarly, although females of most birds and other animals can secure all the sperm they need to fertilize their eggs from a single mate, polyandry is common, even in socially monogamous birds. Polyandrous females may secure any of a spectrum of genetic and material benefits, ranging from superior genes for their offspring to greater amounts of parental assistance from their partners. Because females usually control how many males can mate with them, both female choice and conflict between the sexes play major roles in the evolution of mating systems.

4. Ecological factors that affect males' ability to monopolize receptive females are also important influences on mating system diversity. In particular, the distribution of females may affect the profitability of different kinds of territorial mating tactics for males. When females or the resources they need are clumped in space, female defense or resource defense polygyny becomes more likely. If, however, females are widely dispersed or male density is high, males may engage in nonterritorial scramble competition for mates, or they may acquire mates by displaying at a lek. Many aspects of lek polygyny are still not completely understood, such as why it is that males of most lekking species gather in groups to display to females.

Suggested Reading

Steve Emlen and Lew Oring's now classic paper on the ecology of mating systems changed the analysis of mating systems [368]. For an especially thorough examination of a single species with multiple mating systems, read *Dunnock Behaviour and Social Evolution* [291]. Jacob Höglund and Rauno Alatalo [543] have written a useful book on leks. Malte Andersson's book on sexual selection explores mating system evolution in depth [31], while Dave Ligon focuses exclusively on avian mating systems [719]. Extra-pair copulations by birds are the focus of two fine review articles [482, 1278].

12 The Evolution of Parental Care

A lthough most human beings think that both parents should contribute to child care, in a great many animal species neither the
mother nor the father lifts a finger for their offspring, while in others, just one parent—either the female or, much more rarely, the male—
takes full responsibility for her or his progeny. This diversity in parental
behavior has been hinted at in previous chapters, in which we mentioned
the biparental care provided by the prairie vole, the strictly maternal care
provided by most other mammals, and the fathers-only approach of the
spotted sandpiper and red phalarope. Variety of this sort, as my readers
know by now, always interests evolutionary biologists.

The key to explaining the variety in parental behavior lies with the
cost–benefit approach used by behavioral ecologists. The benefits of
parental care are obvious, having to do with the improved survival of the
assisted offspring. But the costs of being a devoted parent must also be
considered if one is to deal with the major evolutionary questions that
surround parental behavior. We have already seen how important it is to
think about fitness costs if we want to explain why only some gulls mob
predators hunting for their chicks (see Chapter 6). So let's begin this chapter with another look at both the benefits *and* the costs of helping offspring survive.

◀ *Parental care is full of*
puzzles for the evolutionary
biologist. Why did this adult
brown booby permit its downy
chick to kill its sibling earlier
in the breeding season?
Photograph by the author

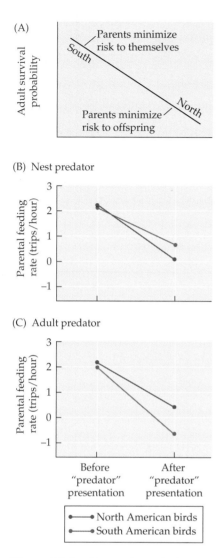

Figure 12.1   Parental care is the evolutionary product of fitness costs and benefits. (A) The reactions of parent birds to predatory threats to themselves and their offspring should vary in relation to the annual mortality rates of adults, which differ between South and North America. Shorter-lived North American birds are predicted to do more to reduce risks to their offspring; longer-lived South American birds are predicted to do more to promote their own survival. (B) The rate at which parents feed their nestlings falls more sharply in North American birds than in South American birds in response to an apparent risk of nest predation by jays, which can find nests by watching where parent birds go. (C) The rate at which parents feed their nestlings falls more sharply in South American birds than in North American birds in response to seeing a hawk capable of killing the parents. After Ghalambor and Martin [446].

The Cost–Benefit Analysis of Parental Care

In many bird species, parents of both sexes work tirelessly to bring food to a brood of all but helpless offspring ensconced in the relative safety of a hidden nest. Without parentally supplied food, the baby birds would soon be dead. But there are risks attendant upon the parents' feeding trips. Predators can use their comings and goings to and from the nest to find it and feast on its occupants; alternatively, predators can lurk near the nest to intercept the parents as they return to their offspring with food. How does a parental bird balance the costs and benefits of its activities?

Cameron Ghalambor and Tom Martin predicted that parent birds should adjust their provisioning behavior adaptively in accordance with two key factors: the nature of the predator (whether it consumes nestlings or adults) and the annual mortality rate for breeding adults [446]. In birds with a generally low adult mortality rate, parents ought to minimize the risk of getting themselves killed by a predator because they will have many more chances to reproduce in the future if a predator does not get them now. In breeding birds with a high annual mortality rate, however, parents should be less concerned with their own safety and more sensitive to the risks that their nestlings may confront from nest-raiding predators; these parents will have relatively few chances to reproduce in the future, so they can gain the most by doing what they can for their current brood.

Ghalambor and Martin knew that birds that breed in North America tend to have shorter lives and produce larger clutches of eggs than their close relatives that breed in South America. So they matched up five pairs of these relatives, including, for example, two members of the same genus, the short-lived North American robin and the longer-lived Argentinian rufous-bellied thrush. When the ornithologists played tapes of Steller's jays to Arizonan robins and tapes of plush-capped jays to their Argentinian counterparts, the robins and thrushes both greatly reduced their visits to their nests for some time so as not to reveal to these nest predators where their nests were hidden. However, the robins curtailed their activity around the nest more strongly than the thrushes, presumably because they had more to gain by protecting their current crop of nestlings from the keen-eyed jays, given their relatively low probability of reproducing in subsequent years. When the ornithologists placed a stuffed sharp-shinned hawk, a killer of adult birds, near active nests and played recorded sharp-shin calls, the parent birds in the sampled species again reduced their visits to their nests for some time. In this round of tests, however, the potentially long-lived Argentinian birds delayed their return to the nest longer than the corresponding Arizonan species (Figure 12.1) [446]. This case illustrates how the evolution of parental care has been fine-tuned by the nature of the costs, as well as the benefits, of the behavior.

Why More Care by Mothers than Fathers?

Although in many species of birds, including robins, both fathers and mothers help their progeny survive, across the animal kingdom as a whole, females are much more likely to be maternal than males are to be paternal. Thus, for example, in the treehoppers of the insect subfamily Membracinae, females of some species, but never males, stand watch over their eggs day and night to protect them against predatory or parasitic insects that would destroy their brood. In some cases, females even stay on to protect their nymphs until they have become adults. Chung-Ping Lin and his co-workers have explored the evolutionary history of egg-guarding parental care by mapping the trait on a phylogeny of the Membracinae that was derived from molecular comparisons among the genera in this group [725]. Their work indicates that while maternal

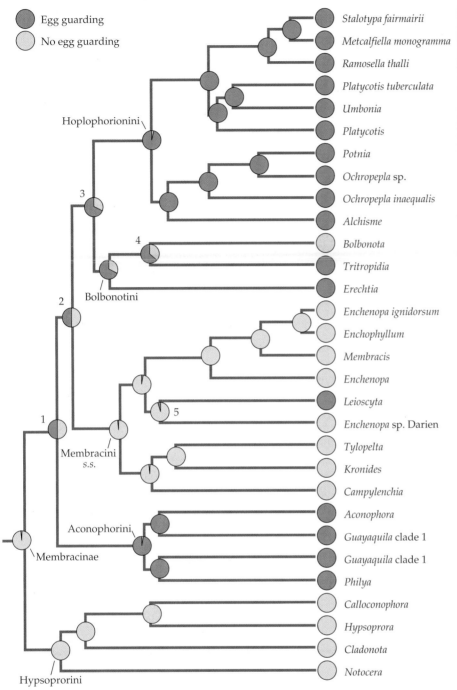

- ● Egg guarding
- ○ No egg guarding

Stalotypa fairmairii
Metcalfiella monogramma
Ramosella thalli
Platycotis tuberculata
Umbonia
Platycotis
Potnia
Ochropepla sp.
Ochropepla inaequalis
Alchisme
Bolbonota
Tritropidia
Erechtia
Enchenopa ignidorsum
Enchophyllum
Membracis
Enchenopa
Leioscyta
Enchenopa sp. Darien
Tylopelta
Kronides
Campylenchia
Aconophora
Guayaquila clade 1
Guayaquila clade 1
Philya
Calloconophora
Hypsoprora
Cladonota
Notocera

Hoplophorionini

3

4

Bolbonotini

2

1

Membracini
s.s.

5

Aconophorini

Membracinae

Hypsoprorini

Figure 12.2   Parental care is provided by females in the Membracinae. The genera that exhibit maternal care have been placed on a molecular phylogeny of this subfamily of treehoppers. The circles at the bases of the lineages represent ancestral species; the degree to which a given circle is blue reflects the probability that females of this ancestral species cared for their eggs. The photograph shows a female treehopper in the genus *Guayaquila* standing guard over a cluster of eggs covered by a secretion applied by the female. After Lin et al. [725]; photograph by Chung-Ping Lin.

care has probably originated three different times in the Membracinae (Figure 12.2), paternal care has never evolved in these insects.

Discussion Question

12.1   On the basis of Figure 12.2, how many times has maternal care been lost after evolving in the Membracinae? If you wanted to employ the comparative method to examine the evolved function of maternal care in this group, what genera would be of special interest to you, and why?

Figure 12.3 Paternal care in fishes and opportunities for polygyny. (Left) A male Randall's jawfish holds his mate's eggs in his mouth. Mouth brooding limits a male to one clutch at a time. (Right) In contrast, male sticklebacks caring for a nest with a clutch of eggs can attract additional females, which add their eggs to the nest. This male is aerating the eggs in his nest, at the base of the aquatic plant, by drawing water through the nest. Left, photograph by Roger Steene; right, photograph by David Thompson.

The treehoppers therefore offer a clear example of the general rule that maternal care evolves more readily than paternal care. One intuitively appealing explanation for the female domination of parental caregiving takes the following form: because females (unlike males) have already invested so much energy in making eggs, they have a special incentive to make sure that their large initial gametic investment is not wasted. Females therefore continue to provide parental care after the eggs have been produced and fertilized. This hypothesis comes to grief, however, when we observe that females of a substantial number of species, including spotted sandpipers and many fishes, abruptly terminate their parental investment after laying their large and expensive-to-produce eggs, which they leave totally in the care of their partners (Figure 12.3). These species show that a considerable initial investment in offspring does not automatically make it advantageous for females to invest still *more* in their brood. Instead, each increment of maternal care is subject to selection. When the costs of providing an additional unit of care exceed the fitness benefits gained from the act, females that do not make the added investment will leave more surviving descendants, on average, than females that provide the extra care.

So we must find another explanation for the general pattern of female-only parental care. David Queller comes to the rescue by pointing out that if the cost–benefit ratio for parental care were usually lower for females than for males, as it may well be, then we would expect females to provide care more often than males [991]. Let's assume for the sake of simplicity that one standard unit of parental care invested in a current offspring reduces the future reproductive output of a male and a female by the same amount. Let's also assume that we are looking at a species in which females sometimes mate with more than one male in a breeding season. In this case, the average benefit to a male from caring for a brood of offspring will be reduced to the extent that some of "his" offspring were actually fathered by other males. For example, if his paternity is, on average, 80 percent, then for every five offspring assisted, the male's investment can yield at best only four descendants, whereas all five youngsters could advance the female's reproductive success. In other words, when cuckolded paternal males waste some of their costly parental care on nonrelatives, they become less likely than their mates to experience a favorable benefit-to-cost ratio for parenting.

Not only are the benefits from paternal care likely to be less than the benefits from a comparable amount of maternal care, but the costs of parental care are likely to be greater for males than for females. As we noted when discussing sexual selection theory in Chapter 10, males generally attempt to acquire many mates because those that succeed leave a great many descendants. Time and energy spent in caring for offspring cannot be invested in mating effort, and therefore, sexual selection should often block the evolution of paternal care. Imagine a lek of black grouse in which the top male fertilizes most of the eggs of the 20 or so females that come to the lek to mate (see Chapter 11). Because regular attendance at the lek is one of the main correlates of male mating success, a grouse with a reasonable chance of becoming an alpha male would pay a heavy price in terms of lost offspring if he took time off from lekking to help care for some of his offspring [991]. In most species, the *potential* reproductive rate of certain elite males far exceeds that of females (see pp. 322–323), making the cost of parental care greater for males than for females, and thereby decreasing the likelihood that paternal care will evolve [656].

Discussion Questions

12.2   Here's another way of looking at the effect of differences between males and females in the probability that their putative offspring are truly their genetic offspring. If a female lays a fertilized egg or gives birth to a baby, this offspring will definitely have 50 percent of her genes. In contrast, a male that mated with this female may or may not be the father of that offspring. Thus, the argument goes, males have less to gain by parental behavior, and this tilts the equation in favor of nonpaternal males that seek out multiple partners rather than limiting themselves to caring for the offspring of one or a few mates. But let's check the logic of this argument by imagining a hypothetical species in which males have a fairly low probability of siring the offspring of any given mate—say, a 40 percent chance. Further, imagine that there are two hereditarily different male phenotypes in the population, a paternal type and a nonpaternal type. The average paternal male mates with two females (each with an average of 10 eggs), whereas the average nonpaternal male mates with five females (which enables him to fertilize a proportion of 50 eggs). In addition, let's say that the paternal male boosts the survival chances of the eggs under his care to 50 percent versus 10 percent for the unprotected offspring of nonpaternal males. Which behavior is the adaptation here? Show your math. What point does this example make about the evolution of male parental care?

12.3   Among certain monkeys and apes with prolonged parental care, females live longer than males in species in which females provide most or all of the parenting, but males live longer than females in species in which males make the major contribution to offspring care [24]. In other words, adults of the parental sex tend to live longer than the nonparental sex. Does this finding indicate that parental care provides a fitness benefit for caretakers in the form of improved survival? Someone then claims that the longer life span of the parental sex has been selected for because primate young are very slow to develop, and therefore parents must live long enough to get their offspring to the age of independence in order to maintain a stable population. What do you have to say about that hypothesis? Do you have an alternative explanation for the observed pattern?

Exceptions to the Rule

The general rule that males are not parental has many exceptions. If evolutionary theory is to help us explain these unusual cases, then paternal care systems have to make sense in terms of their relative fitness costs and benefits to males. Male-only parental care is especially common among fishes (see Figure 12.3), despite the fact that male fish make vast quantities of sperm, like most other male animals. Because some male fish could potentially have many more offspring than the most fecund female, they would seem, at first glance, to have much to lose by diverting time and energy from mating effort to parental effort. Upon mature reflection, however, we can see that there does not have to be a trade-off between parental care and mate attraction in mating systems in which female choice revolves around male care of eggs. Stickleback females are actually drawn to egg-guarding males because these males are demonstrating their commitment to parenting, and perhaps also because the more eggs in a male's nest, the more the dilution effect (see Chapter 6) can help keep their offspring safe [1060].

In fact, stickleback males can care for up to ten clutches of eggs over the 2 weeks or so that it takes the eggs to hatch. In contrast, an average female stickleback can produce only seven clutches of eggs during this period, even without taking time out to guard her eggs [227]. Parenting would be far less advantageous for stickleback females than for males, given that a female would be caring for only one clutch at a time. Furthermore, while the female was so engaged, she would not be able to forage freely, and so would not grow as rapidly as she might otherwise. This loss in growth could be especially damaging in those species in which female fecundity increases exponentially with increasing body size. In other words, for each unit of growth lost by being parental, a female may pay a particularly stiff price in loss of future egg production. Males that are parental also grow more slowly than they would otherwise, but since they must remain in a territory to attract mates in any case, the decrease in their growth *resulting exclusively from parental care* is slight.

The costs of parental behavior for males and females have been directly measured for a mouth-brooding cichlid, St. Peter's fish, in which either the male or the female may care for their young by orally incubating the fertilized eggs. Both sexes lose weight while mouth brooding, since they find it difficult to eat with a mouth full of eggs or fry (baby fish). Furthermore, the interval between spawnings increases for parental fish of both sexes compared with individuals whose clutches are experimentally removed (Figure 12.4). However, the mean inter-spawn interval is greater for brooding females (11 days) than for brooding males (7 days). Moreover, parental females produce fewer young in their next clutch than do nonparental females, whereas parental males are just as able to fertilize complete clutches of eggs in their next spawning as are nonparental males. Although both sexes pay a price for parental behavior, the costs of brood care to females are greater in terms of reduced fecundity [61]. Thus, in fishes, paternal behavior may evolve because males lose less from parenting than females do, which creates a more favorable benefit-to-cost ratio for paternal behavior than for maternal care of offspring [486].

Figure 12.4   Parental care costs female St. Peter's fish more than it costs males. (A) Females that have cared for a clutch of eggs are much slower to produce a new batch of mature eggs than are females that have not provided care to their previous clutch. (B) Parental males also spawn less often than nonparental males, but the difference between the two groups is less than for females. After Balshine-Earn [61].

St. Peter's fish

Figure 12.5   Male water bugs provide uniparental care. A male belostomatid broods eggs glued onto his back by his mate. Photograph by Bob Smith.

Why Do Male Water Bugs Do All the Work?

Although *exclusive* paternal care of young is common among fishes, the trait is very rare among other animals, vertebrates and invertebrates alike [222, 1179]. Among the exceptional paternal insects are water bugs of the genus *Lethocerus*, which guard and moisten clutches of eggs that females glue onto the stems of aquatic vegetation above the waterline (see Figure 1.15) [1128]. Males in some other genera of water bugs (e. g., *Abedus* and *Belostoma*) permit their mates to lay eggs directly on their backs (Figure 12.5), after which the male assumes responsibility for their welfare. A brooding male *Abedus* spends hours perched near the water surface, pumping his body up and down to keep well-aerated water moving over the eggs. Clutches that are experimentally separated from a male attendant do not develop, demonstrating that male parental care is essential for offspring survival in this case.

Bob Smith has explored both the history and the adaptive value of these unusual paternal behaviors [1127]. Since the closest relatives of the Belostomatidae, the family that contains these paternal water bugs, are the Nepidae, a family of insects without male parental care, we can be confident that the brooding species evolved from nonpaternal ancestors (Figure 12.6). Whether out-of-water brooding and back brooding evolved independently from such an ancestor, or whether one preceded the other, is not known, although some evidence suggests that back brooding came later. In particular, female *Lethocerus* sometimes lay their eggs on the backs of other individuals, male or female, when they cannot find suitable vegetation for that purpose. This unusual behavior indicates how the transition from out-of-water brooding to back brooding might have occurred. Females with the tendency to lay their eggs on the backs of their mates could have reproduced in temporary ponds and pools where emergent aquatic vegetation was scarce or absent.

Figure 12.6   Evolution of brood care by males in the Nepoidea, the group that includes the belostomatid water bugs. The illustrations, drawn to scale, show the largest representatives of each group. Paternal care is widespread throughout the family Belostomatidae, but no species in the family Nepidae exhibits this trait. After Smith [1127].

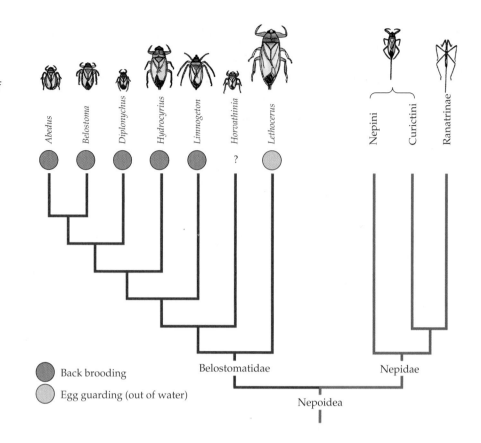

But why do the eggs of water bugs require brooding? Huge numbers of aquatic insects lay eggs that do perfectly well without a caretaker of either sex. However, Smith notes that belostomatid eggs are much larger than the standard aquatic insect egg, with a correspondingly large requirement for oxygen, which is needed to sustain the high metabolic rates underlying embryonic development. But the relatively low surface to volume ratio of a large aquatic egg leads to an oxygen deficit inside the egg. Since oxygen diffuses through air much more easily than through water, laying eggs out of water can solve that problem. But this solution creates another problem, which is the risk of desiccation that the eggs face when they are high and dry. The solution, brooding by males that moisten the eggs repeatedly, sets the stage for the evolutionary transition to back brooding at the air–water interface.

Wouldn't things be simpler if belostomatids simply laid small eggs with large surface-to-volume ratios? To explain why some water bugs produce eggs so large that they need to be brooded, Smith points out that water bugs are among the world's largest insects, almost certainly because they specialize in grasping and stabbing large vertebrates, including fish, frogs, and tadpoles. Water bugs, like all other insects, grow in size only during the immature stages. After the final molt to adulthood, no additional growth occurs. As an immature insect molts from one stage to the next, it acquires a new flexible cuticular skin that permits an expansion of size, but no immature insect grows more than 50 or 60 percent per molt. One way for an insect to grow large, therefore, would be to increase the number of molts before making the final transition to adulthood. However, no member of the belostomatid family molts more than six times. This observation suggests that these insects are locked into a five- or six-molt sequence, just as spotted sandpipers evidently cannot lay more than four eggs per clutch. If a water bug is to grow large enough to kill a frog in just five or six molts, then the first instar (the nymph that hatches from the egg) must be large, because it will get to undergo only five or so 50-percent expansions. Thus, water bug development is an example of the "panda principle" (see p. 295) in that evolutionary modifications of body size have to be layered on what has already evolved in this lineage. In order for the first-instar nymph to be large, the egg must be large. In order for large eggs to develop quickly, they must have access to oxygen, which is where male brooding comes into play Thus, male brooding is an ancillary evolutionary development whose foundation lies in selection for large body size in conjunction with prey specialization [1127].

Female water bugs, however, could conceivably provide care for their own eggs after laying them on exposed aquatic vegetation. Why is it that the males do the brooding, never the females? Here the situation parallels the fish story closely. First, male water bugs with one clutch of eggs sometimes attract a second female, perhaps because a male bug with a partial load of eggs is effectively advertising his capacity for parental care, just like a stickleback [1179]. Second, just as is also true for some fishes, the costs of parental care may be disproportionately great for females in terms of lost fecundity. In order to produce large clutches of large eggs, female belostomatids require far more prey than males do. Because brooding limits mobility and thus access to prey, parental care probably has greater fitness costs for females than for males, biasing selection in favor of male parental care.

Discriminating Parental Care

No matter which parent provides care, an adaptationist would not expect a parent or parents to provide assistance freely to young animals that are not their own genetic offspring. But can parental animals always identify their own progeny? Consider the Mexican free-tailed bat, which migrates to certain caves in the American Southwest, where pregnant females form colonies in the millions. After giving birth to a single pup, a mother bat leaves her offspring clinging to the roof of the cave in a crèche that may contain 4000 pups per square meter (Figure 12.7). When the female returns to nurse her infant, she flies back to the spot where she last nursed her pup and is promptly besieged by a host of hun-

Figure 12.7   Mexican free-tailed bat mothers recognize their pups, despite the fact that they leave their infants in a dense mass of baby bats when they leave their cave to forage outside. When a female returns to nurse her pup, she has to relocate one individual among thousands. Photograph by Gary McCracken.

gry bat babies [805]. Given the swarms of pups, early observers believed that mothers could not possibly identify their own offspring and instead had to provide milk on a first-come, first-served basis.

But do Mexican free-tailed bat mothers really nurse indiscriminately? To find out, Gary McCracken captured female bats and the pups nursing from them and took blood samples from both [805]. He then analyzed the samples using starch-gel electrophoresis, a technique with much the same goal as DNA fingerprinting (see Box 11.1) that can be used to determine whether two individuals have the same variant form of a given enzyme, and thus the same allele of the gene that codes for that enzyme. If female bats are indiscriminate care providers, then the enzyme variants of females and the pups they nurse should often be different. But if females tend to nurse their own offspring, then females and pups should tend to share the same alleles. McCracken focused on the gene for superoxide dismutase, an enzyme represented by six different forms in the population he sampled. Despite the chaos of the bat colony, the enzyme data indicated that females found their own pups at least 80 percent of the time. More recent direct observational studies indicate that females probably do better than that, almost always recognizing their own pups through vocal and olfactory signals [55, 807]. Thus, mother free-tailed bats clearly deliver their parental care primarily to their own pups.

Discussion Questions

12.4   Gary McCracken found that although female Mexican free-tailed bats usually feed their own pups, they do make some "mistakes," which they could have avoided by leaving the pup in a spot by itself instead of in a crèche with hundreds of other babies [805]. Does this mean that the parental behavior of this species is not adaptive? Use a cost–benefit approach to develop alternative hypotheses to account for these "mistakes."

12.5   In some cases, males or females do care for young other than their own, as when certain male fish take over and protect egg masses being brooded by other males or when female ducks acquire ducklings that have just left someone else's nest (Figure 12.8) (see also Chapter 11 on male parental care in species in which extra-pair copulations occur). Devise alternative hypotheses to explain this phenomenon. Under what circumstances might adoptions actually raise the caregiver's reproductive success? Under what other circumstances might adopters be forced into helping nongenetic offspring as a cost of achieving some other goal.

Figure 12.8   Adoption by a female goldeneye duck. This mother escorts a mob of marked ducklings, some of whom were determined by their tags to have come from broods produced by other females. Photograph by Bruce Lyon.

If the ability to recognize one's genetic offspring is advantageous in proportion to the risk of misdirected parental care, then we would expect convergent evolution of parent–offspring recognition in colonial mammals in addition to Mexican free-tailed bats. A comparative test is provided by the subantarctic fur seal, whose females give birth on crowded island beaches. Females remain with their pups for about a week before leaving for an oceanic fishing trip that can last as much as 3 weeks. Playback experiments demonstrate that baby seals take no more than 5 days to learn their mother's call, while the females are also quick learners. When a mother seal returns to the beach, therefore, she calls out, and her infant calls back, leading to their reunion in less than 15 minutes as a rule [207].

Yet another comparative test of the hypothesis that offspring recognition functions to prevent misdirected parental care takes advantage of variation among swallows in the risk of making caregiving mistakes. Although both the bank swallow and the rough-winged swallow nest in clay banks, the bank swallow is colonial, whereas the rough-winged swallow nests by itself. Individual fledglings of the colonial bank swallow produce highly distinctive vocalizations, giving their parents a potential cue to use when making decisions about which individuals to feed, allowing them to distinguish between their own offspring and other fledglings that sometimes wind up in the wrong nests begging for food [814]. Bank swallow parents rarely make mistakes, despite the high density of nests in their colonies. The solitary rough-winged swallow, on the other hand, never has a chance in nature to feed another's fledglings, and so would not be expected to show the same skillful recognition of offspring as its cousin, the bank swallow. Indeed, when Mike and Inger Beecher transferred fledglings between distant rough-wing burrows, they found that the parents readily accepted the transplants. Rough-winged swallows will even act as foster parents for fledgling bank swallows [76].

Two other swallow species, the highly colonial cliff swallow and the less social barn swallow, should also differ in their chick recognition attributes. As expected, cliff swallow chicks produce calls containing about sixteen times as much variation as the corresponding calls of barn swallow chicks (Figure 12.9) [815]. Therefore, it should be easier for cliff swallows to recognize their young than for barn swallows to discriminate among barn swallow chicks. In operant conditioning experiments that required adults of both species to discriminate between pairs of chick calls, cliff swallows reached 85 percent accuracy significantly faster than barn swallows. These results suggest that the acoustical perceptual systems of the cliff swallow, as well

Figure 12.9 Call distinctiveness enables offspring recognition by parents. The chicks of cliff swallows, a colonial species, produce highly structured and distinctive calls, helping their parents recognize them as individuals. The calls of barn swallow chicks, a less colonial species, are much less structured and more similar. The call frequencies of both species lie between 1 and 6 kHz; the durations of the calls range from 0.7 to 1.3 seconds for the cliff swallows and from 0.4 to 0.8 seconds for the barn swallows. After Medvin et al. [815].

Cliff swallow   Barn swallow

Sound frequency

Bluegill

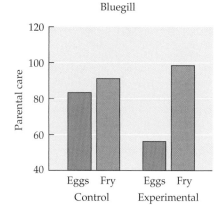

Figure 12.10   Reactions of nest-defending bluegill males to potential egg and fry predators under two conditions. Experimental males had been exposed to clear containers holding smaller male bluegills, mimicking the presence of rivals who might fertilize some of the eggs in the defender's nest; control males were not subjected to this treatment. "Parental care" was quantified using a formula based on the number of displays and bites directed at a plastic bag holding a predatory pumpkinseed sunfish. After Neff [866].

as its calls, have evolved to promote accurate offspring recognition [737], just as is true of subantarctic fur seals.

Discussion Questions

12.6   Territorial male bluegill sunfish defend the eggs and fry in their nests against predatory fish such as largemouth bass (see Figure 13.7). Figure 12.10 shows how intensely males defended their nests in an experiment in which some territorial bluegills were exposed to potential cuckolds during the spawning season. Bryan Neff put sneaker males (see Figure 10.25) in plastic jars near the nests of his experimental subjects to provide the cues associated with a high risk of cuckoldry; he measured male brood defense by quantifying how intensely bluegill dads charged and threatened a predator of bluegill eggs and fry, a pumpkinseed sunfish, which Neff placed near bluegill nests in a clear plastic bag [866]. How do you interpret the results shown in Figure 12.10? What is puzzling about them? Does it help to know that bluegill males can apparently evaluate the paternity of fry, but not eggs, by the olfactory cues they offer?

12.7   When male baboons intervene in disputes between two juveniles, they tend to take the side of their genetic offspring (Figure 12.11). How do you suppose the team of researchers that published these results secured this information?

(A)

(B)

Figure 12.11   Male baboons intervene on behalf of their own offspring when young baboons start fighting with one another. (A) The adult male cradles his offspring, protecting it against an aggressive youngster. (B) Of fifteen fathers whose behavior was monitored, twelve were more likely to help their own offspring than an unrelated juvenile. A, photograph by Joan Silk; B, after Buchan et al. [166].

Why Adopt Genetic Strangers?

The cases described thus far support the prediction that parents should recognize their own young and discriminate against others when the probability of being exploited by someone else's offspring is high. And yet some colonial, ground-nesting gulls occasionally adopt unrelated chicks. Although researchers initially reported that adults consistently rejected older, mobile chicks when they were experimentally transplanted between nests [825], attacks by adults on these transferred youngsters apparently occurred because of the frightened behavior of the displaced chicks [469]. When juveniles *voluntarily* leave their natal nest—which they sometimes do if they have been poorly fed by their parents (Figure 12.12)—they do not flee from potential adopters, but instead beg for food and crouch submissively when threatened. They have a good chance of being adopted, even at the ripe old age of 35 days [555], and if they are taken in, they are more likely to survive than if they had remained with the genetic parents that were failing to supply them with enough food [162].

The adoption of strangers qualifies as a Darwinian puzzle, for here we have parents apparently failing to act in the best interests of their genes. As usual, resolving the paradox requires thinking about the costs, not just the benefits, of a trait of interest. And learned recognition of offspring carries costs as well as benefits, notably the risk of making a mistake and not feeding, or even attacking and killing, one's own offspring. Rather than erring on the side of harming their genetic offspring, gulls have evolved a readiness to feed any chicks in their nest that beg confidently when approached by an adult [951]. Sometimes this rule of thumb permits a genetic outsider to steal food from a set of foster parents by slipping into a nest with other youngsters of its age and size [650]. When adoption occurs, the adoptive parents lose about 0.5 chicks of their own on average; however, adoption is rare, with fewer than 10 percent of adult

Figure 12.12   Why seek adoptive parents? Gull chicks that abandoned their natal nest in search of foster parents weighed much less than average for chicks their age that were known to survive to day 12. But these potential adoptees were sometimes adopted by nonrelatives, and as a result they weighed more on average at day 11 than the subset of gull nestlings known to have died or disappeared by this time. After Brown [162].

ring-billed gulls taking in a stranger in any year [162]. The modest average annual fitness cost of a rule of thumb that results in occasional adoptions has to be weighed against the cost of rejecting one's own genetic offspring that would arise if parent gulls were more cautious about feeding chicks in their nests.

The argument here is that a less than perfectly accurate rule of thumb can be selectively advantageous if the proximate mechanism responsible for that rule has a more favorable cost–benefit ratio than alternative psychological mechanisms that would result in different decision rules. This hypothesis has been used to explain why male western bluebirds adopt broods under some circumstances. When Janis Dickinson and Wes Weathers removed some nesting males, most of the experimentally widowed females soon attracted replacement partners, about half of whom fed the chicks of the females they had joined, even though at least some of those chicks had surely been fathered by the original male [331]. The observation that the paternal replacement males were those that found a female while she was still laying eggs (and thus both sexually receptive and potentially fertile) suggests that male western bluebirds have a proximate all-or-none decision-making mechanism that regulates their parental caregiving. If the male joins a female during her fertile phase, he exhibits all-out parental care; if he joins her after her fertile period is over, then the rule is "no parental care at all." Likewise, in the white-browed scrub-wren, a polyandrous species with alpha and beta males, the rule for the subordinate is to be paternal only if he has copulated with the female [1289]. In both bluebirds and beta scrub-wrens, males apparently use proximate decision rules that motivate them to help broods that probably contain some of their genetic offspring, as do male dunnocks (see pp. 388–389). If bluebirds or scrub-wrens could run DNA fingerprinting tests on their putative offspring, then they could do a perfect job of discriminating between their own genetic progeny and those of another male. But since this technology is obviously out of their reach, selection can favor behavioral decision-making rules that are better than the *possible* alternatives, even though these rules do not always produce the "perfect" response.

The use of another generally adaptive rule of thumb may explain why breeding adults of various species devote themselves to the young of specialized brood parasites, such as certain cuckoos and cowbirds, that never rear their own offspring but instead always place their eggs in the nests of other birds. Parental songbirds often appear to distribute the food they bring to the nest in a biased fashion, preferentially feeding their larger, more actively begging nestlings (see p. 299). This kind of bias enables parents to avoid wasting hard-earned resources on weak or sickly offspring unlikely to reach reproductive age, but this potentially advantageous decision rule can be exploited by brood parasites. A young cuckoo or cowbird tends to hatch a day or two sooner than the young of the host species, which enables the parasite to get a head start on its competitors in the nest. Because adult brood parasites are also generally larger birds than the species they victimize (see Figure 9.28), their babies soon become much larger than the hosts' own youngsters. When a strapping young cowbird stretches up higher than its nestmates to greet a returning parent bird, odds are that the parasite will get fed (Figure 12.13) [717].

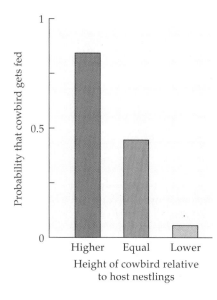

Figure 12.13   Effective begging depends on how high a baby bird can reach relative to its nestmates. Brown-headed cowbird nestlings generally can reach higher than their hosts' offspring, which is why they have a higher probability of getting fed. After Lichtenstein and Sealy [717].

12.8 Interspecific brood parasitism is very rare in birds, with only about 1 percent of all species practicing the trait [689]. Make a prediction about which group of birds, those with precocial young or those with altricial young, would be more likely to evolve into specialist brood parasites. (In altricial species, the eggs are small in relation to parental body weight, but the hatchlings are initially completely dependent on food supplied to them by a parent. In precocial species, the eggs are relatively large, but the youngsters can move about and feed themselves shortly after hatching.) Check your prediction against [758].

The History of Interspecific Brood Parasitism

But how did cowbirds, cuckoos, and the like come to specialize in parasitizing other bird species? In the case of cuckoos, one phylogenetic reconstruction based on molecular data indicates that specialized parasitism has arisen three times over the evolutionary history of this group (Figure 12.14) [39]. Given this phylogeny and the overall rarity of specialist brood parasitic bird species, Oliver Krüger and Nick Davies hypothesized that the ancestor of the current parasitic cuckoos was a "standard" bird whose adults cared for their own offspring. By taking advantage of detailed information on the natural history of many of the 136 species of cuckoos worldwide, of which 83 are nonparasitic while 53 let other birds brood their young, Krüger and Davies were able to demonstrate that the ancestral state was represented not only by parental care, but also by the occupation of small home ranges and the absence of migration. The next stage of evolution involved species that provided parental care for their offspring but possessed relatively large ranges and a tendency to migrate. From lineages of this sort came the modern brood parasites, which also have large home ranges and are generally migratory [676]. Brood parasites presumably gain by roaming widely in search of appropriate hosts throughout an extensive home range; this readiness to move while hunting for ephemeral sources of hosts may have led to the evolution of migratory tendencies in this group.

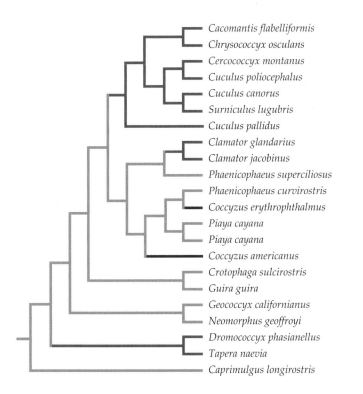

Figure 12.14   Specialized brood parasitism by cuckoos has evolved three times, based on a phylogeny of this family of birds, which contains both brood parasites and standard parental, nesting species. The blue branches indicate those lineages whose modern descendants are specialist parasites (The red branches are for two species that are occasional parasites). After Aragón et al. [39].

But what about the transition between parental behavior and specialized brood parasitism? This shift could have taken place gradually, with an intermediate phase when the parasites of the day targeted nesting adults of their own species, with the shift to members of one or more other species occurring later. Alternatively, specialized interspecific parasitism may have arisen abruptly with the exploitation of adults of another species right from the start. The first, gradualist scenario leads to the prediction that females of some of today's birds should lay their eggs in nests of their *own* species, and indeed, parasitism of this sort has been documented in about 200 species and may occur in many others as well [292, 973].

A possible hint of a very early stage in the evolution of this kind of brood parasitism comes from a study of wood ducks. In this species, suitable nest cavities in trees are scarce, with the result that two females sometimes lay eggs in the same nest before one duck evicts the other. The "winner" then cares for the eggs of the other female along with her own, having made the "loser" an involuntary parasite in the process [1084]. If the losers were to produce more offspring than they would have if they had retained the nest site, then selection could favor variants that voluntarily deposited and abandoned their eggs in another's nest—a behavior that is fairly common in another hole-nesting duck, the Barrow's goldeneye [352].

Another voluntary brood parasite is the coot, a common water bird, in which "floater" females that lack nests or territories of their own lay their eggs in the nests of other coots, apparently in an effort to make the best of a bad situation, since they cannot brood their own eggs themselves. But some fully territorial females with nests of their own also regularly pop surplus eggs into the nests of unwitting neighbors. Since there are limits to how many young one female coot can rear with her partner, even a territorial female can boost her fitness a little by surreptitiously enlisting the parental care of other pairs [756]. The exploitative nature of this behavior is revealed by the finding that older, larger females select younger, smaller ones to receive their eggs, presumably because this kind of host cannot fight back effectively when visited by a brood parasite [757].

The gradual shift hypothesis yields the prediction that when intraspecific brood parasites first began to exploit other species as hosts, they should have selected other related species with similar nestling food requirements. Currently, most specialized brood parasites take advantage of species that are *not* closely related to them, but this observation could simply reflect the fact that most brood parasites have been evolving for many millions of years since the onset of their *interspecific* parasitic behavior. Thus, to check this prediction, we need to find brood parasites that have a relatively recent origin. The familiar cowbirds of the Americas are one such group, its parasitic species having originated "only" 3 to 4 million years ago, whereas the parasitic cuckoos evolved at least 60 million years before the present [292]. The living cowbird species believed to be closest to the ancestral brood parasite does indeed parasitize a single host species that belongs to the same genus that it does; the next closest species to the ancestral parasite parasitizes other birds belonging to its own family (Figure 12.15) [699]. These data, if they have been interpreted properly, provide support for the gradual shift hypothesis.

Similarly, female widowbirds in the family Viduidae parasitize finch species that belong to the family Estrildidae. The two families are very closely related (Figure 12.16), which may be why both parasites and hosts share a number of important features in common, especially bright white eggs and an unusual form of begging behavior by nestlings, in which the baby birds turn the head nearly upside down and shake it from side to side, rather than stretching upward in the manner of most other nestling songbirds. Assuming that the

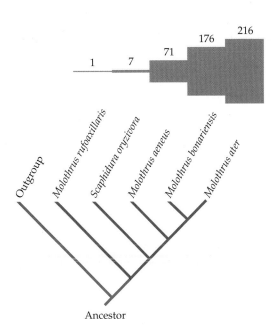

Figure 12.15 Evolution of brood parasitism among cowbirds. The phylogeny depicts the evolutionary relationships among cowbirds as determined by molecular genetic analyses. Above the phylogeny, the number of hosts parasitized by each current living species of cowbird is shown. (The "outgroup" is a species of cowbird that does not engage in brood parasitism.) The pattern suggests that the first brood parasitic cowbird victimized only a single closely related host species, with increasingly generalized brood parasitism probably evolving subsequently. After Lanyon [699].

ancestral parasitic widowbird also possessed these attributes, sensory exploitation (see Chapter 9) could account for the success the offspring of the original brood parasite had after hatching in the nest of an estrildid finch host [1142].

On the other hand, the very large majority of living brood parasites take advantage of unrelated species much smaller than they are [1119]. Some persons interpret this finding to mean that the original brood parasites may have made an abrupt shift to one or more unrelated smaller host species because such a shift was more likely to succeed, given that, as noted above, brood parasite nestlings that become larger than their hosts' offspring are more likely to be fed, another form of sensory exploitation that works to a parasite's advantage.

The importance of the size disparity between host and parasite has been demonstrated by creating some experimental brood parasites. When Tore Slagsvold shifted blue tit eggs into great tit nests, the experimentally produced brood parasitic blue tit nestlings, which are smaller than those of great tits, did

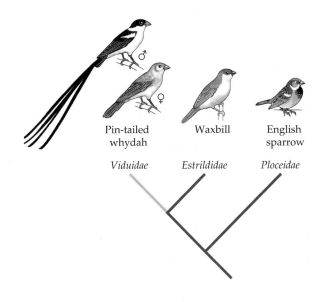

Figure 12.16 Widowbirds parasitize closely related species. Widowbirds (in the family Viduidae) parasitize nests of finches in the family Estrildidae (the family most closely related to the Viduidae). Adult male widowbirds, like the male pin-tailed whydah (*Vidua macroura*) shown here, look nothing like their hosts, but female pin-tailed whydahs and their nestlings do resemble adult and nestling waxbills (*Estrilda astrild*), an estrildid finch parasitized by this widowbird.

Figure 12.17 The size of an experimental "brood parasite" nestling relative to its host species determines its survival chances. Larger great tit nestlings survived well when transferred to the nests of smaller blue tits, whereas blue tits did poorly in great tit nests. After Slagsvold [1119].

Blue tit

Great tit

very poorly. In the reciprocal experiment, however, most of the great tit chicks cared for by blue tit parents survived to fledge (Figure 12.17) [1119]. These findings suggest that unless the original mutant interspecific brood parasites happened to deposit their eggs in the nests of a smaller host species, the likelihood of success (from the parasite's perspective) was not great. Therefore, proponents of both the gradual and abrupt shift scenarios for the evolution of interspecific brood parasitism have at least some supportive evidence to which they can point, which is why the issue is still unresolved.

Discussion Question

12.9 When great tits are experimentally reared in blue tit nests, they survive quite well, as just noted. But many of these experimental brood parasites did not associate with members of their own species when they became adults, and they often failed to mate with a great tit (see Figure 3.6). The reproductive success of these individuals was consequently low [1120]. Explain these results in terms of the proximate mechanism of imprinting (see Chapter 3) and discuss both the negative *and* positive effects that this developmental process could have had on the evolution of interspecific brood parasitism.

Why Accept a Parasite's Egg?

Whatever the origin and subsequent history of interspecific brood parasitism, it is true that in order for a parasite nestling to take advantage of a host species' parental decision rules, the egg containing the parasite has to hatch. Why, then, don't host species take immediate action against the eggs of parasites? Some birds do, by recognizing the foreign egg and removing it from the nest or by abandoning the nest along with the parasite's egg. However, each of these options has its disadvantages [1322]. Parent birds that made incubation of eggs dependent on their learned recognition of the eggs they had laid might sometimes abandon or destroy some of their own eggs by mistake, treating them as if they were foreign eggs. Indeed, reed warblers sometimes do throw out some of their own eggs while trying to get rid of a cuckoo's egg [293]. If brood parasites victimize only a very small minority of the host population, then even a

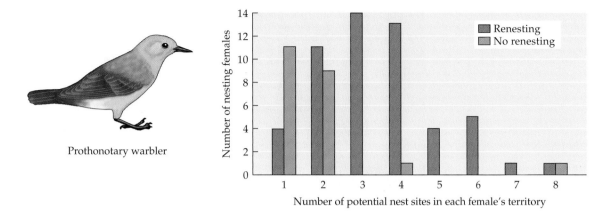

Figure 12.18   The probability that a female prothonotary warbler will nest again in her territory is a function of the number of potential nest sites in her territory. When only a few good nest holes are present, a female rarely makes a second nesting attempt on her territory; if relatively many sites are available, a female usually does renest there. After Petit [940].

small risk of costly recognition errors by the host can make accepting parasite eggs the better parental option [744].

Acceptance of the parasite egg is even more likely to be adaptive when the host is a small species unable to grasp and remove large cowbird or cuckoo eggs [1030]. Such small-billed birds have two choices: abandon the clutch, either by leaving the site or by building a new nest on top of the old one, or stay put and continue brooding the clutch along with the parasite egg. The abandonment option imposes heavy penalties on the hosts, which must find a new nest site, build a new nest, and lay a new clutch of eggs. The costs of this option are especially high for hole-nesting species because suitable tree holes are generally rare. Perhaps the willingness of a female to start over after finding a cowbird egg in her nest is affected by how many other good nest holes are in her territory. Indeed, when female prothonotary warblers do not have several such sites on their territory, they often do not make a second breeding attempt there (Figure 12.18). These females may tolerate a parasite in their first nest for lack of an alternative [940]. Likewise, yellow warblers tend to accept foreign eggs when their nests are parasitized near the end of the breeding season when little time remains to rear a new brood from scratch [1077].

Even if host birds could throw out or cover up a brood parasite's eggs without risk of error, the parasite might make this option unprofitable by returning to the nest to destroy or consume the host's eggs or young if it found that its offspring had been harmed (Figure 12.19). This "Mafia hypothesis" has been tested by examining the interactions between European magpies and parasitic great spotted cuckoos [1137]. As predicted by the hypothesis, magpie nests from which cuckoo eggs had been ejected suffered a significantly higher rate of predation than nests with accepted cuckoo eggs (87 percent versus 12 percent in one sample). Furthermore, when researchers removed the cuckoo egg from a nest that was apparently being checked by a cuckoo, and replaced the magpie's eggs as well with plasticine imitations, the cuckoo approached the nest after the researchers had finished and violently pecked the false magpie eggs. In nature, magpies that lose their clutch have to renest, which exposes them to all the negative effects that delays in breeding have in a seasonal environment. In this light, it is not surprising that acceptor magpies actually have somewhat higher reproductive success, as measured in terms of fledglings produced, than do ejectors of cuckoo eggs [1137].

The approach we have taken thus far is to examine the proposition that the costs of refusing to incubate a brood parasite's egg, or to feed a parasitic nestling after it hatched, could actually outweigh the benefits of refusal. Another complementary way to look at the interaction between host and parasite is to apply

Figure 12.19   Egg removal by a cuckoo. Here a female European cuckoo destroys a reed warbler's egg prior to laying her own egg in this nest. Cuckoos could punish birds that harmed their eggs by destroying the entire brood in an uncooperative victim's nest. Photograph by Ian Wyllie.

the perspective of evolutionary arms race theory. Whenever there are two parties in conflict with each other, they exert reciprocal selection pressure on each other, with an adaptive advance made by one leading in subsequent generations to an adaptive counter-response by the other. We have seen this phenomenon in the interplay between ultrasound-producing bats and ultrasound-detecting moths (Chapter 4), in the sexual conflict that occurs between males and females of many species (Chapter 10), and in the attempts of parasitic cuckoo nestlings to manipulate the behavior of their adoptive parents (Chapter 9). You may recall that the young European cuckoo produces vocalizations that closely mimic those of an entire brood of reed warblers (see Figure 9.28), perhaps because successful mimics in the past were able to overcome the adaptive tendencies of their hosts to withhold food from chicks that do not sound like members of their own species [297].

The arms race approach helps us make sense of the interaction between Horsfield's bronze-cuckoo and its host, the superb fairy-wren. If breeding fairy-wrens find an egg in their nest before they have begun to deposit their own eggs there, they almost always build over the intruder's egg, and they abandon nests altogether if a cuckoo drops an egg in after the wrens have begun to incubate their own complete clutch. So the wrens have defenses against cuckoos. Adult female cuckoos, however, are adept at slipping in to lay an egg at just the right moment; namely, when a nest has only a partial clutch of fairy-wren eggs. In these cases, the cuckoo egg is almost always accepted and incubated along with the hosts' eggs. But when the cuckoo chick hatches and pushes its wren nestmates out of the nest to their deaths, the fairy-wrens abandon the nest about 40 percent of time, leaving the baby cuckoo to starve to death. In the remainder of cases, however, the fairy-wrens keep caring for the sole occupant of their nest, and so waste their time and energy rearing the killer of their offspring.

If the semi-acceptance of Horsfield's bronze-cuckoo chicks stems from the parasite's ability to combat the evolved defenses of its hosts, then we would predict that these chicks must have some special way of stimulating parental care—which they do. Like European cuckoo chicks, they produce begging calls that sound very much like those made by their host's offspring (Figure 12.20) [696]. Naomi Langmore and her colleagues believe that the vocal mimicry of Horsfield's bronze-cuckoo chicks is an evolved response to the host's discriminatory abilities. In support of this conclusion, they point to the fact that superb fairy-wrens *always* abandon nests that have been parasitized by another species, the shining bronze-cuckoo, whose nestlings produce vocalizations very different from those of the host chicks. (I suspect that you can guess how often shining bronze-cuckoos make the mistake of laying an egg in a superb fairy-wren nest.)

Figure 12.20   A product of an evolutionary arms race? Chicks of Horsfield's bronze-cuckoo, a specialist brood parasite on Australian fairy-wrens, mimic the calls of their host's chicks very closely, which may help them overcome the defenses of fairy-wren host parents. In contrast, the chicks of another bronze-cuckoo species, which rarely parasitizes fairy-wren nests, not only do not look like fairy-wren chicks, but lack a good facsimile of the begging calls of fairy-wren nestlings. After Langmore et al. [696].

12.10 Fairy-wrens are the only birds that abandon the chicks of brood parasites. Why don't other victimized species do the same? After all, many other birds victimized by cuckoos can identify and take action against the parasite's eggs, which they do by learning the distinctive visual features of their own eggs and then rejecting those that do not match. But the same species that are extremely good at learning to recognize egg features usually completely fail to recognize a cuckoo chick [743]. At first glance, this failure appears to be maladaptive, but consider the consequences of learned chick recognition for birds that were successfully parasitized in their first year of breeding by a single cuckoo chick that took over the nest and eliminated all the host's youngsters, as is their habit. If this kind of learning occurred, how would the host adults respond to their own chicks in the next breeding attempt? (What do you predict about the response of fairy-wrens to their young after having once mistakenly reared a parasitic cuckoo?) How does this case illustrate the importance of cost–benefit analyses in behavioral ecology as well as the value of considering the proximate mechanisms by which individuals achieve behavioral goals?

12.11 When a yellow warbler's nest has been parasitized by a cowbird, odds are that the cowbird will return within a few days to try to remove a yellow warbler egg from the nest. Because yellow warblers can sometimes rear some of their own offspring along with the parasite, the loss of an egg from a parasitized clutch adds to the damage already done by the addition of the parasite's egg. However, female warblers can prevent the loss of an egg if they increase the time they spend sitting on the nest, in which case returning cowbirds are generally unable to get at their eggs. But female warblers that increase their egg-guarding time have less time to forage and so must be fed by their partners—which their partners can do. But alas, more comings and goings at the nest by the food-bearing male make the nest's location more conspicuous to nest predators. Given this trade-off between dealing with parasitic cowbirds versus predators that eat eggs and chicks, should yellow warbler females always guard their nests to the maximum extent? In this context, why did a team of researchers expose some nesting yellow warblers to tapes of cowbird vocalizations near their nests, which also received a warm but nonviable cowbird egg while the incubating female warbler was temporarily away from the nest [1184]? Put this example in the context of evolutionary arms race theory.

The Evolution of Parental Favoritism

Even when parents invest only in their own progeny, they rarely distribute their care in a completely democratic fashion. Parental birds, for example, regularly use a decision rule that steers food toward some nestlings, often those that beg most energetically, while depriving others, typically those that beg less vigorously. This uneven distribution of food takes an extreme form in the great egret, a bird in which brothers and sisters fight for possession of the fish their parents bring to the nest (Figure 12.21) [831]. The dominant individuals in a brood often (though not always) bludgeon a sibling to death or push it out of the nest, thereby monopolizing the food supplied by their parents. How can siblicide possibly advance a parent's fitness when it leads to the outright loss of one or even two members from a brood of three or four?

Perhaps parents do *not* gain from siblicide, which may have evolved because of the fitness advantages enjoyed by offspring able to dispose of their siblings. The parental perspective on sibling conflict need not be the same as that of their siblicidal offspring, which we can illustrate with a hypothetical case. Imagine a family with two offspring, each of which will have three surviving youngsters of its own on average. Now imagine that if one of the two eliminates its sibling, the siblicidal offspring boosts its output to five surviving offspring, thanks to its ability to hog the food its parents supply after the death of its brother or sis-

Figure 12.21   Sibling aggression in the great egret. Two chicks fight viciously in a battle that may eventually result in the death of one of them. The adult egret yawns while the fight goes on below. Photograph by Douglas Mock, from Mock et al. [833].

ter. Although the killer kid loses three nephews or nieces, which its sib would have produced were it not for its premature death, this genetic cost is more than compensated for, from the vantage point of the siblicidal individual, by the two extra offspring of its own. (Nephews and nieces share one-fourth of their genes with their uncles and aunts, whereas parents and offspring have one-half of their genes in common [see p. 450]. You can do the math here.) But the parent of a murderous offspring loses the three grandoffspring that its dead progeny would have had, a figure that is not matched by the two extra grandoffspring coming from the successful siblicidal survivor. This is the kind of situation predicted to lead to **parent–offspring conflict**, a concept developed by Bob Trivers after he realized that some actions can advance the fitness of an offspring while reducing the reproductive success of its parent, and vice versa [1213].

In some animals in which siblicide occurs, parents can and apparently do resist their progeny's siblicidal behavior. Evidence for this claim comes from studies of colonially nesting seabirds called boobies, some of which exhibit "early siblicide," in which an older "A" chick disposes of a younger "B" chick in the first few days of its sibling's short and unhappy life [745]. The A chick's ability to kill its brother or sister stems in part from the pattern of egg laying and incubation in boobies. In these birds, females lay one egg, begin incubat-

Figure 12.22   Early siblicide in the brown booby.
A very young chick is dying in front of its parent, which
continues to incubate the larger, siblicidal chick that has
forced its younger brother or sister out of the nest and
into the sun. Photograph by the author.

ing it at once, and then some days later lay a second egg. Because the first egg
hatches sooner than the second, the A chick is relatively large by the time the B
chick comes on the scene. In early siblicidal species, the A chick immediately
begins to manhandle the B chick, soon forcing it out of the nest scrape, where
it dies of exposure and starvation (Figure 12.22).

Early siblicide is standard practice for the masked booby, but not the blue-
footed booby, whose A chicks engage in siblicide less often and generally later
in the nesting period. If, however, you give a pair of blue-footed booby chicks
the chance to be cared for by masked booby parents, which tolerate early sibling aggression, the A chick often quickly kills the B chick under the vacant gaze
of its substitute parents. In contrast, blue-footed booby parents appear to keep
their A chick under control during its initial days with its sibling. If so, then
when masked booby chicks are given to blue-footed booby parents, the foster
parents should sometimes be able to prevent them from immediately killing
their siblings. The actual results of such a switch match the predicted ones (Figure 12.23) [745], providing evidence that parents can evolve to interfere with
lethal sibling rivalries, should it be in their interest to do so. (However, the
researchers who did this work did not actually watch parent–offspring interactions, and so the means by which adult masked boobies intervene in sibling disputes is still unknown.)

Observations of egrets reveal that parental intervention definitely does not
occur when two juvenile egrets go at one another. Indeed, lethal sibling battles
are actually promoted by earlier parental decisions about when to begin incubating eggs and what to put in them in the first place. Thus, as soon as a female
egret lays her first egg, incubation begins, just as is true for boobies. Because 1
or 2 days separate the laying of each egg in a three-egg clutch, the young hatch
out asynchronously, with the firstborn getting a head start in growth. As a result,
this chick will be much larger than the third-born chick, which helps ensure that
the senior chick will monopolize the small fish its parents bring to the nest. The

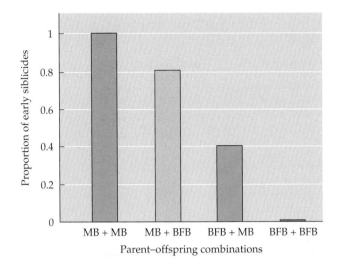

Figure 12.23 Parent boobies can control siblicide to some extent. The rate of early siblicide by masked booby (MB) chicks declines when they are placed in nests with intervention-prone blue-footed booby (BFB) foster parents. Conversely, the rate of early siblicide by blue-footed booby chicks rises when they are given laissez-faire masked boobies as foster parents. After Lougheed and Anderson [745].

senior chick is not only bigger, but also more aggressive, because, judging from what happens in the cattle egret, the first-laid eggs receive relatively large quantities of androgens—a hormonal aggression facilitator. The unequal feeding rates that result further exaggerate the size differences among siblings, creating a runt of the litter that often dies from the combined effects of starvation and assault [1074].

Discussion Question

12.12 As noted above and elsewhere, some female birds make adjustments in the androgens they supply to their eggs in accordance with various factors. In the case of cattle egrets, the egg's position in the laying sequence is linked to its infusion with androgens. In the case of canaries, the ability of the father to sing an attractive song influences the mother's willingness to add male sex hormones to the eggs [450] (see Chapter 10). In other birds, females adjust the amount of feeding effort in relation to the quality of the offspring's father. Female blue tits, for example, provide less food for the offspring of partners whose crown feathers have been manipulated so that they reflect less ultraviolet light [576]. Why can these different kinds of allocation decisions all be considered examples of parental favoritism, and what do the three examples have in common with respect to how maternal decisions advance the fitness of the mother?

Thus, parent egrets not only tolerate siblicide, they actually promote it. Why? Perhaps because parental interests are served by having the chicks themselves eliminate those members of the brood that are especially unlikely to survive to reproduce. In most years, food will be moderately scarce, making it impossible for the adults to find enough to feed all three offspring. In good years, they can supply a large brood with plenty to eat, and it is then that the ability to lay three eggs pays off. Under conditions of food scarcity, however, a reduction in the brood accomplished by siblicide saves the parents the time and energy that would otherwise be wasted on offspring with little or no chance of reaching adulthood, even if their siblings left them alone.

One way to test this hypothesis is to create unnaturally synchronous broods of cattle egrets [834]. In one experiment, synchronous broods were formed by putting chicks that had hatched on the same day in the same nest; normal asynchronous broods were assembled by bringing chicks together that differed in age by the typical 1.5-day interval. A category of exaggeratedly asynchronous broods was also created by putting chicks that had hatched 3 days apart into

TABLE 12.1 *The effect of hatching asynchrony on parental efficiency in cattle egrets*

	Mean survivors per nest	Food brought to nest per day (ml)	Parental efficiency[a]
Synchronous brood	1.9	68.3	2.8
Normal asynchronous brood	2.3	53.1	4.4
Exaggerated asynchronous brood	2.3	65.1	3.5

Source: Mock and Ploger [834]

[a]The number of surviving chicks divided by the volume of food brought to the nest per day × 100

the same nests. If the normal hatching interval is optimal in promoting efficient brood reduction, then the number of offspring fledged per unit of parental effort should be highest for the normal asynchronous broods. This prediction was confirmed. Members of synchronous broods not only fought more and survived less well, but required more food per day than normal broods, resulting in low parental efficiency (Table 12.1). The same result has been recorded for the blue-footed booby, in which experimental synchronous broods of two fought more and required much more food than control broods composed of an asynchronously hatched pair of chicks [906].

Thus, cattle egret parents and others like them know (unconsciously) what they are doing when they manipulate the hormone content of their eggs and incubate them in ways that ensure differences in size and fighting ability among their chicks. Sibling rivalry and siblicide actually help parents deliver their care only to offspring that have a good chance of eventually reproducing while keeping their food delivery costs to a minimum. Although cases of this sort represent extreme examples of parental favoritism, even those nesting birds that bias their allocation of food resources toward vigorous offspring are really practicing infanticide, speeding the demise of those progeny unlikely to reproduce even if they were well fed.

Discussion Question

12.13 In species like boobies and egrets, parental decisions about incubation and hormone allocation put their second or third offspring at great risk of being destroyed. If the second or third egg laid is destined to produce a chick that will die within a few days, why don't parents save the energy that goes into the superfluous egg, which would also save energy for their favored offspring? One possibility goes under the label of the insurance hypothesis: adults invest in a backup egg in order to have a replacement for a favored first-laid egg should something happen to that egg or to the nestling itself after the egg hatches [220]. How would you test this hypothesis experimentally?

How to Evaluate the Reproductive Value of Offspring

Parents sometimes bias food delivery to their offspring based on signals provided by their progeny, such as the vigorousness of their begging calls, as we saw in Chapter 9. But what about the bright red lining of a baby bird's mouth [1057], which is conspicuously displayed by many nestling songbirds as they reach up to solicit food from a returning parent (Figure 12.24)? Because the red color of the mouth lining is generated by carotenoid pigments in the blood, and because carotenoids are believed to contribute to immune function, a bright

Figure 12.24   An honest signal of condition? The red mouth gape of nestling lark buntings is exposed when the birds beg for food from their parents. The brightness of a baby bird's gape could reveal something about the strength of its immune system. Photograph by Bruce Lyon.

red gape could signal a healthy nestling that will make good use of extra food. Parents that preferentially fed those members of their brood that had bright red gapes would be investing in healthy nestlings that were more likely to fledge and eventually become reproducing adults than their sickly nestmates. If parents do indeed make adaptive parental decisions of this sort, then nestlings made ill by injection of a foreign material should have paler mouth linings than those that have not had their immune systems challenged. Furthermore, parents should feed offspring with artificially reddened gapes more than they feed offspring with unaltered mouth coloration. When Nicola Saino's research team tested both predictions in the barn swallow, the results were positive (Figure 12.25).

Figure 12.25   The color of the mouth gape affects the amount of food that nestling barn swallows are given by their parents. After experimenters colored the gapes of some nestlings with two drops of red food coloring, they received more food. In contrast, nestlings that received two drops of yellow food coloring or water were not fed more. After Saino et al. [1057].

On the other hand, there are alternative explanations for bright gape coloration [221], including the straightforward possibility that young birds gain by having colored gapes simply because these make the begging bird's maw more visible to its parents, especially in species in which the nest is placed in a dark tree hollow or other cavity [529]. Philipp Heeb and his co-workers found that when they painted the mouths of great tit nestlings, those birds with artificial yellow gapes were fed more often in relatively dark nest boxes than red-painted birds, whose mouths were less visible under low light conditions. In nest boxes with clear Plexiglas windows on top, however, the red-mouthed nestlings suffered no begging handicap, as shown by their ability to achieve the same weight as their yellow-mouthed companions [529]. The fact that the mouths of some nestling birds strongly reflect ultraviolet radiation is also consistent with the hypothesis that mouth colors are designed simply to help parents find a receptacle into which food can be stuffed because ultraviolet signals stand out strongly against the visual background of the typical nest [577].

Perhaps, therefore, we should withhold judgment on whether nestlings are communicating something about their physical condition to their parents via the color of their mouth linings. The idea, however, that an offspring's potential to survive and reproduce (its **reproductive value**) influences parental decisions continues to be actively explored by adaptationists, as we noted when discussing the possibility that acoustical begging vocalizations provide honest signals of offspring condition (see p. 300). Parents that use these calls to favor some individuals over others may be able to avoid costly investments in progeny of low reproductive value.

Along similar lines, Bruce Lyon and his colleagues suspected that the long, orange-tipped plumes on the backs and throats of baby coots might be the cues used by their parents to determine which individuals to feed and which to ignore. Coots produce large clutches of eggs, but soon after the young begin to hatch, the adult birds often begin what to humans seems an especially unpleasant process of brood reduction. As some babies swim up to beg for food from a parent, the parent may not only refuse to provide something to eat, but may also deliver a volley of pecks at the head of its youngster. As a result, some brood members soon stop begging, and eventually expire head-down in the water.

To test the link between chick ornaments and parental care decisions, Lyon and company trimmed the thin orange tips from these special feathers on half of the chicks in a brood, while leaving the other members of the brood untouched. The unaltered orange-plumed chicks were fed more frequently by their parents, and they grew more rapidly as well (Figure 12.26) [759]. In control broods, in which *all* of the chicks had had their orange feathers trimmed, the youngsters were fed as often, and survived as well, as broods consisting only of untouched orange-feathered chicks. This result shows that the parents of the experimental mixed broods discriminated against the orange-deprived chicks because they were not as strongly ornamented as their feather-intact broodmates, not because the parents failed to recognize them as their offspring [759].

One wonders whether the orange ornaments of baby coots signal something about offspring health, thereby enabling parents to make adaptive decisions about parental care. The general message provided by coots, boobies, and egrets is that parents do not always treat each offspring the same, but instead often help some survive at the expense of others. Cases of this sort remind us that selection acts not on variation in the number of offspring produced, but on the number that survive to reproduce and pass on the hereditary traits of their parents.

(A)

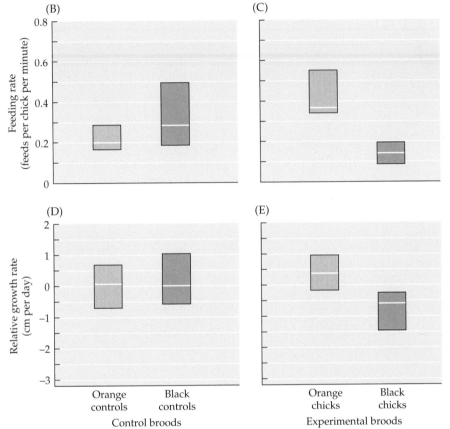

Figure 12.26   The effect of orange feather ornaments of baby coots on parental care. (A) Baby coots have unusually colorful feathers near the head. (B) Control groups composed *entirely* of either unaltered (orange) chicks or chicks that had had the orange tips trimmed from their ornamental feathers (black) were fed at the same mean rate. (C) In experimental broods composed of half orange and half black chicks, the ornamented individuals received more frequent feedings from parent birds. (D) The relative growth rates of chicks in both control groups were the same, but (E) ornamented chicks grew faster in mixed broods compared with the experimentally altered chicks. A, photograph by Bruce Lyon; B–E, after Lyon et al. [759].

Discussion Questions

12.14   Magpie nestlings are increasingly likely to survive as they get older (Figure 12.27A) [1004]. Given that fact, use a cost–benefit approach to explain why magpie parents change their defensive behavior in response to a biologist's approach to the nest over the nesting period (Figure 12.27B).

Figure 12.27 Nestling survival probability and defense of the nest by parent magpies. (A) The probability that nestlings of different ages will be alive at fledging. (B) The relation between nestling age and the intensity of nest defense by adult magpies, as measured by how close adult magpies came to a human observer (the higher the score, the closer the magpies came). After Redondo and Carranza [1004].

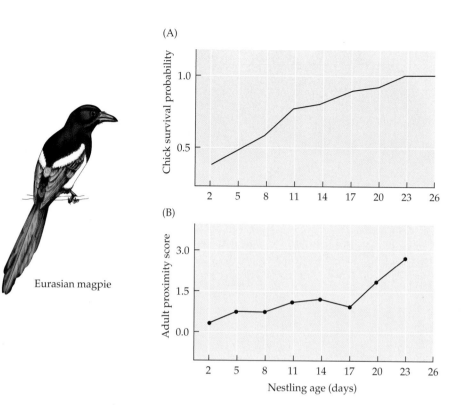

Eurasian magpie

12.15 Use the concept of reproductive value to make predictions about the escape decisions of incubating mallard ducks. These birds, which nest amid dense aquatic vegetation, are faced with a trade-off when a potential predator approaches. They can improve the odds of saving their own skin by quickly flushing from the nest, but their noisy departure will often give the nest location away, with the likely loss of all the eggs within. Or they can sit tight, remaining as inconspicuous as possible, improving the odds that the predator will pass by the nest and its contents, but also increasing their personal risk of being killed. Make predictions about how mallards will respond in relation to the number of eggs being incubated, the mean size of the eggs in the clutch, and their stage of development. Check your predictions against data in [9].

Summary

1. The time, energy, and resources that parents devote to their offspring have costs, including reduced fecundity in the future and fewer opportunities to mate in the present, as well as the obvious benefit of improved survival of the assisted progeny. A cost–benefit approach helps explain why females are more likely than males to provide parental care by focusing on the costs to males of helping broods of mixed paternity, costs that females rarely experience because they are usually genetically related to all the offspring in their broods.

2. Exceptions to the general rule occur, even to the extent of cases in which males take exclusive care of their young. These cases too have been analyzed by a cost–benefit approach. Thus, paternal care in fishes may often be favored because males caring for eggs laid in their territories can be more attractive to potential mates than are males without eggs to brood. In contrast, the costs of parental care to females may include large reductions in growth rate and consequent losses of fecundity.

3. An evolutionary approach to parental care yields the expectation that parents will be able to identify their own offspring when the risk of investing in genetic strangers is high. As expected, offspring recognition is widespread, particularly in colonial species in which opportunities for misdirected parental care are frequent. However, adults of many species do sometimes adopt nongenetic offspring, including specialist brood parasites, with consequent losses of fitness. Multiple hypotheses exist to account for these puzzling cases, including the possibility that highly discriminating host adults could lose fitness by sometimes mistakenly rejecting their own offspring.

4. Another Darwinian puzzle is the indifference shown by some animals to lethal aggression among their young offspring. Cases like these may be explained as part of a parental strategy to let the offspring themselves identify which individuals are most likely to survive, and therefore which youngsters will provide a payoff for continuing parental investment. The more general principle is that selection rarely favors completely even treatment of offspring because some youngsters are more likely than others to survive to reproduce.

Suggested Reading

Good books on parental care include *The Evolution of Parental Care* by Timothy Clutton-Brock [222] and *Mother Nature* by Sarah Hrdy [571]. A vast literature exists on brood parasites and their interactions with their hosts; for a superb review, see *Cuckoos, Cowbirds and Other Cheats* by Nick Davies [292]. Avian siblicide is the subject of an excellent article by Doug Mock and his co-workers [833] as well as a fine book by Mock written for a general audience [832].

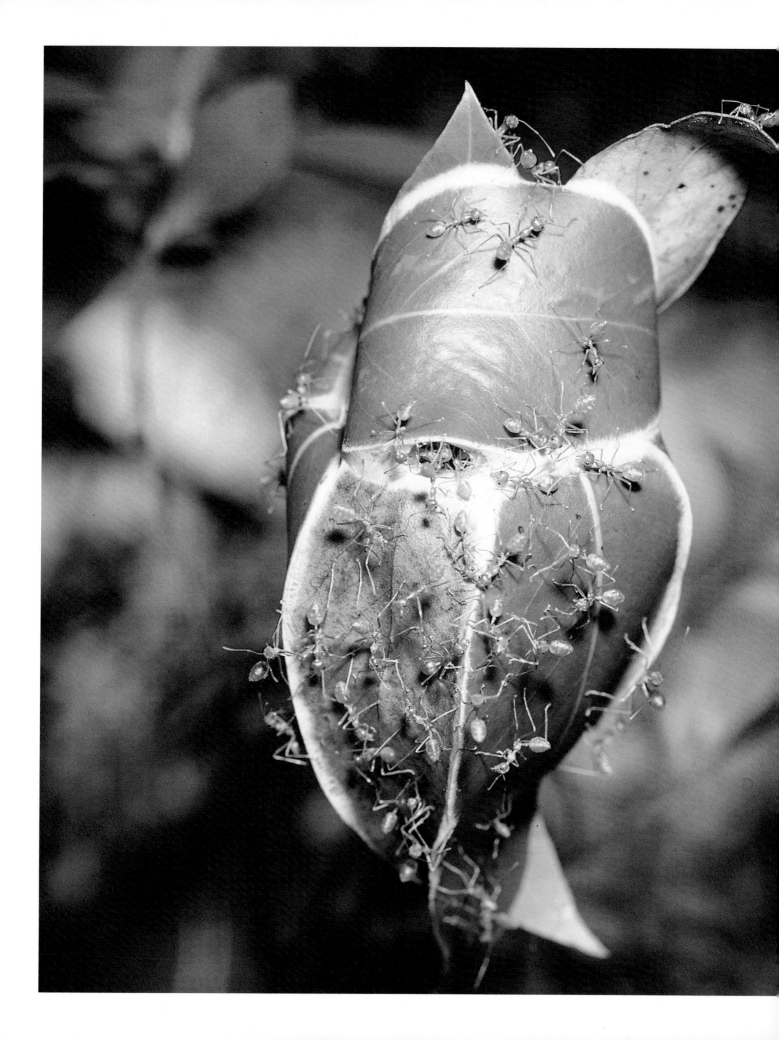

13 | *The Evolution of Social Behavior*

*D*on't reach for the Raid the next time you find a paper wasp colony under an eave on your house (see Figure 3.10). The wasps will not sting you, provided you approach the nest careful-ly—and I have noticed that most people do not need too much encour-agement to be cautious around wasps. But if you overcome your fear of these insects, I am confident you will be fascinated by their behavior. In this opinion I join the great evolutionary biologist W. D. Hamilton, whose work is featured prominently in this chapter. Hamilton had this to say about paper wasps [505]:

> Social wasps are among the least loved of insects … Yet, where sta-tistics will not alter a general impression, another approach might. Every schoolchild, perhaps as part of religious training, ought to sit watching a *Polistes* wasp nest for just an hour … I think few will be unaffected by what they see. It is a world human in its seeming motivations and activities far beyond all that seems reasonable to expect from an insect: constructive activity, duty, rebellion, mother care, violence, cheating, cowardice, unity in the face of a threat—all these are there.

The quasi-human melodramas that take place at a paper wasp nest involve a community of females both competing and cooperating as they

◀ *Weaver ants form superbly cooperative societies based on a sterile worker caste. One of the many products of their altruism is a magnificent leaf nest woven together with silk provided by the colony's larvae. Photograph by the author* rear the eggs and grubs that occupy the cells of the nest. If I were to tell you that only one of the several females at the nest might be the mother of all the eggs and grubs housed there, and that these youngsters were being regularly fed with food collected by females other than their mother, I hope you would be at least mildly surprised. Although parental care can evolve by natural selection when the benefits of the behavior exceed its costs, as we saw in the previous chapter, it is hard to imagine how an adult's fitness could be increased by behaving parentally toward someone else's youngsters. Yet helpers at nests are found not just in paper wasps, but in a host of other insects, as well as in some birds and mammals. These self-sacrificing individuals pose a wonderful Darwinian puzzle whose solution has engaged some of the best evolutionary biologists in the world, including W. D. Hamilton and Charles Darwin himself.

This chapter focuses on how altruism and other helpful acts of social organisms can be analyzed from an adaptationist perspective. But first we must ask a more basic question: Why do any animals live in groups instead of leading solitary lives?

The Costs and Benefits of Social Life

You may think that the reason so many animals live with others of their species is that social creatures are higher up the evolutionary scale, and so better adapted than animals that live solitary lives. You might hold this belief because you know that humans are highly social, and you would like to think that we, and perhaps a few other highly social species, represent the crowning achievements of the evolutionary process. But if you believed these things, you would be mistaken, because natural selection does not aim for preset endpoints (see Chapter 1). Instead, in each and every species, generation after generation, relatively social and relatively solitary types compete unconsciously with one another in ways that determine who leaves more offspring on average. In some species, the more social individuals have won out, but in a large majority, it is the solitary types that have consistently left more surviving descendants on average.

But how can living alone ever be superior to living together? Under some conditions, the cost–benefit equation for solitary life is better than that for a more social existence (Table 13.1). For example, in most social species, animals have to expend time and energy jockeying for social status. Those that do not occupy the top positions regularly have to signal their submissive state to their supe-

TABLE 13.1 *Some potential costs and benefits of social living*

Costs	Benefits
Greater conspicuousness of clumped individuals to predators	Defense against predators via the dilution effect or via mutual defense (see Chapter 6)
Greater transmission of disease and parasites among group members	Opportunities to receive assistance from others in dealing with pathogens
More competition for food among group members	Improved foraging via the information center effect (see Chapter 7)
Time and energy expended by subordinates in dealing with more dominant companions	Subordinates are granted permission to remain safely within the group
Greater male vulnerability to cuckoldry	Opportunity for some males to cuckold others
Greater female vulnerability to egg tossing, egg dumping, and other forms of reproductive interference by companions	Opportunity to toss the eggs of others, dump eggs in others' nests, and interfere with the reproduction of competitors

(A) Helpers

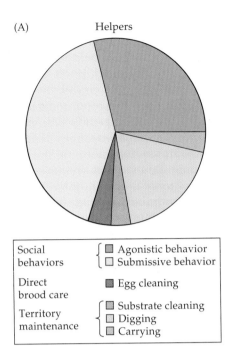

Social behaviors	{	☐ Agonistic behavior ☐ Submissive behavior
Direct brood care		■ Egg cleaning
Territory maintenance	{	☐ Substrate cleaning ☐ Digging ☐ Carrying

(B)

Figure 13.1 The energy budget of subordinate, nonbreeding "helpers" that associate with breeding pairs in the cichlid fish *Neolamprologus pulcher*. (A) The largest proportion of the subordinate fish's energy is expended on the performance of submissive behaviors—specifically, a tail-quivering display. Most of the remainder of the subordinate fish's energy budget is spent attacking intruders and removing sand and debris from the area defended by the breeding pair and their helpers. (B) A subordinate helper quivers his tail as the dominant fish approaches from behind. A, after Taborsky and Grantner [1178]; B, photograph by Michael Taborsky.

riors if they are to be permitted to remain in the group. All the kowtowing can take up a major share of a social subordinate's life (Figure 13.1) [1178]. In fact, even small social groups are rife with subtle and not-so-subtle competition, as we shall document in the pages ahead.

Discussion Question

13.1 In an African cichlid fish known for its cooperative breeding, helpers live with a breeding pair within the group's communal territory. In Figure 13.2, the locations occupied by five of these helpers are shown over a 3-day period; on day 3, the largest helper (1) was removed by researchers but its territory outline remains in the figure [1271]. How would you interpret these data in light of the possibility that helpers are competing with one another while helping a breeding pair? What benefit might helpers derive from achieving dominance over other helpers?

Figure 13.2 Effect of removal of the top-ranked subordinate helper in a cooperatively breeding group of cichlid fish (*Neolamprologus brichardi*). The removal occurred on day 3 in an aquarium that housed a breeding pair and five helpers. (The different colored dots represent the different fish and show where these individuals were seen on a given day.) After Werner et al. [1271].

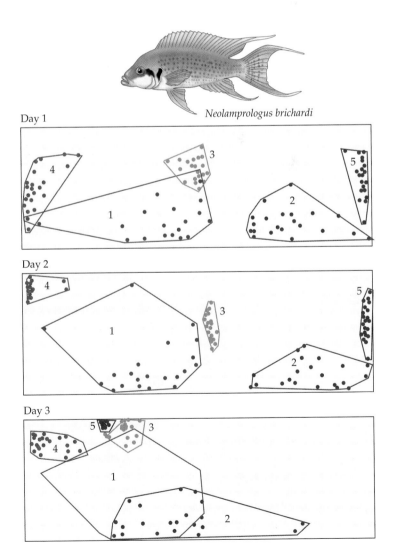

Neolamprologus brichardi

Day 1

Day 2

Day 3

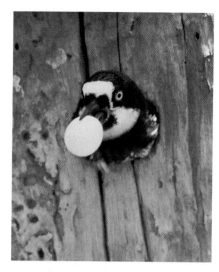

Figure 13.3   Reproductive interference in a social animal. One member of a breeding group of acorn woodpeckers removes an egg from a nest that she shares with several companions. Photograph courtesy of Walt Koenig.

Social groups also offer opportunities for reproductive interference. Breeding males that live in close association with more attractive rivals may lose their mates to these individuals. Breeding females may wind up incubating eggs dumped in their nest by brood parasites of their own species [566, 838]. Such costly reproductive penalties come with social living in the acorn woodpecker, a bird that forms breeding groups containing as many as three females and four males. The several females all lay their eggs in the same tree hole nest, perhaps because any female that tried to keep a nest to herself would have all her eggs destroyed by her vindictive companions [655]. Even when several females agree to use the same nest, the first eggs laid are almost always removed by another female member of the group (Figure 13.3) [860]. Eventually these "cooperatively breeding" females all lay eggs on the same day, at which time they finally stop tossing one another's eggs out and incubate the clutch. By this time, however, more than a third of the eggs laid by the woodpeckers may have been destroyed. Having your eggs tossed is an unadulterated cost of social living for females of this species.

In addition to these direct reproductive costs, sociality has two other potential disadvantages. The first is heightened competition for food, which occurs in animals as different as colonial fieldfares (Figure 13.4) [1302] and prides of lions, whose females are often pushed from their kills by hungry males [1066]. The second is increased vulnerability to parasites and pathogens, which plague social species of all sorts [16] (but see [1033]). The fact that some social animals have evolved special responses to these enemies may enable those animals to reduce the damage they cause, but cannot totally eliminate the burden they impose. Thus, honey bees warm their hives in response to an infestation by a fungal pathogen, which apparently helps kill the heat-sensitive fungus, but at the price of time and energy expended by the heat-producing workers [1149]. Similarly, termites can reduce the lethal effect of a different fungal invader in their nest mounds because unexposed colony members can acquire some immunity to the fungus simply by associating with others of their group that have already developed an immune response to the pathogen [1210]. Even so, the existence of special responses to fungal infection speaks to the high probability that a colony will become infected, perhaps because it offers such a large target. Moreover, the antifungal mechanisms themselves do not come for free, but require physiological expenditures on the part of the termites.

If social living carries a heightened risk of infection by parasites or pathogens, then the larger the group, the greater the risk. This prediction holds for cliff swallows, which pack their nests side by side in colonies composed of anywhere from a handful of birds to several thousand pairs. The more swallows nesting together, the greater the chance that at least one bird will be infested with swallow bugs, which can then readily spread from one nest to another. Charles

Figure 13.4   Competition for food is a cost of sociality in the fieldfare, a songbird that nests in loose colonies in woodlands. The larger the colony, the lower the survival rate of nestlings, due to increased juvenile mortality caused largely by starvation. After Wiklund and Andersson [1302].

Fieldfare

Figure 13.5   Effect of parasites on cliff swallow nestlings. The much larger nestling on the right came from an insecticide-treated nest; the stunted baby of the same age on the left occupied a nest infested with swallow bugs. From Brown and Brown [155].

and Mary Brown found that nestlings in large colonies were heavily assaulted by blood-sucking, growth-stunting swallow bugs [155]. The Browns demonstrated that the bugs were guilty of harming the nestlings by fumigating a sample of nests in an infested colony while leaving other control nests untreated. The nestlings doused with insecticide weighed much more, and were more likely to survive, than those plagued by parasites (Figure 13.5).

The parasites and fungi that make life miserable for swallows and other social creatures demonstrate that if sociality is to evolve, the assorted costs of living together must be outweighed by compensatory benefits. Cliff swallows may join others to take advantage of the improved foraging that comes from following companions to good feeding sites (see Chapter 7) [156, 473], while other animals, such as egg-brooding male emperor penguins, save thermal energy by huddling shoulder-to-shoulder during the brutal Antarctic winter [25]. Still others, such as lionesses, join forces to fend off enemies of their own species, including infanticidal males [1152].

The most widespread fitness benefit for social animals, however, probably is improved protection against predators (Figure 13.6) [16]. Many studies have

Figure 13.6   Social living with defensive benefits? The members of this dense school of small (5 cm long) striped catfish living on a coral reef near Sulawesi have almost certainly joined forces to improve their chances of survival. Schooling in this and other species can enhance the survival of individual fish either by intimidating some predators through the collective size of the massed school or by amplifying their defenses, if the fish are protected by spines or chemical repellents. Photograph by Roger Steene.

Figure 13.7   Mutual defense in a society of bluegills. Each colonial male defends a territory bordered by the nest sites of other males, while bass (above), bullhead catfish (left), snails, and pumpkinseed sunfish (right foreground) roam the colony in search of eggs. Drawing courtesy of Mart Gross.

shown that animals in groups gain by diluting the risk of being captured, or by spotting danger sooner, or by ganging up on their enemies (see Chapter 6). Thus, in the maomao, a reef fish with a wonderful Latin name (*Abudefduf abdominalis*), individual males in large nesting groups chase other egg-eating fish only about one-fourth as often as males in small nesting aggregations. And when a nest-defending male is removed from a small group, his eggs are attacked by a predator sooner than the eggs of a male removed from a large group, indicating that maomao males definitely derive mutual antipredator benefits by nesting together [1220].

Males in nesting colonies of the bluegill sunfish also cooperate in driving egg-eating bullhead catfish away from their nests at the bottom of a freshwater lake (Figure 13.7) [485]. If the social behavior of the bluegill has indeed evolved in response to predation, then closely related species that nest alone should suffer less from predation. As predicted, the solitary pumpkinseed sunfish, a member of the same genus as the bluegill, has powerful biting jaws and so can repel egg-eating enemies on its own, whereas bluegills have small, delicate mouths good only for inhaling small, soft-bodied insect larvae [485]. Pumpkinseed sunfish are in no way inferior to or less well adapted than bluegills because they are solitary; they simply gain less through social living, which makes solitary nesting the adaptive tactic for them.

The Evolution of Helpful Behavior

Animals that live together have the potential to assist one another, and they often do, as maomao and bluegill males demonstrate. Until the mid-1960s, biologists took helpful behavior of this sort more or less for granted because they assumed that animals should help one another for the benefit of the species as a whole. But when George C. Williams pointed out the defects of this assumption (see p. 19), helpful actions, especially altruistic ones, suddenly became more interesting to evolutionary biologists.

TABLE 13.2 *The direct reproductive success of individuals that engage in different kinds of social interactions*

Type of interaction	Effect on direct reproductive success of	
	Social donor	Social recipient
Mutualism (Cooperation)	+	+
Reciprocity	+ (delayed)	+
Altruism	–	+
Selfish behavior	+	–
Spiteful behavior[a]	–	–

[a]You should not be surprised that spiteful behavior is almost never observed in nature; you should be surprised that altruism is not uncommon despite the loss of reproductive success experienced by altruists.

There are a number of ways in which animals can behave toward one another in social interactions, which have different payoffs for the donor and the recipient of the behavior (Table 13.2, Figure 13.8). Helping sometimes generates immediate returns for both parties, in which case they are said to be engaged in a **mutualism.** When one lioness drives a wildebeest into a lethal ambush set by her fellow pride members [1145], the cooperative driver will usually get some meat, even if she did not personally pull the antelope down and strangle it herself. Likewise, if several male bluegills succeed in fending off a bullhead catfish that has entered their part of the nesting colony, the eggs in all the males' nests are more likely to survive to hatch. When both helper and recipient enjoy reproductive gains from their interaction, their mutualism, or cooperation, requires no special evolutionary explanation.

This is not to say that mutualism is uninteresting. Consider the coalitions of male lions that form to oust rival males living with a pride of females. When cooperating males are successful, they may gain sexual access to a large group of receptive females. When Craig Packer and his associates analyzed lion coali-

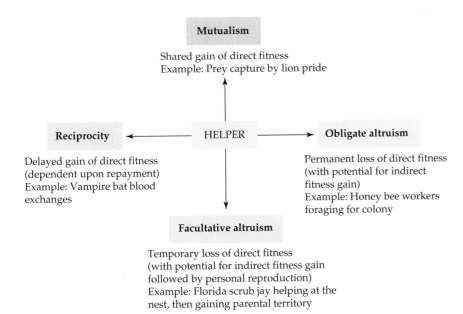

Mutualism
Shared gain of direct fitness
Example: Prey capture by lion pride

Reciprocity
Delayed gain of direct fitness
(dependent upon repayment)
Example: Vampire bat blood
exchanges

HELPER

Obligate altruism
Permanent loss of direct fitness
(with potential for indirect
fitness gain)
Example: Honey bee workers
foraging for colony

Facultative altruism
Temporary loss of direct fitness
(with potential for indirect fitness gain
followed by personal reproduction)
Example: Florida scrub jay helping at the
nest, then gaining parental territory

Figure 13.8   The different categories of helping behavior. Cooperative helpers can be placed into four groups based on the fitness consequences of their actions.

Figure 13.9 Cooperation among competitors. Yearling male lazuli buntings range in color from dull brownish to bright blue and orange. (Their plumage scores range from less than 16 to more than 32.) Bright yearling males permit dull males, but not males of intermediate brightness, to settle on territories neighboring their own. As a result, brownish males often pair off with females in their first year whereas yearling males of intermediate plumage typically remain unpaired. After Greene et al. [474]; photographs courtesy of Erick Greene.

tions, they found that partnerships of two or three males shared access to the females fairly evenly [918]. Even so, some males in these groups do not do as well as others. Why do the disadvantaged males tolerate their situation? Probably because they could do no better alone. If they went solo, their chance of acquiring and defending a pride would be next to zero because one male has little chance against two or three rivals. Thus, some males may be forced to cooperate with domineering companions if they are to have any chance of mating [653].

Likewise, subordinate yearling male lazuli buntings, which have dull brown plumage, engage in an interesting mutualism with brightly colored, dominant yearling males (Figure 13.9). The bright males aggressively drive other males with bright or intermediate plumage away from top-quality territories with good shrub cover, but they tolerate the presence of dull-plumaged males. These low-ranking individuals are permitted to claim territories in good habitat right next to their brightly colored companions. One hypothesis for the bright males' surprising acceptance of their brownish neighbors is that bright males are unlikely to lose paternity to such neighbors, which lack the attributes that make females eager to engage in extra-pair copulations. If this hypothesis is true, then brightly colored males should not be cuckolded as often as duller ones. In keeping with this prediction, in the nests sampled by Erick Greene and his co-workers, bright males lost paternity on average at half the rate of their dull neighbors. The more frequently victimized duller buntings usually cared for between one and two extra-pair young, which were probably the genetic offspring of their brightly plumaged next-door neighbors [474].

Given the costs of trying to raise a family next to dominant male lazuli buntings, why do dull males accept the offer to live near them? Perhaps because the subordinate buntings at least get to hold high-quality territories, which enables them to acquire a social mate more often than males of intermediate plumage brightness. Those in-between males are often pushed by dominant rivals into habitat so poor that no female will join them. Although drab yearling males may often rear the chicks of other males, they also produce some of their own on occasion, achieving a modest amount of reproductive success, unlike most yearlings of intermediate plumage and in-between social status, which have to wait another full year to breed [474].

Discussion Question

13.2   Given the differences in reproductive success for the three categories of male lazuli buntings, how can we account for the evolutionary persistence of males with dull and, especially, intermediate plumage?

The fact that both dull and bright yearling neighbors gain some fitness from their interaction means that their social arrangement constitutes a mutualism. But what about the male coalitions of the long-tailed manakin studied by David McDonald, in which only one of two cooperative males appears to reproduce? In this bird, males form pairs and sing loud duets over and over to attract females to a display court [401, 808]. Visiting females land on the pair's display perch, often a horizontal section of liana that lies a foot or so above the ground. In response, the two males dart in and land close to the prospective mate before performing an astonishing cartwheel display (Figure 13.10). After a series of these moves, the males may perform the "butterfly flight," in which they flutter slowly back and forth in front of the female, displaying their beautiful plumage and coordinated flight capacity. Should a female visitor start jumping excitedly on the perch in response to these displays, one member of the duo discreetly leaves, while the remaining male stays to copulate with her. The female then flies off, after which the mated male calls for his display partner, who hurries back to resume his duties.

By marking the males at display perches, McDonald and his manakin watchers found that each site has only one mating male. This alpha manakin may

Figure 13.10   Cooperative courtship of the long-tailed manakin. The two males are in the cartwheeling portion of their dual display to a female, who is perched on the vine to the right.

Figure 13.11   Cooperation with an eventual payoff. After the death of his alpha male partner, the beta male long-tailed manakin (now an alpha) copulates about as frequently as his predecessor did, presumably because the females attracted to the duo in the past continue to visit the display arena when receptive. After McDonald and Potts [808].

have several display companions, but not one of them gets to reproduce, not even the alpha's favorite colleague, a beta male, who in turn is dominant to any other part-time cooperators [808]. How can it be adaptive for the subordinates to work so hard on behalf of the sexually monopolistic alpha male? By patiently following males year after year, McDonald established that whenever an alpha male disappeared, the beta male took over, after which a lower-ranking individual moved up to become the new alpha's main nonbreeding display partner. Therefore, by cooperating with the alpha male, a beta individual establishes his claim to be next in line, keeping other (probably younger) birds at bay for years. When a beta male becomes an alpha, he usually gets to mate with many of the same females that copulated with the previous alpha (Figure 13.11) [808]. Thus, beta males form a mutualism with their exclusionary partners because this is the only way to join a queue to become a reproducing alpha male—eventually.

Discussion Question

13.3   In several ant species, two or more unrelated females may join forces to found a colony after they have mated. The females may cooperate in digging the nest and producing the first generation of workers, but then they start fighting until only one is left alive [93]. How can it pay to join such an association? What prediction can you make about the survival rates and average productivity of colonies founded by a single female? Under what conditions would it be accurate to call this social system a mutualism? Develop at least one cost–benefit hypothesis to account for the timing of the switch from cooperation to aggressive behavior. If the behavior of the two queens is the product of natural selection, not group selection, what prediction can you make about the interactions between them during the colony establishment phase prior to the fight-to-the-death phase?

The Reciprocity Hypothesis

The study of long-tailed manakins shows us that some superficially self-sacrificing actions actually advance the reproductive chances of helpful individuals. Another possible case of this sort involves the meerkat, a small African mammal that forages in groups. From time to time, one meerkat will stop digging for insects in the soil and climb a tree or a termite mound to look around for approaching predators (Figure 13.12) [226]. Should a goshawk come swooping in, the elevated sentinel is usually the first to give an alarm, which sends all the still-foraging meerkats dashing for cover. One explanation for this behavior is that sentinels help others at personal cost now because they will be repaid later by their teammates when they take their turns at being lookouts. Bob Trivers called this kind of social relationship "reciprocal altruism" (it is also known as **reciprocity**) because helped individuals eventually return the favors they receive [1211]. If the initial cost of helping is modest, but the benefit from receiving the returned favor is great, then selection can favor making the initial gesture. Imagine, for example, that a meerkat lookout has a 2 percent chance of being killed for every 100 hours spent scanning for enemies. But imagine that for every hour he spends on his perch, another companion will pay him back. If having others keep an eye open for danger for 100 hours improves the helpful lookout's chance of survival by any amount more than 2 percent, then the benefit is greater than the cost, and reciprocity can spread through a population.

However, consider an alternative explanation for sentinel behavior. Perhaps the lookouts are sated and do not need to forage for food, so they climb a tree in order to better spot danger to themselves. Rather than offering costly assistance to others in their band, the

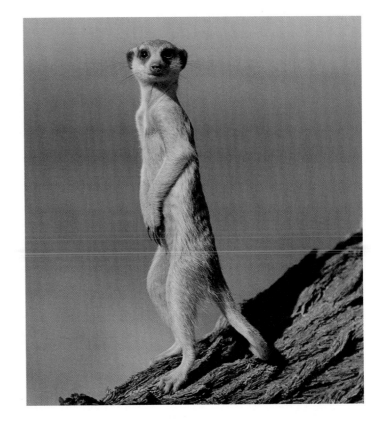

Figure 13.12 **A meerkat sentinel** on the alert for approaching predators. Photograph by Nigel J. Dennis.

"sentinels" could be securing personal fitness benefits, especially if an approaching goshawk is likely to chase one of the lookout's fleeing companions rather than the alert sentinel. (Note that this argument requires that the sated meerkats be safer on a lookout perch than in a burrow. Moreover, the sentinel must be less likely to be attacked if its companions are running for cover than if they are not. Finally, this hypothesis also requires that the signaler's companions gain more by dashing for a burrow than by remaining frozen in place in an effort to avoid detection by the onrushing predator.)

How can we test the reciprocity hypothesis against the personal safety alternative? The reciprocity hypothesis predicts that meerkats should follow a regular rotation of sentinel duty and that sentinels should run some risk of predation. However, in reality, sentinel duty is established haphazardly, and lookouts are usually closer to an escape burrow than are their fellows, suggesting that lookouts do not put themselves in special danger. The personal safety hypothesis also receives support from the finding that solitary meerkats spend about the same proportion of each day in sentinel behavior as do the members of a band. Moreover, when meerkats are given supplemental food, which reduces the cost of taking time out to look around for predators, they increase the amount of time spent on a lookout perch. Thus, what initially appears to be a rotation of lookouts may actually be the product of individuals spending as much time as possible during the day in a relatively safe position [226].

This is not to say that reciprocity is absent from nature [915, 1304]. In many primates, for example, individuals spend considerable time carefully grooming the fur of a companion (Figure 13.13). The groomer helps the groomee by removing parasites and debris, but gains no immediate benefit from its actions, cashing in only when the animal it helped returns the favor. In bands of baboons, pairs of females do take turns grooming and being groomed, as predicted by the reciprocity hypothesis [1109].

Figure 13.13   Reciprocity in a social primate. A vervet monkey grooms a companion. The groomee will return the favor at a later date. Photograph by Dorothy Cheney.

Figure 13.14   Experimental demonstration of reciprocity in cotton-top tamarins. (A) A double-compartmented cage with a pull tool that one cotton-top subject (the actor, on the right) could use to drag food down the tray toward its companion (on the left, reaching for a food reward). (B) The proportion of trials during which a cotton-top reciprocated by pulling food to a helpful companion (which had been trained to always pull a food item within reach of the other monkey) and an unhelpful companion (which had been trained never to pull a food item within reach of another individual). A, photograph courtesy of Marc Hauser; B, after Hauser et al. [522].

The capacity for reciprocity also appears to exist in another primate, the cotton-top tamarin, as Marc Hauser and his fellow researchers demonstrated experimentally in the following way [522]. They constructed a special cage with separate compartments for each of two monkeys (Figure 13.14), one of whom had access to a pull bar that could be used to drag food to within reach of either the puller or his companion (depending on the placement of the food item by a researcher). The question was, would a cotton-top repay a puller that used the tool to deliver food to it? Hauser and company conditioned one monkey to always pull the food within reach of a companion. This invariant puller was then paired with a genetically unrelated individual, which we shall call the actor. The actor and the trained altruist were given opportunities to take turns pulling food for each other over 24 test trials. The actor repaid the trained-to-pull-every-time companion somewhere between a third and a half of the time, much more so than when the actor was paired with a "defector" monkey that had been trained *never* to use the pull tool to deliver food to a cagemate. In other words,

(A)

(B)

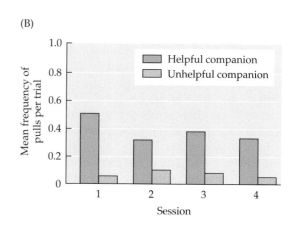

when paired with an apparently helpful companion, cotton-tops reciprocated, but when given an opportunity to assist an unhelpful companion, the tamarins withheld their assistance.

Although at least some primates appear to have the capacity for reciprocity, the behavior is not particularly common among animals generally, perhaps because a population composed of reciprocal altruists would be vulnerable to invasion by individuals who accept help but then forget about the payback. "Defectors" reduce the fitness of "helpers" in such a system, which ought to make reciprocity less likely to evolve. The problem can be illustrated with a game theory model called the **prisoner's dilemma** (Figure 13.15), which is based on a human situation (see also Figure 6.33). Imagine that a crime has been committed by two persons who agreed not to squeal on each other if caught. The police have brought them in for interrogation and have put them in separate rooms. The cops have enough evidence to convict them both on lesser charges, but need to have the criminals implicate each other in order to jail them for a more serious crime. They therefore offer each suspect freedom if he will squeal on his pal. If suspect A accepts the offer ("Defect") while B maintains their agreed-upon story ("Cooperate"), A gets his freedom (the maximum reward) while B gets hit with the maximum punishment—say, 10 years in prison (the "sucker's payoff"). If together they maintain their agreement ("Cooperate + Cooperate"), then the police will have to settle for conviction of both on the lesser charge, leading to, say, a 1-year prison term for each suspect. And if each one fingers the other, the police will use this evidence against both and renege on their offer of freedom for the snitch, so that both A and B will be punished quite severely with, say, a 5-year prison sentence each.

In a setting in which the payoffs for the various responses are ranked "Defect while other player cooperates" > "Both cooperate" > "Both defect" > "Cooperate while other player defects," the optimal response for suspect A is always to defect, never to cooperate. Under these circumstances, if suspect B maintains their joint innocence, A gets a payoff that exceeds the reward he could achieve by cooperating with a cooperative B; if suspect B squeals on A, defection is still the superior tactic for A, because he suffers less punishment when both players defect than if he cooperates while his companion squeals on him. By the same token, suspect B will always come out ahead, on average, if he defects and points the finger at his buddy.

This model predicts, therefore, that reciprocal cooperation should never evolve. How, then, can we account for the cases of reciprocity that have been observed in nature? One answer comes from examining scenarios in which two players interact repeatedly, not just once. Robert Axelrod and W. D. Hamilton have shown that when this condition applies, individuals that use the simple

Figure 13.15   The prisoner's dilemma. The diagram lays out the payoffs for player A that are associated with cooperating or not cooperating with player B. Defection is the adaptive choice for player A given the conditions specified here (if the two individuals will interact only once).

decision rule "Do unto individual X as he did unto you the last time you met" can reap greater overall gains than cheaters who accept assistance but do not return the favor [46]. When multiple interactions are possible, the rewards for back-and-forth cooperation add up, exceeding the short-term payoff from a single defection. In fact, the potential accumulation of rewards can even favor individuals who "forgive" a fellow player for an occasional defection because that tactic can encourage maintenance of a long-term relationship with its additional payoffs [1259].

Vampire bats appear to meet the required conditions for adaptive multiple-play reciprocity. These animals must find scarce vertebrate victims from which to draw the blood meals that are their only food. After an evening of foraging, the bats return to a roost where individuals known to one another regularly assemble. A bat that has had success on a given evening can collect a large amount of blood, so much that it can afford to regurgitate a life-sustaining amount to a companion who has had a run of bad luck. Under these circumstances, the cost of the gift to the donor is modest, but the potential benefit to the recipient is high, since vampire bats die if they fail to get food three nights in a row. Thus, a cooperative, blood-transferring vampire bat is really buying insurance against starvation down the road. Individuals that establish durable "give and take" relationships with one another are better off over the long haul than those cheaters that accept one blood gift but then renege on repayment, thereby ending a potentially durable cooperative arrangement that could involve many more meal exchanges [1304].

Altruism and Indirect Selection

Reciprocity is really a special kind of mutualism in which the helpful individual endures a short-term loss until its help is reciprocated, at which time it earns a net increase in fitness. In contrast, there are some cases in which a donor really does permanently lose opportunities to produce offspring of its own as a result of helping another individual. In evolutionary biology, this kind of helpful behavior is called **altruism** (see Table 13.2). Altruistic actions, if they exist, are an especially exciting Darwinian puzzle for adaptationists because they violate the "rule" that traits cannot spread over evolutionary time if they lower an individual's reproductive success relative to that of other individuals (see Chapter 1).

In order to explain how altruism could evolve, W. D. Hamilton developed a special explanation that did not rest on for-the-good-of-the-group arguments [502]. Instead, Hamilton's theory was based on the premise that individuals reproduce with the unconscious goal of propagating their alleles more successfully than other individuals. Personal reproduction contributes to this ultimate goal in a direct fashion. But helping genetically similar individuals—that is, one's relatives—survive to reproduce can provide an indirect route to the very same end.

To understand why, the concept of the **coefficient of relatedness** comes in handy. This term refers to the probability that an allele in one individual is present in another because both individuals have inherited it from a recent common ancestor. Imagine, for example, that a parent has the genotype A/a, and that a is a rare form of the A gene. Any offspring of this parent will have a 50 percent chance of inheriting the a allele because any egg or sperm that the parent donates to the production of an offspring has one chance in two of bearing the a allele. The coefficient of relatedness (r) between parent and offspring is therefore $\frac{1}{2}$, or 0.5.

The coefficient of relatedness varies for different categories of relatives. For example, an uncle and his sister's son have one chance in four of sharing an allele by descent because the man and his sister have one chance in two of having this allele in common, and the sister has one chance in two of passing that allele on to any given offspring. Therefore, the coefficient of relatedness for an uncle and his nephew is $\frac{1}{2} \times \frac{1}{2} = \frac{1}{4}$, or 0.25. For two cousins, the r value falls to $\frac{1}{8}$, or 0.125. In contrast, the coefficient of relatedness between an individual and another, unrelated individual is 0.

With knowledge of the coefficient of relatedness between altruists and the individuals they help, we can determine the fate of a rare "altruistic" allele that is in competition with a common "selfish" allele. The key question is whether the altruistic allele becomes more abundant if its carriers forgo reproduction and instead help relatives reproduce. Imagine that an animal could potentially have one offspring of its own, or alternatively invest its efforts in the offspring of its siblings, thereby helping three nephews or nieces survive that would have otherwise died. A parent shares half its genes with an offspring; the same individual shares one-fourth of its genes with each nephew or niece. Therefore, in this example, personal reproduction yields $r \times 1 = 0.5 \times 1 = 0.5$ genetic units contributed directly to the next generation, whereas altruism directed at relatives yields $r \times 3 = 0.25 \times 3 = 0.75$ genetic units passed on indirectly in the bodies of relatives. In this example, the altruistic tactic is adaptive because it results in more shared alleles being transmitted to the next generation.

Discussion Question

13.4   If an altruistic act increases the genetic success of the altruist, then in what sense is this kind of altruism actually selfish? In everyday English, words like "altruism" and "selfishness" carry with them an implication about the motivation and intentions of the helpful individual. Why might everyday usage of these words get us into trouble when we hear them in an evolutionary context? Consider the proximate–ultimate distinction here. If an individual inadvertently helped another at reproductive cost to itself, could the behavior be called altruistic under the evolutionary definition?

Another way of looking at this matter is to compare the genetic consequences for individuals who aid others at random versus those who direct their aid to close relatives. If aid is delivered randomly, then no one form of a gene is likely to benefit more than any other. But if close relatives aid one another selectively, then any rare family alleles they possess may survive better, causing those alleles to increase in frequency compared with other forms of the gene in the population at large. When one thinks in these terms, it becomes clear that a kind of natural selection can occur when genetically different individuals differ in their effects on the reproductive success of close relatives. Jerry Brown calls this form of selection **indirect selection**, which he contrasts with **direct selection** for traits that promote success in personal reproduction (Figure 13.16A) [158].

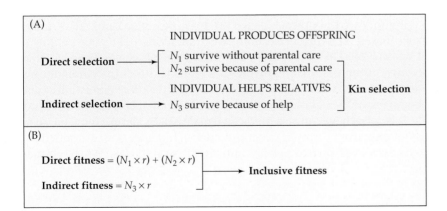

Figure 13.16   The components of selection and fitness. (A) Direct selection acts on variation in individual reproductive success. Indirect selection acts on variation in the effects individuals have on their relatives' reproductive success. (B) Direct fitness is measured in terms of personal reproductive output; indirect fitness is measured in terms of genetic gains derived by helping relatives reproduce. Inclusive fitness is the sum of the two measures and represents the total genetic contribution of an individual to the next generation. After Brown [158].

A brief digression is necessary here to deal with yet another term, **kin selection**, which was originally defined by John Maynard Smith to embrace the evolutionary effects of *both* parental aid given to descendant kin (offspring) *and* altruism directed to **nondescendant kin** (relatives other than offspring). Biologists have long recognized that parents can improve the survival chances of their offspring, and that parental care spreads when the resulting increase in the survival of the aided offspring more than compensates a parent for the loss of opportunities to produce additional offspring in the future (see Chapter 12). In genetic terms, parents can gain via parental investment because they share 50 percent of their genes with each offspring. By the same token, however, individuals can promote the survival of certain of their genes by helping relatives other than offspring. Altruism can be favored by the component of kin selection that Brown calls indirect selection, and the use of this term keeps the focus clearly on the distinction between parental effects on offspring and an aid-giver's effects on nondescendant kin [158]. The term *kin selection* is, however, widely used by evolutionary biologists as a synonym for *indirect selection*, and readers should be aware that when they see the term, odds are that it is being used to refer to selection for altruism supplied to relatives other than offspring.

Indirect Selection and the Alarm Call of Belding's Ground Squirrel

Having laid the necessary theoretical groundwork, let's use it in analyzing an alarm-calling behavior that might be an altruistic act, but keeping in mind that what looks like altruism might actually be a case of mutualism or reciprocity with direct fitness benefits for the helpful individual [480]. You will recall that African meerkats stand on guard to warn of approaching predators, but that their behavior appears to improve their own survival more than that of the other meerkats they alert to danger. Does the same story apply to Belding's ground squirrel? This North American rodent produces a staccato alarm whistle (Figure 13.17) when a coyote or badger approaches, which sends other nearby ground squirrels rushing for safety.

Paul Sherman collected the evidence needed to evaluate alternative hypotheses on the adaptive value of the ground squirrel's alarm call [1094]. He found that alarm-calling squirrels are tracked down and killed by weasels, badgers, and coyotes at a higher rate than noncallers, a discovery that eliminates the possibility that alarm calling confers direct fitness on the signaler by confusing or deterring predators. Moreover, the possibility that alarm calling evolved as a form of reciprocity is also unlikely, because the probability that an individual will give an alarm call is not correlated with familiarity or length of association between the caller and the animals that benefit from its signal [1095]. Remember that reciprocity is more likely to evolve when reciprocators form long-term associations.

Sherman's observation that adult female squirrels with relatives nearby are more than twice as likely as males to give costly alarm calls is consistent with both a parental care hypothesis (based on direct selection) and an altruism hypothesis (based on indirect selection). You may recall that female Belding's ground squirrels tend to settle near their mothers, whereas males disperse some distance away from their natal burrow (see Figure 8.10). If the parental care hypothesis is correct, we would expect females to give more alarm calls than males because only female squirrels live near their offspring. If the altruism hypothesis is correct, we can also predict a female bias in alarm calling because females not only live near their genetic offspring, but are also surrounded by other female relatives, such as sisters, aunts, and female cousins. When self-sac-

Figure 13.17   A Belding's ground squirrel gives an alarm call after spotting a terrestrial predator. Photograph by George Lepp, courtesy of Paul Sherman.

rificing females warn their nondescendant kin, they could be compensated for the personal risks they take by an increased probability that kin *other than offspring* will survive to pass on shared genes, resulting in indirect fitness gains for the altruists. Females with offspring living nearby as well as females with *only* nondescendant kin as neighbors are in fact more likely to call upon detecting a predator than are females who lack relatives in their neighborhood. These findings suggest that both direct and indirect selection contribute to the maintenance of alarm calling behavior in this species [1094].

The Concept of Inclusive Fitness

Because fitness gained through personal reproduction (**direct fitness**) and fitness achieved by helping nondescendant kin survive (**indirect fitness**) can both be expressed in identical genetic units, we can sum up an individual's total contribution of genes to the next generation, creating a quantitative measure that can be called **inclusive fitness** (Figure 13.16B). Note that an individual's inclusive fitness is not calculated by adding up that animal's genetic representation in its offspring plus that in all of its other relatives. Instead, what counts is an individual's own effects on gene propagation (1) directly in the bodies of its surviving offspring *that owe their existence to the parent's actions*, not to the efforts of others, and (2) indirectly via nondescendant kin *that would not have existed except for the individual's assistance*. For example, if the animal we mentioned earlier successfully reared one of its own offspring and also adopted three of its sibling's progeny, then its direct fitness would be $1 \times 0.5 = 0.5$, and its indirect fitness would be $3 \times 0.25 = 0.75$; the union of these two figures provides a measure of the animal's inclusive fitness ($0.5 + 0.75 = 1.25$).

The concept of inclusive fitness, however, is not used to secure absolute measures of the lifetime genetic contribution of individuals, but rather to help us compare the evolutionary (genetic) consequences of two alternative hereditary traits [990]. In other words, inclusive fitness becomes important as a means to determine the relative genetic success of two or more competing behavioral strategies. If, for example, we wish to know whether an altruistic strategy is superior to one that promotes personal reproduction, we can compare the inclusive fitness consequences of the two traits. In order for an altruistic trait to be adaptive, the inclusive fitness of altruistic individuals has to be greater than it would have been if those individuals had tried to reproduce personally. In other words, a rare allele "for" altruism will become more common only if the indirect fitness gained by the altruist is greater than the direct fitness it loses as a result of its self-sacrificing behavior. This statement is often presented as **Hamilton's rule**: a gene for altruism will spread only if $r_b B > r_c C$. Spelling this out, we calculate the indirect fitness gained by multiplying the extra number of relatives that exist thanks to the altruist's actions (B) by the mean coefficient of relatedness between the altruist and those extra individuals (r_b); we calculate the direct fitness lost by multiplying the number of offspring not produced by the altruist (C) by the coefficient of relatedness between parent and offspring (r_c). For example, if the genetic cost of an altruistic act were the loss of one offspring ($1 \times r_c = 1 \times 0.5 = 0.5$ genetic units), but that altruistic act led to the survival of three nephews that would have otherwise perished ($3 \times r_b = 3 \times 0.25 = 0.75$ genetic units), the altruist would experience a net gain in inclusive fitness, thereby increasing the frequency of any distinctive allele associated with its altruistic behavior.

13.5 Let's say that in calculating the inclusive fitness of a male in a coalition of lions, you measured his direct fitness by multiplying by 0.5 the number of offspring produced by the male, and then added as his indirect fitness the total number of all offspring produced by the other members of his coalition times the mean value of *r* between those offspring and the male in question. Your calculation of his inclusive fitness would be challenged on what grounds?

Inclusive Fitness and the Pied Kingfisher

The value of Hamilton's rule can be illustrated by Uli Reyer's study of the pied kingfisher [1013]. These attractive African birds nest colonially in tunnels in banks by large lakes. Some year-old males are unable to find a mate, and instead become *primary helpers* that bring fish to their mother and her nestlings while attacking predatory snakes and mongooses that threaten the nest. Are these males propagating their genes as effectively as possible by helping to raise their siblings? They do have other options: they could help unrelated nesting pairs in the manner of *secondary helpers*, or they could simply sit out the breeding season, waiting for next year in the manner of *delayers*.

To learn why primary helpers help, we need to know the costs and benefits of their actions. Primary helpers work harder than delayers and the more laid-back secondary helpers (Figure 13.18). The greater sacrifices of primary helpers translate into a lower probability of their surviving to return to the breeding grounds the next year (just 54 percent return) compared with secondary helpers (74 percent return) or delayers (70 percent return). Furthermore, only two in three surviving primary helpers find a mate in their second year and reproduce personally, whereas 91 percent of returning secondary helpers succeed in breeding. Many one-time secondary helpers breed with the female they helped the preceding year (10 of 27 in Reyer's sample), suggesting that improved access to a potential mate is the ultimate payoff for their initial altruism.

These data enable us to calculate the direct fitness cost to the altruistic primary helpers in terms of reduced personal reproduction in their second year of life. For simplicity's sake, we shall restrict our comparison to solo primary helpers that help their parents rear siblings in the first year and then breed on their own in the second year, if they survive and find a mate, versus second-

Pied kingfisher

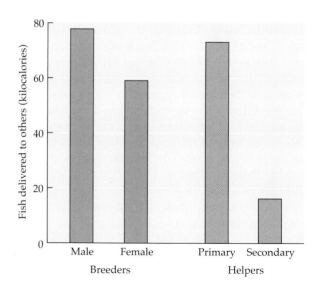

Figure 13.18 **Altruism and relatedness in pied kingfishers.** Primary helpers deliver more calories per day in fish to a nesting female and her offspring than do secondary helpers, which are not related to the breeders they assist. After Reyer [1013].

TABLE 13.3 *Calculations of inclusive fitness for male pied kingfishers*

Behavioral tactic	First year			Second year				
	y	r	f_1	o	r	s	m	f_2
Primary helper	1.8 × 0.32 = 0.58			2.5 × 0.50 × 0.54 × 0.60 = 0.41				
Secondary helper	1.3 × 0.00 = 0.00			2.5 × 0.50 × 0.74 × 0.91 = 0.84				
Delayer	0.0 × 0.00 = 0.00			2.5 × 0.50 × 0.70 × 0.33 = 0.29				

Source: Reyer [1013]

Symbols: y = extra young produced by helped parents; o = offspring produced by breeding ex-helpers and delayers; r = coefficient of relatedness between the male and y, and between the male and o; f_1 = fitness in first year (indirect fitness for the primary helper); f_2 = direct fitness in second year; s = probability of surviving into the second year; m = probability of finding a mate in the second year.

ary helpers that help nonrelatives with no other helpers present in the first year and then reproduce on their own in the second year, if they survive and find a mate.

Primary helpers throw themselves into helping their parents produce offspring *at the cost of having less chance of reproducing personally in the next year*. Although primary helpers do better than delayers in the second year (0.41 versus 0.29 units of direct fitness), secondary helpers do better still (0.84 units of direct fitness) because they have a higher survival rate and a greater probability of securing a partner (Table 13.3).

But is the cost to primary helpers of 0.43 lost units of direct fitness (0.84 − 0.41 = 0.43) in the second year offset by a gain in indirect fitness during the first year? To the extent that these males increase their parents' reproductive success, they create siblings that would not otherwise exist, indirectly propagating their genes in this fashion. In Reyer's study, the parents of a primary helper gained an *extra* 1.8 offspring, on average, when their son was present. Some primary helpers assisted their genetic mother and father, in which case the extra 1.8 siblings were full brothers and sisters, with a coefficient of relatedness of 0.5. But in other cases, one parent had died and the other had remated, so that the offspring produced were only half-siblings (r = 0.25). The average coefficient of relatedness for sons helping a breeding pair was thus between one-fourth and one-half (r = 0.32). Therefore, the average gain for primary helper sons was 1.8 extra sibs × 0.32 = 0.58 units of indirect fitness, a figure higher than the mean direct fitness loss experienced in their second year of life.

Reyer used Hamilton's rule to establish that primary helpers sacrifice future personal reproduction in year 2 in exchange for *increased* numbers of nondescendant kin in year 1 [1013]. Because these added siblings carry some of the helpers' alleles, they provide indirect fitness gains that more than offset the loss in direct fitness that primary helpers experience in their second year relative to secondary helpers.

Discussion Questions

13.6 Given the results of our calculations of inclusive fitness for male pied kingfishers, why are there ever any delayers? Is it maladaptive to be a delayer? Can you use conditional strategy theory to analyze this case?

13.7 Another lion problem: Let's say that a lion pride typically consists of 10 reproductively mature females. Imagine that a male working by himself has a 30 percent chance of acquiring and defending a pride for 1 year. However, a pair of males has a 80 percent chance of holding a pride for this same period. (1) Assuming that all the females mate with the male or males that control their pride and produce one youngster apiece during the year, and that both males in a two-male pride sire an equal number of young, should two unrelated males get together to secure this harem of females? (2) What if the males are cousins? (3) Now imagine that the males are half-brothers, but the dominant male manages to get 80 percent of all matings, and so 80 percent of the offspring are his. Should the subordinate join a coalition with his dominant half-brother? In all your calculations of inclusive fitness, identify the direct and indirect fitness components. (This question is courtesy of Mike Beecher, to whom any complaints should be directed.)

Inclusive Fitness and Helpers at the Nest

In the pied kingfisher, primary helpers raise their fitness indirectly through their increased production of nondescendant kin, whereas secondary helpers raise their fitness directly by increasing their future chances of reproducing personally. Primary helpers demonstrate that altruism can be adaptive; secondary helpers show that helping need not be altruistic, but instead may generate direct fitness benefits to helpers. Thus, this one species offers support for two very different adaptationist hypotheses on the evolution of helping behavior. These hypotheses can be tested for other cases of helpers at the nest, which are found in a variety of other birds, as well as fishes, mammals, and insects [158, 229, 994, 1177].

Each case of helpers at the nest presents a separate puzzle that deserves to be analyzed in light of a full range of hypotheses, including the possibility that caring for another's offspring is a *nonadaptive* side effect of other adaptive traits. As Ian Jamieson has pointed out, helping may have *originated* in some bird species as an incidental by-product of genetic or ecological changes that made it adaptive for young adults to delay their dispersal from their natal territory [587, 588]. If these stay-at-home birds were exposed to the nestlings being cared for by their parents, the begging behavior of the baby birds might have activated parental behavior in the young nonbreeding adults. This behavior could then be maintained over evolutionary time as a by-product of two adaptive traits, delayed dispersal and the tendency to care for one's own offspring, even if feeding someone else's young reduced the fitness of nonbreeding helpers.

This nonadaptive by-product hypothesis assumes that selection could not eliminate the helper's tendency to feed its parents' offspring without also destroying the capacity of the stay-at-home bird to invest in its own nestlings at a later date. This proposition is testable. One of its key predictions is that the underlying mechanisms of parental care should be no different in species with helpers than in species without helpers. The group of birds known as jays provides the necessary comparative test. In the Mexican jay (*Aphelocoma ultramarina*) and Florida scrub jay (*Aphelocoma coerulescens*), some nonbreeding birds help their parents rear additional siblings (Figure 13.19). The western scrub jay (*Aphelocoma californica*) is a member of the same genus, but lacks helpers at the nest. In this species, only breeding individuals have high prolactin levels, while nonbreeders have low levels of this hormone, which appears to regulate parental care in many birds. In contrast, nonbreeding helpers in the Mexican jay and Florida scrub jay have prolactin levels that match those of their breeding parents [1072, 1237]. Moreover, the prolactin in nonbreeding Mexican jays rises to peak levels *before* there are young to feed in the nest (Figure 13.20), suggesting that selection has favored nonbreeding individuals of this species that are capable of becoming hormonally primed to rear their siblings [1237]. These results are at odds with the nonadaptive by-product explanation for helping.

But if helping is adaptive, do helpers derive inclusive fitness gains via the direct or indirect route, or both? In the Mexican jay and Florida scrub jay, some

Figure 13.19   Cooperation among scrub jay relatives. Helpers at the nest in the Florida scrub jay provide food for the young, defense for the territory, and protection against predators. Based on a drawing by Sarah Landry, from Wilson [1315].

stay-at-home helpers inherit their natal territories from their parents—a direct fitness benefit. Furthermore, parents with helpers rear more offspring than parents without helpers, which generates an indirect fitness benefit for the helpers as well (Table 13.4) [158, 1338]. However, the apparent increase in the number of offspring fledged by pairs with helpers might arise strictly because helpers live with parents that have better territories, which provide more food or superior nesting sites. The hypothesis that territory quality, not helpers at the nest, is responsible for differences among breeding pairs in the number of fledglings produced was tested by Ronald Mumme. He captured and removed the nonbreeding helpers from some randomly selected breeding pairs of Florida scrub jays, while leaving other helpers untouched. The experimental removal of

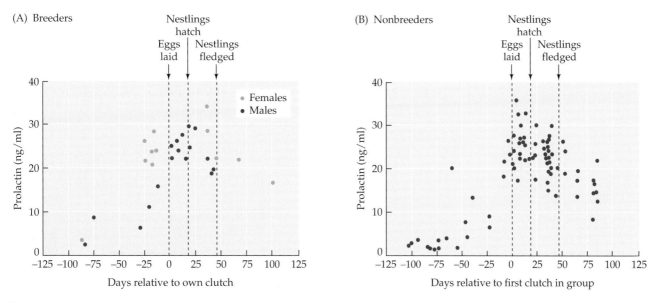

Figure 13.20   Seasonal changes in prolactin concentrations in breeders and nonbreeding helpers at the nest in the Mexican jay. Nonbreeding birds in a group exhibit the same pattern of increased prolactin production prior to the hatching of eggs as do breeding adults. After Brown and Vleck [160].

TABLE 13.4 *Effect of Florida scrub jay helpers at the nest on the reproductive success of their parents and on their own inclusive fitness*

	Parents without breeding experience[a]	Parents with breeding experience
Average number of fledglings produced with no helpers	1.03	1.62
Average number of fledglings produced with helpers	2.06	2.20
Increased reproductive success due to help	1.03	0.58
Average number of helpers	1.70	1.90
Indirect fitness gained per helper	0.60	0.30

Source: Emlen [365]

[a]Includes pairs in which one parent has reproduced, which is why some pairs in this category acquire a helper at the nest.

helpers reduced the reproductive success of the experimental pairs by about 50 percent, as measured by the number of offspring known to be alive 60 days after hatching (Figure 13.21). Helpers apparently really do help in this species [859].

In fact, helper scrub jays also improve the chances that their parents will live to breed again another year, as do helpers in the pied kingfisher. Improved parental survival means that the helpers are responsible for still more siblings in the future; these extra siblings yield an average of about 0.30 additional indirect fitness units for helpers [861]. Thus, the total indirect fitness gains from helping at the nest can potentially exceed its costs in terms of lost direct fitness, especially if the young birds have almost no chance of reproducing personally. When very few openings are available for dispersing young adults, helping is more likely to be the adaptive option for individuals that have the potential to be either altruistic or to reproduce on their own. Individuals of this sort can be said to be facultative altruists (see Figure 13.8) because they are not locked into the helper role.

Whether saturated nesting habitats contribute to the maintenance of helping at the nest is also testable. If young birds remain on their natal territories because they cannot find suitable nesting habitat, then yearlings given an opportunity to claim good open territories should promptly become breeders. Jan Komdeur did the necessary experiment with the Seychelles warbler, a drab little brown bird that has played a big role in testing evolutionary hypotheses

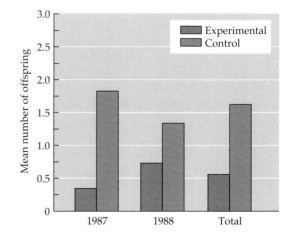

Figure 13.21 **Helpers at the nest help parents raise more siblings** in the Florida scrub jay. The graph shows numbers of offspring alive after 60 days in experimental nests that lost their helpers and in unmanipulated control nests during a 2-year experiment. After Mumme [859].

about helping at the nest. When Komdeur transplanted 58 birds from one island (Cousin) to two other nearby islands with no warblers, he created vacant territories on Cousin, and helpers at the nest there immediately stopped helping in order to move into open spots and begin breeding. Since the islands that received the transplants initially had many more suitable territorial sites than warblers, Komdeur expected that the offspring of the transplanted adults would also leave home promptly in order to breed elsewhere on their own. They did, providing further evidence that young birds help only when they have little chance of making direct fitness gains by dispersing [658].

Moreover, the sophisticated conditional strategy that controls the dispersal decisions made by young Seychelles warblers is sensitive to the quality of their natal territory. Breeding birds occupy sites that vary in size, vegetational cover, and insect supplies. By using these variables to divide warbler territories into categories of low, medium, and high quality, Komdeur showed that young helpers on good territories were likely to survive there while also increasing the odds that their parents would reproduce successfully. Young birds whose parents had prime sites often stayed put, securing both direct and indirect fitness gains in the process. In contrast, young birds on poor natal territories had little chance of making it to the next year, nor could they have a positive effect on the reproductive success of their parents. They left home and tried to find a breeding opportunity of their own [657].

Discussion Question

13.8   Helpers at the nest have been found in only about 3 percent of all birds [521]. One attribute of this small minority of birds that has often been linked to the evolution of helpers is the delayed dispersal of juveniles, as we have just illustrated for Florida scrub jays and Seychelles warblers. But another factor that might have promoted the evolution of helping is a very low adult mortality rate. These two ideas have sometimes been presented as competing hypotheses, but how might they both reflect the same ecological pressure that makes helping at the nest an adaptive making-the-best-of-a-bad-job option for young birds?

The flexibility of behavior exhibited by Seychelles warblers is not unique to that species. Consider how young female white-fronted bee-eaters make adaptive conditional decisions about reproducing (Figure 13.22). This African bird nests in loose colonies in clay banks. Like male pied kingfishers, young female white-fronted bee-eaters can choose to breed, or to help a breeding pair at their nest burrow, or to sit out the breeding season altogether. If an unpaired, dominant, older male courts her, a young female almost always leaves her family and natal territory to nest in a different part of the colony, particularly if her mate has a group of helpers to assist in feeding the offspring they will produce. Her choice usually results in high direct fitness payoffs. But if young, subordinate males are the only potential mates available to her, the young female will usually refuse to set up housekeeping. Young males come with few or no helpers, and when they try to breed, they are often harassed by their fathers, who may force their sons to abandon their mates and return home to help rear their siblings.

A female that opts not to pair off under unfavorable conditions may choose to slip an egg into someone else's nest, or to become a helper at the nest in her natal territory—provided that the breeding pair there are her parents, to whom she is closely related. If one or both of her parents have died or moved away, she is unlikely to help rear the chicks there, which are at best half-siblings, and instead will simply wait, conserving her energy for a better time in which to reproduce [364]. Thus, although daughter bee-eaters have the potential to become helpers at the nest, they choose this option only when the indirect fitness benefits of helping are likely to be substantial.

Figure 13.22 Conditional reproductive tactics of female white-fronted bee-eaters. Females of this species have many options, of which helping at the nest is only one. Females select a given option depending on their special circumstances. After Emlen et al. [364].

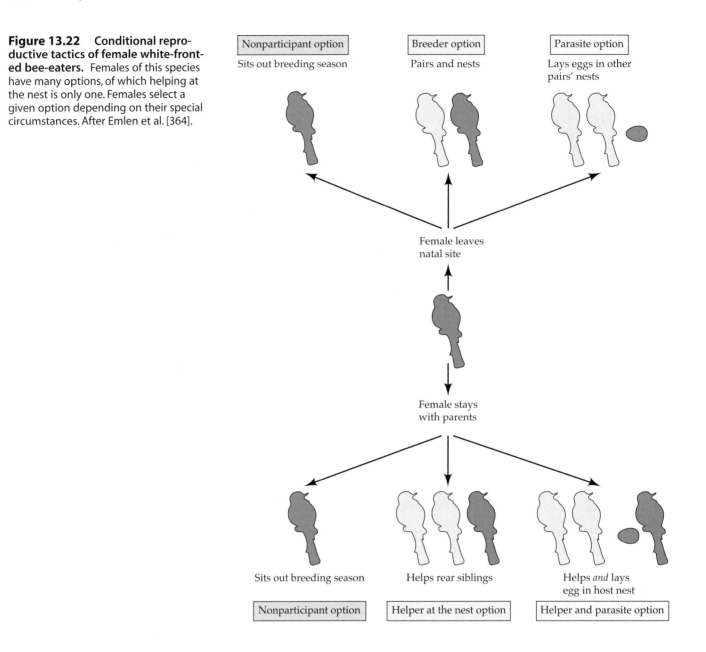

Insect Helpers at the Nest

Although some birds provide impressive examples of adaptive helping at the nest, the phenomenon also occurs in highly sophisticated forms in certain insects, including the paper wasps mentioned at the beginning of this chapter. Paper wasp colonies usually consist of one or more reproductively active females and a number of helpers at the nest, or *workers*, which are always females, never males. In order to generate a set of female helpers early in the nesting cycle, egg-laying female wasps fertilize some eggs by releasing sperm from a sperm storage organ as the eggs pass down the oviduct. These diploid eggs will develop into females, whereas to produce a son, a queen need only lay an unfertilized (haploid) egg. All the Hymenoptera (the wasps, bees, and ants) can use this haplodiploid system of sex determination to produce daughters and sons at the appropriate times during the breeding cycle (Figure 13.23).

The degree to which hymenopteran queens can control the fertilization of their eggs has been measured in the honey bee. In this species, worker (female)

(A)

(B)

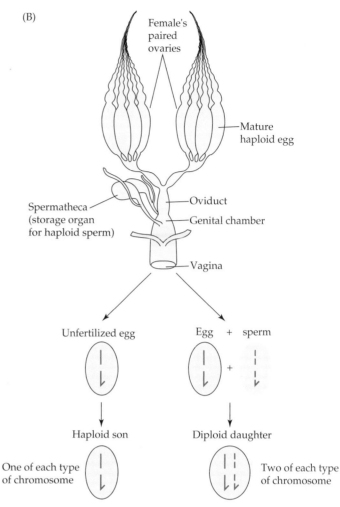

Figure 13.23   Haplodiploid sex determination in Hymenoptera. (A) When a haploid male bee copulates with a diploid adult female, he provides her with a quantity of haploid sperm, which are stored in the female bee's spermatheca. (B) When the female "decides" to produce a son, she releases a mature haploid egg from an ovary and lays it as an unfertilized egg, which will develop into a male. In order to produce a daughter, the female fertilizes the mature egg with sperm from her spermatheca as the egg passes down the oviduct. For simplicity's sake, only two chromosomes are shown here. Photograph by the author.

larvae are reared in smaller brood cells than are drone (male) larvae. Direct genetic analysis of over 100 eggs laid in small cells revealed that they were all diploid, and so had been fertilized before oviposition occurred, whereas all the eggs removed from large cells were haploid, and so had not been fertilized. Queen bees apparently have complete control over the sex of their progeny [1003]. The same applies to the fire ant, whose queens make extraordinarily efficient use of the sperm stored within their bodies by releasing just three sperm on average to fertilize an egg when they are producing a daughter [1217].

Paper wasp queens make daughters early in the nesting cycle because many will remain at the nest to help rear more offspring, which later in the season will include both brothers and sisters that are destined to reproduce, not help at the nest. Nonetheless, the helper daughters of paper wasps have functional ovaries (unlike most honey bee workers, for example) and so are capable of reproducing. Even if they have not mated, they could lay unfertilized eggs and so have sons, contributing to their direct fitness if these offspring were to survive. But helper paper wasp daughters do not often lay eggs, and when they do, their mother may consume those eggs rather than let them develop. Therefore, helper daughters typically forgo direct fitness in order to assist their mother in the production of siblings. We can explain their altruism in terms of indirect fitness gains to the extent that their actions yield more surviving, reproducing brothers and sisters than their mother could have produced by herself. Given the higher rate of loss of nests built and defended by a single paper wasp to raiding female wasps and nest predators [429, 883], we can be quite confident that helpers usually do boost the reproductive success of the queen they help. Our confidence in this conclusion is further enhanced by the results of some experimental studies involving the removal of subordinate females wasps by researchers. In one such study, nests that lost helpers did not survive as well as those that retained them; in another, the role of subordinates in producing extra brood was documented experimentally [1102, 1195].

Discussion Question

13.9   In some species of ants, workers either possess or can develop functional ovaries, and these individuals may go on to mate and lay eggs. But these workers can be prevented from reproducing by their colonymates. For example, in *Dinoponera quadriceps*, the dominant reproducing female smears a potential competitor with a chemical from her stinger, after which lower-ranking workers immobilize the unlucky pretender for days at a time (Figure 13.24A) [844]. Likewise, workers of *Harpegnathos saltator* punish nestmates that are developing their ovaries by holding the offenders in a firm grip (Figure 13.24B), preventing them from doing anything. These tactics inhibit further development of an immobilized ant's ovaries [718]. Why, in evolutionary terms, do the workers of these two species treat their colonymates in this fashion? What do you suppose workers will do if the fertility of a dominant queen drops? (See [271] for the results of a study in which the fertility of the queen was lowered experimentally by application of a hormone.)

The composition of paper wasp societies varies greatly. Although some groups are of the sort just described, composed of a mother and her altruistic daughters, others are made up of various combinations of relatives and genetic strangers. Because of this variety and the many different reproductive tactics exhibited by female paper wasps [1146], these social insects have been favorite subjects for evolutionary biologists interested in the possible evolutionary causes of altruism. As indicated, some females at multiple-female nests, whether relatives of the queen or not, exhibit altruism by tending the eggs and larvae of the single dominant female. In some species, subordinates are effectively sterile because the queen monopolizes almost all the egg laying done at the nest. Queens of the paper wasp *Polistes dominulus*, for example, are responsible for more than 95 percent of all the offspring at their nests [994]. Should a subordinate of this species lay an egg, the queen sniffs it out and eats it. Despite this indignity, subordinates usually stay on. When colonies are composed of related females, indirect selection theory can potentially explain why subordinate relatives of a queen give up reproducing in favor of helping. This kind of explanation will not work, however, for colonies in which *unrelated* females band together to rear brood even though the queen does almost all the egg laying.

The puzzle posed by altruistic subordinates unrelated to the queen has been tackled with the help of what can be called the **transactional**

(A)

(B)

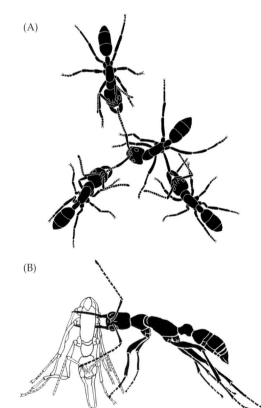

Figure 13.24   Conflict within ant colonies over reproduction. In two ant species, individuals that are about to reproduce are detected by colonymates and physically prevented from doing anything for hours or days. (A) The queen has smeared the ant in the center with her sting. Three workers grasp the would-be reproductive by assorted appendages, preventing her from moving. (B) The ant in black has grabbed and immobilized a nestmate whose ovaries had begun to develop. After holding her nestmate captive for three or four days, a worker may turn her prisoner over to another worker to continue the imprisonment. A, from Monnin et al. [844]; B, drawing by Malu Obermayer, from Liebig et al. [718].

Figure 13.25   A test of a hypothesis based on transactional theory. When a queen of *Polistes fuscatus* is not closely related to the other females that join her to set up a colony (foundresses), she apparently concedes some reproductive chances to her companions, which reduces the degree to which a single female monopolizes reproduction at the nest. After Reeve et al. [1008].

theory of social behavior. This theory views animal societies as arenas in which dominants and subordinates "negotiate" their reproductive rights within the group [607, 1005]. One derivative of transactional theory is a *concessions model*, in which a powerful dominant member of the group concedes a certain amount of reproduction to lower-ranking individuals to make it advantageous for them to stay in the group, thereby boosting the productivity and survival chances of the colony and thus the fitness of the concession-making animal. When unrelated females help a dominant queen paper wasp, they cannot derive indirect fitness from their assistance, as the queen's relatives do, so these females should require some direct fitness payoffs for staying and helping. One would predict, therefore, that queens allied with nonrelatives should make larger concessions to keep helpers on hand than queens fortunate enough to have a bevy of relatives available. Indeed, when dominant females of the wasp *P. fuscatus* associate with unrelated helpers, the reproductive differences among dominants and subordinates are less than when all the females in a colony of this species are related to one another (Figure 13.25) [1008].

Alternative transactional hypotheses are, however, available for cases in which subordinates assist unrelated queens. Perhaps instead of accepting concessions from a queen, subordinate helpers reproduce to the extent that they can by circumventing the controlling dominant female, something they might be quite good at. In other words, there might be a tug-of-war going on within a colony, whose members may be attempting to take as many opportunities as possible to reproduce rather than merely taking whatever a dominant queen is willing to offer.

The fact is that there are numerous examples of insect societies that do not neatly fit the concessions model. In one Australian social bee, for example, nesting groups that are composed of two unrelated individuals exhibit a *greater* degree of reproductive skew (with the dominant taking the lion's share) than is found in duos made up of close relatives [693]. This result is not consistent with expectations from a concessions hypothesis, but does match predictions from the alternative tug-of-war scenario, in which dominants are confronted by subordinates trying to secure as much reproductive success as possible. When the members of a social unit spend time fighting with one another, they should have less time to contribute to colony productivity. Thus, under a tug-of-war scenario, we would expect groups of unrelated individuals to produce fewer offspring overall than groups of relatives, which have an indirect fitness incentive to moderate their conflicts with their sisters. The Australian bees behave in ways consistent with the tug-of-war hypothesis.

Subordinates unrelated to a dominant group member might in some instances tolerate a reduction in personal reproduction because (1) they have little chance of founding a successful nest elsewhere on their own and (2) they have some chance of inheriting the nest where they are helping upon the demise of the current queen, at which time their reproductive payoff could be substantial. (Note the similarity between this hypothesis and those proposed to account for secondary helpers in pied kingfishers and beta display partners in long-tailed manakins.) This kind of facultative altruism (see Figure 13.8) can be considered yet another kind of social transaction, with some individuals helping others in exchange for permission to remain at the colony where they have some chance of eventually becoming a dominant breeding female. In support of this hypothesis, observations of 28 nests of *P. dominulus* yielded 13 records of a change in ownership within the nesting season, 10 of which were accomplished by resident helpers [994]. Much the same thing occurs in a tropical social wasp in a different subfamily in which one dominant female typically takes total charge of reproduction, while an assortment of subordinates help whether they are related to the queen or not. In effect, the nonreproducers, which have their own dominance hierarchy, are lined up in a queue awaiting the demise of the queen, at which point the top subordinate will have a chance to become an independent reproducer. Moreover, because this now dominant individual inherits a coterie of workers, her offspring are likely to survive even if she dies before completion of a brood cycle [1163]. The point is that an abundance of alternative hypotheses exist to account for helping at the nest in the Hymenoptera, and the evidence suggests that many of them are valid for one or another member of this group.

Discussion Questions

13.10   Figure 13.26 contains data on the relative reproductive success of alpha and beta males of the white-browed scrub-wren in two different kinds of cooperative breeding associations [1290]. Is the observed pattern consistent with transactional theory?

13.11   Coalitions of male lions appear to be of two types: large groups of closely related individuals and smaller groups of nonrelatives. Use two hypotheses based on transactional theory to make predictions about the degree of reproductive monopolization that should be observed in the two kinds of coalitions. In one hypothesis, dominants (which are in total control of the group) concede reproduction to subordinates to retain their services as group members; in another hypothesis, dominants (which lack complete control) and subordinates are engaged in a tug-of-war over who gets to reproduce. Check the actual data presented by Craig Packer and colleagues [918].

White-browed scrub-wren

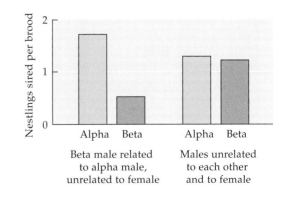

Figure 13.26   The effect of relatedness on equality of reproductive opportunities in the cooperatively breeding white-browed scrub-wren. After Whittingham et al. [1290].

Figure 13.27   Eusocial insects have sterile castes. (Left) In many ant species, several kinds of sterile females exist, each suited for a particular role within the colony. Here a tiny worker ant perches on the head of a giant soldier. (Right) In a nasute termite colony, the sterile soldiers have glands in their heads that can spray enemies with sticky repellent fluids. Left, photograph by Mark Moffett; right, photograph by the author.

The Evolution of Eusocial Behavior

Although the evolution of helping at the nest by reproductively capable females is a worthy problem, perhaps even more intriguing is the sterility exhibited by helpers in some social insect colonies (Figure 13.27). Unlike the facultative altruists discussed already, these helpers have very little or no chance to reproduce personally and so can be considered obligate altruists (see Figure 13.8). The anatomy and behavior of these sterile castes with their undeveloped ovaries often reveal extreme specialization for self-sacrifice. Honey bee workers, for example, have a barbed sting designed to catch in the skin of vertebrate enemies threatening their hive, the better to deter them, even though it means the death of the hive defender (Figure 13.28). Soldier ants of some species have

Figure 13.28   Suicidal sacrifice by a worker bee. When a honey bee stings a vertebrate, she dies after leaving her stinger and the associated poison sac attached to the body of the victim. Photograph of Bernd Heinrich's knee by Bernd Heinrich.

immense jaws, the better to pierce the bodies of predators, including fellow ants, that have invaded their colonies. Other ant species have grenade soldiers, whose suicidal mission is to rush at predatory insects invading their colony while simultaneously constricting their abdominal muscles so violently that they burst a large abdominal gland, spilling a disabling glue over their enemies [791].

The **eusocial** insects—those with a sterile worker caste—caught Darwin's eye, and his solution to the problem of their evolution was closely related to the indirect fitness hypothesis already discussed at length in this chapter. Darwin noted that social insect colonies are extended families, so that when sterile members of the group help others survive to reproduce, the helpers (even if they die in the process) are helping to maintain family traits—including the ability of the reproducing members of the family to generate some sterile helpers [285].

Discussion Question

13.12 Eusociality has a history as well as potential adaptive value. Figure 13.29 shows a phylogeny of a portion of the bee species in the genus *Lasioglossum*, with the social system of the species superimposed on the evolutionary tree. What is the minimum number of times that eusociality has evolved in this group? How many times has eusociality been lost, either completely or partially? (Partial loss of eusociality has occurred in those species that are labeled "polymorphic," which means that in these cases some populations are eusocial and others are not.) How is this phylogeny relevant to the common view that complex social systems are generally superior, and more recently evolved, than simpler ones?

The next major advance on Darwin's explanation came 120 years later, when W. D. Hamilton developed his now famous genetic cost–benefit analysis of worker altruism. Remember that according to Hamilton's rule, altruism can evolve when the altruist's loss in personal reproduction (C) times the degree of relatedness of a parent to its offspring (r_c) is less than the added number of reproducing relatives that owe their existence to the altruist (B) times the degree of relatedness between the altruist and the helped individuals (r_b). Hamilton realized that if r_b, the relatedness of the altruist to the relatives it helped, was particularly high, then the indirect fitness side of the equation would be increased. He was also the first to point out that r_b for sisters could indeed be unusually high in the Hymenoptera because of the haplodiploid system of sex determination in this group [502].

The potentially very close genetic similarity between sister bees, ants, and wasps arises because male Hymenoptera, as noted earlier, are haploid (have only one set of chromosomes, not two). Therefore, all the haploid sperm a male ant, bee, or wasp makes are chromosomally (and thus genetically) identical. So if a female ant, bee, or wasp mates with just one male, all the sperm she receives will have the same set of genes. When the female uses those sperm to fertilize her eggs, all her diploid daughters will carry the same set of paternal chromosomes and genes, which make up 50 percent of their total genotype. The other set of chromosomes carried by daughter hymenopterans comes from their mother. The mother's haploid eggs are not genetically uniform because she is diploid; gamete formation by a parent with two sets of chromosomes involves the production of a cell with just one set drawn at random from those in that parent. An egg made by a female bee, ant, or wasp will share, on average, 50 percent of the alleles carried within her other eggs. Thus, when a queen bee's eggs unite with genetically identical sperm, the resulting offspring will share, on average, 75 percent of their alleles: 50 percent of their father's alleles and 0–50 percent of their mother's alleles (Figure 13.30).

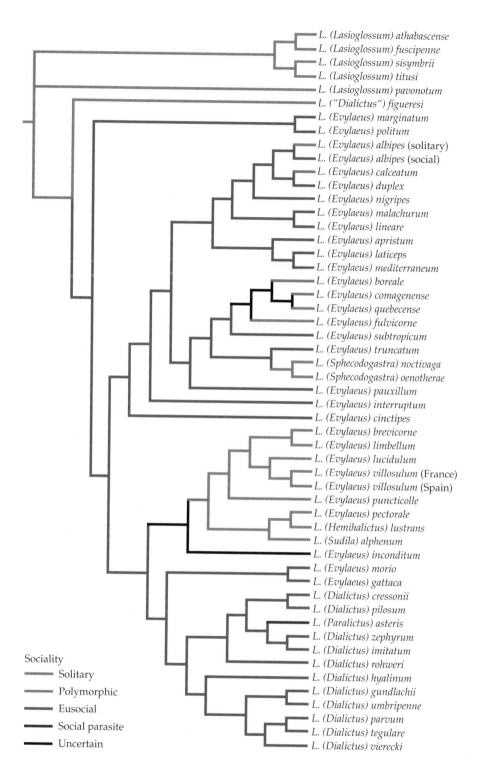

Lasioglossum oenotherae

Sociality
— Solitary
— Polymorphic
— Eusocial
— Social parasite
— Uncertain

Figure 13.29   Eusociality has an evolutionary history. Among the many species of bees belonging to the genus *Lasioglossum*, some are eusocial, others are solitary, and still others are polymorphic (exhibit both patterns in different populations). The social parasitic species exploits established colonies of other bees. The photograph shows two females of *L. (Sphecodogastra) oenotherae*, a solitary species whose closest relatives are mostly eusocial bees. After Danforth et al. [284]; photograph by Bryan Danforth.

(A) Mother–daughter genetic relatedness

(B) Sister–sister genetic relatedness

Figure 13.30   Haplodiploidy and the evolution of eusociality in the Hymenoptera. The degree of genetic relatedness of a female wasp (A) to her daughters and (B) to her sisters. For the sake of simplicity, only two chromosomes are shown. A queen's sons develop from unfertilized eggs with only one copy of each chromosome, so that mothers share 50 percent of their genes with their sons.

Under the haplodiploid system of sex determination, hymenopteran sisters may therefore have a coefficient of relatedness of 0.75, higher than the 0.5 figure for a mother and her daughters and sons. As a consequence of this genetic fact, $(r_c \times C)$ should be less than $(r_b \times B)$ more often in the Hymenoptera than in other groups, all other things being equal, which would facilitate the evolution of eusociality in these insects. If sisters really are especially closely related, indirect selection could more easily favor hymenopterans that, so to speak, put all their eggs (alleles) in a sister's basket rather than reproducing personally. Perhaps not coincidentally, the Hymenoptera have the greatest number of eusocial species of any insect family.

Testing the Haplodiploid Hypothesis

The haplodiploid hypothesis for eusociality generates many testable predictions. First, if a female worker in a eusocial bee, ant, or wasp colony is to cash in on her potentially high degree of relatedness to her fellow females, she should bias her help toward reproductively competent sisters rather than toward male siblings. Although sister hymenopterans share up to 75 percent of their genes, a sister shares only 25 percent of her genes with her haploid brother (see Figure 13.30). Males do not receive any of the paternal genes their sisters possess. The remaining half of the genome that sisters and brothers both receive from their mother ranges from 0 to 100 percent identical, averaging 50 percent; 50 percent of a half means that a sister shares, on average, only one-fourth of her genes with her brothers ($r = 0.25$).

Bob Trivers and Hope Hare recognized that these brother–sister inequalities in r should lead to conflict between queens and workers over how workers allocate their assistance to their siblings [1215]. Because a queen donates 50 percent of her genes to each offspring, whether male or female, she gains no genetic advantage by biasing offspring production toward daughters rather than sons, or vice versa. Imagine a hypothetical population of, say, an ant species in which queens did tend to make more of one sex than the other. In this situation, any mutant queens that did the opposite and had more offspring belonging to the rarer sex would be handsomely repaid in grandoffspring. If males were scarce, for example, then a queen that used her parental capital to generate sons would create offspring with an abundance of potential mates, and thus many more opportunities to reproduce than a comparable number of daughters. The greater fitness of son-producing queens would effectively add more males to the next generation, moving the sex ratio back toward equality. If over time, the sex ratio overshot and became male-biased, then daughter-producing queens would gain the upper hand, shifting the sex ratio back the other way. When the investment ratio for sons and daughters is 1:1, there is no advantage for a son-producing specialist or a daughter-producing specialist. Therefore, an equal investment strategy is favored by frequency-dependent selection (see p. 217) acting on queens.

But let's think about the nature of selection on workers in species in which workers share three times as many genes on average with their sisters as with their brothers. Under these circumstances, workers that were able to bias the sex ratio of their siblings toward sisters would gain greater representation for their genes than those that went along with the sex ratio advantageous for their mother. Thus, conflict would arise if workers did what they could to ensure that their colony put out more reproducing females than males.

If queen and workers in large eusocial colonies come into conflict about the allocation of resources to the queen's larval sons and daughters, the workers should hold the upper hand because the queen cannot possibly monitor the activities of thousands of workers. In these cases, workers could conceivably withhold food from brother larvae in order to nourish sister larvae instead. If workers indeed attempt to maximize their own inclusive fitness, then the combined *weight* of all the adult female reproductives (a measure of the total resources devoted to the production of females) raised by the colony's workers should be three times as much as the combined weight of all the adult male reproductives. When Trivers and Hare surveyed the literature, they found the expected 3:1 investment ratio [1215].

13.13   Precisely why would a worker ant gain if the colony invested three times as much in the production of reproductively capable females as males? Illustrate your answer with a case in which male and female offspring cost exactly the same amount to produce, so that 100 units of investment (such as food for larvae) will yield the same number of adult males as females. Consider the population-wide sex ratio and explain why it might not pay workers to force their colony to produce only female reproductives even though the workers would share many more genes with those individuals than with their brothers.

Hymenopteran workers should bias their production of reproductives toward females *only* if their mother mates just once. Queens that copulate with two or more haploid males have two or more sperm genotypes with which to fertilize their eggs. Daughters with different fathers (i.e., half-sisters) will not be closely related at all. Only when females have the same father will they share 75 percent of their genes (see Figure 13.30). In reality, in many hymenopterans, queens do mate with several males. Moreover, sometimes a group of queens occupies the same nest, another factor that reduces the relatedness between workers and the queens of the future that they are rearing. These facts have great significance for the evolution of worker behavior [115, 882].

We can illustrate this point by considering a species of *Formica* ant whose queens may be either monogamous or polyandrous. Liselotte Sündstrom realized that this species provided a wonderful opportunity to find out whether workers did indeed take r_b into account when allocating food to future reproductive sisters and brothers. They did. The daughters of single-mating mothers heavily biased their investment toward producing sister queens. But workers in colonies with multiply mated queens behaved quite differently. For them, brothers were as genetically valuable as sisters, and they did *not* bias the colony's production toward females [1164].

Ulrich Mueller has also shown that worker hymenopterans alter their investment in colonymates according to their coefficient of relatedness [857]. He experimentally manipulated colonies of a eusocial bee, removing the foundress queen from some nests but leaving her alone in others. When a colony has its foundress queen, the usual asymmetry in relatedness persists between workers and their sisters ($r = 0.75$) and their brothers ($r = 0.25$). Under these conditions, a bias toward investment in female progeny is expected under the haplodiploid hypothesis. But in colonies from which the foundress has been removed, a daughter assumes reproductive leadership. Under these conditions, her sister–workers are now helping her produce nieces ($r = 0.375$) and nephews ($r = 0.375$), rather than additional siblings. Thus, the relatedness asymmetry disappears, and workers ought to treat male progeny more favorably in these colonies. In fact, workers in the experimental colonies did invest more in males (the combined weight of which equaled 63 percent of the total weight of all reproductives) than did workers in colonies that retained their foundress queen (in which males constituted 43 percent of the total weight).

13.14   If a female of a social wasp species could help to produce more sisters with an *r* of 0.75, why would she ever reproduce personally, given that reproducers are related to their offspring by just 0.5? In addition, use Hamilton's rule to explain why some "worker" paper wasps produced early in the year leave their natal nests to wait for opportunities to "adopt" a nest containing unrelated individuals rather than becoming helpers at the nest of their mother [1146].

13.15   As we have seen, helpers and queens of social insects can have disputes about any number of things, despite being members of one family (e.g., [535]). Why, for example, might a worker with a once-mated mother let her mother produce daughters, but attempt to produce sons herself (assuming that in this social species, workers had functional ovaries)? On the other side of the coin, why would an un-mated future queen ant be very aggressive toward egg-laying workers in the period before she mated, but then back off after she had mated? (Queens of some ants lay eggs in their natal colony both before and after mating [270].)

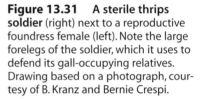

Figure 13.31   A sterile thrips soldier (right) next to a reproductive foundress female (left). Note the large forelegs of the soldier, which it uses to defend its gall-occupying relatives. Drawing based on a photograph, courtesy of B. Kranz and Bernie Crespi.

If haplodiploidy makes it easier for eusocial systems to originate, then sterile altruists should also appear in haplodiploid groups other than the Hymenoptera. The thrips are haplodiploid insects, and as predicted, a sterile self-sacrificing caste has evolved in some gall-forming thrips species [206]. Foundress female thrips induce the formation of a gall in plant tissue and then fill the protective hollow interior of the gall with their offspring, some of which form a soldier caste. These soldiers possess enlarged, spiny forelegs, the better to grasp and stab enemies that would enter the gall and eat their siblings (Figure 13.31). Direct measurements of the degree of relatedness between soldiers and the individuals they protect reveal an r value well over 0.5 for a species whose foundress females mate with only one male. Thus, members of this species exhibit the altruism that Hamilton predicted would be more likely to evolve in association with the haplodiploid system of sex determination.

The haplodiploid hypothesis is based on the premise that exceptionally close relatedness between helper and recipient promotes the evolution of eusociality. If this is true, then other mechanisms that result in extremely close genetic relatedness among members of social groups should also be associated with sterile caste formation and extraordinary self-sacrificing behavior. Both inbreeding and clonal or asexual reproduction can result in very high coefficients of relatedness among family members. In those species of aphids in which mothers reproduce asexually, for example, daughters get carbon copies of their mother's genotype, so the r value for sisters is 1.0. This fact makes it easier to understand why in some clonal species, an aphid that has been captured by a deadly ladybird beetle immediately smears its killer with an alarm pheromone. When the beetle resumes its search for additional victims, the odor it now carries encourages clonemates of the deceased aphid to drop off their perch, an effective anti-ladybird response [842].

A number of other species of asexual aphids have converged anatomically and behaviorally with the gall-forming thrips [585, 1153]. These aphids also live in galls made by a foundress female, and regular females live with their soldier sisters, a polyphenism that arises because of a developmental switch mechanism that these aphids possess (see p. 86.). Aphid soldiers use their powerful spiny forelegs (or thickened swordlike mouthparts) to dispatch enemies such as syrphid fly larvae when those predatory insects attempt to enter the gall to feast on the soldiers' sisters (Figure 13.32). Certain aphid soldiers do more than simply stab their enemies; they inject a poisonous insecticidal protein through their piercing tubular mouthparts into the body of the predator, thereby hastening its demise [685].

In some aphid species, soldiers are obligatorily sterile and so never reproduce; indeed, many die young in defense of their sisters. In experiments conducted by William Foster, an average of about 20 soldiers of *Pemphigus spyrothecae* fell in battle while dispatching one syrphid fly larva. In the absence of soldiers, however, one predatory larva could eat all of the 100 nonsoldier aphids that Foster brought together for his tests [402]. In additional experiments, all the aphids occupying a sample of galls were removed before being returned to their homes in reassembled groups that either included some soldiers or were

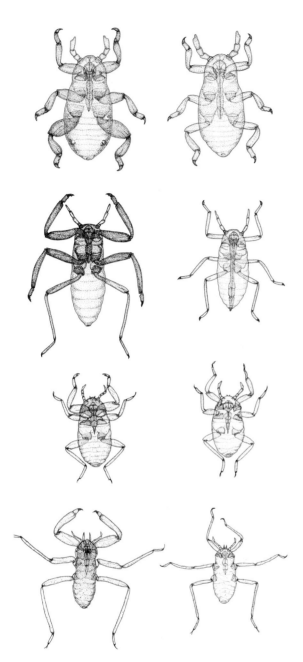

Figure 13.32   Altruism in aphids. Four species of aphids in which obligately sterile soldiers (left) with enlarged grasping legs and short stabbing beaks protect their more delicate colonymates (right), which have the potential to reproduce when mature. The species were drawn at different scales by Christina Thalia Grant. After Stern and Foster [1153].

without defenders. The soldierless galls were ten times as likely to be attacked by syrphids or other insect predators as those with soldiers present [403]. Thus, in nature, soldier aphids do not die in vain, because they derive an indirect fitness payoff when the beneficiaries of their actions survive to reproduce.

Discussion Question

13.16   We have examined the evolution of sterile workers primarily from the perspective of fitness gains and losses incurred by the workers themselves. But what if we view the issue from the standpoint of selection on the mothers of workers and sacrificial soldiers [1272]? How might queen bees, ants, and wasps and foundress aphids gain direct fitness benefits by "forcing" some of their offspring to be nonreproductive helpers—even if this reduced the inclusive fitness of those helpers? If social insect workers behave the way they do because of the suppressive effects of their mothers, can they really be said to be altruistic?

Because of the nature of asexual reproduction in aphids, it had always been simply assumed that apparent family groups in a gall were indeed a clone. But when Patrick Abbot and his colleagues used DNA fingerprinting technology to examine that assumption in *Pemphigus obesinymphae*, they made the surprising discovery that, on average, about 40 percent of the inhabitants of a given gall were intruders from nearby galls. These aphids had left the cottonwood leaf gall in which they were born and had moved in with the neighbors to take advantage of someone else's food resources and the protection offered by someone else's soldiers. When these colonies of mixed parentage were experimentally challenged by a pseudo-predator, a fruit fly larva, aphids living in their natal colony attacked just as if the grub were a syrphid larva. But although about 40 percent of the aphids in mixed colonies were outsiders, they contributed a mere 2 percent to the soldier workforce that responded to the potential predator (Figure 13.33). Instead of helping with defense, the intruders were free to develop relatively rapidly, reaching the reproductive stage long before the more slowly maturing soldiers, which in this species move from the defensive form to the reproductive form over time. The selfish behavior of the genetically unrelated gall freeloaders reduced the benefits gained by altruistic soldiers defending their natal colony [1].

Eusociality in the Absence of Very Close Relatedness

Although the studies described above are generally consistent with the contention that a high coefficient of relatedness can facilitate the evolution of altruism, many unresolved issues remain. Note, for example, that altruistic soldier aphids are extremely rare, appearing in only 1 percent of the thousands of the asexually reproducing aphid species worldwide. Even among the many gall-forming aphids, soldiers are uncommon, having evolved in only about 50 species [1153]. Moreover, as we have just seen, soldier altruism in aphids can

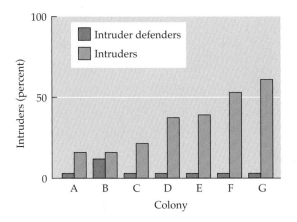

Figure 13.33 Selfish behavior of aphid clone invaders. Gall-defending aphid colonies may contain intruders that have come from elsewhere. Although these outsiders may make up a substantial proportion of the group, almost no newcomers take part in the defense of the gall. Instead, defenders are supplied almost exclusively by the colony's native population. After Abbot et al. [1].

persist even in species in which colonies contain numerous unrelated intruders, a situation that greatly lowers the average relatedness of colonymates.

Similarly, among the eusocial Hymenoptera, sisters are often *not* especially closely related [30], for reasons already discussed. Direct measurements of r for female offspring of two species of polyandrous eusocial wasps produced mean r values no greater than 0.40 [1035]. Likewise, in *Polistes* wasp colonies, the average r of nestmates almost never reaches the 0.75 maximum value, and often is less than 0.5 [1157]. These results could arise if the current queen were a fairly recent replacement for the preceding one, or if several females simultaneously contributed to brood production, or, as noted earlier, if the sole queen had mated with more than one male, all phenomena known to occur in some social insects.

The point is that the haplodiploid system of sex determination does not guarantee that workers in eusocial hymenopteran societies will be very closely related. Of course, it is possible that high coefficients of relatedness were a necessary precondition for the origins of eusocial systems long ago. Subsequently, eusociality may have been retained even as high r values were lost over time. In the honey bee, for example, polyandry, with the consequent decline in r among workers, may have spread secondarily after eusociality had evolved because of the benefits of having high genetic diversity within a colony to promote disease resistance [1180] or worker specialization and task efficiency [920]. Thus, the highly eusocial lifestyle of the now polyandrous honey bee may currently be maintained by selection pressures that differ from those that were responsible for the origin of eusociality in this species. But we can say with certainty that the haplodiploid method of sex determination is not absolutely essential for either the origin or the maintenance of a complex social system. The termites, for example, have diploid males and diploid females, but they are as eusocial as honey bees and paper wasps (see Figure 13.27, right).

Another diplo-diploid, but definitely eusocial, organism is the bizarre-looking naked mole-rat [170, 1098]. This little hairless, sausage-shaped mammal (Figure 13.34) lives in a complex maze of burrows under the African plains that can house more than 200 individuals. The impressive size of their subterranean home stems from extraordinary cooperation among chain gangs of colony members, which work together to move tons of earth to the surface each year while burrowing about in search of edible tubers. Yet when it comes to reproducing, breeding is restricted to a single big "queen" and several "kings" that live in a centrally located nest chamber. Females other than the queen do not even ovulate. Instead, they serve as sterile helpers at the nest, consigned to specialized support roles for the queen and kings, as are most of the males in the colony [688].

Naked mole-rat workers are very closely related to their fellow workers in those colonies in which their parents are siblings [1009]. When a brother and

Figure 13.34 A mammal with an effectively sterile caste. Naked mole-rats live in large colonies made up of many workers who serve a queen and one or a few breeding males. Photograph by Raymond Mendez.

sister mate, their inbred offspring are likely to have coefficients of relatedness above 0.5 because both parents share rare family alleles as a result of having the same mother and father. Nonetheless, not every mole-rat king and queen are siblings, or even cousins; some reproducing mole-rats evidently prefer to pair off with a nonrelative [214]. Moreover, mole-rat colonies produce some especially fat individuals that apparently leave home to found a new colony elsewhere, presumably with an unrelated individual of the opposite sex from another group [138]. The fact that mole-rat workers in some colonies are not extraordinarily closely related tells us again that the evolution of eusociality is not utterly dependent on exceptionally high r values. This point is reinforced by the finding that the breeding pairs at the head of the eusocial colonies of the Damaraland mole-rat, a relative of the naked mole-rat, have a mean coefficient of relatedness of 0.02—no inbreeding here. As a result, the mean relatedness among members of the same colony is very close to 0.5, the standard coefficient of relatedness for siblings of outbred unions [173]. In the naked mole-rat, the Damaraland mole-rat, and many other species, helpers usually care for siblings, and that is apparently enough to accommodate the evolution of complex sociality.

The Ecology of Eusociality

Hamilton's rule that altruism can evolve when $(r_c \times C)$ is less than $(r_b \times B)$ contains more elements than r_b. In fact, altruism can spread even when r_b is close to 0, provided that C (the number of offspring the helper gave up to be an altruist) is also very low. In other words, if the ecology of the species is such that young dispersing adults have little chance of reproducing successfully, then nondispersers that remain at their natal site to help their relatives are likely to secure sufficient indirect fitness to make helping at the nest the adaptively superior option—an argument that we presented earlier when discussing avian helpers at the nest.

For many social animals, especially social insects, the chance that a dispersing individual will ever succeed in building something like the natal colony is vanishingly small (Figure 13.35) [16]. Consider that a foundress female of a eusocial ambrosia beetle takes about half a year to gnaw her way just 5 centimeters into a eucalyptus tree [633]. Most foundresses die long before they get to this point, which is only the first phase of building a tunnel complex in the tree. Once a large tunnel network is established, however, a colony can persist for decades, with helper daughters assured of a safe home in which to assist their mother in rearing reproductive males and new foundresses, a very few of which may disperse successfully, raising the indirect fitness of their stay-at-home sisters.

One of the main functions of helpers is the defense of their mother's valuable nest against predators and potential nest usurpers of their own species. Defense of a fortress nest where colony members can safely feed occurs in the ambrosia beetle just mentioned, as well as in a few thrips and many termite species. Likewise, soldier aphids tend to belong to species whose galls are unusually hard and durable [1015]. But defense of a large or long-lasting nest is not the only way in which sterile workers can increase the reproduction of their relatives. David Queller and Joan Strassmann argue that in many of the eusocial ants, bees, and wasps, the sterile workers' most important service is gathering food for their larval relatives [992]. Whereas fortress defenders such as termites and aphids typically live amid a wealth of digestible plant material, the standard ant, bee, or wasp must roam far from the nest site in search of scarce food. In so doing, it runs a gauntlet of predators. Because mortality rates are high under these conditions, a female that lived and foraged alone would often die before her brood achieved independence. If, however, a nesting female can enlist the aid of others, care will continue to be provided for her young even if she should die prematurely. Under these circumstances, helpers related to the primary reproductive female gain considerable indirect fitness by bringing the brood along the rest of the way to adulthood [387].

Thus, although certain genetic factors may give the evolution of altruism a boost, ecological factors that increase the positive effect of helping on the survival of relatives are equally important. Nonetheless, our understanding of complex sociality is still incomplete. For example, there are seven species of African mole-rats, all of which are burrowing, parental animals. For all these species, the costs of leaving a safe underground burrow would seem to be high, and the benefits of helping great, given the value of the burrow, the care required by helpless youngsters, and the advantages of communal tunneling. Therefore, we might expect to observe eusocial life in all seven species. However, direct evidence of eusociality is available for only two species, the naked mole-rat and the Damaraland mole-rat (although some persons believe that the other mole-rats may also exhibit some elements of eusociality) [170]. More generally still, worker castes are extremely rare in burrowing rodents, of which there are many species [591]. In fact, joint occupation of a burrow by several adults has been reported for only a handful of rodents [688]. These facts pose uncomfortable questions for the argument that fortress nests promote the evolution of eusociality [591]. The use of subterranean tunnels and the associated high costs of

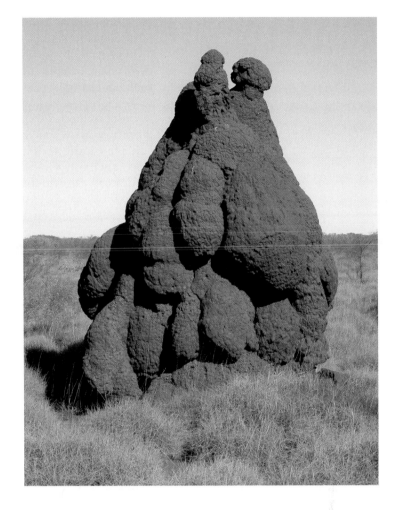

Figure 13.35   Fortress living space may provide an incentive against dispersal in many eusocial insects. The probability that a young spinifex termite will found a colony that eventually builds a mound of the size shown here is exceedingly low. Millions of workers laboring over many years in Western Australia produced this safe home made of red clay as hard as concrete. Photograph by the author.

dispersal must be only part of the ecological story behind the evolution of the eusocial mammals.

In general, it is easier to offer a tentative explanation for why a species *has* evolved a particular trait than for why a species has *not* evolved a particular trait. For example, researchers have accounted convincingly for the social life of the Florida scrub jay. But what about its close relative, the western scrub jay, a nonsocial species? Why haven't members of this species become social? There must be years when young western scrub jays have little chance of finding a suitable vacant territory. Why haven't they evolved the ability to remain as helpers at the nest under these conditions? Perhaps one or another transactional model will come to the rescue [1005], but much more remains to be learned about the genetic and ecological bases of altruism and social living before we can close the book on this great evolutionary puzzle.

Summary

1. In animal societies, individuals often tolerate the close presence of other members of their species despite the reproductive interference, increased competition for limited resources, and heightened risk of disease that social living entails. Under some ecological circumstances, the advantages of sociality (often improved defense against predators) are great enough to outweigh the many and diverse costs of social living. The common view that social life is always evolutionarily superior to solitary life is incorrect.

2. Animals that live together may help one another in various ways. Some cooperative acts may immediately elevate the personal reproductive success of both cooperators (mutualism). Still others may be performed at some cost that will be more than repaid when the recipient pays the helper back at a later date (reciprocity). Finally, some helpful actions are considered altruistic because they reduce the reproductive success of the helper while at the same time raising the reproductive output of another individual.

3. Mutualism and reciprocity can spread through a population via the action of direct natural selection. If, however, a helper really does permanently reduce its direct fitness while helping to raise the fitness of another, its altruism poses an evolutionary puzzle. One solution may be that the direct fitness costs of certain kinds of altruism are outweighed by the indirect fitness gains generated when an individual increases the number of its surviving nondescendant kin.

4. An allele for altruism can spread in competition with an alternative form of a gene that promotes personal reproduction provided that altruistic individuals increase their relatives' reproductive success enough that the indirect fitness they gain more than compensates for any reduction in their direct fitness. As expected, the overwhelming majority of cases of altruism found in nature fit this description. Self-sacrificing helpers, which may even be sterile, almost always assist close relatives, thereby boosting their inclusive fitness (the sum of their indirect and direct fitness).

5. Although the indirect fitness gained by helping is increased if the coefficient of relatedness between helper and beneficiary is high, sterile castes can evolve even when the degree of relatedness is not great among group members, provided that the ecology of the species is such that helpers can improve the direct fitness of relatives, especially if the odds are against successful personal reproduction by individuals dispersing from a safe natal nest. But many questions about extreme altruism remain unanswered.

Suggested Reading

W. D. Hamilton's work [502] initiated a revolution in the understanding of social behavior; see also reviews by Richard Alexander [16], Mary Jane West-Eberhard [1272], and Steve Emlen [366]. I have relied heavily on Jerry Brown's *Helping and Communal Breeding in Birds* as a guide for understanding the kinds of selection that affect the evolution of social behavior [158]. Kern Reeve and Laurent Keller describe ways to test transactional theory and alternative explanations for why subordinate members of social groups vary in how much they reproduce [1005]. Uli Reyer's study of the social pied kingfisher is superb [1013], as are the papers by Jan Komdeur and colleagues on the Seychelles warbler [658, 660]. Excellent reviews of social behavior are available for ants [553] and other social insects [821, 1314], lions [916], acorn woodpeckers [654], and naked mole-rats [1098].

14 | *The Evolution of Human Behavior*

Humans are an animal species with an evolutionary history. Yes, we are unusual and quite wonderful in our own way, but so are kittiwakes and hangingflies. Having applied evolutionary thinking to kittiwakes and hangingflies, let's see if we can do the same for humans. If our evolution has been shaped by natural selection, then there should be some connection between our behavioral attributes and the ultimate goal of passing on our genes to the next generation. Admittedly, this proposition is far from universally accepted, especially by those who believe that our behavior is essentially "cultural" as opposed to "biological." And there is no doubt that our cultural traditions influence our behavior in many striking ways. If, for example, I had been born and reared by tribesmen in Papua New Guinea, I would consider it natural to be seen in public completely naked except for a long, thin, hollow gourd covering my penis (Figure 14.1). Having not grown up in Papua New Guinea, I have no desire whatsoever to wander around in a similar outfit here in Tempe, Arizona, or anywhere else, for that matter. The pervasive effects of our cultures on our behavior obviously make it especially challenging to apply evolutionary theory to the human species. But because human cultures are the product of an evolved brain, it seems highly likely that an understanding of our evolution would cast light on why we do what we do, whether in New

◀ *How can an adaptationist approach be applied to humans when our behavior varies so much from culture to culture? Photograph by Bill Bachmann*

Figure 14.1 Cultural traditions are powerful influences on human behavior. Men in many parts of New Guinea, including this Dani man from Irian Jaya, have traditionally worn large penis sheaths, but little else. Photograph by G. J. James.

Guinea or New York. Therefore, we shall use natural selection theory to tackle some of the Darwinian puzzles associated with our behavior. Our first step is to analyze the possible adaptive basis of my generous donation of blood to a blood bank the other day. Could this action really have anything to do with human evolution?

The Adaptationist Approach to Human Behavior

When I gave blood, I was not alone. Those of us hooked up to our little plastic blood bags were behaving altruistically because we were taking a tiny risk of a medical complication of some sort in order to provide a substantial fitness benefit to an unknown recipient in need of a transfusion. You may recall from the preceding chapter that one ultimate hypothesis to account for helpful, self-sacrificing acts is that they generate indirect fitness for the helper. Blood donation, however, cannot produce such gains when the helper and the recipient are not related, as is usually the case. Moreover, the action can hardly be a standard case of reciprocity with eventual repayment to the donor, given that most people who receive a blood transfusion do not know who the donor was.

Some persons have claimed, therefore, that blood donation is immune to evolutionary analysis and is instead a kind of "pure" altruism practiced for no possible fitness advantage. According to Peter Singer, "Common sense tells us that people who give blood do so to help others, not for a disguised benefit to themselves" [1114]. Maybe so, but because Richard Alexander was intrigued by the challenge of explaining blood donation from an evolutionary perspective, he kept at it. Perhaps, said Alexander, the donor is repaid, not by the recipient of his blood, but by the donor's everyday companions, who may be impressed by, and wish to cooperate with, someone "so altruistic that he is willing to give up a most dear possession for a perfect stranger" [18].

In Alexander's view, blood donation occurs at the proximate level because our psychological mechanisms make us feel better when we do any of a wide range of conspicuous good deeds, perhaps especially for people unrelated to us. In effect, Alexander is proposing that our brains have a special system that rewards us for certain actions as a means of encouraging us to behave adaptively. Such a mechanism would be similar to the human brain's face recognition center, which, as you may recall (see pp. 128–129), gives us the adaptive ability to identify individuals visually very quickly and accurately. The putative "cheap altruism" mechanism in our brains motivates some of us in Western societies to give blood occasionally or to exhibit other low-cost forms of culturally approved charity. In the past, when our psychological systems were evolving, our hominid ancestors could not donate blood or give used clothing to the Salvation Army, but they could potentially help their companions in various small ways. In a social environment in which an individual's reproductive success surely depended on building and maintaining good relationships with other people, those persons who performed visible acts of charity, which everyone knew were unlikely to be repaid, may have acquired a reputation for generosity, stimulating others to join them in beneficial alliances and thereby generating a fitness payoff for their behavior.

This **indirect reciprocity** hypothesis for blood donation, according to which the donor gains fitness from others who see his good deed, yields many testable predictions. For one thing, unpaid blood donors should almost always let their friends know that they have given blood. Bobbi Low and Joel Heinen report that students at the University of Michigan are significantly more likely to donate to fundraising drives if they receive a pin or tag that advertises their participation. Such devices are routinely distributed by the American Red Cross during their blood drives [747].

A second prediction is that people will be reluctant to broadcast their refusal to give blood. In one interview study in Germany, 34 percent of the subjects said that although they had never given blood, they were nevertheless willing to do so, whereas only 17 percent said that they simply would not be donors [1018]. In an American study, 30 percent of the students given a survey form checked a box indicating that they probably would *not* give blood in a drive to be held in the following week; the other 70 percent left the box blank, which meant, according to the instructions they received, that they probably would give blood the next week. In reality, a mere 17 percent of the students actually gave blood in the following week, showing that many of those who failed to check the "not a donor" box did not wish to acknowledge, even on an anonymous survey form, that they were not going to participate in the blood drive [213].

A third prediction is that when people have a choice of other people to help, they will prefer to assist those who are known to be generous. To test this prediction, Klaus Wedekind and Manfred Milinski devised an experimental game that involved many rounds of giving and (potentially) receiving. Players were supplied with a certain amount of cash at the outset of the game, and were offered the opportunity to give some of it to other players. A donor could specify that a recipient, unknown to the donor, would receive one Swiss franc (or two), in which case the recipient actually received four francs, with the extra money coming from the experimenters. Thus, the donor paid a small price in order to give the recipient a substantial benefit. As each player made his decision whether to give up a franc or two, he knew the past record of giving by each potential recipient, but again, he did not know the identity of this individual, nor did he know whether this person had given him money in a previous round. Even so, individuals with a record of giving were more likely to receive cash donations from the other players, as predicted by the indirect reciprocity hypothesis for altruism [1259].

Milinski and some others have used a similar kind of game to look at the effects of giving to charity. They found that individuals known (under a pseudonym) to donate both to members of their group of game players and to UNICEF (the highly regarded children's fund of the United Nations) not only received more money in return from their fellow players, but also received more votes in a student council election (admittedly staged by the experimenters). In other words, public donations appeared to boost the social reputation of the generous person, making it more likely that others would believe that he or she had the best interests of the group in mind and so was deserving of financial and political support [823].

A key assumption underlying research of the sort we have been describing is that it pays to have a good reputation. This proposition is itself open to evaluation, which Linda Mealey and her co-workers supplied via an experiment in which they showed a set of photographs to a group of subjects, who were asked to rate the attractiveness of the faces they saw. Each of the 36 images shown to the evaluators came with a short negative, positive, or neutral story about the character of the person being viewed. For example, one person might be described negatively as an embezzler while another was portrayed positively as someone who returned a lost wallet containing $250 to its owner. A week later the subjects were asked back into the laboratory to rate the facial attractiveness of a set of 72 images that contained the 36 photographs they had already seen. Although the participants in the study were told the data they were providing was designed to check on the reliability of their attractiveness ratings, in reality the goal of the research was to see whether the supposed social character of a person affected their memory of his face. In fact, the faces of persons of supposed bad character were remembered more often than the faces that had been linked to a positive or a neutral story [813]. In other words, humans apparently have a memory bias that could help us identify, remember, and avoid social cheats.

This finding is complemented by more recent work on the contribution of different parts of the brain to our social behavior. Ralph Adolphs and Anthony Damasio have show that one distinctive structure, the amygdala, apparently directly influences our evaluation of the trustworthiness of individuals. The amygdala helps arouse particularly strong negative emotional feelings in response to faces thought to be untrustworthy [4]. Thus, the amygdala and other parts of the brain contain proximate mechanisms whose properties suggest that people have exerted selection on one another favoring individuals able to avoid being labeled a bad person. One means to this end is be concerned about one's reputation and to achieve recognition for one's generosity rather than one's untrustworthiness.

Discussion Question

14.1   Almost all of us have, at one time or another, been approached by a person in search of a donation who gave us something, a pamphlet or pen or ribbon, prior to or during his or her request for money [212]. Why, in evolutionary terms, is this such a common tactic by those in search of charity?

The Sociobiology Controversy

The studies I have just outlined have not established a direct link between acquiring a reputation for generosity and an increase in reproductive success. Indeed, it would be difficult to do so in modern environments, given the size of the social units in which we currently operate and the access we have to effective birth control. In order to test the prediction that indirect reciprocity confers fitness benefits on generous individuals, we would presumably have to examine the relationship between these variables in a traditional society, one in which people live in relatively small groups where everyone's behavior is under constant scrutiny and where no one has access to modern birth control technology. But I have no doubt that even if studies of this sort showed that persons whose reputations had been enhanced by their generosity to nonrelatives did indeed have higher fitness, a great many people would still object to the conclusion that our capacity for generosity was the product of natural selection. Even W. D. Hamilton, whose research on inclusive fitness was founded on the principle that individuals were expected to behave in the interests of their genes, wrote that he had a "dislike for the idea that my own behaviour or the behaviour of my friends illustrates my own theory of sociality or any other. I like always to imagine that I and we are above all that, subject to far more mysterious laws" [504].

Although Hamilton exhibited some discomfort with evolutionary analyses of human behavior, others have been downright antagonistic to the approach, as illustrated by the response to the 1975 publication of E. O. Wilson's *Sociobiology: The New Synthesis*. This book contained a final chapter on human behavioral evolution [1315], and this one chapter generated a furious controversy. Wilson's critics, including some of his own colleagues at Harvard University, accused him of promoting a discredited view of human behavior that, if accepted, would cause great societal harm [20]. Wilson and others effectively answered these charges [1316], and now, three decades after the publication of *Sociobiology*, human sociobiology is flourishing under the new labels of human behavioral ecology, evolutionary psychology, and evolutionary anthropology [11]. Although this particular controversy has largely faded, many people would still like to ignore or discredit the evolutionary analysis of human behavior, and these individuals often raise the very same criticisms that Wilson's opponents published in 1975. Three common misunderstandings of the evolutionary approach to human behavior follow, with a rebuttal for each one.

"We humans don't do things just because we want to raise our inclusive fitness." Some opponents of evolutionary studies of human behavior have pointed out that, although humans desire a great many things, the wish to maximize our inclusive fitness rarely, if ever, is at the top of our list [1031]. If you had asked Picasso why he wanted to paint, or Bill why he wanted to marry Jane, Picasso and Bill would not have explained that they wanted to increase their genetic success. But if a baby cuckoo could talk, it would not tell you that it rolled its host's eggs out of the nest "because I want to propagate as many copies of my genes as possible." Neither cuckoos nor humans need to be aware of the ultimate reasons for their activities in order to behave adaptively. The human brain's costly decision-making mechanisms were shaped by natural selection to enhance our fitness, not to provide us with the capacity to monitor the reproductive consequences of each and every action. It is enough that proximate mechanisms, like a well-developed sex drive, motivate individuals to do things, like copulate, that are correlated with fitness—the production of offspring. On the proximate level, we learn a language, we enjoy sweet foods, we fall in love, we have sex, and we derive satisfaction from our charitable actions because we possess physiological mechanisms that facilitate these behaviors. Because honey tastes good, we want to eat it, and when we do, we acquire useful calories that may contribute to our survival and reproductive success without our ever being aware of the evolved function of our fondness for sweets.

"But not all human behavior is biologically adaptive!"   Over the years, critics have claimed that an array of cultural practices, such as circumcision, prohibitions against eating perfectly edible foods, and the use of birth control devices to limit fertility, seem unlikely to advance individual fitness. If some humans do things that reduce their fitness, these persons argue, then sociobiology cannot tell us anything about human behavior. Note, however, that the claims of these critics are based on the assumption that natural selection theory requires that every aspect of every organism be currently adaptive [463]. This assumption is incorrect, as noted before (see Table 6.1). The adaptationist approach is not used to assert that every trait is an adaptation. Instead, evolutionary biologists use natural selection theory to identify interesting puzzles, produce plausible hypotheses, and attempt to test alternative explanations. If, for example, an adaptationist were to examine religiously motivated celibacy, he or she would be fully aware of the possibility that the trait was a *maladaptive* by-product of certain brain modules that control other generally adaptive abilities. Nonetheless, a sociobiologist might still try to produce a testable hypothesis on how acceptance of celibacy could paradoxically enable priests to leave more copies of their genes than if they were heterosexually active. Needless to say, this would be a challenge, but perhaps not an insuperable one for someone aware of indirect selection.

Even if twenty adaptationist hypotheses on celibacy were developed, there is no guarantee that any would withstand testing. This is as it should be. T. H. Huxley, the great defender of Darwinian theory, wrote, "There is a wonderful truth in [the] saying [that] next to being right in this world, the best of all things is to be clearly and definitely wrong, because you will come out somewhere"

[580]. If our sociobiological hypotheses about the celibate priesthood were incorrect, effective tests would tell us so, which would enable us to remove some ideas from the table.

"Evolutionary approaches to human behavior are based on a politically reactionary doctrine that supports social injustice and inequality."   The original critics of sociobiology denounced it as politically dangerous because they feared it would provide scientific cover for immoral social policies of the sort advanced by racist and fascist demagogues in the past [20, 21]. According to this view, the claim that such and such a trait is adaptive implies that it is both genetically determined and good, and therefore cannot and should not be changed. After all, if one claims that male dominance is adaptive, isn't this saying that the status quo is desirable and that feminist claims fly in the face of what is genetically fixed and morally necessary?

Scientific findings can be employed in ways that often surprise, and even horrify, the investigator. Einstein's basic research on the relation between energy and matter contributed to the development of atomic weapons, much to his dismay; Darwin's theory of evolution has been misunderstood and misused by some persons to defend the principle that the rich are evolutionarily superior beings, as well as to promote unabashedly racist plans for the "improvement of the human species" by selective breeding.

We can hope that political perversions of evolutionary theory have been so discredited that they will not happen again. The critical point here, however, is that sociobiologists attempt to explain why social behavior exists, not to justify any given trait. This distinction is easily understood in cases involving other organisms. Biologists who study infanticide by male langurs or how a small marine copepod adaptively feeds on the eye of the Greenland shark are never accused of approving of infanticide or the blinding of sharks. To say that something is biologically or evolutionarily adaptive means only that it tends to elevate the inclusive fitness of individuals with the trait—nothing more.

Moreover, a hypothesis that a behavioral ability is adaptive does not mean that the characteristic is developmentally inflexible. All sociobiologists understand that development is an interactive process involving both genes and environment. Change the environment, and you will change the gene–environment interactions underlying a behavioral phenotype, with the result that the phenotype may change. A classic example in human biology involves language acquisition, a clearly adaptive ability that rests on vast numbers of extraordinarily complex gene–environment interactions. Change the cultural environment that a baby is exposed to, so that, say, a child of English-speaking parents is raised by adults who speak Urdu, and we all know what will happen. But this ability of infants to adopt the local language requires special genetic information, especially from those genes that code for proteins that promote the development of certain specialized brain modules that underlie language learning. These neural units have evolved specifically in the context of language acquisition, rather than as a side effect of some sort of generalized intelligence, as we can see from the existence of two rare human phenotypes. On the one hand, certain profoundly retarded individuals chatter away, producing completely grammatical sentences with little inherent meaning. On the other hand, some English-speaking individuals with normal to above normal intelligence have great trouble with the rules of grammar, often failing, for example, to add "-ed" to verbs when they wish to speak of past events [955]. This kind of evidence tells us that selection is behind the evolution of the special mechanisms that make language acquisition easy and effective for the vast majority of us.

Discussion Questions

14.2   In 1981, the first vocational school for deaf children was opened in Nicaragua. Although the children at the school had never been taught a sign language, they invented one of their own that became increasingly complex [1087]. This unique language has many of the fundamental properties of any language, including the breakdown of information into discrete units and a grammatical presentation of words (gestures in this case). What is the significance of this case study for discussions about the evolutionary basis of language learning ability in human beings?

14.3   Philip Kitcher states that "socially relevant science," such as sociobiology, demands "higher standards of evidence" because if a mistake is made (a hypothesis presented as confirmed when it is false), the societal consequences may be especially severe. For example, a hypothesis that men are more disposed to seek political power and high status in business and science than women is dangerous because it "threaten(s) to stifle the aspirations of millions" [645]. How might a sociobiologist respond to Kitcher's claim?

Arbitrary Culture Theory

The main alternative to an evolutionary approach to human behavior is the theory that human cultural traditions arise from accidents of history and the almost unlimited inventiveness of the human mind. According to this **arbitrary culture theory**, most of our activities have little or nothing to do with fitness maximization, but instead reflect the more or less arbitrary processes by which traditions originate and persist over time. The contrast between these two approaches can be illustrated by the differences between cultural anthropologists and sociobiologists in their analysis of adoption, which is common in some societies but much less so in others. At one time, for example, an amazing 30 percent of all children became adoptees in Oceania (the islands of the central Pacific Ocean), leading to the claim that this peculiar feature of life in this part of the world is a cultural anomaly immune to evolutionary analysis [1056]. When Joan Silk analyzed data on the relationships between moderately large samples of adopters and adoptees in 11 different cultures in Oceania [1108], however, she found that most adopters were actually caring for their close relatives (Figure 14.2). The highly kin-biased nature of adoption in these societies casts doubt on the arbitrary culture hypothesis while supporting an indirect fitness hypothesis for this form of human altruism.

Nevertheless, a minority of adopters in Oceania do take in the children of strangers, their genetic competitors, who may receive the same love and affec-

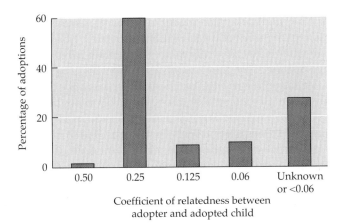

Coefficient of relatedness between adopter and adopted child

Figure 14.2   **The indirect fitness hypothesis for adoption** can be tested by examining the coefficient of relatedness between adoptive parents and their adopted children. In 11 island societies in Oceania, adopter and adoptee were usually close relatives, yielding indirect fitness gains for the adopters. After Silk [1108].

tion that parents typically supply to their own offspring. Are these exceptions to the typical adoption pattern in Oceania impossible to explain from a sociobiological perspective? Not necessarily. Silk suggests that small families in some agricultural cultures might benefit from gaining adoptees, even if they were nonrelatives, because adopted persons can contribute to the family workforce, raising the economic productivity of the family unit and improving the survival chances of the adopters' genetic offspring. This direct fitness hypothesis produces the prediction that small families in Oceania should be more likely to adopt than large ones, a prediction that Silk showed was correct.

An alternative evolutionary hypothesis for adoption among nonrelatives, which occurs in many other societies besides those in Oceania, recognizes that some decisions may be the maladaptive by-product of otherwise adaptive proximate mechanisms. Adoption of a nonrelative might, for example, be a consequence of motivational systems that cause adult humans to want to have children and raise a family. According to this hypothesis, although adults who adopt infant strangers may reduce their fitness, the urge to have and love children is *usually* adaptive. Because such psychological mechanisms tend to elevate fitness, they are maintained in human populations even though they *sometimes* induce people to behave in ways that do not result in fitness gains.

The point is that in the past, selection has favored individuals whose behavior has been guided by desires that steered them toward reproductive success. As noted earlier, these proximate desires, such as a strong sex drive and eagerness to have a family, are what really control our behavior, not dispassionate mathematical calculations of the fitness consequences of various options open to us. Therefore, persons unable to satisfy their proximate wishes will sometimes do things that do not boost their inclusive fitness. For example, a person deprived of the opportunity to have sex may masturbate; a couple unable to have a family may adopt a child who is not genetically related to them.

This side effect hypothesis for adoption generates testable predictions, one of which is that husbands and wives who have lost an only child, or who have been unable to produce children themselves, should be especially likely to adopt nonrelatives. In fact, infertility is by far the most common reason given by Californian couples who want to adopt; moreover, women who have been treated for infertility are significantly more likely to attempt to adopt than are women who have not been treated for this condition [389].

Another prediction from this side effect hypothesis is that adoption should also sometimes occur in birds (see pp. 417–418) and mammals when adults have lost their offspring and fortuitously encounter a substitute (Figure 14.3) [613]. Cardinals have been known to feed goldfish [1269], and a white whale was seen trying to lift a floating log part way out of the water, as if the log were a distressed infant that needed help to reach the surface to breathe [141]. Adopting a goldfish and helping a log did nothing to promote either the cardinal's or the white whale's genes. But these individuals had almost certainly lost their young recently and were employing generally adaptive parental behaviors in an unusual, nonadaptive fashion.

The likelihood of a maladaptive application of the parental drive may be especially high in modern human societies because these environments are so very different from the ones in which our proximate psychological mechanisms evolved. In Western cultures, babies are routinely made available to nonrelatives who do not know the parents of the adoptee, something that almost never occurred in the distant past. Under these novel conditions, our motivational and emotional systems, which evolved long ago, are especially likely to cause us to behave maladaptively and, in so doing, reveal something about the naturally selected features of our psyches.

Figure 14.3 Adoption occurs in nonhuman animals, often when adults have just lost an offspring but encounter a substitute. Here several emperor penguins compete for "possession" of a youngster. Photograph by Kim Westerskov.

The abundance of testable sociobiological hypotheses on adoption (and we have not discussed the possibility that adoption can be a form of charitable behavior that raises the social reputation of adopters) speaks to the productive nature of evolutionary theory. No one evolutionary hypothesis explains every case of adoption, just as no one hypothesis resolves every aspect of a complex anatomical or physiological phenomenon. However, the adoption example demonstrates that an evolutionary approach can more than hold its own against the competing arbitrary culture theory.

Discussion Questions

14.4   Marshall Sahlins has argued that sociobiology is contradicted because people in most cultures do not even have words to express fractions. Without fractions, a person cannot possibly calculate coefficients of relatedness, and without this information (Sahlins claims), people cannot determine how to behave in order to maximize their indirect fitness [1056]. Has Sahlins delivered a knockout blow to sociobiological theory?

14.5   A common view in some circles is that the behavioral differences between men and women stem largely from the effects of cultural pressures to conform to stereotypical male and female roles. Thus, young boys receive certain toys to play with, such as guns and miniature trucks, and young girls get different toys, such as dolls and baby carriages. The sex stereotyping involved in the selection of toys for boys and girls is said to push boys into "masculine roles" while guiding girls into culturally approved "feminine roles." Two researchers gave young male and female monkeys both types of toys and measured the amount of time they spent with each kind of toy. Why did they do this? What prediction based on an arbitrary culture hypothesis were they able to test? Check their results in [15].

Adaptive Mate Preferences

Now that we have seen how evolutionary biologists can produce and then test hypotheses on donating blood and adopting children, let's examine their attempts to explain our reproductive behavior. Tackling reproduction is a challenge because the cultural rules and regulations surrounding human sexual behavior are extraordinarily diverse. There are monogamous, polygynous, and polyandrous societies, some in which you cannot marry an unrelated person who belongs to your clan, others in which adult men can marry prepubescent girls, some in which males and their relatives provide payment for a bride, and others in which women must bring a costly dowry with them to their wedding. Despite all this cultural variety, certain biological facts of life still apply to our species as a whole.

For one thing, women are typical female mammals in that they retain control of reproduction by virtue of their physiological investments in producing eggs, nurturing embryos, and providing infants with breast milk after they are born. Although men are also able and often willing to make large parental investments in their offspring, their reproductive decisions nevertheless take place in a setting defined by female physiology and psychology [276, 440]. For example, consider the reproductive significance for men of variation in female fertility, which is a function of a woman's age, health, and body weight, among other things. Preadolescent and postmenopausal women obviously cannot become pregnant. Women in their twenties are more likely to become pregnant than women in their forties. Healthy women are more fertile than sick ones. Women that are substantially overweight or underweight are less likely to become pregnant than women of average weight [1207].

Given the reality that women differ in their likelihood of conception, many evolutionary biologists have predicted that men should possess psychological mechanisms that enable them to evaluate female fertility with accuracy and that they should attach special value to cues associated with high fertility. One way to check this prediction is to look for a positive correlation between what men consider to be "good looks" and female fertility. Note that the evolutionary expectation here is very different from a frequently cited alternative explanation for male preferences, which is that men and women have been culturally indoctrinated into acceptance of a battery of social norms. These cultural pressures are often said to set nearly impossible standards of feminine beauty as a means of keeping most women on the defensive about their looks [1333].

So are standards of beauty in Western society essentially arbitrary, perhaps even imposed on women by mean-spirited men, or do they reflect a largely unconscious, evolved male interest in the fertility of potential mates? In Western cultures, males generally prefer the following feminine features: full lips, small noses, large breasts, a waist circumference that is substantially smaller than hip circumference (the hourglass figure) [1115], and an intermediate weight rather than extreme thinness or obesity [1207]. Men also prefer certain female scents [243] and vocal stimuli [234, 573]; women that produce the preferred odors and vocalizations are also visually attractive to men. The preferred visual attributes are associated with developmental homeostasis, a strong immune system, good health, high estrogen levels, and, especially, youthfulness [440, 1192], all of which together constitute a recipe for high fertility. Thus, for example, the level of circulating estrogen in a sample of healthy Polish women was related to their body shapes (Figure 14.4). The added estrogen present in the large-breasted, narrow-waisted women in this study may have translated into as much as a threefold increase in the probability of conception for this group compared with the other participants [592].

Thus, many of the physical features that men in Western societies tend to find sexually appealing are linked to a woman's potential to become pregnant. But even highly fertile women can conceive only during the few days each month when they are ovulating. If men evaluate women in terms of their immediate reproductive value, we would expect males to prefer the body odor of a

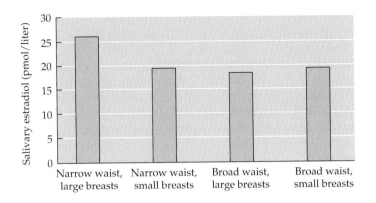

Figure 14.4 Body shape is correlated with fertility in women. Women with features that men often claim they prefer—namely, narrow waists and large breasts—are more fertile than women with other body shapes, based on a sample of healthy Polish women between 24 and 37 years of age who were not taking birth control pills. After Jasiénska et al. [592].

female during her fertile phase over the odor of the same woman during her nonovulatory phase. Devendra Singh and Matthew Bronstad showed that, as predicted, men find the smell of a T-shirt worn by an ovulatory woman to be "more pleasant and sexy" than the scent of a T-shirt worn by the same woman when she is not fertile [1116]. Subsequently, another research group checked the prediction that women's faces should be more attractive during the fertile phase than during the nonfertile phase of the menstrual cycle. Again, the actual results matched the expected ones. Men shown photographs of the face of a woman when she was and when she was not ovulating typically voted for the "fertile face" [1022]. Needless to say, the men were not consciously aware of the reproductive significance of their preferences, nor would they have to be in order to behave in ways that tended to encourage them to pursue women likely to conceive.

Discussion Questions

14.6   The research on male mate preferences outlined above has involved interviews of men in Western cultures. What objection might be raised to the conclusions based on this research? How might those objections be overcome? (You may find the following papers useful: [178, 266, 609, 694, 1279, 1350].) If men from two cultures have different standards of feminine beauty, how might a social scientist explain this finding, and how would an adaptationist account for the result?

14.7   Some persons believe that men in Western societies generally prefer sexual partners who are younger than they are because they have been taught what is an arbitrary cultural convention from an early age. What prediction follows from this hypothesis with respect to the dating preferences of teenage males for females of different ages? What conclusion follows from the data in Figure 14.5 [630]?

Adaptive Mate Preferences of Women

We have presented the argument that because women vary in fertility and because female fertility has a major effect on male reproductive success, men should find cues associated with high fertility sexually attractive. Although men also vary in fertility to some extent, the degree of variation is much less than in women, if only because there is no male equivalent of the menstrual cycle or of menopause. On the contrary, most men of any age can supply plenty of functional sperm on any day of the month to any woman willing to accept their gametes. Not surprisingly, therefore, evolutionary biologists have never bothered to explore whether female mate choice revolves around an assessment of male fertility. Instead, adaptationists have focused on two other factors: the capacity of men to supply good genes to their offspring, and their ability (and

Figure 14.5   Preferred dating partners by age for male teenagers. The age of preferred dates is shown in relation to the age of the male subjects. After Kenrick et al. [630].

willingness) to provide resources for their partner's offspring. Both of these traits appear to vary markedly among men, and both could greatly affect a woman's reproductive success.

From an evolutionary perspective, then, we would expect women to find attractive those male physical attributes that indicate high genetic quality or parental ability. Some studies have reported that women prefer men with "masculine" facial features; namely, a prominent chin and strong cheekbones. In addition, facial symmetry has been identified as a plus, as well as an athletic, muscular upper body [440]. This combination has been linked to high testosterone levels, good current health, and perhaps most importantly, good health during juvenile development. Despite developmental homeostasis (see p. 81), early deficits in nutrition can have some long-lasting negative effects on survival and reproductive success in our species [748]. Male development is at special risk because of the potentially damaging side effects of the male sex hormone, testosterone. Therefore, the ability of a man to develop normally, despite high levels of circulating testosterone, is a possible indicator of a strong immune system capable of overcoming the handicap imposed by the hormone [395]. Considerable debate surrounds the idea that males of our species (and many others) have evolved honest signals of the quality of their immune systems [1021] (see Chapter 10). Nevertheless, one can make the case that women *should* (in fitness terms) find certain masculine signals appealing because strong, healthy partners may be able to pass on these characteristics, assuming that they are heritable, to their male offspring, which would then be sexually attractive to the next generation of women. Moreover, fathers of this sort ought to be able to compete effectively with rivals for dominance within their group, thereby securing the resources that go with high social status; men with political power usually have more material goods to provide their wives and children than do their subordinate companions [440].

You should know, however, that not every study has reported that women find men with the most masculine, big-jawed faces most attractive (Figure 14.6) [1172]. Moreover, few evolutionists claim that handsomeness, however evaluated, necessarily trumps direct evidence of a man's wealth and his willingness to share it with a partner and her offspring. In fact, most persons interested in the evolution of mate choice by women and men have predicted that good looks should be far less important for single women than for single men [178, 465]. According to this majority view, the extremely demanding task of rearing a baby to the age of independence puts a premium on securing a husband willing and able to assist with the children, whereas a man's fitness will be zero unless his sperm get to fertilize an egg, which requires a fertile woman.

The importance for women of having a wealthy partner has been established in studies showing that females in cultures without birth control who secure relatively rich husbands do tend to have higher fitness than females whose partners can provide them with fewer material benefits. Among the Ache of Paraguay, the children of men who were good hunters were in fact more likely to survive to reproductive age than the children of less skillful hunters [621]. Likewise, several studies of traditional societies in Africa and Iran have revealed a positive correlation between a woman's reproductive success and her husband's wealth, as measured by land owned or number of domestic animals in the husband's herds [118, 583, 764]. Even in modern societies, household income is correlated with children's health, with the effect growing larger as children get older. Chronic illnesses in childhood can reduce the earning power of children that reach adulthood, thus perpetuating poverty across generations [198], with all of the reproductive consequences this has for human beings.

Thus, it would make reproductive sense if females tended to put wealth, social status, and political influence ahead of good looks in partners. This evo-

(A)

(B)

Figure 14.6 Testosterone and male facial attractiveness. (A) In this study, digital images of the same male's face were altered to reflect the developmental effects of testosterone from low to high levels (left to right). (B) When young women were asked to judge these photographs for physical attractiveness, they tended to pick images like the one in the middle, whereas when they were asked to rate the images in terms of social dominance, they tended to pick those similar to the right-hand image. After Swaddle and Reierson [1172].

lutionary prediction has been supported by many questionnaire studies. However, even when researchers have found clear differences between the sexes in the value they attach to "good financial prospects" versus "good looks," the absolute measures of importance given to these attributes have not necessarily been especially high for either sex. But the men and women in these studies typically have not had to specify which items among a list of attributes are absolutely essential to their choice of mates, and which would be nice to have in a partner, but not crucial. Therefore, a team of social psychologists led by Norm Li attempted to put constraints on the choices made by the people they interviewed by giving them a limited budget to expend on designing a hypothetical ideal mate. A subject was given a list of traits and told to decide how many of his or her limited supply of "mate dollars" to use when purchasing any one item, such as physical attractiveness, creativity, yearly income, and so on. To secure a partner at the second level of attractiveness, or creativity, or yearly income (the highest level was 10) would require 2 mate dollars; to get someone at the eighth level would require 8 mate dollars. When the persons interviewed had only 20 mate dollars to work with, their investments differed greatly according to their sex. Men devoted 21 percent of their total budget toward the acquisition of a physically attractive partner; women spent 10 percent of the same total amount to the same end. On the other hand, women on this tight budget devoted 17 percent of their money to boost the yearly income of an ideal mate, whereas men invested just 3 percent of their mate dollars on this attribute.

After having specified how they would spend their first 20 mate dollars, the participants were given two additional 20-dollar increments. By the time they reached the third 20 mate dollars, the sexes did not differ markedly with respect to the attributes they were buying. Having already purchased what they really valued, they could and did spend on other attributes. This experiment tells us

Figure 14.7 Men and women from different cultures show similar mate preferences. Men advertise for younger women and women advertise for older men in both (A) the *Arizona Solo* and (B) the *Times* of New Delhi, India. The advertisers indicated their own age and the maximum and minimum ages they would accept in potential partners. After Kenrick and Keefe [631]

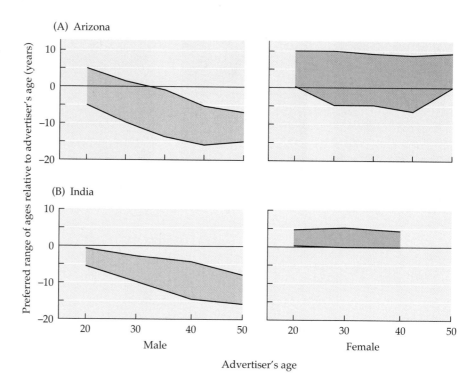

(A) Arizona

(B) India

Male

Female

Advertiser's age

that people view some mate characteristics as essential items and others as mere luxuries, add-ons if you will. The essential elements are not the same for men and women, as predicted by an evolutionary approach [716].

Personal ads, whose cost limits the number of words used by the advertisers, also provide relevant evidence on what people consider fundamentally important in a mate. So, for example, women seeking partners through newspapers are far more likely than men to specify that they are looking for someone who is relatively rich [1256]. In keeping with this goal, women advertisers in both Arizonan and Indian newspapers also often specify an interest in someone older than they are (Figure 14.7) [631]; older men usually have larger incomes than younger men [178].

If women really are highly interested in a partner's wealth and capacity to provide for offspring, then men in their thirties should be most desirable because men of this age have relatively high incomes and are likely to live long enough to invest large amounts in their children over many years. One can calculate the "market value" of men of different ages by using samples of personal ads and dividing the number of women requesting a particular age class of partner in their advertisements by the number of men in that age class who are advertising their availability; this measure thus combines both demand and supply. Men in their late thirties have the highest market value (Figure 14.8) [932].

Figure 14.8 Age and the market value of men. Market value is measured by the number of advertisements in the Personals sections of newspapers by women requesting men of certain age classes divided by the number of men of those age classes announcing their availability. After Pawłowski and Dunbar [932].

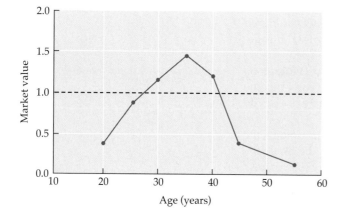

Discussion Question

14.8 If an evolutionary approach to human reproductive behavior is useful, what should the market value curve for women look like? Add your predicted data points to the graph shown in Figure 14.8. Check your prediction against data in [932].

It could be, however, that women's interest in the earning power of potential mates is a purely rational response to the fact that males in almost every culture control their society's economy, making it difficult for a female to achieve material well-being on her own. If this nonevolutionary hypothesis explains why females favor wealthy males, then women who are themselves well off and not dependent on a partner's resources should place much less importance on male earning power. Contrary to this prediction, several surveys have shown that women with relatively high expected incomes actually put *more* emphasis, not less, on the financial status of prospective mates [1208, 1295]. For example, relatively wealthy female undergraduates put higher value on wealth and status in potential long-term mates than do less wealthy females [181].

Mating preferences are one thing, but mating behavior is another. Are women actually more likely to copulate with men who possess the attributes they prefer than with men who lack them? Among the Ache of eastern Paraguay, good hunters with high social status are more likely to have extramarital affairs and produce illegitimate children than poor hunters, suggesting that females in this society find skillful providers sexually attractive [621]. Likewise, in Renaissance Portugal, noblemen were more likely to marry more than once, and more likely to produce illegitimate children, than men of lower social rank. These results are consistent with the prediction that females use possession of resources as a cue when selecting a father for their children [116].

If we bear the imprint of past evolution on our psyches, then women in modern Western societies should also use resource control and its correlates, such as high social status, when deciding which men to accept as sexual partners. In order to study the relationship between male income and copulatory success in modern Quebec, Daniel Perusse secured data from a large sample of respondents on how often they copulated with each of their sexual partners in the preceding year. With this information, Perusse was able to estimate the *number of potential conceptions* (NPC) a male would have been responsible for, had he and his partner(s) abstained from birth control. Male mating success, as measured by NPC, was highly correlated with male income, especially for unmarried men (Figure 14.9). Perusse concluded that many single Canadian men attempt to mate often with more than one woman, but that their ability to do

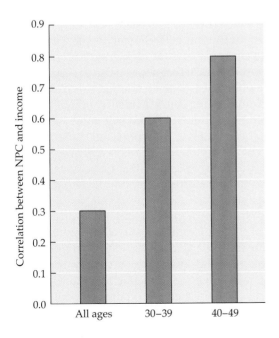

Figure 14.9   Higher income increases male copulatory success. Income is positively correlated with the number of potential conceptions (NPC) in the preceding year for unmarried Canadian men of various age groups, but especially for older men. After Perusse [939].

Figure 14.10 Fertility declines as family income increases in the United States (and many other industrial societies). Shown here is the average number of children per family in relation to the amount of income per family member. The families are grouped by income decile, so that the bottom 10 percent are shown at D1 and the top 10 percent are shown at D10. Data are for households in the United States in 1994 as collected by the Institute for Social Research at the University of Michigan.

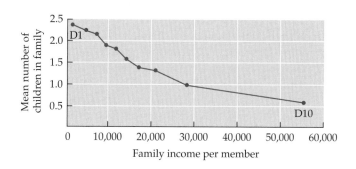

so is heavily influenced by their wealth and status, thanks to the female preference for relatively rich partners. These findings have been replicated in great detail with a much larger random sample of men living in the United States [617]. Thus, the striving for high income and status exhibited by males in North America and elsewhere may be the product of past selection by choosy females, which occurred in environments in which potential conceptions had an excellent chance of being actual ones [939].

Discussion Question

14.9 We have presented the hypothesis that, all other things being equal, women prefer men who bring wealth to the relationship. In most modern cultures, however, high family income does not seem to be positively correlated with the number of children produced (Figure 14.10). Indeed, poor couples often have more surviving children than rich ones. Some persons have concluded that this finding invalidates an evolutionary analysis of human behavior [1235]. How do you suppose an adaptationist might deal with this claim? You might want to contrast aspects of the current human environment with our ancestors' environment in the distant past. Moreover, is it possible that even in the past, restricting the number of offspring produced by resource-rich pairs could have, under some circumstances, promoted fitness? What about right now [119]? (Think about the nature of competition among the upper class and expand your view of fitness to accommodate grandchildren as well as children.)

Conditional Mate Preferences of Men and Women

The kind of research we have reviewed thus far tells us what people want in an ideal mate, but most people are keenly aware that ideal mates are in short supply. Thus, you will not be surprised to learn that not every man is paired off with an exceedingly fertile woman, nor is every woman married to an extremely wealthy, highly parental man with outstanding genes. In reality, although many men evaluate the sexual attractiveness of women in much the same way, their actual mate choices often diverge considerably from what seems to be an evolutionary ideal. The same holds true for women. Why? Perhaps most people are endowed with a conditional reproductive strategy that helps us take real-world conditions and our personal circumstances into account when deciding what mate selection tactics to employ [430].

So, for example, it is entirely possible for a male to find a youthful, fertile woman highly attractive, but to accept a less youthful, less attractive mate given his own limitations as a good catch. Indeed, increasing evidence suggests that humans have a reproductive strategy that is affected by a variety of factors, including a realistic assessment of one's own market value as a mate. A Scottish research team headed by A. C. Little established that women who rate themselves as "highly attractive" show a stronger preference for both relatively masculine and relatively symmetrical faces than women who state that they are of average or low attractiveness (Figure 14.11) [729]. Likewise, Peter Buston and Steve Emlen found that both men and women who consider themselves high-ranking prospects for a long-term relationship expressed a preference for an equally high-ranking partner, whereas those individuals with a lower self-perception of market value were less demanding (Figure 14.12) [181].

Figure 14.11   Women who think they are highly attractive prefer more attractive men. (A) When women were given a choice between a pair of digitally altered photographs of men, one of which was more symmetrical than the other (right versus left), they differed in the degree to which they said they preferred the more symmetrical face. (B) Women who rated themselves highly attractive chose the more symmetrical face nearly 70 percent of the time, whereas women of lower self-rated attractiveness did so less than 60 percent of the time. After Little et al. [729].

Discussion Question

14.10   What fitness benefits might accrue to a person of modest market value in selecting a partner of more or less equal value instead of aiming for a person of much higher fertility or wealth?

The conditional reproductive strategy of females evidently also factors in the potential duration of a relationship. Thus, only women in a long-term partnership with a committed companion feel a change in sexual desire over the menstrual cycle, with a peak occurring around the time of ovulation. An increase in sexual desire at that time increases the probability that the woman will mate and conceive a child with a male who is likely to be there to assist later in rearing that child [954]. The quite different sexual psychology of unattached women reduces the risk of conception at a time when a long-term partner is not available.

The conditional reproductive strategy of men also appears to enable individuals to select different tactics for different kinds of sexual relationships. Short-

Figure 14.12   Self-perception of attractiveness affects mate preferences in both sexes. The degree to which women and men consider themselves attractive long-term mate possibilities is correlated with their mate preferences. Less attractive individuals are willing to settle for less in a partner. The values for mate preferences are mean scores of the subjects' answers to questions about the importance of ten attributes in their dating decisions, such as: "On a scale of 1 to 9, how important is physical attractiveness to you?" The values for self-perception are the mean scores of the respondents' own evaluation of how they would score on these ten attributes. After Buston and Emlen [181].

Figure 14.13   **Sex differences in mate selectivity.**
College men differ from college women in the minimum intelligence that they say they would require in a casual sexual partner. However, men and women have similar standards with respect to the minimum intelligence they say is essential for a marriage partner. "Intelligence" was scored on an IQ scale so that a score of 50 meant that the acceptable individual had an IQ higher than half of the population. After Kenrick et al. [632].

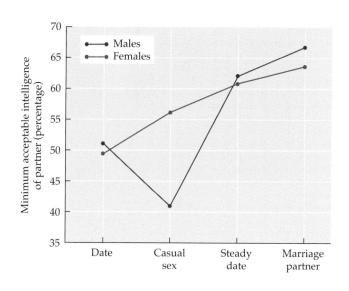

term flings require relatively little commitment on the part of a male because any offspring produced will be cared for by someone else. In contrast, a long-term relationship, as in marriage, requires costly resource transfers to a woman and parental investment for her offspring. A cost–benefit analysis yields the prediction that men should have far lower standards for a one-night stand than for a marriage partner. Indeed, when social psychologist Doug Kenrick surveyed a group of undergraduates about the minimum acceptable level of intelligence they would require in a partner for interactions ranging from a first date to marriage, both men and women adjusted the minimum level upward similarly in relation to degree of long-term commitment involved. But men and women differed strikingly when considering their standards for partners in a casual sexual encounter (Figure 14.13) [632], which tells us that the conditional strategies of the two sexes do not work in exactly the same way.

Thus, game theory, and specifically conditional strategy theory, offers an explanation for why it is that people exhibit so many different condition-dependent reproductive tactics. We, like Seychelles warblers and white-fronted bee-eaters, are able to pick and choose among a restricted range of responses based on an analysis of many variables, such as our social status, our looks, the attributes of potential partners and competitors of the same sex, and the degree of investment required for a sexual relationship.

Sexual Conflict

Because the mate preferences and genetic interests of men and women are not necessarily identical, we would expect to observe a good deal of sexual conflict in our species, as we do in most other animals (see Chapter 10). One significant source of sexual conflict would arise if men are, on average, more interested in acquiring multiple sexual partners than women are, as some have suggested. The basis for this suggestion is that although monogamous men may advance their reproductive success by helping one mate, polygynous men with several wives or an extramarital partner or two can potentially produce still more offspring. Polygyny has almost certainly been an option for men with substantial resources throughout our history as a species, judging from the fact that the acquisition of several wives was culturally sanctioned in 83 percent of all preindustrial societies [863]. Extramarital affairs, which could also have been part of the ancestral reproductive pattern, have the potential to increase the fitness of an adulterous male substantially at very little cost to him, provided his illegitimate children are cared for by his extramarital partners and their husbands.

The special value men place on sexual variety is reflected in their willingness to patronize female prostitutes, whereas women almost never pay men to copulate with them. Males, not females, also support a huge pornography industry in Western societies because men, not women, are willing to pay just to look at nude women. Note that in modern societies these aspects of male sexual behavior are surely *maladaptive*; prostitutes almost universally employ effective birth control or undergo abortions when pregnant, and payment for pornography is unlikely to boost a man's reproductive success. The prostitution and pornography industries take advantage of the male psyche, which evolved prior to modern birth control and the publication of *Playboy* [1176].

David Buss and David Schmitt illuminated these differences between the sexes simply by asking a sample of college undergraduates how many sexual partners they would like to have over different periods of time. The men in his study wanted many more mates than the women did (Figure 14.14A). Moreover, when Buss and Schmitt asked their subjects to evaluate the likelihood that

(A)

Figure 14.14 Sex differences in the desire for sexual variety. (A) Men and women differ in the number of sexual partners that they say they would ideally like to have over different periods of time. (B) Men and women also differ in their estimates of the likelihood that they would agree to have sexual intercourse with an attractive member of the opposite sex after having known that individual for varying lengths of time. After Buss and Schmidt [180].

they would be willing to have sex with a desirable potential mate after having known this person for periods ranging from 1 hour to 5 years, the differences between men and women were also dramatic (Figure 14.14B): "After knowing a potential mate for just 1 hour, men are slightly disinclined to consider having sex, but the disinclination is not strong. For most women, sex after just 1 hour is a virtual impossibility" [180].

The enthusiasm of males for low-cost sexual opportunities has also been documented by two social psychologists who sent confederates, an attractive young man and an attractive young woman, on the following mission. They were to approach strangers of the opposite sex on a college campus, asking some of them, "Would you go to bed with me tonight?" Not one woman agreed to the proposition, but 75 percent of the men said "Yes." Remember that the male subjects had known the woman in question for about a minute [216].

Now, it is possible that all the women in this study who said "No" did so because they sensibly feared becoming pregnant or did not wish to risk injury or disease from a sexual encounter with a male stranger. If so, then homosexual women should have fewer inhibitions about casual sex, since sexual interactions between two women cannot result in pregnancy and are unlikely to lead to physical assault. But homosexual women are no more interested in having multiple partners than are heterosexual females [50].

What are the fitness consequences for women whose partners are able to satisfy their desire for sexual variety? Polygyny carries the risk for a female that her partner will divert resources from her and her children to another woman and her offspring. Over evolutionary history, females of our species have probably gained fitness when they had exclusive access to a male's resources and parental care. In 19th-century Utah, for example, women married monogamously to relatively poor Mormon men had more surviving children on average (6.9) than women married to rich polygynous Mormons (5.5) [527]. Even though the polygynists' wives had lower fitness, the polygynists themselves did much better than monogamous men because they had children by several wives, not just one.

The potential benefits of polygyny to males increase the likelihood of conflict between husbands and wives. On the other side of the coin, some women are receptive to extramarital affairs, which may enable them to acquire additional material goods, or more protection, or better genes for their offspring from their extra-pair partners. If a wealthy or powerful extra-pair partner becomes a woman's primary partner, she may be able to exchange a low-ranking husband for a socially superior one, with all the positive effects on her fitness that trading up affords [1069]. The main point is that because both women and men can potentially elevate their fitness by mating with more than one partner, it is possible for a husband to reduce his wife's fitness, and vice versa.

One result of this fitness conflict between man and woman may have been the evolution of a capacity for sexual jealousy, which is an emotional state that helps individuals detect and interfere with the fitness-damaging sexual behavior of a mate. However, the nature of sexual jealousy should differ between the sexes, according to evolutionary psychologists, because the reproductive harm done by a multiply mating partner differs for men and women. A wife whose husband secures an additional mate (in a polygynous society) or goes on to divorce her in favor of a new partner usually loses some or all of her access to her husband's wealth, and thus the means to support herself and her children. Therefore, a woman's sexual jealousy, so the argument goes, should be focused on the possible loss of an attentive provider and helpmate, which is more likely to occur when a man becomes emotionally involved with another woman. On the other hand, a husband whose wife mates with another man may eventually care unknowingly for offspring fathered by that man. A man's sexual jealousy should therefore revolve around the potential loss of paternity and parental investment arising from a wife's extramarital sexual activity rather than the potential loss of resources and emotional commitment [282].

If this view is correct, then if men and women were asked to imagine their responses to two scenarios, one in which a partner develops a deep friendship with another individual and one in which a partner engages in sexual intercourse with another individual, women should find the first scenario more disturbing than the second, whereas men should be more upset at the thought of a mate copulating with another man. Data from several cultures confirm these predictions [179, 282]. For example, in a study involving Swedish university students, who live in a fairly sexually permissive culture, 63 percent of the women found the prospect of emotional infidelity more troubling, whereas almost exactly the same percentage of the men deemed sexual infidelity more upsetting [1296].

One effect of male sexual jealousy and possessiveness is to reduce the possibility that another male will supply a man's partner with sperm. In this light, marriage, including the honeymoon, may be a cultural institution that serves the function of mate guarding (see p. 345). Everywhere, men aspire to monopolize or restrict sexual access to their mates, although they do not necessarily succeed. Marriage institutionalizes these ambitions. Although one sometimes hears of societies in which complete sexual freedom is the norm, the notion that such cultures actually exist appears to have been a (wistful?) misinterpretation on the part of outside observers. In *all* cultures studied to date, adultery committed by a woman, or even suspicion of it, is considered an offense against her husband and often precipitates violence [279]. Presumably, suspicion of lost paternity explains why pregnant women in the United States are at double the risk of violent domestic assault compared with nonpregnant women [169]. In some other societies, a woman known to have cuckolded her husband may be legally killed by her aggrieved partner [282].

Discussion Questions

14.11   Fortunately, men in the United States are not permitted under any circumstances to kill their wives. Unfortunately, husbands do commit that crime on occasion, with nearly 14,000 homicides of this sort found in an FBI database assembled for the years 1976 to 1994. In some cases, a lovers' triangle was known to be involved [1088]. In this subset of homicides, young women were far more likely to be victims than older ones. Analyze this result as dispassionately as you can in terms of the potential fitness costs and benefits to the killer husband. With these fitness effects in mind, consider the possibility that wife killing in the context of potential infidelity is an evolved adaptation. Contrast this possibility with an alternative explanation; namely, that wife killing occurs as a maladaptive by-product of the psychological mechanisms that inspire violent sexual jealousy in males who think that their partners are unfaithful.

14.12   Mate guarding is an evolved response to sperm competition (see Chapter 10). If sperm competition has been a factor in human evolution, we can make some predictions about the relative investment of men in their testes, the sperm-producing organs, compared with that made by our two closest relatives, chimpanzees and gorillas. Chimpanzee females regularly mate with several males in the same estrous cycle, whereas gorilla females almost never do, since they typically live in bands controlled by a single, powerful male. How large (as a proportion of body size) should the testes of chimpanzee males be relative to gorilla testes? If the testes of men are more similar to those of chimpanzees, what might this tell us about the intensity of sperm competition during our evolutionary past? If, on the other hand, humans resemble gorillas, what conclusion is justified? Check your predictions against [512].

Figure 14.15   Rape occurs in animals other than humans. (Top) In the beetle *Tegrodera aloga*, a male (right) can court a female (left) decorously by repeatedly drawing her antennae into grooves on his head; copulation ensues only if the female responds to this courtship. (Bottom) Alternatively, a male (below) can force a female (above) to mate by running to her, grasping her, throwing her on her side, and inserting his everted genitalia as the female struggles to break free. Photographs by the author.

Coercive Sex

The fact that the murder of women suspected of adultery is still condoned in some parts of the world is one the least attractive manifestations of sexual conflict in our species. Another is the regular occurrence of forced copulation, a phenomenon that is not limited to human beings, by the way (Figure 14.15). Despite the fact that human rapists are often severely punished, rape occurs in every culture studied to date [1193]. This phenomenon has been analyzed by many persons, including Susan Brownmiller in her highly influential book *Against Our Will* [163]. In her view, rapists act on behalf of all men to instill fear in all women, the better to intimidate and control them, thus keeping them "in their place."

Brownmiller's intimidation hypothesis implies that some males are willing to take the substantial risks associated with rape in order to provide a benefit for the rest of male society. This argument suffers from all the logical problems inherent in "for-the-good-of-the-group" hypotheses (with the added difficulty that groups composed of only one sex cannot be the focus of any realistic sort of group selection), but we can test it anyway. If the evolved function of the trait is to subjugate all women, then the rapist element in male society can be predicted to target older, dominant women (or young women who aspire to positions of power) to demonstrate the penalty that comes from stepping outside the traditional subordinate role. This prediction is not supported: most rape victims are young, *poor* women [1194].

An alternative evolutionary hypothesis proposed by Randy and Nancy Thornhill is that rape is an adaptive tactic in a conditional sexual strategy [1194]. According to the Thornhills, sexual selection has favored males with the capacity to commit rape *under some conditions* as a means of fertilizing eggs and leaving descendants. In this view, rape in humans is analogous to forced copulation in *Panorpa* scorpionflies (see p. 337–338), in which males unable to offer nuptial gifts use the low-gain, last-chance tactic of trying to force females to copulate with them. According to the rape-as-adaptive-tactic hypothesis, human males unable to attract willing sexual partners might also use rape as a reproductive option of last resort.

The proposition that rape might serve an adaptive sexual function angered Brownmiller, who wrote, "It is reductive and reactionary to isolate rape from other kinds of violent antisocial behavior and to dignify it with adaptive significance" [164]. This view, however, confuses efforts to explain rape with attempts to excuse or justify the behavior. As noted earlier, when evolutionary biologists examine the adaptive value of a trait, whether it be rape in the human species, forced copulation in scorpionflies, or brood parasitism by cowbirds, their goal is to explain the evolutionary causes of the behavior, not to condone rape, or brood parasitism, or anything else.

The explanatory hypothesis that human rape is an evolved reproductive tactic controlled by a conditional strategy generates the prediction that some raped women will become pregnant, which they do, even in modern societies in which many women take birth control pills [1193]. In fact, copulatory rape apparently is more likely to result in pregnancy than consensual sex [458]. In the past, in the absence of reliable birth control technology and abortion procedures, rapists would have had a still higher probability of fathering children through forced copulation. Furthermore, if rape really is a product of an evolved reproductive mechanism, then rapists should more often target women of high fertility, just as bank swallows and other birds identify fertile (egg-laying) females and try to force those individuals to copulate with them [76, 1141]. In contrast, Brownmiller's view is that rape has nothing to do with reproduction, but is merely another form of violent antisocial behavior driven by a desire of men to dominate women. Note that this hypothesis focuses on the proximate cause of rape, not its ultimate reproductive consequences; it is possible for rapists to be motivated by aggression and for violent rape to generate offspring for the rapists. The notion, however, that rape has no sexual component leads to the prediction that the age distribution of rape victims should be the same as that of women murdered by male assailants. Crime data at odds with this prediction (Figure 14.16).

If rape is an evolved reproductive tactic of last resort, then it should occur more often in populations in which a high proportion of the fertile female population is monopolized by a relatively small part of the male population. Philip Starks and Caroline Blackie checked this prediction by looking at the correlation between rape and divorce rates in different populations. They reasoned that when divorce occurs, some men move on to acquire younger second (and even third) wives, depriving other men of the chance to marry these women and thereby increasing the overall variation in male reproductive success. The more serial monogamists in a population, the smaller the pool of fertile unmarried women relative to unmarried men, and the greater the potential benefits of rape as an alternative reproductive tactic for the excluded males. As Starks and Blackie predicted, the higher the annual divorce rate in the United States over a 36-year period, the higher the frequency of rape (Figure 14.17) [1148].

Figure 14.16   Testing alternative hypotheses for rape. If rape were motivated purely by the intent to attack women violently (a proximate hypothesis), we would expect that the distribution of rape victims would match that of female murder victims. Instead, rape victims are especially likely to be young (fertile) women, a result consistent with ultimate hypotheses proposing that rape is linked to male reproductive tactics. Data on rape victims come from 1974–1975 police reports for 26 U.S. cities. After Thornhill and Thornhill [1194].

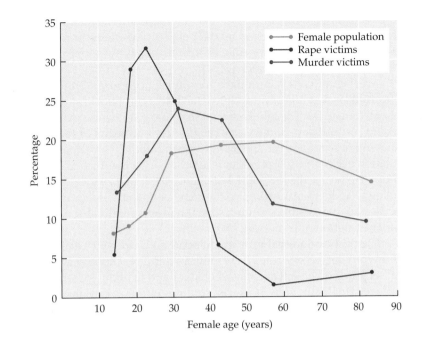

Although these findings suggest that rape may increase the fitness of some men, it is also possible that rape is not adaptive per se, but is instead a *maladaptive* by-product of the male sexual psyche, which causes quick sexual arousal, a desire for variety in sexual partners, and an interest in impersonal sex, all attributes that generate many adaptive (i.e., fitness-enhancing) consequences while also incidentally leading some men to rape some women [1193]. After all, men have been known to engage in many decidedly nonreproductive sexual activities, including masturbation, homosexual rape, and rape of postmenopausal women and prepubertal girls, just as males of many nonhuman species also exhibit sexual activity that cannot possibly result in offspring, such as the copulatory mounting of weaned pups by male elephant seals [1032]. Moreover, the only attempt to estimate the reproductive consequences of rape for men in a traditional society yielded the conclusion that the fitness cost to a rapist exceeded the benefit by a factor of about ten [1123].

Even if coercive copulation usually reduces the fitness of its practitioners, the rape-as-by-product hypothesis would be tenable if the systems motivating male sexual behavior have a net positive effect on fitness. Unfortunately, this hypothesis is difficult to test definitively because it produces many of the same predictions as the rape-as-adaptation hypothesis. One prediction unique to the by-product hypothesis is that rapists will have unusually high levels of sexual activity with consenting as well as nonconsenting partners. Some evidence supports this hypothesis [691, 922], but, as is true of many issues in human sociobiology, more data are required. Nevertheless, in this case as in so many others, the adaptationist approach has generated novel hypotheses that are entirely testable in principle and practice. The evolutionary angle on rape is now available for skeptical scrutiny, and as a result, we may eventually gain a better understanding of the ultimate causes of the behavior. When we do, we will not in any way be obliged to be more understanding of the illegal and immoral activity of rapists [1193].

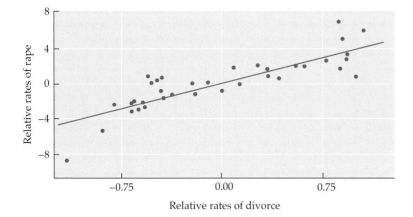

Figure 14.17   Rape as a conditional reproductive tactic. The annual correlation between rape and divorce rates in the United States from 1960 to 1995. After Starks and Blackie [1148].

Discussion Questions

14.13   Discussions of rape are invariably emotionally charged. From an evolutionary perspective, why might women have an especially strong visceral response to the topic and an intense desire to punish rapists? Why might most men also wish to deter rape? Would an understanding of evolutionary theory have led to a revision in a legal ruling by the United States Supreme Court that contained the following claim: "Rape is without doubt deserving of serious punishment; but in terms of moral depravity and of the injury to the person and to the public, it does not compare with murder. … [Rape] does not include … even the serious injury to another person" (quoted in [612]).

14.14   Natalie Angier states that married men have the same probability of fertilizing an egg per copulation with their wives as rapists do when forcing copulation on a victim [37]. In the past, the probability that an offspring of a married man would survive to reproduce was almost certainly much higher than the probability that a rapist's child would reach reproductive age because married men assist their children, whereas rapists do not. Is Angier correct, therefore, in claiming that rape cannot be an adaptive tactic? (Remember that "adaptive" means "reproductively useful.")

Adaptive Parental Care

You may have concluded by now that evolutionary biologists are interested only in human sexual activities. Although it is true that adaptationists have studied the reproductive component of our behavior most intensively, this does not mean that other aspects of human behavior have been ignored. The entire spectrum of human actions is fair game for researchers using an evolutionary approach [439]. As we saw at the outset of this chapter, issues of reciprocity, reputation, and social judgment can be analyzed in terms of their ultimate significance. Now we shall turn to an evolutionary analysis of parental behavior.

As noted before, in our species both males and females can provide parental care for their offspring, but individuals vary with respect to how much care they actually offer a given offspring [1328]. For example, the parents of an infant that produces abnormally high-pitched cries do not respond as rapidly as they would to a baby that cries in a more typical manner [427]. However, even infants with normal crying patterns are occasionally ignored, or even killed, by a parent. Here would seem to be the kind of behavior that must defeat evolutionary analysis.

But maybe not. Remember that an individual's impact on the next generation's gene pool is determined not by the number of babies conceived or born, but by the number of offspring that reach reproductive age. If babies with high-pitched cries have serious birth defects that make it unlikely that they will survive to reproduce, then parents that reduce or end their investment in these children may have more resources to give to other more viable offspring now or in the future. (And babies unable to cry normally often do have serious illnesses or congenital defects [427].) Likewise, if caring for a newborn or carrying a fetus to term threatens to reduce the lifetime reproductive success of a woman, then ending her investment in that offspring can potentially increase, not reduce, her fitness [278]. Throughout most of human history, when single women have attempted to rear a child unassisted, they have probably failed. An evolutionary prediction, therefore, is that single pregnant women should be more likely to end their pregnancies than married women, who have a partner to help them raise the offspring.

However, the evolutionary prediction that single pregnant women will seek abortions more often than married ones should apply more strongly to younger than to older women because young unmarried women have a greater chance of securing a husband and his fitness-enhancing support in the future than older women do. In contrast, older *married* women are expected to terminate pregnancies more often, given the increased probability of medical complications for older pregnant women, which could threaten their ability to care for their other still-dependent children or grandchildren. When the willingness of British women to carry a fetus to term was examined in relation to age and marital status, the results were entirely in accord with these predictions (Figure 14.18) [754]. Much the same pattern has been documented for Swedish women, who are also far more likely to terminate a pregnancy as they approach menopause [1183].

Men, as well as women, can potentially provide or withhold parental investment in their offspring. Because men run the risk of caring for a partner's child that was fathered by someone else, the sociobiological prediction is that

Figure 14.18 Do women employ abortion adaptively? (A) Single women in England and Wales are *less* likely to abort pregnancies as they grow older; married women are *more* likely to abort pregnancies as they age. These differences reflect the different costs and benefits of investing in an offspring when a woman has or does not have a husband's support. (B) The probability that a single woman will undergo an abortion is a function of the age-related probability that she will attract a husband in the future. After Lycett and Dunbar [754].

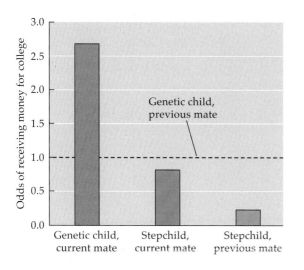

Figure 14.19 Parental favoritism. The odds that a man will give money to a child for college are much higher if the man is the genetic father of the potential recipient than if he is a stepparent to the child. The four categories of offspring examined in this study were genetic offspring living with their father, genetic offspring living with their father's previous mate, and the man's stepchildren living either with him or with a previous mate. The amount given to a genetic offspring living with a previous mate was used as a standard against which the other donations were measured. After Anderson et al. [27].

married men should have evolved psychological mechanisms that protect them against that risk. As we have noted, men that learn of a wife's adultery are likely to react extremely negatively and often seek a divorce, with its consequent reduction or elimination of paternal care [1176]. Male concern about paternity is so obsessive that husbands of rape victims in many cultures may legally divorce their unfortunate wives [163].

Stepfathers are another category of males who are placed in the position of caring for offspring who are not their own. To test the prediction that stepfathers should favor their genetic offspring over stepchildren, Mark Flinn monitored some Trinidadian families in which a stepfather lived with children of his own as well as those his wife had by another man. In these families, the percentage of interactions that involved conflicts of one sort or another was about twice as high for interactions between a stepfather and a stepchild as for those between a father and a genetic offspring [392]. Likewise, in modern American society, a man's genetic offspring by a current mate are far more likely to receive money for college than are his stepchildren with either a current or a previous mate (Figure 14.19) [27].

The same pattern applies to stepmothers, who are less likely to provide for their stepchildren than for their own genetic offspring. Households in which a mother cares for stepchildren, adopted children, or foster children spend less on food than households in which mothers reside with their genetic offspring [196]. Moreover, in blended families, the mother's own genetic offspring receive, on average, one more year of schooling than the stepchildren who reside with her. Note that by studying blended families, researchers have eliminated the possibility that stepchildren receive reduced educational opportunities because women who remarry are on average less capable of providing for the education of *any* children in their household [197].

Stepchildren not only tend to receive fewer resources from their stepparents, but are also at greater risk of being physically assaulted, a discovery made by Martin Daly and Margo Wilson when they tested the evolutionary hypothesis that our evolved psychological mechanisms encourage us to bias our parental care toward our genetic offspring [280]. A by-product of these mechanisms might be a greater tendency of stepparents to maltreat children that are not their own. In keeping with this prediction, Daly and Wilson found that in Hamilton, Ontario, children 4 years old or younger were *40 times more likely* to suffer abuse in families with a stepparent than in families with both genetic parents present! Note that for both categories of Canadian parents, the absolute likelihood that

Figure 14.20 **Child abuse and the relatedness of parents and offspring.** Child abuse is far more likely to occur in households with a stepparent than in households with two genetic parents. After Daly and Wilson [277].

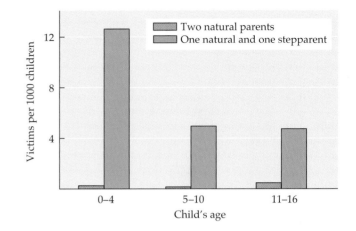

a child would be abused was small (Figure 14.20), but the *relative* risk was far greater for children in households with a stepparent [277]. Daly and Wilson argued that the psychological systems that promote selective parental care in Canada and elsewhere [281] generally lead to increased fitness by encouraging adults to invest in their offspring, but these same aspects of parental psychology may occasionally have maladaptive effects for a very few parents with blended families.

Discussion Question

14.15 Daly and Wilson found that stepchildren incurred an approximately 100-fold increase in the risk of fatal child abuse compared with children reared by two genetic parents [280]. Apply their explanation for the abuse of children by their stepfathers to these cases of infanticide. How does Daly and Wilson's hypothesis differ from the hypothesis for infanticide by male langurs and lions (see Chapter 1)? What would you need to know to test the two alternative explanations for infanticide by stepfathers? Now imagine the following hypothesis: Child abuse by step-parents occurs because the new caretakers typically come into the family some time after the child's birth and so lack the physiological mechanisms, perhaps hormonal ones, that promote tolerance of an infant's difficult behaviors, such as colicky crying or whining. In the absence of these psychological buffering devices, things can go terribly wrong, and they do, albeit very rarely. How does this hypothesis stand up to the other two? What significance would there be in comparing the parental behavior of genetic parents—say, persons serving in the military—who have not seen their child for the first year of its life with the paternal responses of stepfathers who move in with a partner who has a one-year-old toddler?

Helping Children Marry

We have seen that adults care for children in a selective fashion, usually favoring genetic offspring over nonrelatives. But even in families without stepchildren, parents do not always treat their progeny identically when it comes to dishing out parental benefits. An example involves the material sacrifices parents make to help their offspring acquire spouses. In some societies, the groom and his family are called upon to donate resources such as cattle or money or labor—the bridewealth—to the bride's family (Figure 14.21); in other societies, the family of the bride sends their daughter off to marry with a special donation—the dowry—to her new husband or his family.

If bridewealth or dowry payments were purely arbitrary traditions, the accidental products of cultural evolution, then we would predict that the two forms of payments should be equally represented among cultures worldwide. They

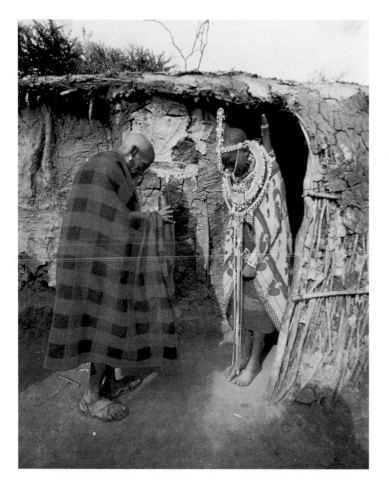

Figure 14.21   Marriage requires bridewealth payments in many traditional African cultures, such as the Masai of Kenya and Tanzania. The father who stands here with his about-to-be-married daughter will receive cattle from her husband-to-be and his relatives. Photograph by Jason Lauré.

are not [436]. (Before reading further, use sexual selection theory to predict whether bridewealth or dowry should be the more common cultural practice worldwide.)

Because males typically compete for access to females, we can predict that bridewealth payments should be far more common than dowries. This is indeed the case. Bridewealth payments have been recorded in 66 percent of the 1267 societies described in the *Ethnographic Atlas* [863], whereas dowry is standard practice in just 3 percent of these societies. Bridewealth payments are found particularly frequently in polygynous cultures, occurring in more than 90 percent of those societies classified under the heading "general polygyny," in which more than 20 percent of married men have more than one wife (Table 14.1). When some males monopolize several females, marriageable females become an especially scarce and valuable commodity. One way to secure multiple wives, and so achieve exceptional reproductive success, is to provide a payment to their parents or other relatives. To do so often requires much wealth.

Rich men in polygynous societies can produce a great many children, which enables wealthy parents to secure more descendants by putting their resources primarily in the hands of a son rather than a daughter, whose direct fitness is limited by the number of embryos she can personally produce. Biased parental investment can even occur posthumously, when wealthy parents pass most of their wealth to a son or sons, enabling these individuals to become successful polygynists [117]. Inheritance rules that favor sons are in fact associated with the practice of polygyny (see Table 14.1) [516].

TABLE 14.1 *The relationship between the mating systems of human cultures, bridewealth payments, and inheritance systems that favor sons*

	Bridewealth payment		Sons favored	
Mating system	No	Yes	No	Yes
Monogamy	62%	38%	42%	58%
Limited polygyny	46%	54%	20%	80%
General polygyny	9%	91%	3%	97%

Source: Hartung [516]

Note: The data are from Murdock's *Ethnographic Atlas* [863] for 112 monogamous cultures, 290 cultures that practice limited polygyny (less than 20 percent of men are polygynous), and 448 that practice general polygyny (more than 20 percent of men are polygynous). When sons are favored, they receive all or almost all of the parental inheritance.

Guy Cowlishaw and Ruth Mace confirmed that this pattern was real after controlling for the nonindependence of cultures—that is, after dealing with the statistical problem of how to treat information from several cultures that may have inherited similar practices as a result of sharing a recent common cultural ancestor. First, they constructed a cultural phylogeny based on linguistic information; this diagram identified which cultures were closely related and which were not. They then superimposed the mating system of each culture on the phylogeny to identify those cultures whose mating systems had changed to polygyny from a monogamous predecessor and vice versa. Independent cultural changes to polygyny were much more likely to be associated with inheritance rules favoring sons than were changes to monogamy [247].

Even in supposedly monogamous Western societies, rich men may have opportunities for unusual copulatory success because, as we noted earlier, their wealth makes them attractive to women. If parents in modern societies retain an ancestrally selected bias that causes them to favor those offspring with the highest reproductive potential, we can predict that even today, very wealthy parents should be inclined to give their sons more inheritance than their daughters—a prediction that has been shown to be correct (Figure 14.22) [1126].

In contrast to the prevalence of biased parental investment in male offspring, societies in which parents give significantly more to their daughters are rare—as one might expect, given that females typically are in demand as marriage partners. However, Lee Cronk has found one tribal society, the Mukogodo of Kenya, in which parents often provide more food and more medical care (from a local Catholic mission clinic) to their daughters than to their sons [261]. The Mukogodo have only recently abandoned their traditional hunting–gathering lifestyle in favor of an economy built on sheep and goat herding. Their herds are small, and their standing with other pastoral tribes in the area is very low. A son of an impoverished Mukogodo family is unlikely to acquire a large enough flock to pay the bride price for a wife from any of the surrounding tribes. In contrast, a Mukogodo daughter has a good chance of marrying a member of a higher-status tribe, since polygyny is standard among these groups and women are in short supply. As a result of the greater ease of marriage for daughters than for sons, the average number of offspring of a Mukogodo daughter is nearly 4, whereas a son's direct fitness is only about 3. This inequality favors families that have more daughters than sons, an outcome that is achieved through greater investment in young girls than in young boys.

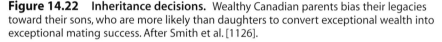

Figure 14.22   Inheritance decisions. Wealthy Canadian parents bias their legacies toward their sons, who are more likely than daughters to convert exceptional wealth into exceptional mating success. After Smith et al. [1126].

This case nicely illustrates the point that human behavior is adaptively flexible, not arbitrarily or infinitely variable. Whatever psychological mechanisms control parental solicitude toward offspring, these evolved systems permit parents to favor sons under some circumstances and daughters under others, while encouraging equal treatment under yet another set of conditions. The option chosen tends to enhance the fitness of parents in their particular environment. The *differences* between the daughter-favoring Mukogodo and those nearby herding cultures in which sons receive preferential treatment are surely not genetic. Instead, these parental differences reflect our ability to use evolved conditional strategies to select among a limited set of options, choosing the one with the highest fitness payoff in a given setting.

Among the Mukogodo and most other tribal groups, men pay a price to acquire a bride. Why do a few contrary cultures sanction payments that help a woman secure a husband? One answer is that in monogamous societies in which males typically invest materially in their children, parents that help their daughters "buy" the right kind of man gain fitness as a result. Males in socially stratified monogamous cultures vary greatly in their status and wealth. In such cultures, women married to elite males should generally enjoy a reproductive advantage because a monogamous husband's wealth will not be divided among a bevy of wives, but instead will go to his only wife and her children. To the extent that wealth and high status translate into reproductive success, a woman's parents may gain by competing with the parents of other families for an "alpha" husband, even if this requires that they offer a material inducement to the right male or to his family. Steven Gaulin and James Boster found that dowry payments occur in substantially less than 0.5 percent of nonstratified societies, whereas 9 percent of stratified societies and 60 percent of *monogamous* stratified cultures permit or encourage dowries [436]. This highly nonrandom set of associations supports the hypothesis that dowry practices are not a random cultural artifact, but part of an adaptive parental strategy. Thus, we have yet another example of how sociobiologists, far from being flummoxed by cultural diversity, can make use of it to test evolutionary hypotheses about human behavior.

Discussion Question

14.16   In the camel-herding Gabbra, an African tribe, the number of camels owned by a household boosts a man's reproductive success more than that of a woman. In the horticultural Chewa, the more land under cultivation, the higher the fertility of an individual, but the effect is the same for men and women (Figure 14.23) [548]. In the polygynous Gabbra, inheritance is male-biased, and in the Chewa, in which individuals belong to their mother's lineage, inheritance is effectively female-biased. Explain why, keeping in mind that certainty of maternity (the probability that a mother's child is genetically related to her) is always greater than the certainty of paternity (the probability that a putative father's child is actually his genetic offspring). In other words, under what conditions would grandparents gain more genetic grandchildren by investing in their daughters than in their sons? How do the data from these two cultures contribute to answering this question?

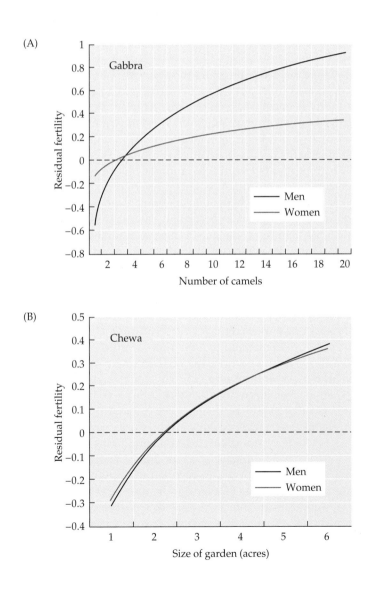

Figure 14.23   Wealth affects male and female fertility differently in two African cultures. In the Gabbra, residual fertility, a measure of reproductive success, increases more rapidly with wealth (number of camels owned) for men than it does for women. In the Chewa, the sexes do not differ in the effect that wealth (size of garden) has on their fertility. After Holden et al. [548].

Applications of Evolutionary Psychology

Although the ability of evolutionary theory to contribute to a purely academic understanding of our behavior is enough to delight many biologists and psychologists, one can also argue that there is a hard-headed, applied side to this kind of approach. For example, criminal investigators working on child abuse cases in North America have surely benefited from an awareness of the statistical relationship between this kind of crime and the presence of a stepparent in the family, a connection first uncovered not by criminologists, but by two evolutionary psychologists. Likewise, I believe that young women have much to gain from an understanding of the evolutionary bases of coercive sex, especially the reality of the reproductive component of the crime, which derives from the evolved sexual psychology of males. If women recognized that coercive sex, including rape, is not practiced solely by classic criminal sociopaths, they might be more on guard against more ordinary men, whose intense sex drive enables them to feel sexual desire even when their "partners" do not.

Moreover, everyone should know that our perfectly natural, highly adaptive drive to have children is directly linked to an exploding human population that stands at 6 billion and counting, much to the detriment of the earth. Our efforts at getting things under control might be advanced if we all knew why we think fat, healthy babies are wonderful and why we believe nothing is too good for our children, even though the effect of having lots of plump babies and affording them a middle-class upbringing has pushed our global life support system toward collapse [936].

Equally in need of correction is our evolved tendency not just to reproduce enthusiastically, but to consume the earth's resources as if there were no tomorrow. This human attribute is driven in part by the production of large families, but perhaps even more so by the desire of human beings to acquire wealth and use it ostentatiously, a key component of the reproductive strategy of men. In addition, people tend to discount the future when it comes to resource use, which is to say that most of us tend to spend what we have now at the expense of setting something aside for the long term. Some persons have linked this aspect of human psychology to the high mortality rates that have characterized most of human evolution until very recently. Given the reality that the odds were against a long life, it would have been generally adaptive for people during our history as a species to consume resources at once and to produce children as soon as they could.

This argument can be tested by comparing the behavior of people in modern communities where life expectancy varies. Margo Wilson and Martin Daly have done the test, taking advantage of the fact that during a period around 1990, male life expectancy in different Chicago neighborhoods varied by about 25 years. If men possess a conditional strategy whose operating rules are affected by the likelihood of early death, then we would predict that individuals in high-mortality neighborhoods should be prepared to take greater risks than those living in places where death from natural causes is likely to be postponed. One dramatic manifestation of risk taking by men is their willingness to engage in extreme violence, which in turn should affect homicide rates. As predicted, the rate at which homicides occurred in a neighborhood was tightly linked to male life expectancy (with the values for this figure adjusted by removal of the effects of homicide itself) (Figure 14.24) [1319]. In other words, when men perceive that their lives may well be short, they are more likely to take up a life of crime and violence—a high-risk tactic, but one with a potentially high immediate payoff for the successful risk taker.

Figure 14.24   Homicide rates are highly correlated with male life expectancy. These data come from a number of neighborhoods in Chicago in which males can expect to live anywhere from under 55 to more than 75 years (after the effects of homicides themselves on life expectancy are removed). The homicide rate is calculated as the number of murders per 100,000 people. After Wilson and Daly [1319].

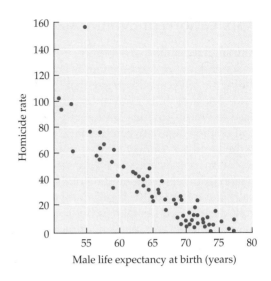

Discussion Question

14.17   Often teenage pregnancies are treated as a pathology of some sort. But can you make a case that women should possess a conditional strategy that takes life expectancy into account when making decisions about the onset of reproduction? If so, then what is the predicted pattern for the mean age of first pregnancy in the different Chicago neighborhoods just mentioned? See [1319] to check your answer.

The point is that not one product of natural selection, whether it be our wish to have children or our desire to have a good reputation, is guaranteed to produce positive results, either for us or for our species as a whole. Take our capacity for cooperation. On the one hand, humans do often band together for goals that seem admirable: making a computer program work better, trying to eliminate polio, striving to reduce white-collar crime, putting on a play or opera—the list is all but endless. But the ability to identify with a group and to form strong bonds with its members can also be employed in profoundly aggressive ways against other groups formed of people with similarly powerful cooperative abilities.

Indeed, Richard Alexander has argued that violent competition among human groups for resources led to the evolution of the extraordinary cooperation and the intensely group-centered sense of morality that people exhibit today [17]. The adaptively unpleasant nature of human morality can be illustrated by examining what the injunction "Thou shalt not kill," has meant to people who have accepted it. This commandment has almost always been interpreted to mean "Thou shalt not kill members of one's own tribe or community or nation—but the destruction of others living outside the band, city, or country will be accepted, even encouraged, if these others pose any threat to the communal welfare or if the resources they control can be taken from them with only modest risk of personal injury" [517]. Richard Wrangham has pointed out that this special kind of morality may well be traced to an ancestral primate in the lineage that gave rise to both the common chimpanzee and *Homo sapiens*. These two species are the only living ones in which males regularly band together to raid other groups with the intent of assaulting, or even killing, their enemies before returning to their home base (Figure 14.25) [1340]. This tribal tendency with its two faces, solicitous support for one's own group and often profound and violent antipathy for all others, is manifested in dozens of ways—some largely benign, as in the allegiance given to a favorite baseball, football, or other sports

Figure 14.25   Male chimpanzees cooperate in attacks on other bands. This species and our own are the only mammals in which male teams leave their home range to assault members of other groups and then return home after the attack. Photograph by Kristin Mosher.

team, and others truly horrific, including the historically frequent and widespread practice of genocide [324].

Discussion Question

14.18   In writing about genocide, Stephen J. Gould reviewed the adaptationist hypothesis that the capacity for large-scale murder evolved as a result of intense competition for resources or mates between small bands during our evolutionary history. Gould dismissed this hypothesis on the grounds that evolutionary theory supposedly has nothing novel to say about this aspect of human behavior. "An evolutionary speculation can only help if it teaches us something we don't know already—if, for example, we learned that genocide was biologically enjoined by certain genes, or even that a positive propensity, rather than a mere capacity, regulated our murderous potentiality. But the observational facts of human history speak against determination and only for potentiality. Each case of genocide can be matched with numerous incidents of social benevolence; each murderous clan can be paired with a pacific clan" [462]. Evaluate Gould's argument critically in the light of what you know about (1) the proximate–ultimate distinction, (2) conditional strategies, and (3) the differences between adaptationist and arbitrary culture hypotheses.

There is no question that the attempts by evolutionary biologists to understand why we behave the way we do are far from complete. But at the very least, evolutionary analyses have demonstrated beyond doubt that just because something is "adaptive," or "natural," or "evolved" does not mean that it is "good" or "moral" or "desirable." Once you know that our genes have the capacity to make us work on their behalf without regard for our welfare or that of most other people, then it is easier to fight against our evolved impulses [1309]. If all of us understood this small but vital point, we would be better off. For one thing, a knowledge of evolutionary outcomes could help us see through the attempts of others to exploit our evolved susceptibility to certain kinds of stratagems [212, 232]. Informed persons could also better avoid the moral certainty and self-righteousness that enables us to demonize our opponents prior to attacking them. When Pogo declared that "We have met the enemy and he is us," he was correct, and those of us who understand the role of natural selection in shaping our evolutionary history know why he was right. Perhaps some of you can use that understanding to help make the human species less an enemy to itself.

Summary

1. Human beings are an evolved animal species. Human sociobiology, a field of study that includes evolutionary psychology and evolutionary anthropology, employs natural selection theory to generate testable hypotheses about the possible adaptive value of our species' behavior.

2. The study of sociobiology has been marked by intense controversy, in part because some persons have misunderstood the goals and foundations of the discipline. Contrary to the views of some critics, sociobiologists are not motivated by a political agenda, nor are sociobiological hypotheses based on the premise that elements of human behavior are genetically determined or morally desirable. Sociobiological hypotheses are designed to explain, not justify, our behavior. Testing these hypotheses can help us understand why we have psychological mechanisms that motivate us to behave in certain ways.

3. Unlike arbitrary culture theory, which proposes that human behavior is the arbitrary product of cultural traditions constrained only by the limits of our imagination, sociobiological theory views human behavior as the product of natural selection. It assumes that human beings have conditional strategies that provide us with limited options that tend to increase, rather than decrease, the fitness of individuals.

4. Many sociobiological hypotheses have been advanced for elements of human reproductive behavior, as illustrated by evolutionary analyses of the different kinds of cues important to men and women in evaluating potential mates. Males prefer those characteristics associated with high fertility; females typically place more emphasis on the resources held by potential partners. The decisions that people make based on these psychological mechanisms would have generally increased the fitness of individuals living in pre-contraceptive environments. The same mechanisms can, however, also lead to behavior that most people would condemn, such coercive sex and marital infidelity.

5. Elements of parental behavior, like elements of mate choice, can be explained in terms of their contribution to individual reproductive success. Even apparently maladaptive actions, such as child abuse by stepfathers, can be better understood as by-products of fitness-maximizing mechanisms that evolved because of their other adaptive consequences, such as the generally adaptive desire to limit one's parental investment to one's genetic offspring.

6. When a claim is made that something is adaptive or natural, this does not mean that it is desirable, moral, or necessary, only that it tends to propagate our genes. Widespread acceptance of this point might enable us to understand that trusting our impulses is more likely to help us pass on our genes than to produce wise decisions that maximize either our personal happiness or the general good.

Suggested Reading

The debate on sociobiology began with E. O. Wilson; see the final chapter of *Sociobiology* [1315] and an attack on the book [21] with Wilson's reply [1316]. Representative books and articles that criticize evolutionary analyses of human behavior include [164, 341, 358, 405, 715]. The misconceptions that many persons have about human sociobiology have been outlined by two of my colleagues at Arizona State, Doug Kenrick [629] and Owen Jones [612]. Most of the classic papers on the evolutionary analysis of human behavior can be found, along with updates and critiques, in *Human Nature*, edited by Laura Betzig [99]. Modern accounts of an evolutionary approach to human behavior include books by Richard Alexander [17, 18], Martin Daly and Margo Wilson [278, 280], and Robert Wright [1342], whose description of evolutionary psychology is extremely readable. The evolution of human sexual behavior and psychology has been explored in various ways by Donald Symons [1176], David Geary [440], and Geoffrey Miller [826]. Readers can also find many interesting articles on human behavior in the journal *Evolution and Human Behavior*. You can participate in on-line research in experimental evolutionary psychology by going to www.alittle.lab.com.

Glossary

Action potential   The neural signal; a self-regenerating change in membrane electrical charge that travels the length of a nerve cell.

Adaptation   A characteristic that confers higher inclusive fitness to individuals than any other existing alternative exhibited by other individuals within the population; a trait that has spread or is spreading or is being maintained in a population as a result of natural selection or indirect selection.

Adaptationist   A behavioral biologist who develops and tests hypotheses on the possible adaptive value of a particular trait. Persons using an **adaptationist approach** test whether a given trait enables individuals to propagate their special genes more effectively than if they were to behave in some other manner.

Adaptive value   The contribution that a trait or gene makes to inclusive fitness.

Allele   A form of a gene; different alleles typically code for distinctive variants of the same enzyme.

Altruism   Helpful behavior that raises the recipient's direct fitness while lowering the donor's direct fitness.

Arbitrary culture theory   The view that human behavior is the arbitrary product of whatever cultural traditions people are exposed to within their society; thus, our actions are not expected to be explicable in evolutionary terms.

Artificial selection   *See* Selection.

Associated reproductive pattern   A seasonal change in reproductive behavior that is tightly correlated with changes in the gonads and hormones in contrast to a **disassociated reproductive pattern** in which the onset of reproductive behavior is apparently not triggered by a sharp change in circulating hormones.

Biological clock   An internal physiological mechanism that enables organisms to time any of a wide assortment of biological processes and activities.

Brood parasite   An animal that exploits the parental care of individuals other than its parents.

By-product hypothesis   An explanation for a maladaptive or nonadaptive attribute that is said to occur as a by-product of a proximate mechanism that has some other adaptive consequence for individuals.

Causal question   In the scientific method, a question about the cause of a natural phenomenon.

Central pattern generator   A group of cells in the central nervous system that produces a particular pattern of signals necessary for a functional behavioral response.

Chase-away selection   *See* Selection.

Circadian rhythm   A roughly 24-hour cycle of behavior that expresses itself independent of environmental changes.

Circannual rhythm   An annual cycle of behavior that expresses itself independent of environmental changes.

Coefficient of relatedness   The probability that an allele present in one individual will be present in a close relative; the proportion of the total genotype of one individual present in the other, as a result of shared ancestry.

Command center   A neural cluster or an integrated set of clusters that has primary responsibility for the control of a particular behavioral activity.

Communication   The cooperative transfer of information from a signaler to a receiver.

Comparative method   A procedure for testing evolutionary hypotheses based on disciplined comparisons among species of known evolutionary relationships.

Concession theory   *See* Transactional theory.

Conditional strategy   *See* Strategy.

Convergent evolution   The independent acquisition over time through natural selection of similar characteristics in two or more unrelated species.

Cooperation   A mutually helpful action.

Cost-benefit approach   A method for studying the adaptive value of alternative traits based on the recognition that phenotypes come with fitness costs and fitness benefits; an adaptation has a better cost-benefit ratio than alternative versions of that trait.

Critical period   A phase in an animal's life when certain experiences are particularly likely to have a potent developmental effect.

Cryptic female choice   The ability of a female in receipt of sperm from more than one male to choose whose sperm get to fertilize her eggs.

Darwinian puzzle   A trait that appears to reduce the fitness of individuals that possess it; traits of this sort attract the attention of evolutionary biologists.

Developmental homeostasis   The capacity of developmental mechanisms within individuals to produce adaptive traits, despite potentially disruptive effects of mutant genes and suboptimal environmental conditions.

Dilution effect   Safety in numbers that comes from swamping the ability of local predators to consume prey.

Diploid   Having two copies of each gene (and two copies of each type of chromosome) in one's genotype.

Direct fitness   *See* Fitness.

Direct selection   *See* Selection.

Display   A stereotyped action used as a communication signal by individuals.

Dissociated reproductive pattern   *See* Associated reproductive pattern.

Divergent evolution   The evolution by natural selection of differences among closely related species that live in different environments and are therefore subject to different selection pressures.

Dominance hierarchy   A social ranking within a group, in which some individuals give way to others, often conceding useful resources to others without a fight.

Entrainment   The re-setting of a biological clock so that an organism's activities are scheduled in keeping with local conditions.

Ethology   The study of the proximate mechanisms and adaptive value of animal behavior.

Eusocial   Refers to groups that contain specialized nonreproducing castes that work for the reproductive members of the society.

Evolutionary psychology   The study of the adaptive value of psychological mechanisms, especially of human beings; a key component of sociobiology.

Evolutionary stable strategy   *See* Strategy.

Evolutionary benefit   That aspect of a trait that tends to raise the inclusive fitness of individuals.

Evolutionary cost   That aspect of a trait that tends to decrease the inclusive fitness of individuals.

Explosive breeding assemblage   The temporary formation of large groups of mating individuals.

Extra-pair copulation   A mating by a male or female with someone other than their primary partner in a seemingly monogamous species.

Female defense polygyny   *See* Polygyny.

Female preference hypothesis   An explanation for the formation of leks in which females prefer to choose mates from a group rather than by inspecting potential mates one by one.

Fertility insurance hypothesis   An explanation for why females might mate with more than one male per breeding cycle with the benefit being an increase in egg fertilization rate.

Fitness   A measure of the genes contributed to the next generation by an individual, often stated in terms of the number of surviving offspring produced by the individual.

>   **Direct fitness**   The genes contributed by an individual via personal reproduction in the bodies of surviving offspring.
>   **Indirect fitness**   The genes contributed by an individual indirectly by helping nondescendant kin, in effect creating relatives that would not have existed without the help of the individual.
>   **Inclusive fitness**   The sum of an individual's direct and indirect fitness.

Fitness benefit   That aspect of a trait that tends to raise the inclusive fitness of individuals.

Fitness cost   That aspect of a trait that tends to reduce the inclusive fitness of individuals.

Fixed action pattern   An innate, highly stereotyped response that is triggered by a well-defined, simple stimulus; once the pattern is activated, the response is performed in its entirety.

Free-running cycle   The cycle of activity of an individual that is expressed in a constant environment.

Frequency-dependent selection   *See* Selection.

Game theory   An evolutionary approach to the study of adaptive value in which the payoffs to individuals associated with one behavioral tactic are dependent on what the other members in the group are doing.

Gene   A segment of DNA, typically one that encodes information about the sequence of amino acids that makes up a protein.

Genetic compatibility   The ability of genes present in some sperm to complement the genes present in some eggs, resulting in an increased likelihood of the development of superior offspring.

Genotype   The genetic constitution of an individual; may refer to the alleles of one gene possessed by the individual or to its complete set of genes.

Good genes theory   The argument that mate choice advances individual fitness because it provides the off-spring of choosy individuals with genes that promote reproductive success by advancing the offspring's chances of survival or reproductive success.

Good parent theory   An explanation for female preferences for males whose appearance or behavior signals that these potential mates are likely to provide above average parental care for their offspring.

Group selection   *See* Selection.

Hamilton's rule   The argument made by W.D. Hamilton that altruism can spread through a population with $r_b > c$ (with r being the coefficient of relatedness between the altruist and the individual helped, b being the fitness benefit received by the helped individual, and c being the cost of altruist in terms of the direct fitness lost by the altruism due to his actions.

Haploid   Having only one copy of each gene in one's genotype, as for example in the sperm and eggs of diploid organisms.

Healthy mates theory   An explanation for female preferences for males whose appearance or behavior signals that these potential mates are unlikely to transmit communicable diseases or parasites to them.

Home range   An area that an animal occupies but does not defend, in contrast to a territory, which is defended.

Honest signals   Signals that convey accurate information about the signaler's real fighting ability or true value as a potential mate.

Hotshot   A male whose attributes are especially appealing to sexually receptive females.

Hot spot   A location whose properties attract sexually receptive females to the male able to hold the site against rival males.

Hypothesis   A tentative explanation that will require testing before acceptance.

Ideal free distribution   The distribution of individuals in space that are free to choose where to go in ways that could maximize their fitness.

Illegitimate receiver   An individual that listens to the signals of others, thereby gaining information that it uses to reduce the fitness of the signaler.

Illegitimate signaler   An individual that produces signals that may deceive others into responding in ways that reduce the fitness of the signal receiver.

Imprinting   A form of learning in which individuals exposed to certain key stimuli, usually early in life, form an association with the object and may later show sexual behavior toward similar objects.

Inbreeding depression   The tendency of inbred organisms to have lower fitness than noninbred members of their species.

Inclusive fitness   *See* Fitness.

Indirect fitness   *See* Fitness.

Indirect reciprocity   *See* Reciprocity.

Indirect selection   *See* Selection.

Individual selection   *See* Selection.

Innate releasing mechanism   A hypothetical neural mechanism thought to control an innate response to a sign stimulus.

Instinct   A behavior pattern that reliably develops in most individuals, promoting a functional response to a releaser stimulus the first time the action is performed.

Interneuron   A nerve cell that relays messages either from receptor neurons to the central nervous system (a sensory interneuron) or from the central nervous system to neurons commanding muscle cells (a motor interneuron).

Kin discrimination   The capacity of an individual to react differently to others based on the degree to which they are genetically related.

Kin selection   *See* Selection.

Learning   A durable and usually adaptive change in an animal's behavior traceable to a specific experience in the individual's life.

Lek   A traditional display site that females visit to select a mate from among the males displaying at their small resource-free territories.

Lek polygyny   *See* Polygyny.

Mate assistance monogamy   *See* Monogamy.

Mate guarding   Actions taken by males (usually) to prevent a sexual partner from acquiring sperm from other males. *See* also Mate guarding monogamy.

Material benefits hypothesis   An explanation for why females of some species might mate with several males per breeding cycle with benefits coming to polyandrous females because they gain access to the material benefits controlled by the several males.

Migration   The regular movement back and forth between two relatively distant locations by animals that use resources concentrated in these different sites.

Mobbing behavior   When prey closely approach and attempt to harass a predator.

Monogamy   A mating system in which one male mates with just one female in a breeding season.

> **Female enforced monogamy**   A monogamous mating system that arises because females are able to prevent their partners from securing additional mates.
> **Mate assistance monogamy**   Monogamy that arises because males gain more fitness by offering parental care for the offspring of his mate than by seeking out additional sexual partners.

Mate guarding monogamy   Monogamy that arises because males gain more fitness by preventing one partner from mating with other males than by seeking out additional sexual partners.

Mutualism   A mutually beneficial relationship or cooperative action.

Natural selection   *See* Selection.

Neuron   A nerve cell.

Nondescendant kin   Relatives other than offspring.

Nuclei   Dense clusters of cell bodies of neurons within nervous systems.

Nuptial gift   A food item transferred by a male to a female just prior to or during copulation.

Operant conditioning   A kind of learning based on trial and error, in which an action or operant becomes more frequently performed if it is rewarded.

Operational sex ratio (OSR)   The ratio of receptive males to receptive females over a given period.

Optimality theory   An evolutionary theory based on the assumption that the attributes of organisms are optimal, i.e., better than others in terms of fitness benefits to costs; the theory is used to generate hypotheses about the possible adaptive value of traits in terms of the net fitness gained by individuals that exhibit these attributes.

Parent-offspring conflict   The clash of interests that occurs when parents can gain fitness by withholding parental care or resources from some offspring in order to invest in others now or later, even though the deprived offspring would gain more fitness by receiving parental care or resources.

Parental investment   Costly parental activities that increase the likelihood of survival for some existing offspring but that reduce the parent's chances of producing offspring in the future.

Phenotype   Any measurable aspect of an individual that arises from an interaction of the individual's genes with its environment.

Phenotype matching   A proximate mechanism of kin discrimination in which an individual's behavior toward another is based on how similar they are in some way, such as odor or appearance.

Pheromone   A volatile chemical released by an individual as a scent signal for another.

Photoperiod   The number of hours of light in 24 hours.

Phylogeny   An evolutionary genealogy of the relationships among a number of species or clusters of species that can be used to develop hypotheses on the evolutionary history of a given trait.

Pleiotropy   The capacity of a gene to have multiple developmental effects on individuals.

Polyandry   A mating system in which a female has several partners in a breeding season.

Polygyny   A mating system in which a male fertilizes the eggs of several partners in a breeding season.

>**Female defense polygyny**   Polygynous males directly defend several mates.
>**Lek polygyny**   Polygynous males attract several mates to a display territory.
>**Resource defense polygyny**   Polygynous males acquire several mates attracted to resources under their control.
>**Scramble competition polygyny**   Polygynous males acquire several widely scattered mates by finding them first.

Polygyny threshold model   An explanation for polygyny based on the premise that females will gain fitness by mating with an already paired male if the resources controlled by that male greatly exceed those under the control of unmated males.

Polyphenism   The occurrence of two or more alternative phenotypes within a species whose differences are induced by key differences in the environments experienced by the members of this species.

Prisoner's dilemma   A game theory construct in which the fitness payoffs to individuals are set such that mutual cooperation between the players generates a lower return than defection, which occurs when one individual accepts assistance from the other but does not return the favor.

Proximate cause   An immediate, underlying cause based on the operation of internal mechanisms possessed by an individual.

Reciprocity   Also known as reciprocal altruism in which a helpful action is repaid at a later date by the recipient of assistance.

>**Indirect reciprocity**   A form of reciprocity in which a helpful action is repaid at a later date by individuals other than the recipient of assistance.

Releaser   A sign stimulus given by an individual as a social signal to another.

Reproductive value   A measure of the probability that a given offspring will reach the age of reproduction or the potential of an individual to leave surviving descendants.

Reproductive success   The number of surviving offspring produced by an individual; direct fitness.

Resource defense polygyny   *See* Polygyny.

Resource holding power   The inherent capacity of an individual to defeat others when competing for useful resources.

Runaway selection   *See* Selection.

Satellite male   A male that waits near another male to intercept females drawn to the signals produced by the other male or attracted by the resources defended by the other male.

Scientific conclusion   In the scientific method, a hypothesis that has been tested and then accepted or rejected on the basis of the test results.

Scientific method   The process of testing hypotheses by examining the validity of predictions derived from them.

Scramble competition polygyny   *See* Polygyny.

Selection   The effects of differences among individuals in their ability to transmit copies of their genes to the next generation.

Artificial selection   A process that is identical with natural selection, except that humans control the reproductive success of alternative types within the selected population.

Chase-away selection   The reciprocal, spiraling effects of males attempting to exploit female mate choice mechanisms while females evolve resistance to these attempts.

Frequency-dependent selection   A form of natural selection in which those individuals that happen to belong to the less common of two types in the population are the ones that are more fit because of their lower frequency in the population.

Group selection   The process that occurs when groups differ in their collective attributes and the differences affect the survival chances of the groups.

Indirect selection   The process that occurs when individuals differ in their effects on the survival of nondescendant kin, creating differences in the indirect fitness of the individuals interacting with this category of kin.

Kin selection   The process that occurs when individuals differ in ways that affect their parental care or helping behavior, and thus the survival of their own offspring, or the survival of nondescendant kin.

Natural selection (direct selection)   The process that occurs when individuals differ in their traits and the differences are correlated with differences in reproductive success. Natural selection can produce evolutionary change when these differences are inherited.

Runaway selection   A form of sexual selection that occurs when female mating preferences for certain male attributes create a positive feedback loop favoring both males with these attributes *and* females that prefer them.

Sexual selection   A form of natural selection that occurs when individuals vary in their ability to compete with others for mates or to attract members of the opposite sex. As with natural selection, when the variation among individuals is correlated with genetic differences, sexual selection leads to genetic changes in the population.

Selfish herd   A group of individuals whose members use others as living shields against predators.

Sensory exploitation   In the evolution of a signal, when a signaler is able to tap into a pre-existing sensitivity or bias in the perceptual system of a receiver, thereby gaining an advantage in transmitting a message to that receiver.

Sex role reversal   A species in which females compete for access to males, which may choose selectively among potential mates.

Sexual dimorphism   A difference between males and females in a species.

Sexual selection   *See* Selection.

Siblicide   The killing of a sibling by a brother or sister.

Sign stimulus   The effective component of an action or object that triggers a fixed action pattern in an animal.

Sociobiology   A discipline that uses evolutionary theory as the foundation for the study of social behavior; often used to refer to studies of this sort involving human beings.

Sperm competition   The competition between males that determines whose sperm will fertilize a female's eggs when both males' sperm have been accepted by that female.

Stimulus filtering   The capacity of nerve cells and neural networks to ignore stimuli that could potentially elicit a response from them.

Strategy   A genetically distinctive set of rules for behavior exhibited by individuals.

Conditional strategy   A set of rules that enables individuals to use different tactics under different environmental conditions; the inherited behavioral capacity to be flexible in response to certain cues or situations.

Evolutionarily stable strategy   That set of rules of behavior that when adopted by a certain proportion of the population cannot be replaced by any alternative strategy.

Synapse   The point of near contact between one nerve cell and another.

Tactic   A behavior pattern that is enabled by an underlying hereditary mechanism of some sort; tactics are often referred to when someone wishes to distinguish between an option available to an individual as opposed to a strategy, which specifies a fixed response to a particular situation.

Territory   An area that an animal defends against intruders.

Test   In the scientific method, actual results that permit one to evaluate a hypothesis by comparing the evidence against the predicted results.

Transactional theory   The view that social units form as a result of the ability of individuals to negotiate for reproductive opportunities with one another; **concession theory** is derived from the transactional approach based on the premise that dominant members of the group concede some reproductive rights to others in return for their cooperation in staying with the group.

Ultimate Cause   The evolutionary, historical reason why something is the way it is.

Bibliography

1 Abbot, P., J. H. Withgott, and N. A. Moran. 2001. Genetic conflict and conditional altruism in social aphid colonies. *Proceedings of the National Academy of Sciences* 98: 12068–12071.

2 Acharya, L., and J. N. McNeil. 1998. Predation risk and mating behavior: The responses of moths to bat-like ultrasound. *Behavioral Ecology* 9:552–558.

3 Adkins-Regan, E. 1981. Hormone specificity, androgen metabolism, and social behavior. *American Zoologist* 21: 257–271.

4 Adolphs, R., D. Tranel, and A. R. Damasio. 1998. The human amygdala in social judgment. *Nature* 393: 470–474.

5 Airey, D. C., H. Castillo-Juarez, G. Casella, E. J. Pollak, and T. J. DeVoogd. 2000. Variation in the volume of zebra finch song control nuclei is heritable: Developmental and evolutionary implications. *Proceedings of the Royal Society of London B* 267: 2099–2104.

6 Airey, D. C., and T. J. DeVoogd. 2000. Greater song complexity is associated with augmented song system anatomy in zebra finches. *NeuroReport* 11: 2339–2344.

7 Alatalo, R. V., and A. Lundberg. 1984. Polyterritorial polygyny in the pied flycatcher *Ficedula hypoleuca*—evidence for the deception hypothesis. *Annales Zoologici Fennici* 21: 217–228.

8 Alberts, S. C., H. E. Watts, and J. Altmann. 2003. Queuing and queue-jumping: long-term patterns of reproductive skew in male savannah baboons, *Papio cynocephalus. Animal Behaviour* 65: 821–840.

9 Albrecht, T., and P. Klvana. 2004. Nest crypsis, reproductive value of a clutch and escape decisions in incubating female mallards *Anas platyrhynchos. Ethology* 110: 603–614.

10 Alcock, J. 1987. Leks and hilltopping in insects. *Journal of Natural History* 21: 319–328.

11 Alcock, J. 2001. *The Triumph of Sociobiology.* New York: Oxford University Press.

12 Alcock, J., and W. J. Bailey. 1995. Acoustical communication and the mating system of the Australian whistling moth *Hecatesia exultans* (Noctuidae: Agaristidae). *Journal of Zoology* 237: 337–352.

13 Alcock, J., and W. J. Bailey. 1997. Success in territorial defence by male tarantula hawk wasps *Hemipepsis ustulata*: The role of residency. *Ecological Entomology* 22: 377–383.

14 Alcock, J., G. C. Eickwort, and K. R. Eickwort. 1977. The reproductive behavior of *Anthidium maculosum* and the evolutionary significance of multiple copulations by females. *Behavioral Ecology and Sociobiology* 2: 385–396.

15 Alexander, G. M., and M. Hines. 2002. Sex differences in response to children's toys in nonhuman primates (*Cercopithecus aethiops sabaeus*). *Evolution and Human Behavior* 23: 467–479.

16 Alexander, R. D. 1974. The evolution of social behavior. *Annual Review of Ecology and Systematics* 5: 325–383.

17 Alexander, R. D. 1979. *Darwinism and Human Affairs.* Seattle, WA: University of Washington Press.

18 Alexander, R. D. 1987. *The Biology of Moral Systems.* Hawthorne, NY: Aldine de Gruyter.

19 Allen, G. R., D. J. Kazmer, and R. F. Luck. 1994. Post-copulatory male behaviour, sperm precedence and multiple mating in a solitary parasitoid wasp. *Animal Behaviour* 48: 635–644.

20 Allen, L., B. Beckwith, J. Beckwith, S. Chorover, D. Culver, et al. 1975. Against "sociobiology." *New York Review of Books* 22 (Nov. 13): 43–44.

21 Allen, L., B. Beckwith, J. Beckwith, S. Chorover, D. Culver, et al. 1976. Sociobiology—another biological determinism. *BioScience* 26: 182–186.

22 Allison, T., A. Puce, and G. McCarthy. 2000. Social perception from visual cues: The role of the STS region. *Trends in Cognitive Sciences* 4: 267–278.

23 Allman, J. 1999. *Evolving Brains.* New York: Scientific American Library.

24 Allman, J., A. Rosin, R. Kumar, and A. Hasenstaub. 1998. Parenting and survival in anthropoid primates: Caretakers live longer. *Proceedings of the National Academy of Sciences* 95: 6866–6869.

25 Ancel, A., H. Visser, Y. Handrich, D. Masman, and Y. L. Maho. 1997. Energy saving in huddling penguins. *Nature* 385: 304–305.

26 Anderson, J. B., and L. P. Brower. 1996. Freeze-protection of overwintering monarch butterflies in Mexico: Critical role of the forest as a blanket and an umbrella. *Ecological Entomology* 21: 107–116.

27 Anderson, K. G., H. Kaplan, and J. Lancaster. 1999. Paternal care by genetic fathers and stepfathers I: Reports from Albuquerque men. *Evolution and Human Behavior* 20: 405–432.

28 Andersson, J., A.-K. Borg-Karlson, and C. Wiklund. 2000. Sexual cooperation and conflict in butterflies: A male-transferred antiaphrodisiac reduces harassment of recently mated females. *Proceedings of the Royal Society of London B* 267: 1271–1275.

29 Andersson, M. 1982. Female choice selects for extreme tail length in a widowbird. *Nature* 299: 818–820.

30 Andersson, M. 1984. The evolution of eusociality. *Annual Review of Ecology and Systematics* 15: 165–189.

31 Andersson, M. 1994. *Sexual Selection.* Princeton, NJ: Princeton University Press.

32 Andersson, M., F. Gotmark, and C. G. Wiklund. 1981. Food information in the black-headed gull, *Larus ridibundus. Behavioral Ecology and Sociobiology* 9: 199–202.

33 Andersson, S., and T. Amundsen. 1997. Ultraviolet colour vision and ornamentation in bluethroats. *Proceedings of the Royal Society of London B* 264: 1587–1591.

34 Andrade, M. C. B. 1996. Sexual selection for male sacrifice in the Australian redback spider. *Science* 271: 70–72.

35 Andrade, M. C. B. 1998. Female hunger can explain variation in cannibalistic behavior despite male sacrifice in redback spiders. *Behavioral Ecology* 9: 33–42.

36 Andrade, M. C. B. 2003. Risky mate search and male self-sacrifice in redback spiders. *Behavioral Ecology* 14: 531–538.

37 Angier, N. 1999. *Woman, An Intimate Geography.* New York: Houghton Mifflin Company.

38 Apollonio, M., M. Festa-Bianchet, F. Mari, E. Bruno, and M. Locati. 1998. Habitat manipulation modifies lek use in fallow deer. *Ethology* 104: 603–612.

39 Aragon, S., A. P. Møller, J. J. Soler, and M. Soler. 1999. Molecular phylogeny of cuckoos supports a polyphyletic origin of brood parasitism. *Journal of Evolutionary Biology* 12: 495–506.

40 Armitage, K. B. 1986. Marmot polygyny revisited: The determinants of male and female reproductive strategies. In *Ecological Aspects of Social Evolution: Birds and Mammals*, ed. Rubenstein, D. I., and R. W. Wrangham. Princeton, NJ: Princeton University Press.

41 Arnold, S. J. 1980. The microevolution of feeding behavior. In *Foraging Behavior: Ecology, Ethological and Psychological Approaches*, ed. Kamil, A., and T. Sargent. New York: Garland STPM Press.

42 Arnold, S. J. 1983. Sexual selection: The interface of theory and empiricism. In *Mate Choice*, ed. Bateson, P. P. G. Cambridge: Cambridge University Press.

43 Arnold, T. W. 1999. What limits clutch size in waders? *Journal of Avian Biology* 30: 216–220.

44 Aspi, J., and A. A. Hoffman. 1998. Female encounter rates and fighting costs of males are associated with lek size in *Drosophila mycetophaga. Behavioral Ecology and Sociobiology* 42: 163–170.

45 Averof, M., and S. M. Cohen. 1997. Evolutionary origins of insect wings from ancestral gills. *Nature* 385: 627–630.

46 Axelrod, R., and W. D. Hamilton. 1981. The evolution of cooperation. *Science* 211: 1390–1396.

47 Bachmann, G. C., and M. A. Chappell. 1998. The energetic cost of begging behaviour in nestling house wrens. *Animal Behaviour* 55: 1607–1618.

48 Backwell, P. R. Y., and M. D. Jennions. 2004. Coalition among male fiddler crabs. *Nature* 430: 417.

49 Badyaev, A. V., K. R. Foresman, and M. V. Fernandes. 2000. Stress and developmental stability: Vegetation removal causes increased fluctuating asymmetry in shrews. *Ecology* 81: 336–345.

50 Bailey, J. M., S. Gaulin, Y. Agyei, and B. A. Gladue. 1994. Effects of gender and sexual orientation on evolutionarily relevant aspects of human mating psychology. *Journal of Personality and Social Psychology* 66: 1081–1093.

51 Baker, A. J., P. M. Gonzalez, T. Piersma, L. J. Niles, I. D. S. do Nascimento, et al. 2004. Rapid population decline in red knots: fitness consequences of decreased refuelling rates and late arrival in Delaware Bay. *Proceedings of the Royal Society of London B* 271: 875–882.

52 Baker, M. C., S. W. Bottjer, and A. P. Arnold. 1984. Sexual dimorphism and lack of seasonal-changes in vocal control regions of the white-crowned sparrow brain. *Brain Research* 295: 85–89.

53 Baker, M. C., and M. A. Cunningham. 1985. The biology of bird-song dialects. *Behavioral and Brain Sciences* 8: 85–133.

54 Bakst, M. R. 1998. Structure of the avian oviduct with emphasis on sperm storage in poultry. *Journal of Experimental Zoology* 282: 618–626.

55 Balcombe, J. P. 1990. Vocal recognition of pups by mother Mexican free-tailed bats, *Tadarida brasiliensis mexicana. Animal Behaviour* 39: 960–966.

56 Balda, R. P. 1980. Recovery of cached seeds by a captive *Nucifraga caryocatactes. Zeitschrift für Tierpsychologie* 52: 331–346.

57 Balda, R. P., and A. C. Kamil. 1992. Long-term spatial memory in Clark's nutcracker, *Nucifraga columbiana. Animal Behaviour* 44: 761–769.

58 Ball, G. F., and J. Balthazart. 2001. Ethological concepts revisited: Immediate early gene induction in response to sexual stimuli in birds. *Brain Behavior and Evolution* 57: 252–270.

59 Ballentine, B., J. Hyman, and S. Nowicki. 2004. Vocal performance influences female response to male bird song: an experimental test. *Behavioral Ecology* 15: 163–168.

60 Balmford, A., J. C. Deutsch, R. J. C. Nefdt, and T. Clutton-Brock. 1993. Testing hotspot models of lek evolution: Data from three species of ungulates. *Behavioral Ecology and Sociobiology* 33: 57–65.

61 Balshine-Earn, S. 1995. The costs of parental care in Galilee St Peter's fish, *Sarotherodon galilaeus. Animal Behaviour* 50: 1–7.

62 Balthazart, J., M. Baillien, T. D. Charlier, C. A. Cornil, and G. F. Ball. 2003. The neuroendocrinology of reproductive behavior in Japanese quail. *Domestic Animal Endocrinology* 25: 69–82.

63 Baptista, L. F., and M. L. Morton. 1988. Song learning in montane white-crowned sparrows: From whom and when. *Animal Behaviour* 36: 1753–1764.

64 Baptista, L. F., and L. Petrinovich. 1984. Social interaction, sensitive phases and the song template hypothesis in the white-crowned sparrow. *Animal Behaviour* 32: 172–181.

65 Baptista, L. F., and L. Petrinovich. 1986. Song development in the white-crowned sparrow: Social factors and sex differences. *Animal Behaviour* 34: 1359–1371.

66 Barrette, C., and D. Vandal. 1990. Sparring, relative antler size, and assessment in male caribou. *Behavioral Ecology and Sociobiology* 26: 383–387.

67 Barta, Z., and L. A. Giraldeau. 2001. Breeding colonies as information centers: a reappraisal of information-based hypotheses using the producer-scrounger game. *Behavioral Ecology* 12: 121–127.

68 Bartlett, T. Q., R. W. Sussman, and J. M. Cheverud. 1993. Infant killing in primates: A review of observed cases with special reference to the sexual selection hypothesis. *American Anthropologist* 95: 958–990.

69 Basolo, A. L. 1990. Female preference predates the evolution of the sword in swordtail fish. *Science* 250: 808–810.

70 Basolo, A. L. 1995. Phylogenetic evidence for the role of a pre-existing bias in sexual selection. *Proceedings of the Royal Society of London B* 259: 307–311.

71 Basolo, A. L. 1998. Evolutionary change in a receiver bias: A comparison of female preference functions. *Proceedings of the Royal Society of London B* 265: 2223–2228.

72 Bass, A. H. 1996. Shaping brain sexuality. *American Scientist* 84: 352–363.

73 Bass, A. H., and M. S. Grober. 2001. Social and neural modulation of sexual plasticity in teleost fish. *Brain Behavior and Evolution* 57: 293–300.

74 Baum, D. A., and A. Larson. 1991. Adaptation reviewed: A phylogenetic methodology for studying character macroevolution. *Systematic Zoology* 40: 1–18.

75 Baylies, M. K., T. A. Bargiello, F. R. Jackson, and M. W. Young. 1987. Changes in abundance or structure of the *Per* gene-product can alter periodicity of the *Drosophila* clock. *Nature* 326: 390–392.

76 Beecher, M. D., and I. M. Beecher. 1979. Sociobiology of bank swallows: Reproductive strategy of the male. *Science* 205: 1282–1285.

77 Beecher, M. D., and J. M. Burt. 2004. The role of social interaction in bird song learning. *Current Directions in Psychological Science* 13: 224–228.

78 Beecher, M. D., and E. Campbell. In press. The role of unshared song in the singing interactions between neighbouring song sparrows. *Animal Behaviour.*

79 Beecher, M. D., S. E. Campbell, and J. C. Nordby. 2000. Territory tenure in song sparrows is related to song sharing with neighbours, but not to repertoire size. *Animal Behaviour* 59: 29–37.

80 Beecher, M. D., S. E. Campbell, J. M. Burt, C. E. Hill, and J. C. Nordby. 2000. Song-type matching between neighbouring song sparrows. *Animal Behaviour* 59: 21–27.

81 Beecher, M. D., P. K. Stoddard, S. E. Campbell, and C. L. Horning. 1996. Repertoire matching between neighbouring song sparrows. *Animal Behaviour* 51: 917–923.

82 Beering, M. 2001. *A Comparison of the Patterns of Dance Language Behavior in House-hunting and Nectar-foraging Honey Bees.* University of California, PhD. thesis. Riverside, CA.

83 Behrmann, M., G. Winocur, and M. Moscovitch. 1992. Dissociation between mental imagery and object recognition in a brain-damaged patient. *Nature* 359: 636–637.

84 Beletsky, L. D., and G. H. Orians. 1989. Territoriality among male red-winged blackbirds. III. Testing hypotheses of territorial dominance. *Behavioral Ecology and Sociobiology* 24: 333–339.

85 Bell, C. P. 1997. Leap-frog migration in the fox sparrow: Minimizing the cost of spring migration. *Condor* 99: 470–477.

86 Bell, C. P. 2000. Process in the evolution of bird migration and pattern in avian ecogeography. *Journal of Avian Biology* 31: 258–265.

87 Bell, D. A., P. W. Trail, and L. F. Baptista. 1998. Song learning and vocal tradition in Nuttall's white-crowned sparrows. *Animal Behaviour* 55: 939–956.

88 Ben-Shahar, Y., A. Robichon, M. B. Sokolowski, and G. E. Robinson. 2002. Influence of gene action across different time scales on behavior. *Science* 296: 741–744.

89 Benkman, C. W. 1990. Foraging rates and the timing of crossbill reproduction. *Auk* 107: 376–386.

90 Bennett, A. T. D., I. C. Cuthill, and K. J. Norris. 1994. Sexual selection and the mismeasure of color. *American Naturalist* 144: 848–860.

91 Bentley, D., and R. R. Hoy. 1974. The neurobiology of cricket song. *Scientific American* 231 (Aug.): 34–44.

92 Berglund, A., G. Rosenqvist, and I. Svensson. 1986. Mate choice, fecundity and sexual dimorphism in two pipefish species (Syngnathidae). *Behavioral Ecology and Sociobiology* 19: 301–307.

93 Bernasconi, G., and J. E. Strassmann. 1999. Cooperation among unrelated individuals: The ant foundress case. *Trends in Ecology and Evolution* 14: 477–482.

94 Berthold, P. 1991. Genetic control of migratory behaviour in birds. *Trends in Ecology and Evolution* 6: 254–257.

95 Berthold, P. 2003. Genetic basis and evolutionary aspects of bird migration. *Advances in the Study of Behavior* 33: 175–229.

96 Berthold, P., A. J. Helbig, G. Mohr, and U. Querner. 1992. Rapid microevolution of migratory behaviour in a wild bird species. *Nature* 360: 668–670.

97 Berthold, P., and F. Pulido. 1994. Heritability of migratory activity in a natural bird population. *Proceedings of the Royal Society of London B* 257: 311–315.

98 Berthold, P., and U. Querner. 1995. Microevolutionary aspects of bird migration based on experimental results. *Israel Journal of Zoology* 41: 377–385.

99 Betzig, L., ed. 1997. *Human Nature, A Critical Reader.* New York: Oxford University Press.

100 Billing, J., and P. W. Sherman. 1998. Antimicrobial functions of spices: Why some like it hot. *Quarterly Review of Biology* 73: 3–49.

101 Birkhead, T. R. 2003. *The Red Canary.* London: Orion Publishing.

102 Birkhead, T. R., N. Chaline, J. D. Biggins, T. Burke, and T. Pizzari. 2004. Nontransitivity of paternity in a bird. *Evolution* 58: 416–420.

103 Birkhead, T. R., and A. P. Møller. 1992. *Sperm Competition in Birds: Evolutionary Causes and Consequences.* London: Academic Press.

104 Birkhead, T. R., and A. P. Møller, eds. 1998. *Sperm Competition and Sexual Selection.* San Diego, CA: Academic Press.

105 Bissoondath, C. J., and C. Wiklund. 1995. Protein content of spermatophores in relation to monandry/polyandry in butterflies. *Behavioral Ecology and Sociobiology* 37: 365–372.

106 Bjorksten, T. A., K. Fowler, and A. Pomiankowski. 2000. What does sexual trait FA tell us about stress? *Trends in Ecology and Evolution* 15: 163–166.

107 Blackledge, T. A. 1998. Signal conflict in spider webs driven by predators and prey. *Proceedings of the Royal Society of London B* 265: 1991–1996.

108 Blackledge, T. A. 1998. Stabilimentum variation and foraging success in *Argiope aurantia* and *Argiope trifasciata* (Araneae: Araneidae). *Journal of Zoology* 246: 21–27.

109 Blackledge, T. A., and J. W. Wenzel. 1999. Do stabilimenta in orb webs attract prey or defend spiders? *Behavioral Ecology* 10: 372–376.

110 Blackledge, T. A., and J. W. Wenzel. 2001. Silk mediated defense by an orb web spider against predatory mud-dauber wasps. *Behaviour* 138: 155–171.

111 Blount, J. D., N. B. Metcalfe, T. R. Birkhead, and P. F. Surai. 2003. Carotenoid modulation of immune function and sexual attractiveness in zebra finches. *Science* 300: 125–127.

112 Boggess, J. 1984. Infant killing and male reproductive strategies in langurs (*Presbytis entellus*). In *Infanticide: Comparative and Evolutionary Perspectives*, ed. Hausfater, G., and S. Hrdy. Chicago: Aldine.

113 Bolles, R. C. 1973. The comparative psychology of learning: The selective association principle and some problems with "general" laws of learning. In *Perspectives in Animal Behavior*, ed. Bermant, G. Glenview, IL: Scott, Foresman & Company.

114 Bollmer, J. L., T. Sanchez, M. D. Cannon, D. Sanchez, B. Cannon, et al. 2003. Variation in morphology and mating system among island populations of Galapagos Hawks. *Condor* 105: 428–438.

115 Boomsma, J. J., and A. Grafen. 1990. Intraspecific variation in ant sex ratios and the Trivers-Hare hypothesis. *Evolution* 44: 1026–1034.

116 Boone, J. L., III. 1986. Parental investment and elite family structure in preindustrial states: A case study of late medieval-early modern Portuguese genealogies. *American Anthropologist* 88: 859–878.

117 Borgerhoff Mulder, M. 1988. Reproductive consequences of sex-biased inheritance. In *Comparative Socioecology of Mammals and Man*, ed. Standen, V., and R. Foley. London: Blackwell.

118 Borgerhoff Mulder, M. 1990. Kipsigis women's preferences for wealthy men: Evidence for female choice in mammals. *Behavioral Ecology and Sociobiology* 27: 255–264.

119 Borgerhoff Mulder, M. 1998. The demographic transition: Are we any closer to an evolutionary explanation? *Trends in Ecology and Evolution* 13: 266–269.

120 Borgia, G. 1985. Bower quality, number of decorations and mating success of male satin bowerbirds (*Ptilonorhynchus violaceus*). *Animal Behaviour* 33: 266–271.

121 Borgia, G. 1986. Sexual selection in bowerbirds. *Scientific American* 254 (June): 92–100.

122 Borgia, G. 1995. Why do bowerbirds build bowers? *American Scientist* 83: 542–547.

123 Borgia, G., and S. W. Coleman. 2000. Co-option of male courtship signals from aggressive display in bowerbirds. *Proceedings of the Royal Society of London B* 267: 1735–1740.

124 Borries, C. 1997. Infanticide in seasonally breeding multimale groups of Hanuman langurs (*Presbytis entellus*) in Ramnagar (South Nepal). *Behavioral Ecology and Sociobiology* 41: 139–150.

125 Borries, C., K. Launhardt, C. Epplen, J. T. Epplen, and P. Winkler. 1999. DNA analyses support the hypothesis that infanticide is adaptive in langur monkeys. *Proceedings of the Royal Society of London B* 266: 901–904.

126 Bouchard, T. J., Jr, and M. McGue. 2003. Genetic and environmental influences on human psychological differences. *Journal of Neurobiology* 54: 4–45.

127 Bouchard, T. J., Jr. 1997. IQ similarity in twins reared apart: Findings and responses to critics. In *Intelligence: Heredity and Environment*, ed. Sternberg, R., and E. Grigorenko. New York: Cambridge University Press.

128 Boughman, J. W. 1998. Vocal learning by greater spear-nosed bats. *Proceedings of the Royal Society of London B* 265: 227–233.

129 Braaten, R. F., and K. Reynolds. 1999. Auditory preference for conspecific song in isolation-reared zebra finches. *Animal Behaviour* 58: 105–111.

130 Bradbury, J. W. 1977. Lek mating behavior in the hammer-headed bat. *Zeitschrift für Tierpsychologie* 45: 225–255.

131 Bradbury, J. W. 1981. The evolution of leks. In *Natural Selection and Social Behavior*, ed. Alexander, R. D., and D. W. Tinkle. New York: Chiron Press.

132 Bradbury, J. W., and R. M. Gibson. 1983. Leks and mate choice. In *Mate Choice*, ed. Bateson, P. Cambridge: Cambridge University Press.

133 Bradbury, J. W., and S. L. Vehrencamp. 1998. *Principles of Animal Communication*. Sunderland, MA: Sinauer Associates.

134 Bradbury, J. W., S. L. Vehrencamp, and R. M. Gibson. 1989. Dispersion of displaying male sage grouse. I. Patterns of temporal variation. *Behavioral Ecology and Sociobiology* 24: 1–14.

135 Brakefield, P. M., and T. G. Liebert. 2000. Evolutionary dynamics of declining melanism in the peppered moth in The Netherlands. *Proceedings of the Royal Society of London B* 267: 1953–1957.

136 Brandt, M., and D. Mahsberg. 2002. Bugs with a backpack: the function of nymphal camouflage in the West African assassin bugs *Paredocla* and *Acanthaspis* spp. *Animal Behaviour* 63: 277–284.

137 Brandt, Y. 2003. Lizard threat display handicaps endurance. *Proceedings of the Royal Society of London B* 270: 1061–1068.

138 Braude, S. 2000. Dispersal and new colony formation in wild naked mole-rats: Evidence against inbreeding as the system of mating. *Behavioral Ecology* 11: 7–12.

139 Breed, M. D., E. Guzmán-Novoa, and G. J. Hunt. 2004. Defensive behavior of honey bees: Organization, genetics, and comparisons with other bees. *Annual Reviews of Entomology* 49: 271–298.

140 Breininger, D. R., V. L. Larson, B. W. Duncan, R. B. Smith, D. M. Oddy, and M. F. Goodchild. 1995. Landscape patterns of Florida scrub jay habitat use and demographic success. *Conservation Biology* 9: 1442–1453.

141 Bremmer, F. 1986. White whales on holiday. *Natural History* 95: 40–49.

142 Brenowitz, E. A. 1991. Evolution of the vocal control system in the avian brain. *Seminars in the Neurosciences* 3: 399–407.

143 Brenowitz, E. A., D. Margoliash, and K. W. Nordeen. 1997. An introduction to birdsong and the avian song system. *Journal of Neurobiology* 33: 495–500.

144 Bretman, A., N. Wedell, and T. Tregenza. 2004. Molecular evidence of post-copulatory inbreeding avoidance in the field cricket *Gryllus bimaculatus*. *Proceedings of the Royal Society of London B* 271: 159–164.

145 Breven, K. A. 1981. Mate choice in the wood frog, *Rana sylvatica*. *Evolution* 35: 707–722.

146 Briskie, J. V., P. R. Martin, and T. E. Martin. 1999. Nest predation and the evolution of nestling begging calls. *Proceedings of the Royal Society of London B* 266: 2153–2159.

147 Brockmann, H. J. 2002. An experimental approach to alternative mating tactics in male horseshoe crabs (*Limulus polyphemus*). *Behavioral Ecology* 13: 232–238.

148 Brockmann, H. J., T. Colson, and W. Potts. 1994. Sperm competition in horseshoe crabs (*Limulus polyphemus*). *Behavioral Ecology and Sociobiology* 35: 153–160.

149 Brockmann, H. J., and D. Penn. 1992. Male mating tactics in the horseshoe crab, *Limulus polyphemus*. *Animal Behaviour* 44: 653–665.

150 Brooks, D. R., and D. A. McLennan. 1991. *Phylogeny, Ecology, and Behavior*. Chicago, IL: University of Chicago Press.

151 Brotherton, P. N. M., and M. B. Manser. 1997. Female dispersion and the evolution of monogamy in the dik-dik. *Animal Behaviour* 54: 1413–1424.

152 Brower, J. V. Z. 1958. Experimental studies of mimicry in some North American butterflies. 1. The monarch, *Danaus plexippus*, and viceroy, *Limenitis archippus*. *Evolution* 12: 3–47.

153 Brower, L. P. 1996. Monarch butterfly orientation: Missing pieces of a magnificent puzzle. *Journal of Experimental Biology* 199: 93–103.

154 Brower, L. P., and W. H. Calvert. 1984. Chemical defence in butterflies. In *The Biology of Butterflies*, ed. Vane-Wright, R. I., and P. R. Ackery. London: Academic Press.

155 Brown, C. R., and M. B. Brown. 1986. Ecto-parasitism as a cost of coloniality in cliff swallows (*Hirundo pyrrhonota*). *Ecology* 67: 1206–1218.

156 Brown, C. R., and M. B. Brown. 1996. *Coloniality in the Cliff Swallow: The Effect of Group Size on Social Behavior*. Chicago, IL: University of Chicago Press.

157 Brown, J. L. 1975. *The Evolution of Behavior*. New York: W. W. Norton.

158 Brown, J. L. 1987. *Helping and Communal Breeding in Birds: Ecology and Evolution*. Princeton, NJ: Princeton University Press.

159 Brown, J. L., and G. H. Orians. 1970. Spacing patterns in mobile animals. *Annual Review of Ecology and Systematics* 1: 239–262.

160 Brown, J. L., and C. M. Vleck. 1998. Prolactin and helping in birds: Has natural selection strengthened helping behavior? *Behavioral Ecology* 9: 541–545.

161 Brown, J. R., H. Ye, R. T. Bronson, P. Dikkes, and M. E. Greenberg. 1996. A defect in nurturing in mice lacking the immediate early gene *fosB*. *Cell* 86: 297–309.

162 Brown, K. M. 1998. Proximate and ultimate causes of adoption in ring-billed gulls. *Animal Behaviour* 56: 1529–1543.

163 Brownmiller, S. 1975. *Against Our Will*. New York: Simon and Schuster.

164 Brownmiller, S., and B. Merhof. 1992. A feminist response to rape as an adaptation in men. *Brain and Behavioral Sciences* 15: 381–382.

165 Bruce, M. J., M. E. Herberstein, and M. A. Elgar. 2001. Signalling conflict between prey and predator attraction. *Journal of Evolutionary Biology* 14: 786–794.

166 Buchan, J. C., S. C. Alberts, J. B. Silk, and J. Altmann. 2003. True paternal care in a multi-male primate society. *Nature* 425: 179–181.

167 Buchanan, K. L., and C. K. Catchpole. 2000. Song as an indicator of male parental effort in the sedge warbler. *Proceedings of the Royal Society of London B* 267: 321–326.

168 Bugoni, L., L. Krause, and M. V. Petry. 2001. Marine debris and human impacts on sea turtles in southern Brazil. *Marine Pollution Bulletin* 42: 1330–1334.

169 Burch, R. L., and G. G. Gallup. 2004. Pregnancy as a stimulus for domestic violence. *Journal of Family Violence* 19: 243–247.

170 Burda, H., R. L. Honeycutt, S. Begall, O. Locker-Grutjen, and A. Scharff. 2000. Are naked and common mole-rats eusocial and if so, why? *Behavioral Ecology and Sociobiology* 47: 293–303.

171 Burger, J., and M. Gochfeld. 2001. Smooth-billed ani (*Crotophaga ani*) predation on butterflies in Mato Grosso, Brazil: risk decreases with increased group size. *Behavioral Ecology and Sociobiology* 49: 482–492.

172 Burke, T. 1989. DNA fingerprinting and other methods for the study of mating success. *Trends in Ecology and Evolution* 4: 139–144.

173 Burland, T. M., N. C. Bennett, J. U. M. Jarvis, and C. G. Faulkes. 2002. Eusociality in African mole-rats: new insights from patterns of genetic relatedness in the Damaraland mole-rat (*Cryptomys damarensis*). *Proceedings of the Royal Society of London B* 269: 1025–1030.

174 Burley, N. T., and R. Symanski. 1998. "A taste for the beautiful": Latent aesthetic mate preferences for white crests in two species of Australian grassfinches. *American Naturalist* 152: 792–802.

175 Burness, G., S. J. Casselman, A. I. Schulte-Hostedde, C. D. Moyes, and R. Montgomerie. 2004. Sperm swimming speed and energetics vary with sperm competition risk in bluegill (*Lepomis macrochirus*). *Behavioral Ecology and Sociobiology* 56: 65–70.

176 Burt, J. M., S. E. Campbell, and M. D. Beecher. 2001. Song type matching as threat: a test using interactive playback. *Animal Behaviour* 62: 1163–1170.

177 Buskirk, R. E., C. Frolich, and K. G. Ross. 1984. The natural selection of sexual cannibalism. *American Naturalist* 123: 612–625.

178 Buss, D. M. 1989. Sex differences in human mate preferences: Evolutionary hypothesis tested in 37 cultures. *Behavioral and Brain Sciences* 12: 1–149.

179 Buss, D. M., R. J. Larsen, D. Westen, and J. Semmelroth. 1992. Sex differences in jealousy: Evolution, physiology, and psychology. *Psychological Science* 3: 251–255.

180 Buss, D. M., and D. P. Schmitt. 1993. Sexual strategies theory: An evolutionary perspective on human mating. *Psychological Review* 100: 204–232.

181 Buston, P. M., and S. T. Emlen. 2003. Cognitive processes underlying human mate choice: The relationship between self-perception and mate preference in Western society. *Proceedings of the National Academy of Sciences* 100: 8805–8810.

182 Butchart, S. H. M., N. Seddon, and J. M. M. Ekstrom. 1999. Yelling for sex: Harem males compete for female access in bronze-winged jacanas. *Animal Behaviour* 57: 637–646.

183 Calsbeek, R., and B. Sinervo. 2002. An experimental test of the ideal despotic distribution. *Journal of Animal Ecology* 71: 513–523.

184 Calvert, W. H., and L. P. Brower. 1986. The location of monarch butterfly (*Danaus plexippus* L.) overwintering colonies in Mexico in relation to topography and climate. *Journal of the Lepidopterists' Society* 40: 164–187.

185 Camhi, J. M. 1984. *Neuroethology.* Sunderland, MA: Sinauer Associates.

186 Cantoni, D., and R. Brown. 1997. Paternal investment and reproductive success in the California mouse, *Peromyscus californicus.* *Animal Behaviour* 54: 377–386.

187 Carde, R. T., R. T. Staten, and A. Mafra-Neto. 1998. Behaviour of pink bollworm males near high-dose, point sources of pheromone in field wind tunnels: insights into mechanisms of mating disruption. *Entomologia Experimentalis et Applicata* 89: 35–46.

188 Carew, T. J. 2000. *Behavioral Neurobiology: The Cellular Organization of Behavior.* Sunderland, MA: Sinauer Associates.

189 Carey, M., and V. Nolan, Jr. 1975. Polygyny in indigo buntings: A hypothesis tested. *Science* 190: 1296–1297.

190 Carey, S. 1992. Becoming a face expert. *Philosophical Transactions of the Royal Society of London B* 335: 95–103.

191 Caro, T. M. 1986. The functions of stotting in Thomson's gazelles: Some tests of the predictions. *Animal Behaviour* 34: 663–684.

192 Caro, T. M. 1986. The functions of stotting: A review of the hypotheses. *Animal Behaviour* 34: 649–662.

193 Caro, T. M. 1995. Pursuit-deterrence revisited. *Trends in Ecology and Evolution* 10: 500–503.

194 Carpenter, S. J., J. M. Erickson, and F. D. Holland. 2003. Migration of a Late Cretaceous fish. *Nature* 423: 70–74.

195 Carter, R. 1998. *Mapping the Brain.* London: Weidenfeld and Nicolson.

196 Case, A., I.-F. Lin, and S. McLanahan. 2000. How hungry is the selfish gene? *Economic Journal* 110: 781–804.

197 Case, A., I.-F. Lin, and S. McLanahan. 2001. Educational attainment of siblings in stepfamilies. *Evolution and Human Behavior* 22: 269–289.

198 Case, A., D. Lubotsky, and C. Paxson. 2002. Economic status and health in childhood: The origins of the gradient. *American Economic Review* 92: 1308–1334.

199 Casselman, S. J., and R. Montgomerie. 2004. Sperm traits in relation to male quality in colonial spawning bluegill. *Journal of Fish Biology* 64: 1700–1711.

200 Catania, K. C. 2000. Cortical organization in insectivora: The parallel evolution of the sensory periphery and the brain. *Brain Behavior and Evolution* 55: 311–321.

201 Catania, K. C., and J. H. Kaas. 1996. The unusual nose and brain of the star-nosed mole. *BioScience* 46: 578–586.

202 Catania, K. C., and J. H. Kaas. 1997. Somatosensory fovea in the star-nosed mole: Behavioral use of the star in relation to innervation patterns and cortical representation. *Journal of Comparative Neurology* 387: 215–233.

203 Catania, K. C., and M. S. Remple. 2002. Somatosensory cortex dominated by the representation of teeth in the naked mole-rat brain. *Proceedings of the National Academy of Sciences* 99: 5692–5697.

204 Catchpole, C. K., and P. J. B. Slater. 1995. *Bird Song, Biological Themes and Variations.* Cambridge: Cambridge University Press.

205 Cézilly, F., and R. G. Nager. 1995. Comparative evidence for a positive association between divorce and extra-pair paternity in birds. *Proceedings of the Royal Society of London B* 262: 7–12.

206 Chapman, T., and B. Crespi. 1998. High relatedness and inbreeding in two species of haplodiploid eusocial thrips (Insecta: Thysanoptera) revealed by microsatellite analysis. *Behavioral Ecology and Sociobiology* 43: 301–306.

207 Charrier, I., N. Mathevon, and P. Jouventin. 2001. Mother's voice recognition by seal pups. *Nature* 412: 873.

208 Cheng, K. 2000. How honeybees find a place: Lessons from a simple mind. *Animal Learning & Behavior* 28: 1–15.

209 Cheng, M. Y., C. M. Bullock, C. Y. Li, A. G. Lee, J. C. Bermak, et al. 2002. Prokineticin 2 transmits the behavioural circadian rhythm of the suprachiasmatic nucleus. *Nature* 417: 405–410.

210 Chilton, G., M. R. Lein, and L. F. Baptista. 1990. Mate choice by female white-crowned sparrows in a mixed-dialect population. *Behavioral Ecology and Sociobiology* 27: 223–227.

211 Christy, J. H. 1995. Mimicry, mate choice, and the sensory trap hypothesis. *American Naturalist* 146: 171–181.

212 Cialdini, R. B. 2001. The science of persuasion. *Scientific American* 284 (Feb.): 76–81.

213 Cioffi, D., and R. Garner. 1998. The effect of response options on decisions and subsequent behavior: Sometimes inaction is better. *Personality and Social Psychology Bulletin* 24: 463–472.

214 Ciszek, D. 2000. New colony formation in the "highly inbred" eusocial naked mole-rat: Outbreeding is preferred. *Behavioral Ecology* 11: 1–6.

215 Clapham, J. 2001. Why do baleen whales migrate? A response to Corkeron and Connor. *Marine Mammal Science* 17: 432–436.

216 Clark, R. D., and E. Hatfield. 1989. Gender differences in receptivity to sexual offers. *Journal of Psychology and Human Sexuality* 2: 39–55.

217 Clayton, D. F. 1997. Role of gene regulation in song circuit development and song learning. *Journal of Neurobiology* 33: 549–571.

218 Clayton, N. C., and J. R. Krebs. 1994. Hippocampal growth and attrition in birds affected by experience. *Proceedings of the National Academy of Sciences* 91: 7410–7414.

219 Clayton, N. S. 1998. Memory and the hippocampus in food-storing birds: A comparative approach. *Neuropharmacology* 37: 441–452.

220 Clifford, L. D., and D. J. Anderson. 2001. Experimental demonstration of the insurance value of extra eggs in an obligately siblicidal seabird. *Behavioral Ecology* 12: 340–347.

221 Clotfelter, E. D., K. A. Schubert, V. Nolan, and E. D. Ketterson. 2003. Mouth color signals thermal state of nestling dark-eyed juncos (*Junco hyemalis*). *Ethology* 109: 171–182.

222 Clutton-Brock, T. H. 1991. *The Evolution of Parental Care.* Princeton, NJ: Princeton University Press.

223 Clutton-Brock, T. H., and S. D. Albon. 1979. The roaring of red deer and the evolution of honest advertisement. *Behaviour* 69: 145–170.

224 Clutton-Brock, T. H., S. D. Albon, R. M. Gibson, and F. E. Guinness. 1979. The logical stag: Adaptive aspects of fighting in red deer. *Animal Behaviour* 27: 211–225.

225 Clutton-Brock, T. H., and P. H. Harvey. 1984. Comparative approaches to investigating adaptation. In *Behavioural Ecology: An Evolutionary Approach*, ed. Krebs, J. R., and N. B. Davies. Oxford: Blackwell.

226 Clutton-Brock, T. H., M. J. O'Riain, P. N. M. Brotherton, D. Gaynor, R. Kansky, et al. 1999. Selfish sentinels in cooperative mammals. *Science* 284: 1640–1644.

227 Clutton-Brock, T. H., and G. A. Parker. 1992. Potential reproductive rates and the operation of sexual selection. *Quarterly Review of Biology* 67: 437–456.

228 Clutton-Brock, T. H., and G. A. Parker. 1995. Sexual coercion in animal societies. *Animal Behaviour* 49: 1345–1365.

229 Clutton-Brock, T. H., A. F. Russell, L. L. Sharpe, A. J. Young, Z. Balmforth, and G. M. McIlrath. 2002. Evolution and development of sex differences in cooperative behavior in meerkats. *Science* 297: 253–256.

230 Cohen, L., S. Lehéricy, F. Chochon, C. Lemer, S. Rivaud, and S. Dehaene. 2002. Language-specific tuning of visual cortex? Functional properties of the Visual Word Form Area. *Brain* 125: 1054–1059.

231 Colantuoni, C., A. E. Purcell, C. M. Bouton, and J. Pevsner. 2000. High throughput analysis of gene expression in the human brain. *Journal of Neuroscience Research* 59: 1–10.

232 Colarelli, S. M., and J. R. Dettmann. 2003. Intuitive evolutionary perspectives in marketing practices. *Psychology & Marketing* 20: 837–865.

233 Collins, J. P., and J. E. Cheek. 1983. Effect of food and density on development of typical and cannibalistic salamander larvae in *Ambystoma tigrinum nebulosum*. *American Zoologist* 23: 77–84.

234 Collins, S. A., and C. Missing. 2003. Vocal and visual attractiveness are related in women. *Animal Behaviour* 65: 997–1004.

235 Conover, M. R. 1994. Stimuli eliciting distress calls in adult passerines and response of predators and birds to their broadcast. *Behaviour* 131: 19–37.

236 Conroy, C. J., and J. A. Cook. 2000. Molecular systematics of a holarctic rodent (Microtus: Muridae). *Journal of Mammalogy* 81: 344–359.

237 Cook, L. M. 2003. The rise and fall of the *Carbonaria* form of the peppered moth. *Quarterly Review of Biology* 78: 399–417.

238 Coombs, W. P., Jr. 1990. Behavior patterns of dinosaurs. In *The Dinosauria*, ed. Weishampel, D. B., P. Dodson, and H. Osmólska. Berkeley: University of California Press.

239 Cooper, W. E. 1995. Foraging mode, prey chemical discrimination, and phylogeny in lizards. *Animal Behaviour* 50: 973–985.

240 Cooper, W. E., V. Pérez-Mellado, T. A. Baird, J. P. Caldwell, and L. J. Vitt. 2004. Pursuit deterrent signalling by the bonaire whiptail lizard *Cnemidophorus murinus*. *Behaviour* 141: 297–311.

241 Cordero, C., and W. G. Eberhard. 2003. Female choice of sexually antagonistic male adaptations: a critical review of some current research. *Journal of Evolutionary Biology* 16: 1–6.

242 Corkeron, P. J., and R. C. Connor. 1999. Why do baleen whales migrate? *Marine Mammal Science* 15: 1228–1245.

243 Cornwell, R. E., L. Boothroyd, D. M. Burt, D. R. Feinberg, B. C. Jones, et al. 2004. Concordant preferences for opposite-sex signals? Human pheromones and facial characteristics. *Proceedings of the Royal Society of London B* 271: 635–640.

244 Coss, R. G., and R. O. Goldthwaite. 1995. The persistence of old designs for perception. *Perspectives in Ethology* 11: 83–148.

245 Court, G. S. 1996. The seal's own skin game. *Natural History* 105(8): 36–41.

246 Cowlishaw, G., and R. I. M. Dunbar. 1991. Dominance rank and mating success in male primates. *Animal Behaviour* 41: 1045–1056.

247 Cowlishaw, G., and R. Mace. 1996. Cross-cultural patterns of marriage and inheritance: A phylogenetic approach. *Ethology and Sociobiology* 17: 97–98.

248 Cox, G. W. 1985. The evolution of avian migration systems between temperate and tropical regions of the New World. *American Naturalist* 126: 452–474.

249 Coyne, J. 1998. Not black and white. *Nature* 396: 35–36.

250 Craig, C. L., and G. D. Bernard. 1990. Insect attraction to ultraviolet-reflecting spider webs and web decorations. *Ecology* 71: 616–623.

251 Craig, P. 1996. Intertidal territoriality and time-budget of the surgeonfish, *Acanthurus lineatus*, in American Samoa. *Environmental Biology of Fishes* 46: 27–36.

252 Crespi, B. J. 2000. The evolution of maladaptation. *Heredity* 84: 623–629.

253 Crews, D. 1975. Psychobiology of reptilian reproduction. *Science* 189: 1059–1065.

254 Crews, D. 1984. Gamete production, sex hormone secretion, and mating behavior uncoupled. *Hormones and Behavior* 18: 22–28.

255 Crews, D. 1991. Trans-seasonal action of androgen in the control of spring courtship behavior in male red-sided garter snakes. *Proceedings of the National Academy of Sciences* 88: 3545–3548.

256 Crews, D. 1992. Behavioral endocrinology and reproduction: An evolutionary perspective. *Oxford Reviews of Reproductive Biology* 14: 303–370.

257 Crews, D., and N. Greenberg. 1981. Function and causation of social signals in lizards. *American Zoologist* 21: 273–294.

258 Crews, D., V. Hingorani, and R. J. Nelson. 1988. Role of the pineal gland in the control of annual reproductive behavioral and physiological cycles in the red-sided garter snake (*Thamnophis sirtalis parietalis*). *Journal of Biological Rhythms* 3: 293–302.

259 Crews, D., and M. C. Moore. 1986. Evolution of mechanisms controlling mating behavior. *Science* 231: 121–125.

260 Cristol, D. A., and P. V. Switzer. 1999. Avian prey-dropping behavior. II. American crows and walnuts. *Behavioral Ecology* 10: 220–226.

261 Cronk, L. 1993. Parental favoritism toward daughters. *American Scientist* 81: 272–279.

262 Cullen, E. 1957. Adaptations in the kittiwake to cliff nesting. *Ibis* 99: 275–302.

263 Cumming, G. S. 1996. Mantis movements by night and the interactions of sympatric bats and mantises. *Canadian Journal of Zoology* 74: 1771–1774.

264 Cumming, J. M. 1994. Sexual selection and the evolution of dance fly mating systems (Diptera: Empididae; Empidinae). *Canadian Entomologist* 126: 907–920.

265 Cummings, M. E., G. G. Rosenthal, and M. J. Ryan. 2003. A private ultraviolet channel in visual communication. *Proceedings of the Royal Society of London B* 270: 897–904.

266 Cunningham, E. J. A. 2003. Female mate preferences and subsequent resistance to copulation in the mallard. *Behavioral Ecology* 14: 326–333.

267 Cunningham, E. J. A., and A. F. Russell. 2000. Egg investment is influenced by male attractiveness in the mallard. *Nature* 404: 74–77.

268 Currie, C. R., J. A. Scott, R. C. Summerbell, and D. Malloch. 1999. Fungus-growing ants use antibiotic-producing bacteria to control garden parasites. *Nature* 398: 701–704.

269 Curtin, R., and P. Dolhinow. 1978. Primate social behavior in a changing world. *American Scientist* 66: 468–475.

270 Cuvillier-Hot, V., R. Gadagkar, C. Peeters, and M. Cobb. 2002. Regulation of reproduction in a queenless ant: aggression, pheromones and reduction in conflict. *Proceedings of the Royal Society of London B* 269: 1295–1300.

271 Cuvillier-Hot, V., A. Lenoir, and C. Peeters. 2004. Reproductive monopoly enforced by sterile police workers in a queenless ant. *Behavioral Ecology* 15: 970–975.

272 Dagg, A. I. 1998. Infanticide by male lions hypothesis: A fallacy influencing research into human behavior. *American Anthropologist* 100: 940–950.

273 Dale, J., R. Montgomerie, D. Michaud, and P. Boag. 1999. Frequency and timing of extrapair fertilisation in the polyandrous red phalarope (*Phalaropus fulicarius*). *Behavioral Ecology and Sociobiology* 46: 50–56.

274 Dale, S., H. Rinden, and T. Slagsvold. 1992. Competition for a mate restricts mate search of female pied flycatchers. *Behavioral Ecology and Sociobiology* 30: 165–176.

275 Dale, S., and T. Slagsvold. 1994. Polygyny and deception in the pied flycatcher: Can females determine male mating status? *Animal Behaviour* 48: 1207–1217.

276 Daly, M., and M. Wilson. 1983. *Sex, Evolution and Behavior.* Boston: Willard Grant Press.

277 Daly, M., and M. Wilson. 1985. Child abuse and other risks of not living with both parents. *Ethology and Sociobiology* 6: 197–210.

278 Daly, M., and M. Wilson. 1988. *Homicide.* Hawthorne, NY: Aldine de Gruyter.

279 Daly, M., and M. Wilson. 1992. The man who mistook his wife for a chattel. In *The Adapted Mind*, ed. Barkow, J., L. Cosmides, and J. Tooby. New York: Oxford University Press.

280 Daly, M., and M. Wilson. 1998. *The Truth about Cinderella.* New Haven, CT: Yale University Press.

281 Daly, M., and M. Wilson. 2001. An assessment of some proposed exceptions to the phenomenon of nepotistic discrimination against stepchildren. *Annales Zoologici Fennici* 38: 287–296.

282 Daly, M., M. Wilson, and S. T. Weghorst. 1982. Male sexual jealousy. *Ethology and Sociobiology* 3: 11–27.

283 Damen, W. G. M., T. Saridaki, and M. Averof. 2002. Diverse adaptations of an ancestral gill: A common evolutionary origin for wings, breathing organs, and spinnerets. *Current Biology* 12: 1711–1716.

284 Danforth, B. N., L. Conway, and S. Q. Ji. 2003. Phylogeny of eusocial *Lasioglossum* reveals multiple losses of eusociality within a primitively eusocial clade of bees (Hymenoptera: Halictidae). *Systematic Biology* 52: 23–36.

285 Darwin, C. 1859. *On the Origin of Species.* London: Murray.

286 Darwin, C. 1871. *The Descent of Man and Selection in Relation to Sex.* London: Murray.

287 Darwin, C. 1892. *The Various Contrivances by which Orchids are Fertilised by Insects.* New York: D. Appleton.

288 Davies, N. B. 1978. Territorial defence in the speckled wood butterfly (*Pararge aegeria*): The resident always wins. *Animal Behaviour* 26: 138–147.

289 Davies, N. B. 1983. Polyandry, cloaca-pecking and sperm competition in dunnocks. *Nature* 302: 334–336.

290 Davies, N. B. 1985. Cooperation and conflict among dunnocks, *Prunella modularis*, in a variable mating system. *Animal Behaviour* 33: 628–648.

291 Davies, N. B. 1992. *Dunnock Behaviour and Social Evolution.* Oxford: Oxford University Press.

292 Davies, N. B. 2000. *Cuckoos, Cowbirds and Other Cheats.* London: T & A D Poyser.

293 Davies, N. B., and M. de L. Brooke. 1988. Cuckoos versus reed warblers: Adaptations and counteradaptations. *Animal Behaviour* 36: 262–284.

294 Davies, N. B., and T. R. Halliday. 1978. Deep croaks and fighting assessment in toads *Bufo bufo*. *Nature* 275: 683–685.

295 Davies, N. B., I. R. Hartley, B. J. Hatchwell, and N. E. Langmore. 1996. Female control of copulations to maximize male help: A comparison of polygynandrous alpine accentors, *Prunella collaris*, and dunnocks, *P. modularis*. *Animal Behaviour* 51: 27–47.

296 Davies, N. B., and A. I. Houston. 1984. Territory economics. In *Behavioural Ecology: An Evolutionary Approach*, ed. Krebs, J. R., and N. B. Davies. Oxford: Blackwell Scientific Publications.

297 Davies, N. B., R. M. Kilner, and D. G. Noble. 1998. Nestling cuckoos *Cuculus canorus* exploit hosts with begging calls that mimic a brood. *Proceedings of the Royal Society of London B* 265: 673–678.

298 Davies, N. B., and A. Lundberg. 1984. Food distribution and a variable mating system in the dunnock, *Prunella modularis*. *Journal of Animal Ecology* 53: 895–912.

299 Davis–Walton, J., and P. W. Sherman. 1994. Sleep arrhythmia in the eusocial naked mole-rat. *Naturwissenschaften* 81: 272–275.

300 Dawkins, R. 1977. *The Selfish Gene.* New York: Oxford University Press.

301 Dawkins, R. 1980. Good strategy or evolutionarily stable strategy? In *Sociobiology: Beyond Nature/Nurture?* ed. Barlow, G. W., and J. Silverberg. Boulder, CO: Westview Press.

302 Dawkins, R. 1982. *The Extended Phenotype.* San Francisco: W.H. Freeman.

303 Dawkins, R. 1986. *The Blind Watchmaker.* New York: W.W. Norton.

304 Dawkins, R. 1989. *The Selfish Gene.* Oxford: Oxford University Press.

305 Dawkins, R., and J. Krebs. 1978. Animal signals: Information or manipulation? In *Behavioural Ecology: An Evolutionary Approach*, ed. Krebs, J. R., and N. B. Davies. Oxford: Blackwell.

306 Dawson, J. W., K. Dawson-Scully, D. Robert, and R. M. Robertson. 1997. Forewing asymmetries during auditory avoidance in flying locusts. *Journal of Experimental Biology* 200: 2323–2335.

307 de Belle, J. S., A. J. Hilliker, and M. B. Sokolowski. 1989. Genetic localization of *foraging* (*for*): A major gene for larval behavior in *Drosophila melanogaster*. *Genetics* 123: 157–163.

308 de Belle, J. S., and M. B. Sokolowski. 1987. Heredity of *rover/sitter*: Alternative foraging strategies of *Drosophila melanogaster* larvae. *Heredity* 59: 73–183.

309 de Renzi, E., and G. di Pellegrino. 1998. Prosopagnosia and alexia without object agnosia. *Cortex* 34: 403–415.

310 DeCoursey, P. J., and J. Buggy. 1989. Circadian rhythmicity after neural transplant to hamster third ventricle: Specificity of suprachiasmatic nuclei. *Brain Research* 500: 263–275.

311 DeHeer, C. J., M. A. D. Goodisman, and K. G. Ross. 1999. Queen dispersal strategies in the multiple-queen form of the fire ant *Solenopsis invicta*. *American Naturalist* 153: 660–675.

312 delBarco-Trillo, J., and M. H. Ferkin. 2004. Male mammals respond to a risk of sperm competition conveyed by odours of conspecific males. *Nature* 431: 446–449.

313 Dennett, D. C. 1995. *Darwin's Dangerous Idea.* New York: Simon & Schuster.

314 Dethier, V. G. 1962. *To Know a Fly.* San Francisco: Holden-Day.

315 Dethier, V. G. 1976. *The Hungry Fly: A Physiological Study of the Behavior Associated with Feeding.* Cambridge, MA: Harvard University Press.

316 Deutsch, C. J., M. P. Haley, and B. J. Le Boeuf. 1990. Reproductive effort of male northern elephant seals: Estimates from mass loss. *Canadian Journal of Zoology* 68: 2580–2593.

317 Deutsch, J. C. 1994. Uganda kob mating success does not increase on larger leks. *Behavioral Ecology and Sociobiology* 34: 451–459.

318 Deviche, P., and P. J. Sharp. 2001. Reproductive endocrinology of a free-living, opportunistically breeding passerine (White-winged Crossbill, *Loxia leucoptera*). *General and Comparative Endocrinology* 123: 268–279.

319 DeVoogd, T. J. 1991. Endocrine modulation of the development and adult function of the avian song system. *Psychoneuroendocrinology* 16: 41–66.

320 DeWitt, T. J., A. Sih, and D. S. Wilson. 1998. Costs and limits of phenotypic plasticity. *Trends in Ecology and Evolution* 13: 77–81.

321 DeWolfe, B. B., L. F. Baptista, and L. Petrinovich. 1989. Song development and territory establishment in Nuttall's white-crowned sparrow. *Condor* 91: 297–407.

322 DeWoody, J. A., D. E. Fletcher, M. Mackiewicz, S. D. Wilkins, and J. C. Avise. 2000. The genetic mating system of spotted sunfish (*Lepomis punctatus*): Mate numbers and the influence of male reproductive parasites. *Molecular Ecology* 9: 2119–2228.

323 Dhondt, A. A., and J. Schillemans. 1983. Reproductive success of the great tit in relation to its territorial status. *Animal Behaviour* 31: 902–912.

324 Diamond, J. M. 1992. *The Third Chimpanzee*. New York: HarperCollins Publishers.

325 Diamond, J. M. 1999. Dirty eating for healthy living. *Nature* 400: 120–121.

326 Diamond, J. M. 2000. Talk of cannibalism. *Nature* 407: 25–26.

327 Dias, P. C. 1996. Sources and sinks in population biology. *Trends in Ecology and Evolution* 11: 326–330.

328 Dickinson, J. L. 1995. Trade-offs between postcopulatory riding and mate location in the blue milkweed beetle. *Behavioral Ecology* 6: 280–286.

329 Dickinson, J. L. 2001. Extrapair copulations in western bluebirds (*Sialia mexicana*): female receptivity favors older males. *Behavioral Ecology and Sociobiology* 50: 423–429.

330 Dickinson, J. L., and R. L. Rutowski. 1989. The function of the mating plug in the chalcedon checkerspot butterfly. *Animal Behaviour* 38: 154–162.

331 Dickinson, J. L., and W. W. Weathers. 1999. Replacement males in the western bluebird: Opportunity for paternity, chick-feeding rules, and fitness consequences of male parental care. *Behavioral Ecology and Sociobiology* 45: 201–209.

332 Dixson, A. F. 1987. Observations on the evolution of genitalia and copulatory behavior in primates. *Journal of Zoology* 213: 423–443.

333 Dloniak, S. M., J. A. French, N. J. Place, M. L. Weldele, S. E. Glickman, and K. E. Holekamp. 2004. Non-invasive monitoring of fecal androgens in spotted hyenas (*Crocuta crocuta*). *General and Comparative Endocrinology* 135: 51–61.

334 Dodson, G. N. 1997. Resource defense mating system in antlered flies, *Phytalmia* spp. (Diptera: Tephritidae). *Annals of the Entomological Society of America* 90: 496–504.

335 Dodson, G. N., and M. W. Beck. 1993. Pre-copulatory guarding of penultimate females by male crab spiders, *Misumenoides formosipes*. *Animal Behaviour* 46: 951–959.

336 Dornhaus, A., and L. Chittka. 2001. Food alert in bumblebees (*Bombus terrestris*): possible mechanisms and evolutionary implications. *Behavioral Ecology and Sociobiology* 50: 570–576.

337 Double, M., and A. Cockburn. 2000. Pre-dawn infidelity: Females control extra-pair mating in superb fairy-wrens. *Proceedings of the Royal Society of London B* 267: 465–470.

338 Doucet, S. M., and R. Montgomerie. 2003. Multiple sexual ornaments in satin bowerbirds: ultraviolet plumage and bowers signal different aspects of male quality. *Behavioral Ecology* 14: 503–509.

339 Doucet, S. M., and R. Montgomerie. 2003. Structural plumage colour and parasites in satin bowerbirds *Ptilonorhynchus violaceus*: implications for sexual selection. *Journal of Avian Biology* 34: 237–242.

340 Doupe, A. J., and M. M. Solis. 1997. Song- and order-selective neurons develop in the songbird anterior forebrain during vocal learning. *Journal of Neurobiology* 33: 694–709.

341 Dover, G. A. 2000. *Dear Mr. Darwin: Letters on the Evolution of Life and Human Nature*. Berkeley, CA: University of California Press.

342 Downes, S. 2001. Trading heat and food for safety: Costs of predator avoidance in a lizard. *Ecology* 82: 2870–2881.

343 Drăgăniou, T. I., L. Nagle, and M. Kreutzer. 2002. Directional female preference for an exaggerated male trait in canary (*Serinus canaria*) song. *Proceedings of the Royal Society of London B* 269: 2525–2531.

344 Drea, C. M., N. J. Place, M. L. Weldele, E. M. Coscia, P. Licht, and S. E. Glickman. 2002. Exposure to naturally circulating androgens during foetal life incurs direct reproductive costs in female spotted hyenas, but is prerequisite for male mating. *Proceedings of the Royal Society of London B* 269: 1981–1987.

345 Drea, C. M., M. L. Weldele, N. G. Forger, E. M. Coscia, L. G. Frank, et al. 1998. Androgens and masculinization of genitalia in the spotted hyaena (*Crocuta crocuta*). 2. Effects of prenatal anti-androgens. *Journal of Reproduction and Fertility* 113: 117–127.

346 Drews, C. 1996. Contests and patterns of injuries in free-ranging male baboons (*Papio cynocephalus*). *Behaviour* 133: 443–474.

347 Dudley, R. 2000. Evolutionary origins of human alcoholism in primate frugivory. *Quarterly Review of Biology* 75: 3–15.

348 Duffy, D. L., and G. F. Ball. 2002. Song predicts immunocompetence in male European starlings (*Sturnus vulgaris*). *Proceedings of the Royal Society of London B* 269: 847–852.

349 Dufour, K. W., and P. J. Weatherhead. 1998. Bilateral symmetry and social dominance in captive male red-winged blackbirds. *Behavioral Ecology and Sociobiology* 42: 71–76.

350 Duncan, J., R. J. Seitz, J. Kolodny, D. Bor, H. Herzog, et al. 2000. A neural basis for general intelligence. *Science* 289: 457–460.

351 Dunn, P. O., and A. Cockburn. 1999. Extrapair mate choice and honest signaling in cooperatively breeding superb fairy-wrens. *Evolution* 53: 938–946.

352 Eadie, J. M., and J. M. Fryxell. 1992. Density dependence, frequency-dependence, and alternative nesting strategies in goldeneyes. *American Naturalist* 140: 621–641.

353 East, M. L., H. Hofer, and W. Wickler. 1993. The erect "penis" is a flag of submission in a female-dominated society: Greetings in Serengeti spotted hyenas. *Behavioral Ecology and Sociobiology* 33: 355–370.

354 Eberhard, W. G. 1990. Animal genitalia and female choice. *American Scientist* 78: 134–141.

355 Eberhard, W. G. 1996. *Female Control: Sexual Selection by Cryptic Female Choice*. Princeton, NJ: Princeton University Press.

356 Eberhard, W. G. 2003. Substitution of silk stabilimenta for egg sacs by *Allocyclosa bifurca* (Araneae: Araneidae) suggests that silk stabilimenta function as camouflage devices. *Behaviour* 140: 847–868.

357 Eggert, A.-K., and S. K. Sakaluk. 1995. Female-coerced monogamy in burying beetles. *Behavioral Ecology and Sociobiology* 37: 147–154.

358 Ehrenreich, B. 2000. How "natural" is rape? *Time* 155: 88.

359 Eising, C. M., and T. G. G. Groothuis. 2003. Yolk androgens and begging behaviour in black-headed gull chicks: an experimental field study. *Animal Behaviour* 66: 1027–1034.

360 Elliott, P. F. 1975. Longevity and the evolution of polygamy. *American Naturalist* 109: 281–287.

361 Ellis, A. W., and A. W. Young. 1996. *Human Cognitive Neuropsychology*. East Sussex, UK: Psychology Press.

362 Emlen, D. J. 2000. Integrating development with evolution: A case study with beetle horns. *BioScience* 50: 403–418.

363 Emlen, D. J. 2001. Costs and the diversification of exaggerated animal structures. *Science* 291: 1534–1536.

364 Emlen, S. T., P. H. Wrege, and N. J. Demong. 1995. Making decisions in the family: An evolutionary perspective. *American Scientist* 83: 148–157.

365 Emlen, S. T. 1978. Cooperative breeding. In *Behavioural Ecology: An Evolutionary Approach*, ed. Krebs, J. R., and N. B. Davies. Oxford: Blackwell.

366 Emlen, S. T. 1991. Evolution of cooperative breeding in birds and mammals. In *Behavioural Ecology: An Evolutionary Approach*, ed. Krebs, J. R., and N. B. Davies. Oxford: Blackwell Scientific.

367 Emlen, S. T., N. J. Demong, and D. J. Emlen. 1989. Experimental induction of infanticide in female wattled jacanas. *Auk* 106: 1–7.

368 Emlen, S. T., and L. W. Oring. 1977. Ecology, sexual selection and the evolution of mating systems. *Science* 197: 215–223.

369 Emlen, S. T., P. H. Wrege, and M. S. Webster. 1998. Cuckoldry as a cost of polyandry in the sex-role-reversed wattled jacana, *Jacana jacana*. *Proceedings of the Royal Society of London B* 265: 2359–2364.

370 Endler, J. A. 1991. Interactions between predators and prey. In *Behavioural Ecology: An Evolutionary Approach*, ed. Krebs, J. R., and N. B. Davies. Oxford: Blackwell Scientific.

371 Esch, H. E., S. W. Zhang, M. V. Srinivasan, and J. Tautz. 2001. Honeybee dances communicate distances measured by optic flow. *Nature* 411: 581–583.

372 Evans, H. E. 1966. *Life on a Little Known Planet*. New York: Dell.

373 Evans, H. E. 1973. *Wasp Farm*. Garden City, NY: Anchor Press.

374 Evans, J. P., and A. E. Magurran. 2000. Multiple benefits of multiple mating in guppies. *Proceedings of the National Academy of Sciences* 97: 10074–10076.

375 Evans, J. P., and A. E. Magurran. 2001. Patterns of sperm precedence and predictors of paternity in the Trinidadian guppy. *Proceedings of the Royal Society of London B* 268: 719–724.

376 Ewer, R. F. 1973. *The Carnivores*. Ithaca, NY: Cornell University Press.

377 Faaborg, J., P. G. Parker, L. DeLay, T. J. de Vries, J. C. Bednarz, et al. 1995. Confirmation of cooperative polyandry in the Galapagos hawk (*Buteo galapagoensis*). *Behavioral Ecology and Sociobiology* 36: 83–90.

378 Fadool, D. A., K. Tucker, R. Perkins, G. Fasciani, R. N. Thompson, et al. 2004. Kv1.3 channel gene-targeted deletion produces "super-smeller mice" with altered glomeruli, inter-acting scaffolding proteins, and biophysics. *Neuron* 41: 389–404.

379 Fain, G. L. 2003. *Sensory Transduction*. Sunderland, MA: Sinauer Associates.

380 Falls, J. B. 1988. Does song deter territorial intrusion in white-throated sparrows (*Zonotrichia albicollis*)? *Canadian Journal of Zoology* 66: 206–211.

381 Farley, C. T., and C. R. Taylor. 1991. A mechanical trigger for the trot-gallop transition in horses. *Science* 253: 306–308.

382 Farner, D. S. 1964. Time measurement in vertebrate photoperi-odism. *American Naturalist* 95: 375–386.

383 Farner, D. S., and R. A. Lewis. 1971. Photoperiodism and reproductive cycles in birds. *Photophysiology* 6: 325–370.

384 Farries, M. A. 2001. The oscine song system considered in the context of the avian brain: Lessons learned from comparative neurobiology. *Brain Behavior and Evolution* 58: 80–100.

385 Ferguson, J. N., L. J. Young, E. F. Hearn, M. M. Matzuk, T. R. Insel, and J. T. Winslow. 2000. Social amnesia in mice lacking the oxytocin gene. *Nature Genetics* 25: 284–288.

386 Fernald, R. D. 1993. Cichlids in love. *The Sciences* 33: 27–31.

387 Field, J., G. Shreeves, S. Sumner, and M. Casiraghi. 2000. Insurance-based advantage to helpers in a tropical hover wasp. *Nature* 404: 869–871.

388 Field, S. A., and M. A. Keller. 1993. Alternative mating tactics and female mimicry as post-copulatory mate-guarding behav-iour in the parasitic wasp *Cotesia rubecula*. *Animal Behaviour* 46: 1183–1189.

389 Fisher, A. P. 2003. Still "not quite as good as having your own"? Toward a sociology of adoption. *Annual Review of Sociology* 29: 335–361.

390 Fisher, J. 1954. Evolution and bird sociality. In *Evolution as a Process*, ed. Huxley, J., A. C. Hardy, and E. B. Ford. London: Allen & Unwin.

391 Fiske, P., P. T. Rintamäki, and E. Karvonen. 1998. Mating suc-cess in lekking males: a meta-analysis. *Behavioral Ecology* 9: 328–338.

392 Flinn, M. V. 1988. Step-parent/step-offspring interactions in a Caribbean village. *Ethology and Sociobiology* 9: 335–369.

393 Foellmer, M. W., and D. J. Fairbairn. 2003. Spontaneous male death during copulation in an orb-weaving spider. *Proceedings of the Royal Society of London B* 270: S183–S185.

394 Follett, B. K., P. W. Mattocks, Jr, and D. S. Farner. 1974. Circadian function in the photoperiodic induction of gonadotropin secretion in the white-crowned sparrow, *Zonotrichia leucophrys gambelli*. *Proceedings of the National Academy of Sciences* 71: 1666–1669.

395 Folstad, I., and A. J. Karter. 1992. Parasites, bright males, and the immunocompetence handicap. *American Naturalist* 139: 603–622.

396 Foltz, D. W., and P. L. Schwagmeyer. 1989. Sperm competition in the thirteen-lined ground squirrel: Differential fertilization success under field conditions. *American Naturalist* 133: 257–265.

397 Ford, E. B. 1955. *Moths*. London: Collins.

398 Forsgren, E., T. Amundsen, A. A. Borg, and J. Bjelvenmark. 2004. Unusually dynamic sex roles in a fish. *Nature* 429: 551–554.

399 Forster, L. M. 1992. The stereotyped behaviour of sexual can-nibalism in *Latrodectus hasselti* Thorell (Araneae: Theridiidae), the Australian redback spider. *Australian Journal of Zoology* 40: 1–11.

400 Forsyth, A., and J. Alcock. 1990. Female mimicry and resource defense polygyny by males of a tropical rove beetle *Leistotrophus versicolor* (Coleoptera: Staphylinidae). *Behavioral Ecology and Sociobiology* 26: 325–330.

401 Foster, M. S. 1977. Odd couples in manakins: A study of social organization and cooperative breeding in *Chiroxiphia linearis*. *American Naturalist* 111: 845–853.

402 Foster, W. A. 1990. Experimental evidence for effective and altruistic colony defence against natural predators by soldiers of the gall-forming aphid *Pemphigus spyrothecae* (Hemiptera: Pemphigidae). *Behavioral Ecology and Sociobiology* 27: 421–439.

403 Foster, W. A., and P. K. Rhoden. 1998. Soldiers effectively defend aphid colonies against predators in the field. *Animal Behaviour* 55: 761–765.

404 Fox, E. A. 2002. Female tactics to reduce sexual harassment in the Sumatran orangutan (*Pongo pygmaeus abelii*). *Behavioral Ecology and Sociobiology* 52: 93–101.

405 Francis, R. C. 2004. *Why Men Won't Ask for Directions: The Seductions of Sociobiology*. Princeton, NJ: Princeton University Press.

406 Francis, R. C., K. K. Soma, and R. D. Fernald. 1993. Social reg-ulation of the brain-pituitary-gonadal axis. *Proceedings of the National Academy of Sciences* 90: 7794–7798.

407 Frank, L. G. 1986. Social organization of the spotted hyena *Crocuta crocuta*. II. Dominance and reproduction. *Animal Behaviour* 34: 1510–1527.

408 Frank, L. G. 1997. Evolution of genital masculinization: Why do female hyaenas have such a large "penis?" *Trends in Ecology and Evolution* 12: 58–62.

409 Frank, L. G., H. E. Holekamp, and L. Smale. 1995. Dominance, demographics and reproductive success in female spotted hyenas: A long-term study. In *Serengeti II: Research, Management, and Conservation of an Ecosystem*, ed. Sinclair, A. R. E., and P. Arcese. Chicago: University of Chicago Press.

410 Frank, L. G., M. L. Weldele, and S. E. Glickman. 1995. Masculinization costs in hyaenas. *Nature* 377: 584–585.

411 Fretwell, S. D., and H. K. Lucas, Jr. 1969. On territorial behav-ior and other factors influencing habitat distribution in birds. I. Theoretical development. *Acta Biotheoretica* 19: 16–36.

412 Friberg, U., and G. Arnqvist. 2003. Fitness effects of female mate choice: preferred males are detrimental for *Drosophila melanogaster* females. *Journal of Evolutionary Biology* 16: 797–811.

413 Frith, C. B., and D. W. Frith. 2001. Nesting biology of the spot-ted catbird, *Ailuroedus melanotis*, a monogamous bowerbird (Ptilonorhynchidae), in Australian Wet Tropics upland rain-forests. *Australian Journal of Zoology* 49: 279–310.

414 Frith, C. B., and D. W. Frith. 2004. *Bowerbirds*. London: Oxford University Press.

415 Frost, W. N., T. A. Hoppe, J. Wang, and L. M. Tian. 2001. Swim initiation neurons in *Tritonia diomedea*. *American Zoologist* 41: 952–961.

416 Froy, O., A. L. Gotter, A. L. Casselman, and S. M. Reppert. 2003. Illuminating the circadian clock in monarch butterfly migration. *Science* 300: 1303–1305.

417 Fu, P., B. D. Neff, and M. R. Gross. 2001. Tactic-specific success in sperm competition. *Proceedings of the Royal Society of London B* 268: 1105–1112.

418   Fuentes, A. 2002. Patterns and trends in primate pair bonds. *International Journal of Primatology* 23: 953–978.

419   Fullard, J. H. 1997. The sensory coevolution of moths and bats. In *Comparative Hearing: Insects*, ed. Hoy, R. R., A. N. Popper, and R. R. Fay. New York: Springer.

420   Fullard, J. H. 2000. Day-flying butterflies remain day-flying in a Polynesian, bat-free habitat. *Proceedings of the Royal Society of London B* 267: 2295–2300.

421   Fullard, J. H., J. W. Dawson, and D. S. Jacobs. 2003. Auditory encoding during the last moment of a moth's life. *Journal of Experimental Biology* 206: 281–294.

422   Fullard, J. H., J. W. Dawson, L. D. Otero, and A. Surlykke. 1997. Bat-deafness in day-flying moths (Lepidoptera, Notodontidae, Dioptinae). *Journal of Comparative Physiology A* 181: 477–483.

423   Fullard, J. H., L. D. Otero, A. Orellana, and A. Surlykke. 2000. Auditory sensitivity and diel flight activity in Neotropical Lepidoptera. *Annals of the Entomological Society of America* 93: 956–965.

424   Fullard, J. H., J. M. Ratcliffe, and A. R. Soutar. 2004. Extinction of the acoustic startle response in moths endemic to a bat-free habitat. *Journal of Evolutionary Biology* 17: 856–861.

425   Fullard, J. H., and J. E. Yack. 1993. The evolutionary biology of insect hearing. *Trends in Ecology and Evolution* 8: 248–252.

426   Funk, D. H., and D. W. Tallamy. 2000. Courtship role reversal and deceptive signals in the long-tailed dance fly, *Rhamphomyia longicauda*. *Animal Behaviour* 59: 411–421.

427   Furlow, F. B. 1997. Human neonatal cry quality as an honest signal of fitness. *Evolution and Human Behavior* 18: 175–194.

428   Fusani, L., M. Gahr, and J. B. Hutchison. 2001. Aromatase inhibition reduces specifically one display of the ring dove courtship behavior. *General and Comparative Endocrinology* 122: 23–30.

429   Gamboa, G. J. 1978. Intraspecific defense: advantage of social cooperation among paper wasp foundresses. *Science* 199: 1463–1465.

430   Gangestad, S. W., and J. A. Simpson. 2000. Trade-offs, the allocation of reproductive effort, and the evolutionary psychology of human mating. *Behavioral and Brain Sciences* 23: 624–644.

431   Gangestad, S. W., J. A. Simpson, A. J. Cousins, C. E. Garver-Apgar, and P. N. Christensen. 2004. Women's preferences for male behavioral displays change across the menstrual cycle. *Psychological Science* 15: 203–207.

432   Garamszegi, L. Z., and M. Eens. 2004. Brain space for a learned task: strong intraspecific evidence for neural correlates of singing behavior in songbirds. *Brain Research Reviews* 44: 187–193.

433   Garcia, J., and F. R. Ervin. 1968. Gustatory-visceral and telereceptor-cutaneous conditioning: Adaptation in internal and external milieus. *Communications in Behavioral Biology (A)* 1: 389–415.

434   Garcia, J., W. G. Hankins, and K. W. Rusiniak. 1974. Behavioral regulation of the milieu interne in man and rat. *Science* 185: 824–831.

435   Garstang, M. 2004. Long-distance, low-frequency elephant communication. *Journal of Comparative Physiology A* 190: 791–805.

436   Gaulin, S. J. C., and J. S. Boster. 1990. Dowry as female competition. *American Anthropologist* 92: 994–1005.

437   Gaulin, S. J. C., and R. W. FitzGerald. 1986. Sex differences in spatial ability: An evolutionary hypothesis and test. *American Naturalist* 127: 74–88.

438   Gaulin, S. J. C., and R. W. FitzGerald. 1989. Sexual selection for spatial-learning ability. *Animal Behaviour* 37: 322–331.

439   Gaulin, S. J. C., and D. H. McBurney. 2001. *Psychology, An Evolutionary Approach*. Upper Saddle River, NJ: Prentice Hall.

440   Geary, D. C. 1998. *Male, Female: The Evolution of Human Sex Differences*. Washington, DC: American Psychological Association.

441   Gentner, T. Q., and S. H. Hulse. 2000. European starling preference and choice for variation in conspecific male song. *Animal Behaviour* 59: 443–458.

442   Getting, P. A. 1983. Mechanisms of pattern generation underlying swimming in *Tritonia*. II. Network reconstruction. *Journal of Neurophysiology* 49: 1017–1035.

443   Getting, P. A. 1989. A network oscillator underlying swimming in *Tritonia*. In *Neuronal and Cellular Oscillators*, ed. Jacklet, J. W. New York: Dekker.

444   Getz, L. L., and C. S. Carter. 1996. Prairie-vole partnerships. *American Scientist* 84: 56–62.

445   Getz, L. L., B. McGuire, and C. S. Carter. 2003. Social behavior, reproduction and demography of the prairie vole, *Microtus ochrogaster*. *Ethology, Ecology and Evolution* 15: 105–118.

446   Ghalambor, C. K., and T. E. Martin. 2001. Fecundity-survival trade-offs and parental risk-taking in birds. *Science* 292: 494–497.

447   Gibson, R. M. 1996. A re-evaluation of hotspot settlement in lekking sage grouse. *Animal Behaviour* 52: 993–1005.

448   Gibson, R. M., A. S. Aspbury, and L. L. McDaniel. 2002. Active formation of mixed-species grouse leks: a role for predation in lek evolution? *Proceedings of the Royal Society of London B* 269: 2503–2507.

449   Gil, D., J. Graves, N. Hazon, and A. Wells. 1999. Male attractiveness and differential testosterone investment in zebra finch eggs. *Science* 286: 126–128.

450   Gil, D., G. Leboucher, A. Lacroix, R. Cue, and M. Kreutzer. 2004. Female canaries produce eggs with greater amounts of testosterone when exposed to preferred male song. *Hormones and Behavior* 45: 64–70.

451   Gilardi, J. D., S. S. Duffey, C. A. Munn, and L. A. Tell. 1999. Biochemical functions of geophagy in parrots: Detoxification of dietary toxins and cytoprotective effects. *Journal of Chemical Ecology* 25: 897–922.

452   Gill, F. B., and L. L. Wolf. 1975. Economics of feeding territoriality in the golden-winged sunbird. *Ecology* 56: 333–345.

453   Gill, F. B., and L. L. Wolf. 1978. Comparative foraging efficiencies of some montane sunbirds in Kenya. *Condor* 80: 391–400.

454   Giraldeau, L. A., and B. Livoreil. 1998. Game theory and social foraging. In *Game Theory and Animal Behavior*, ed. Dugatkin, L. A., and H. K. Reeve. New York: Oxford University Press.

455   Glickman, S. E., L. G. Frank, P. Licht, T. Yalckinkaya, P. K. Siiteri, and J. Davidson. 1993. Sexual differentiation of the female spotted hyena: One of nature's experiments. *Annals of the New York Academy of Sciences* 662: 135–159.

456   Goodall, J. 1988. *In the Shadow of Man*. Boston: Houghton Mifflin.

457   Göth, A., and C. S. Evans. 2004. Social responses without early experience: Australian brush-turkey chicks use specific visual cues to aggregate with conspecifics. *Journal of Experimental Biology* 207: 2199–2208.

458   Gottschall, J. A., and T. A. Gottschall. 2003. Are per-incident rape-pregnancy rates higher than per-incident consensual pregnancy rates? *Human Nature* 14: 1–20.

459   Gould, J. L. 1982. Why do honey bees have dialects? *Behavioral Ecology and Sociobiology* 10: 53–56.

460   Gould, S. J. 1981. Hyena myths and realities. *Natural History* 90: 16–24.

461   Gould, S. J. 1986. Evolution and the triumph of homology, or why history matters. *American Scientist* 74: 60–69.

462   Gould, S. J. 1996. The diet of worms and the defenestration of Prague. *Natural History* 105: 18–24ff.

463   Gould, S. J., and R. C. Lewontin. 1979. The spandrels of San Marco and the Panglossian paradigm: A critique of the adaptationist programme. *Proceedings of the Royal Society of London B* 205: 581–598.

464   Goymann, W., M. L. East, and H. Hofer. 2001. Androgens and the role of female 'hyperaggressiveness' in spotted hyenas (*Crocuta crocuta*). *Hormones and Behavior* 39: 83–92.

465   Grammar, K., B. Fink, A. P. Møller, and R. Thornhill. 2003. Darwinian aesthetics: sexual selection and the biology of beauty. *Biological Reviews* 78: 385–407.

466   Grant, B. S. 1999. Fine tuning the peppered moth paradigm. *Evolution* 53: 980–984.

467   Grant, B. S., D. F. Owen, and C. A. Clarke. 1996. Parallel rise and fall of melanic peppered moths in America and Britain. *Journal of Heredity* 87: 351–357.

468   Grant, B. S., and L. L. Wiseman. 2002. Recent history of melanism in American peppered moths. *Journal of Heredity* 93: 86–90.

469   Graves, J. A., and A. Whiten. 1980. Adoption of strange chicks by herring gulls, *Larus argentatus. Zeitschrift für Tierpsychologie* 54: 267–278.

470   Gray, E. M. 1997. Do red-winged blackbirds benefit genetically from seeking copulations with extra-pair males? *Animal Behaviour* 53: 605–623.

471   Gray, E. M. 1997. Female red-winged blackbirds accrue material benefits from copulating with extra-pair males. *Animal Behaviour* 53: 625–639.

472   Gray, J. R., and P. M. Thompson. 2004. Neurobiology of intelligence: Science and ethics. *Nature Reviews Neuroscience* 5: 471–482.

473   Greene, E. 1987. Individuals in an osprey colony discriminate between high and low quality information. *Nature* 329: 239–241.

474   Greene, E., B. E. Lyon, V. R. Muehter, L. Ratcliffe, S. J. Oliver, and P. T. Boag. 2000. Disruptive sexual selection for plumage colouration in a passerine bird. *Nature* 407: 1000–1003.

475   Greene, E., L. T. Orsak, and D. W. Whitman. 1987. A tephritid fly mimics the territorial displays of its jumping spider predators. *Science* 236: 310–312.

476   Greenspan, R. J. 2004. E Pluribus Unum, Ex Uno Plura: Quantitative and single-gene perspectives on the study of behavior. *Annual Review of Neuroscience* 27: 79–105.

477   Greenwood, P. J. 1980. Mating systems, philopatry, and dispersal in birds and mammals. *Animal Behaviour* 28: 1140–1162.

478   Grether, G. F. 2000. Carotenoid limitation and mate preference evolution: A test of the indicator hypothesis in guppies (*Poecilia reticulata*). *Evolution* 54: 1712–1714.

479   Grether, G. F., S. Kasahara, G. R. Kolluru, and E. L. Cooper. 2004. Sex-specific effects of carotenoid intake on the immunological response to allografts in guppies (*Poecilia reticulata*). *Proceedings of the Royal Society of London B* 271: 45–49.

480   Griffin, A. S., and S. A. West. 2002. Kin selection: fact and fiction. *Trends in Ecology and Evolution* 17: 15–21.

481   Griffin, D. R. 1958. *Listening in the Dark.* New Haven, CT: Yale University Press.

482   Griffith, S. C., I. P. F. Owens, and K. A. Thuman. 2002. Extra pair paternity in birds: a review of interspecific variation and adaptive function. *Molecular Ecology* 11: 2195–2212.

483   Gross, M. R. 1982. Sneakers, satellites, and parentals: Polymorphic mating strategies in North American sunfishes. *Zeitschrift für Tierpsychologie* 60: 1–26.

484   Gross, M. R. 1996. Alternative reproductive strategies and tactics: Diversity within species. *Trends in Ecology and Evolution* 11: 92–98.

485   Gross, M. R., and A. M. MacMillan. 1981. Predation and the evolution of colonial nesting in bluegill sunfish (*Lepomis macrochirus*). *Behavioral Ecology and Sociobiology* 8: 163–174.

486   Gross, M. R., and R. C. Sargent. 1985. The evolution of male and female parental care in fishes. *American Zoologist* 25: 807–822.

487   Grunt, J. A., and W. C. Young. 1953. Consistency of sexual behavior patterns in individual male guinea pigs following castration and androgen therapy. *Journal of Comparative Physiology and Psychology* 46: 138–144.

488   Gubernick, D. J., and T. Teferi. 2000. Adaptive significance of male parental care in a monogamous mammal. *Proceedings of the Royal Society of London B* 267: 147–150.

489   Gurney, M. E., and M. Konishi. 1980. Hormone-induced sexual differentiation of brain and behavior in zebra finches. *Science* 208: 1380–1383.

490   Guzmán-Novoa, E., D. Prieto-Merlos, J. L. Uribe-Rubio, and G. J. Hunt. 2003. Relative reliability of four field assays to test defensive behaviour of honey bees (*Apis mellifera*). *Journal of Apicultural Research* 42: 42–46.

491   Gwinner, E. 1996. Circannual clocks in avian reproduction and migration. *Ibis* 138: 47–63.

492   Gwinner, E., and J. Dittami. 1990. Endogenous reproductive rhythms in a tropical bird. *Science* 249: 906–908.

493   Gwynne, D. T. 1981. Sexual difference theory: Mormon crickets show role reversal in mate choice. *Science* 213: 779–780.

494   Gwynne, D. T. 1983. Beetles on the bottle. *Journal of the Australian Entomological Society* 23: 79.

495   Gwynne, D. T., and L. W. Simmons. 1990. Experimental reversal of courtship roles in an insect. *Nature* 346: 172–174.

496   Hack, M. A. 1998. The energetics of male mating strategies in field crickets. *Journal of Insect Behavior* 11: 853–868.

497   Hahn, T. P. 1995. Integration of photoperiodic and food cues to time changes in reproductive physiology by an opportunistic breeder, the red crossbill, *Loxia curvirostra* (Aves: Carduelinae). *Journal of Experimental Zoology* 272: 213–226.

498   Hahn, T. P. 1998. Reproductive seasonality in an opportunistic breeder, the red crossbill, *Loxia curvirostra. Ecology* 79: 2365–2375.

499   Hahn, T. P., M. E. Pereyra, S. M. Sharbaugh, and G. E. Bentley. 2004. Physiological responses to photoperiod in three cardueline finch species. *General and Comparative Endocrinology* 137: 99–108.

500   Hahn, T. P., J. C. Wingfield, R. Mullen, and P. J. Deviche. 1995. Endocrine bases of spatial and temporal opportunism in arctic-breeding birds. *American Zoologist* 35: 259–273.

501   Halgren, E., A. M. Dale, M. I. Sereno, R. B. H. Tootell, K. Marinkovic, and B. R. Rosen. 1999. Location of human face-selective cortex with respect to retinotopic areas. *Human Brain Mapping* 7: 29–37.

502   Hamilton, W. D. 1964. The genetical theory of social behaviour, I, II. *Journal of Theoretical Biology* 7: 1–52.

503   Hamilton, W. D. 1971. Geometry for the selfish herd. *Journal of Theoretical Biology* 31: 295–311.

504   Hamilton, W. D. 1995. *The Narrow Roads of Gene Land. Vol. 1. Evolution of Social Behaviour.* Oxford: W.H. Freeman.

505   Hamilton, W. D. 1996. Foreword. In *Natural History and the Evolution of Paper Wasps,* ed. Turillazzi, S., and M. J. West-Eberhard, pp. v–vi. Oxford: Oxford University Press.

506   Hamilton, W. D., and M. Zuk. 1982. Heritable true fitness and bright birds: A role for parasites? *Science* 218: 384–387.

507   Hamilton, W. J., and G. H. Orians. 1965. Evolution of brood parasitism in altricial birds. *Condor* 67: 361–382.

508   Hamner, W. M. 1964. Circadian control of photoperiodism in the house finch demonstrated by interrupted-night experiments. *Nature* 203: 1400–1401.

509   Hanby, J. P., and J. D. Bygott. 1987. Emigration of subadult lions. *Animal Behaviour* 35: 161–169.

510   Harbison, H., D. A. Nelson, and T. P. Hahn. 1999. Long-term persistence of song dialects in the mountain white-crowned sparrow. *Condor* 101: 133–148.

511   Harcourt, A. H., and J. Gardiner. 1994. Sexual selection and genital anatomy of male primates. *Proceedings of the Royal Society of London B* 255: 47–53.

512   Harcourt, A. H., P. H. Harvey, S. G. Larson, and R. V. Short. 1981. Testis weight, body weight and breeding system in primates. *Nature* 293: 55–57.

513   Hare, B., M. Brown, C. Williamson, and M. Tomasello. 2002. The domestication of social cognition in dogs. *Science* 298: 1634–1636.

514   Harlow, H. F., and M. K. Harlow. 1962. Social deprivation in monkeys. *Scientific American* 207 (Nov.): 136–146.

515   Harlow, H. F., M. K. Harlow, and S. J. Suomi. 1971. From thought to therapy: Lessons from a primate laboratory. *American Scientist* 59: 538–549.

516   Hartung, J. 1982. Polygyny and inheritance of wealth. *Current Anthropology* 23: 1–12.

517   Hartung, J. 1995. Love they neighbor: The evolution of in-group morality. *Skeptic* 3: 86–99.

518   Harvey, P. H., and M. D. Pagel. 1991. *The Comparative Method in Evolutionary Biology.* London: Oxford University Press.

519 Haskell, D. G. 1999. The effect of predation on begging-call evolution in nestling wood warblers. *Animal Behaviour* 57: 893–901.

520 Hasselquist, D. 1998. Polygyny in great reed warblers: A long-term study of factors contributing to fitness. *Ecology* 79: 2376–2390.

521 Hatchwell, B. J., and J. Komdeur. 2000. Ecological constraints, life history traits and the evolution of cooperative breeding. *Animal Behaviour* 59: 1079–1086.

522 Hauser, M. D., M. K. Chen, F. Chen, E. Chuang, and E. Chuang. 2003. Give unto others: genetically unrelated cotton-top tamarin monkeys preferentially give food to those who altruistically give food back. *Proceedings of the Royal Society of London B* 270: 2363–2370.

523 Hausfater, G. 1975. Dominance and reproduction in baboons (*Papio cynocephalus*): A quantitative analysis. *Contributions in Primatology* 7: 1–150.

524 Hayes, L. D. 2000. To nest communally or not to nest communally: A review of rodent communal nesting and nursing. *Animal Behaviour* 59: 677–688.

525 Hayes, S. A., D. P. Costa, J. T. Harvey, and B. J. Le Boeuf. 2004. Aquatic mating strategies of the male Pacific harbor seal (*Phoca vitulina richardii*): Are males defending the hotspot? *Marine Mammal Science* 20: 639–656.

526 Haynes, K. F., C. Gemeno, K. V. Yeargan, J. G. Millar, and K. M. Johnson. 2002. Aggressive chemical mimicry of moth pheromones by a bolas spider: how does this specialist predator attract more than one species of prey? *Chemoecology* 12: 99–105.

527 Heath, K. M., and C. Hadley. 1998. Dichotomous male reproductive strategies in a polygynous human society: Mating versus parental effort. *Current Anthropology* 39: 369–374.

528 Hedwig, B. 2000. Control of cricket stridulation by a command neuron: efficacy depends on the behavioral state. *Journal of Neurophysiology* 83: 712–722.

529 Heeb, P., T. Schwander, and S. Faoro. 2003. Nestling detectability affects parental feeding preferences in a cavity-nesting bird. *Animal Behaviour* 66: 637–642.

530 Heinrich, B. 1979. *Bumblebee Economics*. Cambridge, MA.: Harvard University Press.

531 Heinrich, B. 1984. *In a Patch of Fireweed*. Cambridge, MA: Harvard University Press.

532 Heinrich, B. 1988. Winter foraging at carcasses by three sympatric corvids, with emphasis on recruitment by the raven, *Corvus corax*. *Behavioral Ecology and Sociobiology* 23: 141–156.

533 Heinrich, B. 1989. *Ravens in Winter*. New York: Summit Books.

534 Heinze, J., and B. Hölldobler. 1993. Fighting for a harem of queens: Physiology of reproduction in *Cardiocondyla* male ants. *Proceedings of the National Academy of Sciences* 90: 8412–8414.

535 Heinze, J., B. Hölldobler, and C. Peeters. 1994. Conflict and cooperation in ant societies. *Naturwissenschaften* 81: 489–497.

536 Heinze, J., B. Hölldobler, and K. Yamauchi. 1998. Male competition in *Cardiocondyla* ants. *Behavioral Ecology and Sociobiology* 42: 239–246.

537 Heistermann, M., T. Ziegler, C. P. van Schaik, K. Launhardt, P. Winkler, and J. K. Hodges. 2001. Loss of oestrus, concealed ovulation and paternity confusion in free-ranging Hanuman langurs. *Proceedings of the Royal Society of London B* 268: 2445–2551.

538 Herberstein, M. E. 2000. Foraging behaviour in orb-web spiders (Araneidae): Do web decorations increase prey capture success in *Argiope keyserlingi* Karsch, 1878? *Australian Journal of Zoology* 48: 217–223.

539 Hill, G. E. 2002. *A Red Bird in a Brown Bag: The Function and Evolution of Colorful Plumage in the House Finch*. Oxford: Oxford University Press.

540 Hitchcock, C. L., and D. F. Sherry. 1990. Long-term memory for cache sites in the black-capped chickadee. *Animal Behaviour* 40: 701–712.

541 Hofer, H., and M. L. East. 2003. Behavioral processes and costs of co-existence in female spotted hyenas: a life history perspective. *Evolutionary Ecology* 17: 315–331.

542 Hogg, J. T. 1984. Mating in bighorn sheep: Multiple creative male strategies. *Science* 225: 526–529.

543 Höglund, J., and R. V. Alatalo. 1995. *Leks*. Princeton, NJ: Princeton University Press.

544 Höglund, J., and A. Lundberg. 1987. Sexual selection in a monomorphic lek-breeding bird: Correlates of male mating success in the great snipe *Gallinago media*. *Behavioral Ecology and Sociobiology* 21: 211–216.

545 Höglund, J., and J. G. M. Robertson. 1990. Spacing of leks in relation to female home ranges, habitat requirements and male attractiveness in the great snipe (*Gallinago media*). *Behavioral Ecology and Sociobiology* 26: 173–180.

546 Högstedt, G. 1983. Adaptation unto death: Function of fear screams. *American Naturalist* 121: 562–570.

547 Hohoff, C., K. Franzen, and N. Sachser. 2003. Female choice in a promiscuous wild guinea pig, the yellow-toothed cavy (*Galea musteloides*). *Behavioral Ecology and Sociobiology* 53: 341–349.

548 Holden, C. J., R. Sear, and R. Mace. 2003. Matriliny as daughter-biased investment. *Evolution and Human Behavior* 24: 99–112.

549 Holekamp, K. E. 1984. Natal dispersal in Belding's ground squirrels (*Spermophilus beldingi*). *Behavioral Ecology and Sociobiology* 16: 21–30.

550 Holekamp, K. E., and P. W. Sherman. 1989. Why male ground squirrels disperse. *American Scientist* 77: 232–239.

551 Holland, B., and W. R. Rice. 1998. Chase-away sexual selection: Antagonistic seduction versus resistance. *Evolution* 52: 1–7.

552 Holland, B., and W. R. Rice. 1999. Experimental removal of sexual selection reverses intersexual antagonistic coevolution and removes a reproductive load. *Proceedings of the National Academy of Sciences* 96: 5083–5088.

553 Hölldobler, B., and E. O. Wilson. 1990. *The Ants*. Cambridge, MA: Harvard University Press.

554 Hölldobler, B., and E. O. Wilson. 1994. *Journey to the Ants*. Cambridge, MA: Harvard University Press.

555 Holley, A. J. F. 1984. Adoption, parent-chick recognition, and maladaptation in the herring gull *Larus argentatus*. *Zeitschrift für Tierpsychologie* 64: 9–14.

556 Holloway, C. C., and D. F. Clayton. 2001. Estrogen synthesis in the male brain triggers development of the avian song control pathway in vitro. *Nature Neuroscience* 4: 170–175.

557 Holmes, W. G. 1986. Identification of paternal half-siblings by captive Belding's ground squirrels. *Animal Behaviour* 34: 321–327.

558 Holmes, W. G., and P. W. Sherman. 1982. The ontogeny of kin recognition in two species of ground squirrels. *American Zoologist* 22: 491–517.

559 Holmes, W. G., and P. W. Sherman. 1983. Kin recognition in animals. *American Scientist* 71: 46–55.

560 Hoogland, J. L. 1998. Why do female Gunnison's prairie dogs copulate with more than one male? *Animal Behaviour* 55: 351–359.

561 Hoogland, J. L., and P. W. Sherman. 1976. Advantages and disadvantages of bank swallow (*Riparia riparia*) coloniality. *Ecological Monographs* 46: 33–58.

562 Hopkins, C. D. 1998. Design features for electric communication. *Journal of Experimental Biology* 202: 1217–1228.

563 Hori, M. 1993. Frequency-dependent natural selection in the handedness of scale-eating cichlid fish. *Science* 260: 216–219.

564 Horn, A. G., M. L. Leonard, and D. M. Weary. 1995. Oxygen consumption during crowing by roosters: Talk is cheap. *Animal Behaviour* 50: 1171–1175.

565 Hosken, D. J., and P. Stockley. 2004. Sexual selection and genital evolution. *Trends in Ecology and Evolution* 19: 87–93.

566   Hotker, H. 2000. Intraspecific variation in size and density of avocet colonies: Effects of nest-distances on hatching and breeding success. *Journal of Avian Biology* 31: 387–398.

567   Houston, D. C., J. D. Gilardi, and A. J. Hall. 2001. Soil consumption by elephants might help to minimize the toxic effects of plant secondary compounds in forest browse. *Mammal Review* 31: 249–254.

568   Howlett, R. J., and M. E. N. Majerus. 1987. The understanding of industrial melanism in the peppered moth (*Biston betularia*) (Lepidoptera: Geometridae). *Biological Journal of the Linnean Society* 30: 31–44.

569   Hrdy, S. B. 1977. Infanticide as a primate reproductive strategy. *American Scientist* 65: 40–49.

570   Hrdy, S. B. 1977. *The Langurs of Abu*. Cambridge, MA: Harvard University Press.

571   Hrdy, S. B. 1999. *Mother Nature: A History of Mothers, Infants, and Natural Selection*. New York: Pantheon.

572   Huang, Z.-Y., and G. E. Robinson. 1992. Honeybee colony integration: Worker-worker interactions mediate hormonally regulated plasticity in division of labor. *Proceedings of the National Academy of Sciences* 89: 11726–11729.

573   Hughes, S. M., F. Dispenza, and G. G. Gallup, Jr. 2004. Ratings of voice attractiveness predict sexual behavior and body configuration. *Evolution and Human Behavior* 25: 295–304.

574   Hume, D. K., and R. D. Montgomerie. 2001. Facial attractiveness signals different aspects of "quality" in women and men. *Evolution and Human Behavior* 22: 93–112.

575   Hunt, J., R. Brooks, M. D. Jennions, M. J. Smith, C. L. Bentsen, and L. F. Bussière. 2004. High-quality male field crickets invest heavily in sexual display but die young. *Nature* 432: 1024–1027.

576   Hunt, S., I. C. Cuthill, A. T. D. Bennett, and R. Griffiths. 1999. Preferences for ultraviolet partners in the blue tit. *Animal Behaviour* 58: 809–815.

577   Hunt, S., R. M. Kilner, N. E. Langmore, and A. T. D. Bennett. 2003. Conspicuous, ultraviolet-rich mouth colours in begging chicks. *Proceedings of the Royal Society of London B* 270: S25–S28.

578   Hunter, M. L., and J. R. Krebs. 1979. Geographic variation in the song of the great tit (*Parus major*) in relation to ecological factors. *Journal of Animal Ecology* 48: 759–785.

579   Hurst, L. D., and J. R. Peck. 1996. Recent advances in understanding of the evolution and maintenance of sex. *Trends in Ecology and Evolution* 11: 46–52.

580   Huxley, T. H. 1910. *Lectures and Lay Sermons*. New York: E. P. Dutton.

581   Hyman, J., M. Hughes, W. A. Searcy, and S. Nowicki. 2004. Individual variation in the strength of territory defense in male song sparrows: Correlates of age, territory tenure, and neighbor aggressiveness. *Behaviour* 141: 15–27.

582   Ijichi, N., H. Shibao, T. Miura, T. Matsumoto, and T. Fukatsu. 2004. Soldier differentiation during embryogenesis of a social aphid, *Pseudoregma bambucicola*. *Entomological Science* 7: 141–153.

583   Irons, W. 1979. Cultural and biological success. In *Evolutionary Biology and Human Social Behavior: An Anthropological Perspective*, ed. Chagnon, N. A., and W. Irons. North Scituate, MA: Duxbury Press.

584   Isbell, L. A., D. L. Cheney, and R. M. Seyfarth. 1993. Are immigrant vervet monkeys, *Cercopithecus aethiops*, at greater risk of mortality than residents? *Animal Behaviour* 45: 729–734.

585   Itô, Y. 1989. The evolutionary biology of sterile soldiers in aphids. *Trends in Ecology and Evolution* 4: 69–73.

586   Jacobs, L. F., S. J. C. Gaulin, D. F. Sherry, and G. E. Hoffman. 1990. Evolution of spatial cognition: Sex-specific patterns of spatial behavior predict hippocampal size. *Proceedings of the National Academy of Sciences* 87: 6349–6352.

587   Jamieson, I. G. 1989. Behavioral heterochrony and the evolution of birds' helping at the nest: An unselected consequence of communal breeding? *American Naturalist* 133: 394–406.

588   Jamieson, I. G. 1991. The unselected hypothesis for the evolution of helping behavior: Too much or too little emphasis on natural selection? *American Naturalist* 138: 271–282.

589   Jamieson, I. G., J. S. Quinn, P. A. Rose, and B. N. White. 1994. Shared paternity among non-relatives is a result of an egalitarian mating system in a communally breeding bird, the pukeko. *Proceedings of the Royal Society of London B* 257: 271–277.

590   Jarvis, E. D., S. Ribeiro, M. L. da Silva, D. Ventura, J. Vielliard, and C. V. Mello. 2000. Behaviourally driven gene expression reveals song nuclei in hummingbird brain. *Nature* 406: 628–632.

591   Jarvis, J. U. M., and N. C. Bennett. 1993. Eusociality has evolved independently in two genera of bathyergid mole-rats—but occurs in no other subterranean mammal. *Behavioral Ecology and Sociobiology* 33: 253–260.

592   Jasiénska, G., A. Ziomkiewicz, P. T. Ellison, S. F. Lipson, and I. Thune. 2004. Large breasts and narrow waists indicate high reproductive potential in women. *Proceedings of the Royal Society of London B* 271: 1213–1217.

593   Jaycox, E. R., and S. G. Parise. 1980. Homesite selection by Italian honey bee swarms, *Apis mellifera ligustica* (Hymenoptera: Apidae). *Journal of the Kansas Entomological Society* 53: 171–178.

594   Jaycox, E. R., and S. G. Parise. 1981. Homesite selection by swarms of black-bodied honey bees, *Apis mellifera caucasia* and *A. m. carnica*. *Journal of the Kansas Entomological Society* 54: 697–703.

595   Jennions, M. D., and P. R. Y. Backwell. 1996. Residency and size affect fight duration and outcome in the fiddler crab *Uca annulipes*. *Biological Journal of the Linnean Society* 57: 293–306.

596   Jennions, M. D., and M. Petrie. 2000. Why do females mate multiply? A review of the genetic benefits. *Biological Reviews* 75: 21–64.

597   Jenssen, T. A., M. B. Lovern, and J. D. Congdon. 2001. Field-testing the protandry-based mating system for the lizard, *Anolis carolinensis*: does the model organism have the right model? *Behavioral Ecology and Sociobiology* 50: 162–171.

598   Jiménez, J. A., K. A. Hughes, G. Alaks, L. Graham, and R. C. Lacy. 1994. An experimental study of inbreeding depression in a natural habitat. *Science* 266: 271–273.

599   Jin, H., and D. F. Clayton. 1997. Localized changes in immediate-early gene regulation during sensory and motor learning in zebra finches. *Neuron* 19: 1049–1059.

600   Jinks, R. N., T. L. Markley, E. E. Taylor, G. Perovich, A. I. Dittel, et al. 2002. Adaptive visual metamorphosis in a deep-sea hydrothermal vent crab. *Nature* 420: 68–70.

601   Johansson, J., H. Turesson, and A. Persson. 2004. Active selection for large guppies, *Poecilia reticulata*, by the pike cichlid, *Crenicichla saxatilis*. *Oikos* 105: 595–605.

602   Johns, T. 1990. *With Bitter Herbs They Shall Eat It: Chemical Ecology and the Origins of Human Diet and Medicine*. Tucson, AZ. University of Arizona Press.

603   Johnsen, A., V. Andersen, C. Sunding, and J. T. Lifjeld. 2000. Female bluethroats enhance offspring immunocompetence through extra-pair copulations. *Nature* 406: 296–299.

604   Johnsen, A., S. Andersson, J. Ornberg, and J. T. Lifjeld. 1998. Ultraviolet plumage ornamentation affects social mate choice and sperm competition in bluethroats (Aves: *Luscinia s. svecica*): A field experiment. *Proceedings of the Royal Society of London B* 265: 1313–1318.

605   Johnsen, A., J. T. Lifjeld, S. Andersson, J. Ornborg, and T. Amundsen. 2001. Male characteristics and fertilisation success in bluethroats. *Behaviour* 138: 1371–1390.

606   Johnson, C. H., and J. W. Hasting. 1986. The elusive mechanisms of the circadian clock. *American Scientist* 74: 29–37.

607   Johnstone, R. A. 2000. Models of reproductive skew: A review and synthesis. *Ethology* 106: 5–26.

608   Jones, A. G., D. Walker, and J. C. Avise. 2001. Genetic evidence for extreme polyandry and extraordinary sex-role reversal in a pipefish. *Proceedings of the Royal Society of London B* 268: 2531–2535.

609   Jones, D., and K. Hill. 1993. Criteria of facial attractiveness in five populations. *Human Nature* 4: 271–296.

610   Jones, I. L., and F. M. Hunter. 1998. Heterospecific mating preferences for a feather ornament in least auklets. *Behavioral Ecology* 9: 187–192.

611   Jones, J. S., and K. E. Wynne-Edwards. 2000. Paternal hamsters mechanically assist the delivery, consume amniotic fluid and placenta, remove fetal membranes, and provide parental care during the birth process. *Hormones and Behavior* 37: 116–125.

612   Jones, O. D. 1999. Sex, culture, and the biology of rape: Toward explanation and prevention. *California Law Review* 87: 827–942.

613   Jouventin, P., C. Barbraud, and M. Rubin. 1995. Adoption in the emperor penguin, *Aptenodytes forsteri. Animal Behaviour* 50: 1023–1029.

614   Just, J. 1988. Siphonoecetinae (Corophiidae). 6: A survey of phylogeny, distribution, and biology. *Crustaceana, Supplement* 13: 193–208.

615   Kalko, E. K. V. 1995. Insect pursuit, prey capture and echolocation in pipistrelle bats (Microchiroptera). *Animal Behaviour* 50: 861–880.

616   Kalmijn, A. J. 1982. Electric and magnetic field detection in elasmobranch fishes. *Science* 218: 916–918.

617   Kanazawa, S. 2003. Can evolutionary psychology explain reproductive behavior in the contemporary United States? *Sociological Quarterly* 44: 291–302.

618   Kannisto, V., K. Christensen, and J. W. Vaupel. 1997. No increased mortality in later life for cohorts born during famine. *American Journal of Epidemiology* 145: 987–994.

619   Kanwisher, N., P. Downing, R. Epstein, and Z. Kourtzi. 2001. Functional neuroimaging of human visual recognition. In *The Handbook of Functional Neuroimaging*, ed. Cabeza, R., and A. Kingstone. Cambridge, MA: MIT Press.

620   Kanwisher, N., J. McDermott, and M. M. Chun. 1997. The fusiform face area: A module in human extrastriate cortex specialized for face perception. *Journal of Neuroscience* 17: 4302–4311.

621   Kaplan, H., and K. Hill. 1985. Hunting ability and reproductive success among male Ache foragers: Preliminary results. *Current Anthropology* 26: 131–133.

622   Keeton, W. T. 1969. Orientation by pigeons: Is the sun necessary? *Science* 165: 922–928.

623   Keil, A., and N. Sachser. 1998. Reproductive benefits from female promiscuous mating in a small mammal. *Ethology* 104: 897–903.

624   Keller, L., and K. G. Ross. 1998. Selfish genes: A green beard in the red fire ant. *Nature* 394: 573–575.

625   Kemp, D. J. 2002. Sexual selection constrained by life history in a butterfly. *Proceedings of the Royal Society of London B* 269: 1341–1345.

626   Kemp, D. J., and C. Wiklund. 2004. Residency effects in animal contests. *Proceedings of the Royal Society of London B* 271: 1701–1711.

627   Kempenaers, B., G. R. Verheyen, and A. A. Dhondt. 1997. Extrapair paternity in the blue tit (*Parus caeruleus*): Female choice, male characteristics, and offspring quality. *Behavioral Ecology* 8: 481–492.

628   Kempenaers, B., G. R. Verheyen, M. van der Broeck, T. Burke, C. van Broeckhoven, and A. A. Dhondt. 1992. Extra-pair paternity results from female preference for high quality males in the blue tit. *Nature* 357: 494–496.

629   Kenrick, D. 1995. Evolutionary theory versus the confederacy of dunces. *Psychological Inquiry* 6: 56–61.

630   Kenrick, D. T., R. C. Keefe, C. Gabrielidis, and J. S. Cornelius. 1996. Adolescents' age preferences for dating partners: Support for an evolutionary model of life-history strategies. *Child Development* 67: 1499–1511.

631   Kenrick, D. T., and R. C. Keefe. 1992. Age preferences in mates reflect sex differences in reproductive strategies. *Behavioral and Brain Sciences* 15: 75-133.

632   Kenrick, D. T., E. K. Sadalla, G. Groth, and M. R. Trost. 1990. Evolution, traits, and the stages of human courtship: Qualifying the parental investment model. *Journal of Personality* 58: 97–116.

633   Kent, D. S., and J. A. Simpson. 1992. Eusociality in the beetle *Australoplatypus incompertus* (Coleoptera: Curculionidae). *Naturwissenschaften* 79: 86–87.

634   Kerverne, E. B. 1997. An evaluation of what the mouse knockout experiments are telling us about mammalian behaviour. *Bioessays* 19: 1091–1098.

635   Kessel, E. L. 1955. Mating activities of balloon flies. *Systematic Zoology* 4: 97–104.

636   Ketterson, E. D., and V. Nolan, Jr. 1999. Adaptation, exaptation, and constraint: A hormonal perspective. *American Naturalist* 154 Supplement: S4–S25.

637   Kettlewell, H. B. D. 1955. Selection experiments on industrial melanism in the Lepidoptera. *Heredity* 9: 323–343.

638   Kilner, R. M., J. R. Madden, and M. E. Hauber. 2004. Brood parasitic cowbird nestlings use host young to procure resources. *Science* 305: 877–879.

639   Kilner, R. M., D. G. Noble, and N. B. Davies. 1999. Signals of need in parent-offspring communication and their exploitation by the common cuckoo. *Nature* 397: 667–672.

640   Kimball, R. T., E. L. Braun, J. D. Ligon, V. Lucchini, and E. Randi. 2001. A molecular phylogeny of the peacock-pheasants (Galliformes: *Polyplectron* spp.) indicates loss and reduction of ornamental traits and display behaviours. *Biological Journal of the Linnean Society* 73: 187–198.

641   King, A. P., and M. J. West. 1983. Epigenesis of cowbird song—a joint endeavour of males and females. *Nature* 305: 704–706.

642   Kirchner, W. H., and A. Grasser. 1998. The significance of odor cues and dance language information for the food search behavior of honeybees (Hymenoptera: Apidae). *Journal of Insect Behavior* 11: 169–178.

643   Kirkpatrick, M. 1982. Sexual selection and the evolution of female choice. *Evolution* 36: 1–12.

644   Kirn, J. R., and T. J. DeVoogd. 1989. The genesis and death of vocal control neurons during sexual differentiation in the zebra finch. *Journal of Neuroscience* 9: 3176–3187.

645   Kitcher, P. 1985. *Vaulting Ambition*. Cambridge, MA: MIT Press.

646   Klein, S. L. 2000. The effects of hormones on sex differences in infection: From genes to behavior. *Neuroscience and Biobehavioral Reviews* 24: 627–638.

647   Klukowski, M., B. Ackerson, and C. E. Nelson. 2004. Testosterone and daily activity period in laboratory-housed mountain spiny lizards, *Sceloporus jarrovi. Journal of Herpetology* 38: 120–124.

648   Klukowski, M., M. Ireland, and C. E. Nelson. 2001. Interaction of testosterone and social stimuli on the feeding behavior of male mountain spiny lizards, *Sceloporus jarrovi. Copeia* 2001: 966–971.

649   Klump, G. M., E. Kretzschmar, and E. Curio. 1986. The hearing of an avian predator and its avian prey. *Behavioral Ecology and Sociobiology* 18: 317–324.

650   Knudsen, B., and R. M. Evans. 1986. Parent-young recognition in herring gulls (*Larus argentatus*). *Animal Behaviour* 34: 77–80.

651   Kodric-Brown, A. 1993. Female choice of multiple male criteria in guppies: Interacting effects of dominance, coloration and courtship. *Behavioral Ecology and Sociobiology* 32: 415–420.

652   Kodric-Brown, A., and J. H. Brown. 1984. Truth in advertising: The kinds of traits favored by sexual selection. *American Naturalist* 124: 309–323.

653   Koenig, W. D. 1981. Coalitions of male lions: Making the best of a bad job? *Nature* 293: 413–414.

654   Koenig, W. D., and R. L. Mumme. 1987. *Population Ecology of the Cooperatively Breeding Acorn Woodpecker*. Princeton, NJ: Princeton University Press.

655   Koenig, W. D., R. L. Mumme, M. T. Stanback, and F. A. Pitelka. 1995. Patterns and consequences of egg destruction among joint-nesting acorn woodpeckers. *Animal Behaviour* 50: 607–621.

656   Kokko, H., and M. Jennions. 2003. It takes two to tango. *Trends in Ecology and Evolution* 18: 103–104.

657   Komdeur, J. 1992. Importance of habitat saturation and territory quality for evolution of cooperative breeding in the Seychelles warbler. *Nature* 358: 493–495.

658   Komdeur, J. 1992. Influence of territory quality and habitat saturation on dispersal options in the Seychelles warbler: An experimental test of the habitat saturation hypothesis for cooperative breeding. *Acta XX Congressus Internationalis Ornithologici* 20: 1325–1332.

659   Komdeur, J. 2001. Mate guarding in the Seychelles warbler is energetically costly and adjusted to paternity risk. *Proceedings of the Royal Society of London B* 268: 2103–2111.

660   Komdeur, J., A. Huffstadt, W. Prast, G. Castle, R. Mileto, and J. Wattel. 1995. Transfer experiments of Seychelles warblers to new islands: Changes in dispersal and helping behavior. *Animal Behaviour* 49: 695–708.

661   Komdeur, J., F. Kraaijeveld-Smit, K. Kraaijeveld, and P. Edelaar. 1999. Explicit experimental evidence for the role of mate guarding in minimizing loss of paternity in the Seychelles warbler. *Proceedings of the Royal Society of London B* 266: 2075–2081.

662   Komers, P. E., and P. N. M. Brotherton. 1997. Female space use is the best predictor of monogamy in mammals. *Proceedings of the Royal Society of London B* 264: 1261–1270.

663   Konishi, M. 1965. The role of auditory feedback in the control of vocalization in the white-crowned sparrow. *Zeitschrift für Tierpsychologie* 22: 770–783.

664   Konishi, M. 1985. Birdsong: From behavior to neurons. *Annual Review of Neuroscience* 8: 125–170.

665   Kramer, M. G., and J. H. Marden. 1997. Almost airborne. *Nature* 385: 403–404.

666   Krebs, J. R. 1971. Territory and breeding density in the great tit, *Parus major* L. *Ecology* 52: 2–22.

667   Krebs, J. R. 1982. Territorial defence in the great tit (*Parus major*): Do residents always win? *Behavioral Ecology and Sociobiology* 11: 185–194.

668   Krebs, J. R., and A. Kacelnik. 1991. Decision-making. In *Behavioural Ecology: An Evolutionary Approach*, ed. Krebs, J. R., and N. B. Davies. Oxford: Blackwell Scientific Publications.

669   Krishnamani, R., and W. C. Mahaney. 2000. Geophagy among primates: Adaptive significance and ecological consequences. *Animal Behaviour* 59: 899–915.

670   Krohmer, R. W. 2004. The male red-sided garter snake (*Thamnophis sirtalis parietalis*): Reproductive pattern and behavior. *ILAR Journal* 45: 65–74.

671   Kroodsma, D. E., and R. A. Canady. 1985. Differences in repertoire size, singing behavior, and associated neuroanatomy among marsh wren populations have a genetic basis. *Auk* 102: 439–446.

672   Kroodsma, D. E., and M. Konishi. 1991. A suboscine bird (Eastern Phoebe, *Sayornis phoebe*) develops normal song without auditory-feedback. *Animal Behaviour* 42: 477–487.

673   Kroodsma, D. E., W. C. Liu, E. Goodwin, and P. A. Bedell. 1999. The ecology of song improvisation as illustrated by North American sedge wrens. *Auk* 116: 373–386.

674   Kroodsma, D. E., and E. H. Miller, eds. 1996. *Ecology and Evolution of Acoustic Communication in Birds*. Ithaca, NY: Cornell University Press.

675   Kroodsma, D. E., J. Sánchez, D. W. Stemple, E. Goodwin, M. L. da Silva, and J. M. E. Vielliard. 1999. Sedentary life style of Neotropical sedge wrens promotes song imitation. *Animal Behaviour* 57: 855–863.

676   Krüger, O., and N. B. Davies. 2002. The evolution of cuckoo parasitism: a comparative analysis. *Proceedings of the Royal Society of London B* 269: 375–381.

677   Kruuk, H. 1964. Predators and anti-predator behaviour of the black-headed gull *Larus ridibundus*. *Behaviour Supplements* 11: 1–129.

678   Kruuk, H. 1972. *The Spotted Hyena*. Chicago: University of Chicago Press.

679   Kruuk, H. 2004. *Niko's Nature: The Life of Niko Tinbergen and His Science of Animal Behavior*. Oxford: Oxford University Press.

680   Kuhn, T. S. 1996. *The Structure of Scientific Revolutions*. Chicago, IL: University of Chicago Press.

681   Kukalová-Peck, J. 1978. Origin and evolution of insect wings and their relation to metamorphosis, as documented by the fossil record. *Journal of Morphology* 158: 53–126.

682   Kurdziel, J. P., and L. L. Knowles. 2002. The mechanisms of morph determination in the amphipod *Jassa*: implications for the evolution of alternative male phenotypes. *Proceedings of the Royal Society of London B* 269: 1749–1754.

683   Kusmierski, R., G. Borgia, R. H. Crozier, and B. H. Y. Chan. 1993. Molecular information on bowerbird phylogeny and the evolution of exaggerated male characters. *Journal of Evolutionary Biology* 6: 737–752.

684   Kusmierski, R., G. Borgia, A. Uy, and R. H. Crozier. 1997. Labile evolution of display traits in bowerbirds indicates reduced effects of phylogenetic constraint. *Proceedings of the Royal Society of London B* 264: 307–313.

685   Kutsukake, M., H. Shibao, N. Nikoh, M. Morioka, T. Tamura, et al. 2004. Venomous protease of aphid soldier for colony defense. *Proceedings of the National Academy of Sciences* 101: 11338–11343.

686   Kvarnemo, C., G. I. Moore, A. G. Jones, W. S. Nelson, and J. C. Avise. 2000. Monogamous pair bonds and mate switching in the Western Australian seahorse *Hippocampus subelongatus*. *Journal of Evolutionary Biology* 13: 882–888.

687   Kvist, A., A. Lindstrom, M. Green, T. Piersma, and G. H. Visser. 2001. Carrying large fuel loads during sustained bird flight is cheaper than expected. *Nature* 413: 730–732.

688   Lacey, E. A., and P. W. Sherman. 1991. Social organization of naked mole-rat colonies: Evidence for divisions of labor. In *The Biology of the Naked Mole-Rat*, ed. Sherman, P. W., J. U. M. Jarvis, and R. D. Alexander. Princeton, NJ: Princeton University Press.

689   Lack, D. 1968. *Ecological Adaptations for Breeding in Birds*. London: Methuen.

690   Laist, D. W. 1987. Overview of the biological effects of lost and discarded plastic debris in the marine environment. *Marine Pollution Bulletin* 18: 319–326.

691   Lalumiére, M. L., L. J. Chalmers, V. L. Quinsey, and M. C. Seto. 1996. A test of the mate deprivation hypothesis of social coercion. *Ethology and Sociobiology* 17: 299–318.

692   Lande, R. 1981. Models of speciation by sexual selection of polygenic traits. *Proceedings of the National Academy of Sciences* 78: 3721–3725.

693   Langer, P., K. Hogendoorn, and L. Keller. 2004. Tug-of-war over reproduction in a social bee. *Nature* 428: 844–847.

694   Langlois, J. H., L. Kalakanis, A. J. Rubenstein, A. Larson, M. Hallam, and M. Smoot. 2000. Maxims or myths of beauty? A meta-analytic and theoretical review. *Psychological Bulletin* 126: 390–423.

695   Langmore, N. E., J. F. Cockrem, and E. J. Candy. 2002. Competition for male reproductive investment elevates testosterone levels in female dunnocks, *Prunella modularis*. *Proceedings of the Royal Society of London B* 269: 2473–2478.

696   Langmore, N. E., S. Hunt, and R. M. Kilner. 2003. Escalation of a coevolutionary arms race through host rejection of brood parasitic young. *Nature* 422: 157–160.

697   Lank, D. B., L. W. Oring, and S. J. Maxson. 1985. Mate and nutrient limitation of egg-laying in a polyandrous shorebird. *Ecology* 66: 1513–1524.

698   Lank, D. B., C. M. Smith, O. Hanotte, T. Burke, and F. Cooke. 1995. Genetic polymorphism for alternative mating behaviour in lekking male ruff *Philomachus pugnax*. *Nature* 378: 59–62.

699   Lanyon, S. M. 1992. Interspecific brood parasitism in blackbirds (Icterinae): A phylogenetic perspective. *Science* 255: 77–79.

700   Latta, S. C., and C. Brown. 1999. Autumn stopover ecology of the blackpoll warbler (*Dendroica striata*) in thorn scrub forest of the Dominican Republic. *Canadian Journal of Zoology* 77: 1147–1156.

701   Lauder, G. V., A. M. Leroi, and M. R. Rose. 1993. Adaptations and history. *Trends in Ecology and Evolution* 8: 294–297.

702 Layne, J. E. 1998. Retinal location is the key to identifying predators in fiddler crabs (*Uca pugilator*). *Journal of Experimental Biology* 201: 2253–2261.

703 Leal, M. 1999. Honest signalling during prey-predator interactions in the lizard *Anolis cristatellus*. *Animal Behaviour* 58: 521–526.

704 Leech, S. M., and M. L. Leonard. 1997. Begging and the risk of predation in nestling birds. *Behavioral Ecology* 8: 644–646.

705 Leitner, S., J. Nicholson, B. Leisler, T. J. DeVoogd, and C. K. Catchpole. 2002. Song and the song control pathway in the brain can develop independently of exposure to song in the sedge warbler. *Proceedings of the Royal Society of London B* 269: 2519–2524.

706 Lema, S. C., and G. A. Nevitt. 2004. Variation in vasotocin immunoreactivity in the brain of recently isolated populations of a death valley pupfish, *Cyprinodon nevadensis*. *General and Comparative Endocrinology* 135: 300–309.

707 Lemaster, M. P., and R. T. Mason. 2001. Evidence for a female sex pheromone mediating male trailing behavior in the red-sided garter snake, *Thamnophis sirtalis parietalis*. *Chemoecology* 11: 149–152.

708 Lemon, W. C., and R. H. Barth. 1992. The effects of feeding rate on reproductive success in the zebra finch, *Taeniopyga guttata*. *Animal Behaviour* 44: 851–857.

709 Lendrem, D. W. 1986. *Modelling in Behavioural Ecology: An Introductory Text*. London: Croom Helm.

710 Leonard, M. L., A. G. Horn, and J. Porter. 2003. Does begging affect growth in nestling tree swallows, *Tachycineta bicolor*? *Behavioral Ecology and Sociobiology* 54: 573–577.

711 Leoncini, I., Y. Le Conte, G. Costagliola, E. Plettner, A. L. Toth, et al. 2004. Regulation of behavioral maturation by a primer pheromone produced by adult worker honey bees. *Proceedings of the National Academy of Sciences* 101: 17559–17564.

712 Lesch, K. P. 1998. Serotonin transporter and psychiatric disorders: Listening to the gene. *Neuroscientist* 4: 25–34.

713 Levey, D. J., and F. G. Stiles. 1992. Evolutionary precursors of long-distance migration: Resource availability and movement patterns in Neotropical landbirds. *American Naturalist* 140: 447–476.

714 Levine, J. D. 2004. Sharing time on the fly. *Current Opinion in Cell Biology* 16: 1–7.

715 Lewontin, R. C., S. Rose, and L. J. Kamin. 1984. *Not In Our Genes*. New York: Random House.

716 Li, N. P., J. M. Bailey, D. T. Kenrick, and J. A. W. Linsenmeier. 2002. The necessities and luxuries of mate preferences: Testing the tradeoffs. *Journal of Personality and Social Psychology* 82: 947–955.

717 Lichtenstein, G., and S. G. Sealy. 1998. Nesting competition, rather than supernormal stimulus, explains the success of parasitic brown-headed cowbird chicks in yellow warbler nests. *Proceedings of the Royal Society of London B* 265: 249–254.

718 Liebig, J., C. Peeters, and B. Hölldobler. 1999. Worker policing limits the number of reproductives in a ponerine ant. *Proceedings of the Royal Society of London B* 266: 1865–1870.

719 Ligon, J. D. 1999. *The Evolution of Avian Mating Systems*. New York: Oxford University Press.

720 Ligon, J. D., R. Kimball, and M. Merola-Zwartjes. 1998. Mate choice by female red junglefowl: The issues of multiple ornaments and fluctuating asymmetry. *Animal Behaviour* 55: 41–50.

721 Lill, A. 1974. Sexual behavior of the lek-forming white-bearded manakin (*Manacus manacus trinitatis* Hartert). *Zeitschrift für Tierpsychologie* 36: 1–36.

722 Lim, M. M., A. Z. Murphy, and L. J. Young. 2004. Ventral striatopallidal oxytocin and vasopressin V1a receptors in the monogamous prairie vole (*Microtus ochrogaster*). *Journal of Comparative Neurology* 468: 555–570.

723 Lim, M. M., X. Wang, D. E. Olazábal, X. Ren, E. F. Terwilliger, and L. J. Young. 2004. Enhanced partner preference in a promiscuous species by manipulating the expression of a single gene. *Nature* 429: 754–757.

724 Lima, S. L., and L. M. Dill. 1990. Behavioral decisions made under the risk of predation: A review and prospectus. *Canadian Journal of Zoology* 68: 619–640.

725 Lin, C. P., B. N. Danforth, and T. K. Wood. 2004. Molecular phylogenetics and evolution of maternal care in Membracine treehoppers. *Systematic Biology* 53: 400–421.

726 Lincoln, G. A., F. Guinness, and R. V. Short. 1972. The way in which testosterone controls the social and sexual behavior of the red deer stag (*Cervus elaphus*). *Hormones and Behavior* 3: 375–396.

727 Lind, J., T. Fransson, S. Jakobsson, and C. Kullberg. 1999. Reduced take-off ability in robins (*Erithacus rubecula*) due to migratory fuel load. *Behavioral Ecology and Sociobiology* 46: 65–70.

728 Lindauer, M. 1961. *Communication among Social Bees*. Cambridge, MA: Harvard University Press.

729 Little, A. C., D. M. Burt, I. S. Penton-Voak, and D. I. Perrett. 2001. Self-perceived attractiveness influences human female preferences for sexual dimorphism and symmetry in male faces. *Proceedings of the Royal Society of London B* 268: 39–44.

730 Lively, C. M., and M. F. Dybdahl. 2000. Parasite adaptation to locally common genotypes. *Nature* 495: 679–681.

731 Lloyd, J. E. 1965. Aggressive mimicry in *Photuris*: Firefly *femmes fatales*. *Science* 149: 653–654.

732 Lloyd, J. E. 1975. Aggressive mimicry in *Photuris* fireflies: Signal repertoires by *femmes fatales*. *Science* 197: 452–453.

733 Lloyd, J. E. 1980. Insect behavioral ecology: Coming of age in bionomics or compleat biologists have revolutions too. *Florida Entomologist* 63: 1–4.

734 Lockard, R. B. 1978. Seasonal change in the activity pattern of *Dipodomys spectabilis*. *Journal of Mammalogy* 59: 563–568.

735 Lockard, R. B., and D. H. Owings. 1974. Seasonal variation in moonlight avoidance by bannertail kangaroo rats. *Journal of Mammalogy* 55: 189–193.

736 Lockhart, D. J., and C. Barlow. 2001. Expressing what's on your mind: DNA arrays and the brain. *Nature Reviews Neuroscience* 2: 63–68.

737 Loesche, P., P. K. Stoddard, B. J. Higgins, and M. D. Beecher. 1991. Signature versus perceptual adaptations for individual vocal recognition in swallows. *Behaviour* 118: 15–25.

738 Loher, W. 1972. Circadian control of stridulation in the cricket *Teleogryllus commodus* Walker. *Journal of Comparative Physiology* 79: 173–190.

739 Loher, W. 1979. Circadian rhythmicity of locomotor behavior and oviposition in female *Teleogryllus commodus*. *Behavioral Ecology and Sociobiology* 5: 383–390.

740 Lohmann, K. J., C. M. F. Lohmann, L. M. Ehrhart, D. A. Bagley, and T. Swing. 2004. Geomagnetic map used in sea-turtle navigation. *Nature* 428: 909–910.

741 Lore, R., and K. Flannelly. 1977. Rat societies. *Scientific American* 236 (May): 106–116.

742 Lorenz, K. Z. 1952. *King Solomon's Ring*. New York: Crowell.

743 Lotem, A. 1993. Learning to recognize nestlings is maladaptive for cuckoo *Cuculus canorus* hosts. *Nature* 362: 743–745.

744 Lotem, A., H. Nakamura, and A. Zahavi. 1995. Constraints on egg discrimination and cuckoo-host co-evolution. *Animal Behaviour* 49: 1185–1209.

745 Lougheed, L. W., and D. J. Anderson. 1999. Parent blue-footed boobies suppress siblicidal behavior of offspring. *Behavioral Ecology and Sociobiology* 45: 11–18.

746 Lovern, M. B., M. M. Holmes, and J. Wade. 2004. The green anole (*Anolis carolinensis*): A reptilian model for laboratory studies of reproductive morphology and behavior. *ILAR Journal* 45: 54–64.

747 Low, B. S., and J. T. Heinen. 1993. Population, resources, and environment: Implications of human behavioral ecology for conservation. *Population and Environment* 15: 7–41.

748 Lummaa, V. 2003. Early developmental conditions and reproductive success in humans: Downstream effects of prenatal famine, birthweight, and timing of birth. *American Journal of Human Biology* 15: 370–379.

749 Lummaa, V., T. Vuorisalo, R. G. Barr, and L. Lehtonen. 1998. Why cry? Adaptive significance of intensive crying in human infants. *Evolution and Human Behavior* 19: 193–202.

750 Lundberg, P. 1985. Dominance behavior, body-weight and fat variations, and partial migration in European blackbirds *Turdus merula*. *Behavioral Ecology and Sociobiology* 17: 185–189.

751 Lundberg, P. 1988. The evolution of partial migration in birds. *Trends in Ecology and Evolution* 3: 172–176.

752 Lung, O., U. Tram, C. M. Finnerty, M. A. Eipper-Mains, J. M. Kalb, and M. F. Wolfner. 2002. The *Drosophila melanogaster* seminal fluid protein Acp62F is a protease inhibitor that is toxic upon ectopic expression. *Genetics* 160: 211–224.

753 Luschi, P., G. C. Hays, C. Del Seppia, R. Marsh, and F. Papi. 1998. The navigational feats of green sea turtles migrating from Ascension Island investigated by satellite telemetry. *Proceedings of the Royal Society of London B* 265: 2279–2284.

754 Lycett, J. E., and R. I. M. Dunbar. 1999. Abortion rates reflect the optimization of parental investment strategies. *Proceedings of the Royal Society of London B* 266: 2355–2358.

755 Lynch, C. B. 1980. Response to divergent selection for nesting behavior in *Mus musculus*. *Genetics* 96: 757–765.

756 Lyon, B. E. 1993. Conspecific brood parasitism as a flexible female reproductive tactic in American coots. *Animal Behaviour* 46: 911–928.

757 Lyon, B. E. 2003. Ecological and social constraints on conspecific brood parasitism by nesting female American coots (*Fulica americana*). *Journal of Animal Ecology* 72: 47–60.

758 Lyon, B. E., and J. M. Eadie. 1991. Mode of development and interspecific avian brood parasitism. *Behavioral Ecology* 2: 309–318.

759 Lyon, B. E., J. M. Eadie, and L. D. Hamilton. 1994. Parental choice selects for ornamental plumage in American coot chicks. *Nature* 371: 240–243.

760 Lyon, B. E., R. D. Montgomerie, and L. D. Hamilton. 1987. Male parental care and monogamy in snow buntings. *Behavioral Ecology and Sociobiology* 20: 377–382.

761 MacDougall-Shackleton, E. A., E. P. Derryberry, and T. P. Hahn. 2002. Nonlocal male mountain white-crowned sparrows have lower paternity and higher parasite loads than males singing local dialect. *Behavioral Ecology* 13: 682–689.

762 MacDougall-Shackleton, S. A., D. F. Sherry, A. P. Clark, R. Pinkus, and A. M. Hernandez. 2003. Photoperiodic regulation of food storing and hippocampus volume in black-capped chickadees, *Poecile atricapillus*. *Animal Behaviour* 65: 805–812.

763 MacDougall, A. K., and R. Montgomerie. 2003. Assortative mating by carotenoid-based plumage colour: a quality indicator in American goldfinches, *Carduelis tristis*. *Naturwissenschaften* 90: 464–467.

764 Mace, R. 1998. The coevolution of human fertility and wealth inheritance strategies. *Philosophical Transactions of the Royal Society of London B* 353: 389–397.

765 Macrae, C. N., K. A. Alnwick, A. B. Milne, and A. M. Schloerscheidt. 2002. Person perception across the menstrual cycle: Hormonal influences on social-cognitive functioning. *Psychological Science* 13: 532–536.

766 Madden, J. R. 2001. Sex, bowers and brains. *Proceedings of the Royal Society of London B* 268: 833–838.

767 Madden, J. R. 2003. Bower decorations are good predictors of mating success in the spotted bowerbird. *Behavioral Ecology and Sociobiology* 53: 269–277.

768 Madden, J. R. 2003. Male spotted bowerbirds preferentially choose, arrange and proffer objects that are good predictors of mating success. *Behavioral Ecology and Sociobiology* 53: 263–268.

769 Maguire, E. A., N. Burgess, J. G. Donnett, R. S. J. Frackowiak, C. D. Frith, and J. O'Keefe. 1998. Knowing where and getting there: A human navigation network. *Science* 280: 921–934.

770 Maguire, E. A., D. G. Gadian, I. S. Johnsrude, C. D. Good, J. Ashburner, et al. 2000. Navigation-related structural change in the hippocampi of taxi drivers. *Proceedings of the National Academy of Sciences* 97: 4398–4403.

771 Mallach, T. J., and P. L. Leberg. 1999. Use of dredged material substrates by nesting terns and black skimmers. *Journal of Wildlife Management* 63: 137–146.

772 Mangel, M., and C. Clark. 1988. *Dynamic Modelling in Behavioral Ecology*. Princeton, NJ: Princeton University Press.

773 Marden, J. H. 1995. How insects learned to fly. *The Sciences* 35: 26–30.

774 Marden, J. H., and M. G. Kramer. 1994. Surface-skimming stoneflies: A possible intermediate stage in insect flight evolution. *Science* 266: 427–430.

775 Marden, J. H., and M. G. Kramer. 1995. Locomotor performance of insects with rudimentary wings. *Nature* 377: 332–334.

776 Marden, J. H., B. C. O'Donnell, M. A. Thomas, and J. Y. Bye. 2000. Surface-skimming stoneflies and mayflies: The taxonomic and mechanical diversity of two-dimensional aerodynamic locomotion. *Physiological and Biochemical Zoology* 73: 751–764.

777 Marden, J. H., and M. A. Thomas. 2003. Rowing locomotion by a stonefly that possesses the ancestral pterygote condition of co-occurring wings and abdominal gills. *Biological Journal of the Linnean Society* 79: 341–349.

778 Marden, J. H., and J. K. Waage. 1990. Escalated damselfly territorial contests and energetic wars of attrition. *Animal Behaviour* 39: 954–959.

779 Maret, T. J., and J. P. Collins. 1994. Individual responses to population size structure: The role of size variation in controlling expression of a trophic polyphenism. *Oecologia* 100: 279–285.

780 Margulis, S. W., and J. Altmann. 1997. Behavioural risk factors in the reproduction of inbred and outbred oldfield mice. *Animal Behaviour* 54: 397–408.

781 Margulis, S. W., W. Saltzman, and D. H. Abbott. 1995. Behavioural and hormonal changes in female naked mole-rats (*Heterocephalus glaber*) following removal of the breeding female from a colony. *Hormones and Behavior* 29: 227–247.

782 Marler, C. A., and M. C. Moore. 1989. Time and energy costs of aggression in testosterone-implanted free-living male mountain spiny lizards (*Sceloporus jarrovi*). *Physiological Zoology* 62: 1334–1350.

783 Marler, C. A., and M. C. Moore. 1991. Supplementary feeding compensates for testosterone-induced costs of aggression in male mountain spiny lizards, *Sceloporus jarrovi*. *Animal Behaviour* 42: 209–219.

784 Marler, P. 1955. Characteristics of some animal calls. *Nature* 176: 6–8.

785 Marler, P. 1970. Birdsong and speech development: Could there be parallels? *American Scientist* 58: 669–673.

786 Marler, P., and M. Tamura. 1964. Culturally transmitted patterns of vocal behavior in sparrows. *Science* 146: 1483–1486.

787 Marra, P. P., K. A. Hobson, and R. T. Holmes. 1998. Linking winter and summer events in a migratory bird by using stable-carbon isotopes. *Science* 282: 1884–1886.

788 Marshall, R. C., K. L. Buchanan, and C. K. Catchpole. 2003. Sexual selection and individual genetic diversity in a songbird. *Proceedings of the Royal Society of London B* 270: S248–S50.

789 Martín, J., and P. López. 2000. Chemoreception, symmetry and mate choice in lizards. *Proceedings of the Royal Society of London B* 267: 1265–1269.

790 Marzluff, J. M., B. Heinrich, and C. S. Marzluff. 1996. Raven roosts are mobile information centres. *Animal Behaviour* 51: 89–103.

791 Maschwitz, U., and E. Maschwitz. 1974. Platzende Arbeiterinnen: Eine neue Art der Feindabwehr bei sozialen Hautflüglern. *Oecologia* 14: 289–294.

792 Massaro, D. W., and D. G. Stork. 1998. Speech recognition and sensory integration. *American Scientist* 86: 236–244.

793 Massaro, M., J. W. Chardine, and I. L. Jones. 2001. Relationships between black-legged kittiwake nest site characteristics and susceptibility to predation by large gulls. *Condor* 103: 793–801.

794 Mateo, J. M. 2002. Kin-recognition abilities and nepotism as a function of sociality. *Proceedings of the Royal Society of London B* 269: 721–727.

795 Mateo, J. M., and W. G. Holmes. 1997. Development of alarm-call responses in Belding's ground squirrels: The role of dams. *Animal Behaviour* 54: 509–524.

796 Mateo, J. M., and R. E. Johnston. 2000. Kin recognition and the "armpit effect": Evidence of self-reference phenotype matching. *Proceedings of the Royal Society of London B* 267: 695–700.

797 Mather, M. H., and B. D. Roitberg. 1987. A sheep in wolf's clothing: Tephritid flies mimic spider predators. *Science* 236: 308–310.

798 Matthews, L. H. 1939. Reproduction in the spotted hyena *Crocuta crocuta* (Erxleben). *Philosophical Transactions of the Royal Society of London B* 230: 1–78.

799 May, M. 1991. Aerial defense tactics of flying insects. *American Scientist* 79: 316–329.

800 Maynard Smith, J. 1974. The theory of games and the evolution of animal conflicts. *Journal of Theoretical Biology* 47: 209–221.

801 Mayr, E. 1961. Cause and effect in biology. *Science* 134: 1501–1506.

802 Mayr, E. 1963. *Animal Species and Evolution*. Cambridge, MA: Harvard University Press.

803 Mays, H. L., and K. R. Hopper. 2004. Differential responses of yellow-breasted chats, *Icteria virens*, to male and female conspecific model presentations. *Animal Behaviour* 67: 21–26.

804 McCandliss, B. D., L. Cohen, and S. Dehaene. 2003. The visual word form area: expertise for reading in the fusiform gyrus. *Trends in Cognitive Sciences* 7: 293–299.

805 McCracken, G. F. 1984. Communal nursing in Mexican free-tailed bat maternity colonies. *Science* 223: 1090–1091.

806 McCracken, G. F., and J. W. Bradbury. 1981. Social organization and kinship in the polygynous bat *Phyllostomus hastatus*. *Behavioral Ecology and Sociobiology* 8: 11–34.

807 McCracken, G. F., and M. K. Gustin. 1991. Nursing behavior in Mexican free-tailed bat maternity colonies. *Ethology* 89: 305–321.

808 McDonald, D. B., and W. K. Potts. 1994. Cooperative display and relatedness among males in a lek-mating bird. *Science* 266: 1030–1032.

809 McGraw, K. J., and D. R. Ardia. 2003. Carotenoids, immunocompetence, and the information content of sexual colors: an experimental test. *American Naturalist* 162: 704–712.

810 McGraw, K. J., and G. E. Hill. 2000. Differential effects of endoparasitism on the expression of carotenoid- and melanin-based ornamental coloration. *Proceedings of the Royal Society of London B* 267: 1525–1531.

811 McGuire, B., and W. E. Bemis. In press. In *Rodent Societies*, ed. Wolff, J. O., and P. W. Sherman. Chicago: University of Chicago Press.

812 McNair, D. B., E. B. Massiah, and M. D. Frost. 2002. Ground-based autumn migration of Blackpoll Warblers at Harrison Point, Barbados. *Caribbean Journal of Science* 38: 239–248.

813 Mealey, L., C. Daood, and M. Krage. 1996. Enhanced memory for faces of cheaters. *Ethology and Sociobiology* 17: 119–128.

814 Medvin, M. B., and M. D. Beecher. 1986. Parent-offspring recognition in the barn swallow (*Hirundo rustica*). *Animal Behaviour* 34: 1627–1639.

815 Medvin, M. B., P. K. Stoddard, and M. D. Beecher. 1993. Signals for parent-offspring recognition: A comparative analysis of the begging calls of cliff swallows and barn swallows. *Animal Behaviour* 45: 841–850.

816 Meire, P. M., and A. Ervynck. 1986. Are oystercatchers (*Haemoptopus ostralegus*) selecting the most profitable mussels (*Mytilus edulis*)? *Animal Behaviour* 34: 1427–1435.

817 Mello, C. V., and S. Ribeiro. 1998. ZENK protein regulation by song in the brain of songbirds. *Journal of Comparative Neurology* 383: 426–438.

818 Melo, L., A. R. Mendes, and M. A. O. Monteriro da Cruz. 2003. Infanticide and cannibalism in wild common marmosets. *Folia Primatologica* 74: 48–50.

819 Mendonca, M. T., D. Daniels, C. Faro, and D. Crews. 2003. Differential effects of courtship and mating on receptivity and brain metabolism in female red-sided garter snakes (*Thamnophis sirtalis parietalis*). *Behavioral Neuroscience* 117: 144–149.

820 Meshaka, W. E. 1996. Anuran Davian behavior: a darwinian dilemma. *Florida Scientist* 59: 74–75.

821 Michener, C. D. 1974. *The Social Behavior of the Bees: A Comparative Study*. Cambridge, MA: Harvard University Press.

822 Michl, G., J. Török, S. C. Griffith, and B. C. Sheldon. 2002. Experimental analysis of sperm competition mechanisms in a wild bird population. *Proceedings of the National Academy of Sciences* 99: 5466–5470.

823 Milinski, M., D. Semmann, and H. J. Krambeck. 2002. Donors to charity gain in both indirect reciprocity and political reputation. *Proceedings of the Royal Society of London B* 269: 881–883.

824 Milius, S. 2004. Where'd I put that. *Science News* 165: 103–105.

825 Miller, D. E., and J. T. Emlen, Jr. 1975. Individual chick recognition and family integrity in the ring-billed gull. *Behaviour* 52: 124–144.

826 Miller, G. F. 2000. *The Mating Mind*. New York: Doubleday.

827 Miller, L. A. 1971. Physiological responses of green lacewings (*Chrysopa*, Neuroptera) to ultrasound. *Journal of Insect Physiology* 17: 491–506.

828 Miller, L. A. 1983. How insects detect and avoid bats. In *Neuroethology and Behavioral Physiology*, ed. Huber, F., and H. Markl. Berlin: Springer-Verlag.

829 Miller, L. A., and J. Olesen. 1979. Avoidance behavior in green lacewings. I. Behavior of free flying green lacewings to hunting bats and ultrasound. *Journal of Comparative Physiology* 131: 113–120.

830 Mills, M. G. L. 1990. *Kalahari Hyaenas: Comparative Behavioural Ecology of Two Species*. London: Unwin Hyman.

831 Mock, D. W. 1984. Siblicidal aggression and resource monopolization in birds. *Science* 225: 731–733.

832 Mock, D. W. 2004. *More Than Kin and Less Than Kind: The Evolution of Family Conflict*. Cambridge, MA: Harvard University Press.

833 Mock, D. W., H. Drummond, and C. H. Stinson. 1990. Avian siblicide. *American Scientist* 78: 438–449.

834 Mock, D. W., and B. J. Ploger. 1987. Parental manipulation of optimal hatch asynchrony in cattle egrets: An experimental study. *Animal Behaviour* 35: 150–160.

835 Moffat, S. D., E. Hampson, and M. Hatzipantelis. 1998. Navigation in a "virtual" maze: Sex differences and correlation with psychometric measures of spatial ability in humans. *Human Behavior and Evolution* 19: 73–87.

836 Moiseff, A., G. S. Pollack, and R. R. Hoy. 1978. Steering responses of flying crickets to sound and ultrasound: Mate attraction and predator avoidance. *Proceedings of the National Academy of Sciences* 75: 4052–4056.

837 Møller, A. P. 1988. Female choice selects for male sexual tail ornaments in the monogamous swallow. *Nature* 332: 640–642.

838 Møller, A. P. 1991. Density-dependent extra-pair copulations in the swallow *Hirundo rustica*. *Ethology* 87: 316–329.

839 Møller, A. P. 1992. Female swallow preference for symmetrical male sexual ornaments. *Nature* 357: 238–240.

840 Møller, A. P. 1999. Asymmetry as a predictor of growth, fecundity and survival. *Ecology Letters* 2: 149–156.

841 Møller, A. P., and J. P. Swaddle. 1997. *Asymmetry, Developmental Stability, and Evolution*. New York: Oxford University Press.

842 Mondor, E. B., and B. D. Roitberg. 2004. Inclusive fitness benefits of scent-marking predators. *Proceedings of the Royal Society of London B* 271: S341–S343.

843 Money, J., and A. A. Ehrhardt. 1972. *Man and Woman, Boy and Girl*. Baltimore, MD: Johns Hopkins University Press.

844 Monnin, T., F. L. W. Ratnieks, G. R. Jones, and R. Beard. 2002. Pretender punishment induced by chemical signalling in a queenless ant. *Nature* 419: 61–65.

845 Montgomery, H. E., E. S. Adams, G. C. Bachmann, M. L. da Silva, M. E. Hauber, et al. 1998. Human gene for physical performance. *Nature* 393: 221–222.

846 Mooney, R., W. Hoese, and S. Nowicki. 2001. Auditory representation of the vocal repertoire in a songbird with multiple song types. *Proceedings of the National Academy of Sciences* 98: 12778–12783.

847 Moore, J., and R. Ali. 1984. Are dispersal and inbreeding avoidance related? *Animal Behaviour* 32: 94–112.

848 Moore, M. C., and B. Kranz. 1983. Evidence for androgen independence of male mounting behavior in white-crowned sparrows (*Zonotrichia leucophrys gambelii*). *Hormones and Behavior* 17: 414–423.

849 Moreno, J., J. P. Veiga, P. J. Cordero, and E. Mínguez. 1999. Effects of paternal care on reproductive success in the polygynous spotless starling *Sturnus unicolor*. *Behavioral Ecology and Sociobiology* 47: 47–53.

850 Morley, J., and S. Balshine. 2002. Faithful fish: territory and mate defence favour monogamy in an African cichlid fish. *Behavioral Ecology and Sociobiology* 52: 326–331.

851 Morley, R., and A. Lucas. 1997. Nutrition and cognitive development. *British Medical Bulletin* 53: 123–124.

852 Morrow, E. H., and G. Arnqvist. 2003. Costly traumatic insemination and a female counter-adaptation in bed bugs. *Proceedings of the Royal Society of London B* 270: 2377–2381.

853 Morrow, P. A., T. E. Bellas, and T. Eisner. 1976. Eucalyptus oils in defensive oral discharge of Australian sawfly larvae (Hymenoptera: Pergidae). *Oecologia* 24: 193–206.

854 Morse, D. H. 2004. A test of sexual cannibalism models, using a sit-and-wait predator. *Biological Journal of the Linnean Society* 81: 427–437.

855 Moss, C. 1988. *Elephant Memories*. New York: William Morrow.

856 Mouritsen, H., and B. J. Frost. 2002. Virtual migration in tethered flying monarch butterflies reveals their orientation mechanisms. *Proceedings of the National Academy of Sciences* 99: 10162–10166.

857 Mueller, U. G. 1991. Haplodiploidy and the evolution of facultative sex ratios in a primitively eusocial bee. *Science* 254: 442–444.

858 Muller, M. N., and R. Wrangham. 2002. Sexual mimicry in hyenas. *Quarterly Review of Biology* 77: 3–16.

859 Mumme, R. L. 1992. Do helpers increase reproductive success? An experimental analysis in the Florida scrub jay. *Behavioral Ecology and Sociobiology* 31: 319–328.

860 Mumme, R. L., W. D. Koenig, and F. A. Pitelka. 1983. Reproductive competition in the communal acorn woodpecker: Sisters destroy each other's eggs. *Nature* 306: 583–584.

861 Mumme, R. L., W. D. Koenig, and F. L. W. Ratnieks. 1989. Helping behaviour, reproductive value, and the future component of indirect fitness. *Animal Behaviour* 38: 331–343.

862 Munn, C. A. 1991. Macaw biology and ecotourism, or "when a bird in the bush is worth two in the hand". In *New World Parrots in Crisis*, ed. Beisinger, S. R., and N. F. R. Snyder. Washington, D.C.: Smithsonian Institution Press.

863 Murdock, G. P. 1967. *Ethnographic Atlas*. Pittsburgh, PA: Pittsburgh University Press.

864 Myerscough, M. R. 2003. Dancing for a decision: a matrix model for nest-site choice by honeybees. *Proceedings of the Royal Society of London B* 270: 577–582.

865 Nealen, P. M., and D. J. Perkel. 2000. Sexual dimorphism in the song system of the Carolina wren *Thryothorus ludovicianus*. *Journal of Comparative Neurology* 418: 346–360.

866 Neff, B. D. 2003. Decisions about parental care in response to perceived paternity. *Nature* 422: 716–719.

867 Nelson, D. A. 1999. Ecological influences on vocal development in the white-crowned sparrow. *Animal Behaviour* 58: 21–36.

868 Nelson, D. A. 2000. Song overproduction, selective attrition and song dialects in the white-crowned sparrow. *Animal Behaviour* 60: 887–898.

869 Nelson, D. A., and J. A. Soha. 2004. Male and female white-crowned sparrows respond differently to geographic variation in song. *Behaviour* 141: 53–69.

870 Nelson, R. J. 2000. *An Introduction to Behavioral Endocrinology*. Sunderland, MA.: Sinauer Associates.

871 Newcomer, S. D., J. A. Zeh, and D. W. Zeh. 1999. Genetic benefits enhance the reproductive success of polyandrous females. *Proceedings of the National Academy of Sciences* 96: 10236–10241.

872 Newman, E. A., and P. H. Hairline. 1982. The infrared "vision" of snakes. *Scientific American* 20 (Mar.): 116–127.

873 Newton, P. N. 1986. Infanticide in an undisturbed forest population of hanuman langurs, *Presbytis entellus*. *Animal Behaviour* 34: 785–789.

874 Nieh, J. C. 1998. The role of a scent beacon in the communication of food location by the stingless bee, *Melipona panamica*. *Behavioral Ecology and Sociobiology* 43: 47–58.

875 Nieh, J. C. 1999. Stingless-bee communication. *American Scientist* 87: 428–435.

876 Nieh, J. C. 2004. Recruitment communication in stingless bees (Hymenoptera, Apidae, Meliponini). *Apidologie* 35: 159–182.

877 Nieh, J. C., L. S. Barreto, F. A. L. Contrera, and V. L. Imperatriz-Fonseca. 2004. Olfactory eavesdropping by a competitively foraging stingless bee, *Trigona spinipes*. *Proceedings of the Royal Society of London B* 271: 1633–1640.

878 Nieh, J. C., F. A. L. Contrera, J. Rangel, and V. L. Imperatriz-Fonseca. 2003. Effect of food location and quality on recruitment sounds and success in two stingless bees, *Melipona mandacaia* and *Melipona bicolor*. *Behavioral Ecology and Sociobiology* 55: 87–94.

879 Nijhout, H. F. 2003. Development and evolution of adaptive polyphenisms. *Evolution & Development* 5: 9–18.

880 Noë, R., and A. A. Sluijter. 1990. Reproductive tactics of male savanna baboons. *Behaviour* 113: 117–170.

881 Nolen, T. G., and R. R. Hoy. 1984. Phonotaxis in flying crickets: Neural correlates. *Science* 226: 992–994.

882 Nonacs, P. 1986. Ant reproductive strategies and sex allocation theory. *Quarterly Review of Biology* 61: 1–21.

883 Nonacs, P., and H. K. Reeve. 1995. The ecology of cooperation in wasps: Causes and consequences of alternative reproductive decisions. *Ecology* 76: 953–967.

884 Nordby, J. C., S. E. Campbell, and M. D. Beecher. 1999. Ecological correlates of song learning in song sparrows. *Behavioral Ecology* 10: 287–297.

885 Norris, D. R., P. P. Marra, T. K. Kyser, T. W. Sherry, and L. M. Ratcliffe. 2004. Tropical winter habitat limits reproductive success on the temperate breeding grounds in a migratory bird. *Proceedings of the Royal Society of London B* 271: 59–64.

886 Nottebohm, F., and A. P. Arnold. 1976. Sexual dimorphism in vocal control areas of songbird brain. *Science* 194: 211–213.

887 Nowicki, S., D. Hasselquist, S. Bensch, and S. Peters. 2000. Nestling growth and song repertoire size in great reed warblers: Evidence for song learning as an indicator mechanism in mate choice. *Proceedings of the Royal Society of London B* 267: 2419–2424.

888 Nowicki, S., S. Peters, and J. Podos. 1998. Song learning, early nutrition, and sexual selection in birds. *American Zoologist* 38: 179–190.

889 Nowicki, S., W. A. Searcy, and M. Hughes. 1998. The territory defense function of song in song sparrows: A test with the speaker occupation design. *Behaviour* 135: 615–628.

890 Nowicki, S., W. A. Searcy, and S. Peters. 2002. Brain development, song learning and mate choice in birds: a review and experimental test of the "nutritional stress hypothesis." *Journal of Comparative Physiology A* 188: 1003–1114.

891 Nowicki, S., W. A. Searcy, and S. Peters. 2002. Quality of song learning affects female response to male bird song. *Proceedings of the Royal Society of London B* 269: 1949–1954.

892 Nunn, C. L., J. L. Gittleman, and J. Antonovics. 2000. Promiscuity and the primate immune system. *Science* 290: 1168–1170.

893 O'Neill, K. M. 1983. Territoriality, body size, and spacing in males of the bee wolf *Philanthus basilaris* (Hymenoptera; Sphecidae). *Behaviour* 86: 295–321.

894 Olendorf, R., T. Getty, K. Scribner, and S. K. Robinson. 2004. Male red-winged blackbirds distrust unreliable and sexually attractive neighbours. *Proceedings of the Royal Society of London B* 271: 1033–1038.

895 Olson, D. J., A. C. Kamil, R. P. Balda, and P. J. Nims. 1995. Performance of four seed-caching corvid species in operant tests of nonspatial and spatial memory. *Journal of Comparative Psychology* 109: 173–181.

896 Olson, V. A., and I. P. F. Owens. 1998. Costly sexual signals: Are carotenoids rare, risky, or required? *Trends in Ecology and Evolution* 13: 510–514.

897 Orians, G. H. 1962. Natural selection and ecological theory. *American Naturalist* 96: 257–264.

898 Orians, G. H. 1969. On the evolution of mating systems in birds and mammals. *American Naturalist* 103: 589–603.

899 Oring, L. W. 1985. Avian polyandry. *Current Ornithology* 3: 309–351.

900 Oring, L. W., M. A. Colwell, and J. M. Reed. 1991. Lifetime reproductive success in the spotted sandpiper (*Actitis macularia*): Sex differences and variance components. *Behavioral Ecology and Sociobiology* 28: 425–432.

901 Oring, L. W., R. C. Fleischer, J. M. Reed, and K. E. Marsden. 1992. Cuckoldry through stored sperm in the sequentially polyandrous spotted sandpiper. *Nature* 359: 631–633.

902 Oring, L. W., and M. L. Knudson. 1973. Monogamy and polyandry in the spotted sandpiper. *The Living Bird* 11: 59–73.

903 Orr, M. R. 1992. Parasitic flies (Diptera: Phoridae) influence foraging rhythms and caste division of labor in the leaf-cutter ant, *Atta cephalotes* (Hymenoptera: Formicidae). *Behavioral Ecology and Sociobiology* 30: 395–402.

904 Orrell, K. S., and T. A. Jenssen. 2002. Male mate choice by the lizard *Anolis carolinensis*: a preference for novel females. *Animal Behaviour* 63: 1091–1102.

905 Osborne, K. A., A. Robichon, E. Burgess, S. Butland, R. A. Shaw, et al. 1997. Natural behavior polymorphism due to a cGMP-dependent protein kinase of *Drosophila*. *Science* 277: 834–836.

906 Osorno, J. L., and H. Drummond. 1995. The function of hatching asynchrony in the blue-footed booby. *Behavioral Ecology and Sociobiology* 37: 265–274.

907 Östlund-Nilsson, S., and M. Holmlund. 2003. The artistic three-spined stickleback (*Gasterosteus aculeatus*). *Behavioral Ecology and Sociobiology* 53: 214–220.

908 Östlund, S., and I. Ahnesjö. 1998. Female fifteen-spined sticklebacks prefer better fathers. *Animal Behaviour* 56: 1177–1183.

909 Ostner, J., and M. Heistermann. 2003. Intersexual dominance, masculinized genitals and prenatal steroids: comparative data from lemurid primates. *Naturwissenschaften* 90: 141–144.

910 Outlaw, D. C., G. Voelker, B. Mila, and D. J. Girman. 2003. Evolution of long-distance migration in and historical biogeography of *Catharus* thrushes: A molecular phylogenetic approach. *Auk* 120: 299–310.

911 Owens, D. D., and M. J. Owens. 1979. Communal denning and clan associations in brown hyenas (*Hyaena brunnea*, Thunberg) in the central Kalahari Desert. *African Journal of Ecology* 17: 35–44.

912 Owens, D. D., and M. J. Owens. 1996. Social dominance and reproductive patterns in brown hyenas, *Hyaena brunnea*, of the central Kalahari desert. *Animal Behaviour* 51: 535–551.

913 Owens, I. P. F., and R. V. Short. 1995. Hormonal basis of sexual dimorphism in birds: Implications for sexual selection theory. *Trends in Ecology and Evolution* 10: 44–47.

914 Owings, D. H., and R. G. Coss. 1977. Snake mobbing by California ground squirrels: Adaptive variation and ontogeny. *Behaviour* 62: 50–69.

915 Packer, C. 1977. Reciprocal altruism in *Papio anubis*. *Nature* 265: 441–443.

916 Packer, C. 1986. The ecology of sociality in felids. In *Ecological Aspects of Social Evolution*, ed. Rubenstein, D. I., and R. W. Wrangham. Princeton, NJ: Princeton University Press.

917 Packer, C. 1994. *Into Africa*. Chicago: University of Chicago Press.

918 Packer, C., D. A. Gilbert, A. E. Pusey, and S. J. O'Brien. 1991. A molecular genetic analysis of kinship and cooperation in African lions. *Nature* 351: 562–565.

919 Packer, C., D. Scheel, and A. E. Pusey. 1990. Why lions form groups: Food is not enough. *American Naturalist* 136: 1–19.

920 Page, R. E., G. E. Robinson, M. K. Fondrk, and M. E. Nasr. 1995. Effects of worker genotypic diversity on honey-bee colony development and behavior (*Apis mellifera* L.). *Behavioral Ecology and Sociobiology* 36: 387–396.

921 Page, T. L. 1985. Clocks and circadian rhythms. In *Comprehensive Insect Physiology, Biochemistry, and Pharmacology*, ed. Kerkut, G. A., and L. I. Gilbert. New York: Pergamon Press.

922 Palmer, C. T. 1991. Human rape: Adaptation or by-product? *Journal of Sex Research* 28: 365–386.

923 Palombit, R. A., R. M. Seyfarth, and D. L. Cheney. 1997. The adaptive value of "friendships" to female baboons: Experimental and observational evidence. *Animal Behaviour* 54: 599–614.

924 Panhuis, T. M., and G. S. Wilkinson. 1999. Exaggerated male eye span influences contest outcome in stalk-eyed flies (Diopsidae). *Behavioral Ecology and Sociobiology* 46: 221–227.

925 Papaj, D. R., and R. H. Messing. 1998. Asymmetries in physiological state as a possible cause of resident advantage in contests. *Behaviour* 135: 1013–1030.

926 Parker, G. A. 1970. Sperm competition and its evolutionary consequences in the insects. *Biological Reviews* 45: 526–567.

927 Parker, G. A., R. R. Baker, and V. G. F. Smith. 1972. The origin and evolution of gamete dimorphism and the male-female phenomenon. *Journal of Theoretical Biology* 36: 529–553.

928 Partecke, J., A. von Haeseler, and M. Wikelski. 2002. Territory establishment in lekking marine iguanas, *Amblyrhynchus cristatus*: support for the hotshot mechanism. *Behavioral Ecology and Sociobiology* 51: 579–587.

929 Patricelli, G. L., J. A. C. Uy, and G. Borgia. 2003. Multiple male traits interact: attractive bower decorations facilitate attractive behavioural displays in satin bowerbirds. *Proceedings of the Royal Society of London B* 270: 2389–2395.

930 Patricelli, G. L., J. A. C. Uy, and G. Borgia. 2004. Female signals enhance the efficiency of mate assessment in satin bowerbirds (*Ptilonorhynchus violaceus*). *Behavioral Ecology* 15: 297–304.

931 Patricelli, G. L., J. A. C. Uy, G. Walsh, and G. Borgia. 2002. Male displays adjusted to female's response. *Nature* 415: 279–280.

932 Pawłowski, B., and R. I. M. Dunbar. 1999. Impact of market value on human mate choice decisions. *Proceedings of the Royal Society of London B* 266: 281–285.

933 Peakall, R. 1990. Responses of male *Zaspilothynnus trilobatus* Turner wasps to females and the sexually deceptive orchid it pollinates. *Functional Ecology* 4: 159–167.

934 Pelli, D. G., B. Farell, and D. C. Moore. 2003. The remarkable inefficiency of word recognition. *Nature* 423: 752–756.

935 Pengelly, E. T., and S. J. Asmundson. 1974. Circannual rhythmicity in hibernating animals. In *Circannual Clocks*, ed. Pengelley, E. T. New York: Academic Press.

936 Penn, D. J. 2003. The evolutionary roots of our environmental problems: Toward a Darwinian ecology. *Quarterly Review of Biology* 78: 275–301.

937 Pennisi, E. 2000. Fruit fly genome yields data and a validation. *Science* 287: 1374.

938 Perrigo, G., W. C. Bryant, and F. S. vom Saal. 1990. A unique neural timing system prevents male mice from harming their own offspring. *Animal Behaviour* 39: 535–539.

939   Perusse, D. 1993. Cultural and reproductive success in industrial societies: Testing the relationship at the proximate and ultimate levels. *Behavioral and Brain Sciences* 16: 267–283.

940   Petit, L. J. 1991. Adaptive tolerance of cowbird parasitism by prothonotary warblers: A consequence of site limitation? *Animal Behaviour* 41: 425–432.

941   Petrie, M. 1992. Peacocks with low mating success are more likely to suffer predation. *Animal Behaviour* 44: 585–586.

942   Petrie, M. 1994. Improved growth and survival of offspring of peacocks with more elaborate trains. *Nature* 371: 585–586.

943   Petrie, M., and T. Halliday. 1994. Experimental and natural changes in the peacock's (*Pavo cristatus*) train can affect mating success. *Behavioral Ecology and Sociobiology* 35: 213–217.

944   Petrie, M., T. Halliday, and C. Sanders. 1991. Peahens prefer peacocks with elaborate trains. *Animal Behaviour* 41: 323–332.

945   Pfennig, D. W., and J. P. Collins. 1993. Kinship affects morphogenesis in cannibalistic salamanders. *Nature* 362: 836–838.

946   Pfennig, D. W., G. J. Gamboa, H. K. Reeve, J. S. Reeve, and I. D. Ferguson. 1983. The mechanism of nestmate discrimination in social wasps (*Polistes*, Hymenoptera: Vespidae). *Behavioral Ecology and Sociobiology* 13: 299–305.

947   Pfennig, D. W., P. W. Sherman, and J. P. Collins. 1994. Kin recognition and cannibalism in polyphenic salamanders. *Behavioral Ecology* 5: 225–232.

948   Phillips, R. A., R. W. Furness, and F. M. Stewart. 1998. The influence of territory density on the vulnerability of Arctic skuas *Stercorarius parasiticus* to predation. *Biological Conservation* 86: 21–31.

949   Picciotto, M. R. 1999. Knock-out mouse models used to study neurobiological systems. *Critical Reviews in Neurobiology* 13: 103–149.

950   Pierce, G. J., and J. G. Ollason. 1987. Eight reasons why optimal foraging theory is a complete waste of time. *Oikos* 49: 111–118.

951   Pierotti, R., and E. C. Murphy. 1987. Intergenerational conflicts in gulls. *Animal Behaviour* 35: 435–444.

952   Pietrewicz, A. T., and A. C. Kamil. 1977. Visual detection of cryptic prey by blue jays (*Cyanocitta cristata*). *Science* 195: 580–582.

953   Pietsch, T. W., and D. B. Grobecker. 1978. The compleat angler: Aggressive mimicry in the antennariid anglerfish. *Science* 201: 369–370.

954   Pillsworth, E. G., M. G. Haselton, and D. M. Buss. 2004. Ovulatory shifts in female sexual desire. *Journal of Sex Research* 41: 55–65.

955   Pinker, S. 1994. *The Language Instinct*. New York: W. Morrow & Co.

956   Pinxten, R., E. de Ridder, and M. Eens. 2003. Female presence affects male behavior and testosterone levels in the European starling (*Sturnus vulgaris*). *Hormones and Behavior* 44: 103–109.

957   Piper, W. H., D. C. Evers, M. W. Meyer, K. B. Tischler, J. D. Kaplan, and R. C. Fleischer. 1997. Genetic monogamy in the common loon (*Gavia immer*). *Behavioral Ecology and Sociobiology* 41: 25–32.

958   Pitcher, T. 1979. He who hesitates lives: Is stotting anti-ambush behavior? *American Naturalist* 113: 453–456.

959   Pitkow, L. J., C. A. Sharer, X. L. Ren, T. R. Insel, E. F. Terwilliger, and L. J. Young. 2001. Facilitation of affiliation and pair-bond formation by vasopressin receptor gene transfer into the ventral forebrain of a monogamous vole. *Journal of Neuroscience* 21: 7392–7396.

960   Pitnick, S., and F. Garcia-Gonzalez. 2002. Harm to females increases with male body size in *Drosophila melanogaster*. *Proceedings of the Royal Society of London B* 269: 1821–1828.

961   Pitnick, S., T. A. Markow, and G. S. Spicer. 1995. Delayed male maturity is a cost of producing large sperm in *Drosophila*. *Proceedings of the National Academy of Sciences* 92: 10614–10618.

962   Pizzari, T., and T. R. Birkhead. 2000. Female feral fowl eject sperm of subdominant males. *Nature* 405: 787–789.

963   Place, N. J., and S. E. Glickman. 2004. Masculinization of female mammals: Lessons from nature. In *Hypospadias and Genital Development*, ed. Baskin, L. New York: Kluwer Academic/Plenum Publishers.

964   Plaistow, S., and M. T. Siva-Jothy. 1996. Energetic constraints and male mate securing tactics in the damselfly *Calopteryx splendens xanthosoma* (Charpentier). *Proceedings of the Royal Society of London B* 263: 1233–1238.

965   Platt, J. R. 1964. Strong inference. *Science* 146: 347–353.

966   Platzen, D., and R. D. Magrath. 2004. Parental alarm calls suppress nestling vocalization. *Proceedings of the Royal Society of London B* 271: 1271–1276.

967   Plomin, R., and I. Craig. 1997. Human behavioural genetics of cognitive abilities and disabilities. *BioEssays* 19: 1117–1124.

968   Plomin, R., D. W. Fulker, R. Corley, and J. C. DeFries. 1997. Nature, nurture, and cognitive development from 1 to 16 years: A parent-offspring adoption study. *Psychological Science* 8: 442–447.

969   Polak, M., L. L. Wolf, W. T. Starmer, and J. S. F. Barker. 2001. Function of the mating plug in *Drosophila hibisci* Bock. *Behavioral Ecology and Sociobiology* 49: 196–205.

970   Porter, R. H., V. J. Tepper, and D. M. White. 1981. Experiential influences on the development of huddling preferences and "sibling" recognition in spiny mice. *Developmental Psychobiology* 14: 375–382.

971   Powell, A. N., and C. L. Collier. 2000. Habitat use and reproductive success of western snowy plovers at new nesting areas created for California least terns. *Journal of Wildlife Management* 64: 24–33.

972   Powell, G. V. N., and R. D. Bjork. 2004. Habitat linkages and the conservation of tropical biodiversity as indicated by seasonal migrations of three-wattled bellbirds. *Conservation Biology* 18: 500–509.

973   Power, H. W. 1998. Quality control and the important questions in avian conspecific brood parasitism. In *Parasitic Birds and Their Hosts: Studies in Coevolution*, ed. Rothstein, S. I., and S. K. Robinson. New York: Oxford University Press.

974   Powzyk, J. A., and C. B. Mowry. 2003. Dietary and feeding differences between sympatric *Propithecus diadema diadema* and *Indri indri*. *International Journal of Primatology* 24: 1143–1162.

975   Pravosudov, V. V., and N. S. Clayton. 2002. A test of the adaptive specialization hypothesis: Population differences in caching, memory, and the hippocampus in black-capped chickadees (*Poecile atricapilla*). *Behavioral Neuroscience* 116: 515–522.

976   Preston-Mafham, R., and K. Preston-Mafham. 1993. *The Encyclopedia of Land Invertebrate Behaviour*. Cambridge, MA: MIT Press.

977   Prete, F. R. 1995. Designing behavior: A case study. *Perspectives in Ethology* 11: 255–277.

978   Pribil, S., and W. A. Searcy. 2001. Experimental confirmation of the polygyny threshold model for red-winged blackbirds. *Proceedings of the Royal Society of London B* 268: 1643–1646.

979   Proctor, H. C. 1991. Courtship in the water mite *Neumania papillator*: Males capitalize on female adaptations for predation. *Animal Behaviour* 42: 589–598.

980   Proctor, H. C. 1992. Sensory exploitation and the evolution of male mating behaviour: A cladistic test. *Animal Behaviour* 44: 745–752.

981   Pruett-Jones, S., and M. Pruett-Jones. 1994. Sexual competition and courtship disruptions: Why do male bowerbirds destroy each other's bowers? *Animal Behaviour* 47: 607–620.

982   Prum, R. O. 1999. Development and evolutionary origin of feathers. *Journal of Experimental Zoology* 285: 291–306.

983   Pryke, S. R., and S. Andersson. 2003. Carotenoid-based epaulettes reveal male competitive ability: experiments with resident and floater red-shouldered widowbirds. *Animal Behaviour* 66: 217–224.

984   Puce, A., T. Allison, and G. McCarthy. 1999. Electrophysiological studies of human face perception. III: Effects of top-down processing of face-specific potentials. *Cerebral Cortex* 9: 445–458.

985 Pulido, F., P. Berthold, G. Mohr, and U. Querner. 2001. Heritability of the timing of autumn migration in a natural bird population. *Proceedings of the Royal Society of London B* 268: 953–959.

986 Purseglove, J. W., E. G. Brown, C. L. Green, and S. R. J. Robbins. 1981. *Spices*. London: Longman.

987 Pusey, A. E., and C. Packer. 1987. The evolution of sex-biased dispersal in lions. *Behaviour* 101: 275–310.

988 Pusey, A. E., and C. Packer. 1994. Infanticide in lions. In *Infanticide and Parental Care*, ed. Parmigiani, S., and F. S. vom Saal. Chur, Switzerland: Harwood Academic Press.

989 Pusey, A. E., and M. Wolf. 1996. Inbreeding avoidance in animals. *Trends in Ecology and Evolution* 11: 201–206.

990 Queller, D. C. 1996. The measurement and meaning of inclusive fitness. *Animal Behaviour* 51: 229–232.

991 Queller, D. C. 1997. Why do females care more than males? *Proceedings of the Royal Society of London B* 264: 1555–1557.

992 Queller, D. C., and J. E. Strassmann. 1998. Kin selection and social insects. *BioScience* 48: 165–175.

993 Queller, D. C., J. E. Strassmann, and C. R. Hughes. 1993. Microsatellites and kinship. *Trends in Ecology and Evolution* 8: 285–288.

994 Queller, D. C., F. Zacchi, R. Cervo, S. Turillazzi, M. T. Henshaw, et al. 2000. Unrelated helpers in a social insect. *Nature* 405: 784–787.

995 Quillfeldt, P. 2002. Begging in the absence of sibling competition in Wilson's storm-petrels, *Oceanites oceanicus*. *Animal Behaviour* 64: 579–587.

996 Quinn, J. S., G. E. Woolfenden, J. W. Fitzpatrick, and B. N. White. 1999. Multi-locus DNA fingerprinting supports genetic monogamy in Florida scrub-jays. *Behavioral Ecology and Sociobiology* 45: 1–10.

997 Quinn, V. S., and D. K. Hews. 2000. Signals and behavioural responses are not coupled in males: Aggression affected by replacement of an evolutionarily lost colour signal. *Proceedings of the Royal Society of London B* 267: 755–758.

998 Racey, P. A., and J. D. Skinner. 1979. Endocrine aspects of sexual mimicry in spotted hyenas *Crocuta crocuta*. *Journal of Zoology* 187: 315–326.

999 Rachlow, J. L., E. V. Berkeley, and J. Berger. 1998. Correlates of male mating strategies in white rhinos (*Ceratotherium simum*). *Journal of Mammalogy* 79: 1317–1324.

1000 Ralls, K., K. Brugger, and J. Ballou. 1979. Inbreeding and juvenile mortality in small populations of ungulates. *Science* 206: 1101–1103.

1001 Ramirez, M. I., J. G. Azcarate, and L. Luna. 2003. Effects of human activities on monarch butterfly habitat in protected mountain forests, Mexico. *Forestry Chronicle* 79: 242–246.

1002 Ratcliffe, J. M., M. B. Fenton, and B. G. Galef. 2003. An exception to the rule: common vampire bats do not learn taste aversions. *Animal Behaviour* 65: 385–389.

1003 Ratnieks, F. L. W., and L. Keller. 1998. Queen control of egg fertilization in the honey bee. *Behavioral Ecology and Sociobiology* 44: 57–62.

1004 Redondo, T., and J. Carranza. 1989. Offspring reproductive value and nest defense in the magpie (*Pica pica*). *Behavioral Ecology and Sociobiology* 25: 369–378.

1005 Reeve, H. K., and L. Keller. 2001. Tests of reproductive-skew models in social insects. *Annual Review of Ecology and Systematics* 46: 347–385.

1006 Reeve, H. K., and P. W. Sherman. 1993. Adaptation and the goals of evolutionary research. *Quarterly Review of Biology* 68: 1–32.

1007 Reeve, H. K., and P. W. Sherman. 2001. Optimality and phylogeny. In *Adaptationism and Optimality*, ed. Orzack, S. H., and E. Sober. Cambridge: Cambridge University Press.

1008 Reeve, H. K., P. T. Starks, J. M. Peters, and P. Nonacs. 2000. Genetic support for the evolutionary theory of reproductive transactions in social wasps. *Proceedings of the Royal Society of London B* 267: 75–79.

1009 Reeve, H. K., D. F. Westneat, W. A. Noon, P. W. Sherman, and C. F. Aquadro. 1990. DNA "fingerprinting" reveals high levels of inbreeding in colonies of the eusocial naked mole-rat. *Proceedings of the National Academy of Sciences* 87: 2496–2500.

1010 Reid, J. M., P. Monaghan, and G. D. Ruxton. 2002. Males matter: The occurrence and consequences of male incubation in starlings (*Sturnus vulgaris*). *Behavioral Ecology and Sociobiology* 51: 255–261.

1011 Reinhardt, K., R. Naylor, and M. T. Siva-Jothy. 2003. Reducing a cost of traumatic insemination: female bedbugs evolve a unique organ. *Proceedings of the Royal Society of London B* 270: 2371–2375.

1012 Reppert, S. M., H. S. Zhu, and R. H. White. 2004. Polarized light helps monarch butterflies navigate. *Current Biology* 14: 155–158.

1013 Reyer, H.-U. 1984. Investment and relatedness: A cost/benefit analysis of breeding and helping in the pied kingfisher. *Animal Behaviour* 32: 1163–1178.

1014 Reynolds, J. D., and M. R. Gross. 1990. Costs and benefits of female mate choice: Is there a lek paradox? *American Naturalist* 136: 230–243.

1015 Rhoden, P. K., and W. A. Foster. 2002. Soldier behaviour and division of labour in the aphid genus *Pemphigus* (Hemiptera, Aphididae). *Insectes Sociaux* 49: 257–263.

1016 Rhodes, G., F. Proffitt, J. M. Grady, and A. Sumich. 1998. Facial symmetry and the perception of beauty. *Psychonomic Bulletin and Review* 5: 659–669.

1017 Richner, H., and P. Heeb. 1996. Communal life: Honest signaling and the recruitment center hypothesis. *Behavioral Ecology* 7: 115–118.

1018 Riedel, S., A. Hinz, and R. Schwarz. 2000. Attitude towards blood donation in Germany: Results of a representative survey. *Infusion Therapy and Transfusion Medicine* 27: 196–199.

1019 Rintamäki, P. T., R. V. Alatalo, J. Höglund, and A. Lundberg. 1995. Male territoriality and female choice on black grouse leks. *Animal Behaviour* 49: 759–767.

1020 Robert, D., J. Amoroso, and R. R. Hoy. 1992. The evolutionary convergence of hearing in a parasitoid fly and its cricket host. *Science* 258: 1135–1137.

1021 Roberts, M. L., K. L. Buchanan, and M. R. Evans. 2004. Testing the immunocompetence handicap hypothesis: a review of the evidence. *Animal Behaviour* 68: 227–239.

1022 Roberts, S. C., J. Havlicek, J. Flegr, M. Hruskova, A. C. Little, et al. 2004. Female facial attractiveness increases during the fertile phase of the menstrual cycle. *Proceedings of the Royal Society of London B* 271: S270–S272.

1023 Robinson, G. E. 1998. From society to genes with the honey bee. *American Scientist* 86: 456–462.

1024 Robinson, G. E. 1999. Integrative animal behavior and sociogenomics. *Trends in Ecology and Evolution* 14: 202–205.

1025 Robinson, G. E. 2004. Beyond nature and nurture. *Science* 304: 397–399.

1026 Rodd, F. H., K. A. Hughes, G. F. Grether, and C. T. Baril. 2002. A possible non-sexual origin of mate preference: are male guppies mimicking fruit? *Proceedings of the Royal Society of London B* 269: 475–481.

1027 Roeder, K. D. 1963. *Nerve Cells and Insect Behavior*. Cambridge, MA: Harvard University Press.

1028 Roeder, K. D. 1970. Episodes in insect brains. *American Scientist* 58: 378–389.

1029 Roeder, K. D., and A. E. Treat. 1961. The detection and evasion of bats by moths. *American Scientist* 49: 135–148.

1030 Rohwer, S., and C. D. Spaw. 1988. Evolutionary lag versus bill-size constraints: A comparative study of the acceptance of cowbird eggs by old hosts. *Evolutionary Ecology* 2: 27–36.

1031 Rose, M. 1998. *Darwin's Spectre*. Princeton, NJ: Princeton University Press.

1032 Rose, N. A., C. J. Deutsch, and B. J. Le Boeuf. 1991. Sexual behavior of male northern elephant seals: III. The mounting of weaned pups. *Behaviour* 119: 171–192.

1033   Rosengaus, R. B., A. B. Maxmen, L. E. Coates, and J. F. A. Traniello. 1998. Disease resistance: A benefit of sociality in the dampwood termite *Zootermopsis angusticollis* (Isoptera: Termopsidae). *Behavioral Ecology and Sociobiology* 44: 125–134.

1034   Rosenqvist, G. 1990. Male mate choice and female-female competition for mates in the pipefish *Nerophis ophidion*. *Animal Behaviour* 39: 1110–1116.

1035   Ross, K. G. 1986. Kin selection and the problem of sperm utilization in social insects. *Nature* 323: 798–800.

1036   Rowe, L., G. Arnqvist, A. Sih, and J. Krupa. 1994. Sexual conflict and the evolutionary ecology of mating patterns: Water striders as a model system. *Trends in Ecology and Evolution* 9: 289–293.

1037   Rowley, I., and G. Chapman. 1986. Cross-fostering, imprinting, and learning in two sympatric species of cockatoos. *Behaviour* 96: 1–16.

1038   Royle, N. J., I. R. Hartley, and G. A. Parker. 2002. Begging for control: when are offspring solicitation behaviours honest? *Trends in Ecology and Evolution* 17: 434–440.

1039   Rubenstein, D. R., C. P. Chamberlain, R. T. Holmes, M. P. Ayres, J. R. Waldbauer, et al. 2002. Linking breeding and wintering ranges of a migratory songbird using stable isotopes. *Science* 295: 1062–1065.

1040   Rubenstein, D. R., and K. A. Hobson. 2004. From birds to butterflies: animal movement patterns and stable isotopes. *Trends in Ecology and Evolution* 19: 256–263.

1041   Ruegg, K. C., and T. B. Smith. 2002. Not as the crow flies: a historical explanation for circuitous migration in Swainson's thrush (*Catharus ustulatus*). *Proceedings of the Royal Society of London B* 269: 1375–1381.

1042   Ruffieux, L., J. M. Elouard, and M. Sartori. 1998. Flightlessness in mayflies and its relevance to hypotheses on the origin of insect flight. *Proceedings of the Royal Society of London B* 265: 2135–2140.

1043   Runcie, M. J. 2000. Biparental care and obligate monogamy in the rock-haunting possum, *Petropseudes dahli*, from tropical Australia. *Animal Behaviour* 59: 1001–1008.

1044   Rutowski, R. L. 1998. Mating strategies in butterflies. *Scientific American* 279 (July): 64–69.

1045   Ryan, M. J. 1985. *The Túngara Frog*. Chicago: University of Chicago Press.

1046   Ryan, M. J., J. H. Fox, W. Wilczynski, and A. S. Rand. 1990. Sexual selection for sensory exploitation in the frog *Physalaemus pustulosus*. *Nature* 343: 66–67.

1047   Ryan, M. J., and A. Keddy-Hector. 1992. Directional patterns of female mate choice and the role of sensory biases. *American Naturalist* 139: S4–S35.

1048   Ryan, M. J., and A. S. Rand. 1990. The sensory basis of sexual selection for complex calls in the túngara frog, *Physalaemus pustulosus* (sexual selection for sensory exploitation). *Evolution* 44: 305–314.

1049   Ryan, M. J., and A. S. Rand. 1999. Phylogenetic influence on mating call preferences in female tungara frogs, *Physalaemus pustulosus*. *Animal Behaviour* 57: 945–956.

1050   Ryan, M. J., M. D. Tuttle, and L. K. Taft. 1981. The costs and benefits of frog chorusing behavior. *Behavioral Ecology and Sociobiology* 8: 273–278.

1051   Ryan, M. J., and W. E. Wagner, Jr. 1987. Asymmetries in mating behavior between species: Female swordtails prefer heterospecific males. *Science* 236: 595–597.

1052   Rydale, J., H. Roininen, and K. W. Philip. 2000. Persistence of bat defence reactions in high Arctic moths (Lepidoptera). *Proceedings of the Royal Society of London B* 267: 553–557.

1053   Rydell, J., and R. Arlettaz. 1994. Low-frequency echolocation enables the bat *Tadarida teniotis* to feed on tympanate insects. *Proceedings of the Royal Society of London B* 257: 175–178.

1054   Sacks, O. W. 1985. *The Man Who Mistook His Wife for a Hat and Other Clinical Tales*. New York: Summit Books.

1055   Sadowski, J. A., A. J. Moore, and E. D. Brodie, III. 1999. The evolution of empty nuptial gifts in a dance fly, *Empis snoddyi* (Diptera: Empididae): Bigger isn't always better. *Behavioral Ecology and Sociobiology* 45: 161–166.

1056   Sahlins, M. 1976. *The Use and Abuse of Biology*. Ann Arbor: University of Michigan Press.

1057   Saino, N., P. Ninni, S. Calza, R. Martinelli, F. de Bernardi, and A. P. Møller. 2000. Better red than dead: Carotenoid-based mouth coloration reveals infection in barn swallow nestlings. *Proceedings of the Royal Society of London B* 267: 57–61.

1058   Sandberg, R., and F. R. Moore. 1996. Migratory orientation of red-eyed vireos, *Vireo olivaceus*, in relation to energetic condition and ecological context. *Behavioral Ecology and Sociobiology* 39: 1–10.

1059   Sandercock, B. K., and A. Jaramillo. 2002. Annual survival rates of wintering sparrows: Assessing demographic consequences of migration. *Auk* 119: 149–165.

1060   Sargent, R. C. 1989. Allopaternal care in the fathead minnow, *Pimephales promelas*: Stepfathers discriminate against their adopted eggs. *Behavioral Ecology and Sociobiology* 25: 379–386.

1061   Sargent, T. D. 1976. *Legion of Night, The Underwing Moths*. Amherst, MA: University of Massachusetts Press.

1062   Sato, T. 1994. Active accumulation of spawning substrate: A determinant of extreme polygyny in a shell-brooding cichlid fish. *Animal Behaviour* 48: 669–678.

1063   Sax, A., H. Hoi, and T. R. Birkhead. 1998. Copulation rate and sperm use by female bearded tits, *Panurus biarmicus*. *Animal Behaviour* 56: 1199–1294.

1064   Schaller, G. B. 1964. *The Year of the Gorilla*. Chicago: University of Chicago Press.

1065   Schamel, D., D. M. Tracy, and D. B. Lank. 2004. Male mate choice, male availability and egg production as limitations on polyandry in the red-necked phalarope. *Animal Behaviour* 67: 847–853.

1066   Scheel, D., and C. Packer. 1991. Group hunting behaviour of lions: A search for cooperation. *Animal Behaviour* 41: 711–722.

1067   Schieb, J. E., S. W. Gangestad, and R. Thornhill. 1999. Facial attractiveness, symmetry and cues of good genes. *Proceedings of the Royal Society of London B* 266: 1913–1917.

1068   Schlaepfer, M. A., M. C. Runge, and P. W. Sherman. 2002. Ecological and evolutionary traps. *Trends in Ecology and Evolution* 17: 474–480.

1069   Schmitt, D. P., T. K. Shackelford, J. Duntley, W. Tooke, and D. M. Buss. 2001. The desire for sexual variety as a key to understanding basic human mating strategies. *Personal Relationships* 8: 425–455.

1070   Schneider, J. M., and Y. Lubin. 1997. Infanticide by males in a spider with suicidal maternal care, *Stegodyphus lineatus*. *Animal Behaviour* 54: 305–312.

1071   Schneider, J. S., M. K. Stone, K. E. Wynne-Edwards, T. H. Horton, J. Lydon, et al. 2003. Progesterone receptors mediate male aggression toward infants. *Proceedings of the National Academy of Sciences* 100: 2951–2956.

1072   Schoech, S. J. 1998. Physiology of helping in Florida scrub-jays. *American Scientist* 86: 70–77.

1073   Schwabl, H. 1983. Auspragung und Bedeutung des Teilzugverhaltnes einer sudwestdeutschen Population der Amsel *Turdus merula*. *Journal für Ornithologie* 124: 101–116.

1074   Schwabl, H., D. W. Mock, and J. A. Gieg. 1997. A hormonal mechanism for parental favouritism. *Nature* 386: 231.

1075   Schwagmeyer, P. L. 1994. Competitive mate searching in the 13-lined ground squirrel: Potential roles of spatial memory? *Ethology* 98: 265–276.

1076   Schwagmeyer, P. L. 1995. Searching today for tomorrow's mates. *Animal Behaviour* 50: 759–767.

1077   Sealy, S. G. 1995. Burial of cowbird eggs by parasitized yellow warblers: An empirical and experimental study. *Animal Behaviour* 49: 877–889.

1078   Searcy, W. A., S. Nowicki, M. Hughes, and S. Peters. 2002. Geographic song discrimination in relation to dispersal distances in song sparrows. *American Naturalist* 159: 221–230.

1079   Seeley, T. D. 1977. Measurement of nest cavity volume by the honey bee (*Apis mellifera*). *Behavioral Ecology and Sociobiology* 2: 201–227.

1080   Seeley, T. D. 1995. *The Wisdom of the Hive*. Cambridge, MA: Harvard University Press.

1081 Seeley, T. D., and S. C. Buhrman. 1999. Group decision making in swarms of honey bees. *Behavioral Ecology and Sociobiology* 45: 19–32.

1082 Seeley, T. D., and P. K. Visscher. 1988. Assessing the benefits of cooperation in honeybee foraging: Search costs, forage quality, and competitive ability. *Behavioral Ecology and Sociobiology* 22: 229–237.

1083 Seeley, T. D., and P. K. Visscher. 2003. Choosing a home: how the scouts in a honey bee swarm perceive the completion of their group decision making. *Behavioral Ecology and Sociobiology* 54: 511–520.

1084 Semel, B., and P. W. Sherman. 2001. Intraspecific parasitism and nest-site competition in wood ducks. *Animal Behaviour* 61: 787–803.

1085 Semple, S., K. McComb, S. Alberts, and J. Altmann. 2002. Information content of female copulation calls in yellow baboons. *American Journal of Primatology* 56: 43–56.

1086 Senar, J. C., J. Figuerola, and J. Pascual. 2002. Brighter yellow blue tits make better parents. *Proceedings of the Royal Society of London B* 269: 257–261.

1087 Senghas, A., S. Kita, and A. Özyürek. 2004. Children creating core properties of language: Evidence from an emerging sign language in Nicaragua. *Science* 305: 1779–1782.

1088 Shackelford, T. K., D. M. Buss, and V. A. Weekes-Shackelford. 2003. Wife killings committed in the context of a lovers triangle. *Basic and Applied Social Psychology* 25: 137–143.

1089 Shealer, D. A., and J. A. Spendelow. 2002. Individual foraging strategies of kleptoparasitic roseate terns. *Waterbirds* 25: 436–441.

1090 Sheldon, B. C. 1993. Sexually-transmitted disease in birds: occurrence and evolutionary significance. *Philosophical Transactions of the Royal Society of London B* 339: 491–497.

1091 Shellman-Reeve, J. S. 1999. Courting strategies and conflicts in a monogamous, biparental termite. *Proceedings of the Royal Society of London B* 266: 137–144.

1092 Shelly, T. E. 2001. Lek size and female visitation in two species of tephritid fruit flies. *Animal Behaviour* 62: 33–40.

1093 Sherman, G., and P. K. Visscher. 2002. Honeybee colonies achieve fitness through dancing. *Nature* 419: 920–922.

1094 Sherman, P. W. 1977. Nepotism and the evolution of alarm calls. *Science* 197: 1246–1253.

1095 Sherman, P. W. 1981. Kinship, demography and Belding's ground squirrel nepotism. *Behavioral Ecology and Sociobiology* 8: 251–259.

1096 Sherman, P. W. 1988. The levels of analysis. *Animal Behaviour* 36: 616–618.

1097 Sherman, P. W., and G. A. Hash. 2001. Why vegetable dishes are not very spicy. *Evolution and Human Behavior* 22: 147–163.

1098 Sherman, P. W., J. U. M. Jarvis, and R. D. Alexander, eds. 1991. *The Biology of the Naked Mole-Rat*. Princeton, NJ: Princeton University Press.

1099 Sherry, D. F. 1984. Food storage by black-capped chickadees: Memory of the location and contents of caches. *Animal Behaviour* 32: 451–464.

1100 Sherry, D. F., M. R. L. Forbes, M. Kjurgel, and G. O. Ivy. 1993. Females have a larger hippocampus than males in the brood-parasitic brown-headed cowbird. *Proceedings of the National Academy of Sciences* 90: 7839–7843.

1101 Shorey, L. 2002. Mating success on white-bearded manakin (*Manacus manacus*) leks: male characteristics and relatedness. *Behavioral Ecology and Sociobiology* 52: 451–457.

1102 Shreeves, G., M. A. Cant, A. Bolton, and J. Field. 2003. Insurance-based advantages for subordinate cofoundresses in a temperate paper wasp. *Proceedings of the Royal Society of London B* 270: 1617–1622.

1103 Shuster, S. M. 1989. Male alternative reproductive strategies in a marine isopod crustacean (*Paracerceis sculpta*): the use of genetic markers to measure differences in the fertilization success among alpha, beta, and gamma-males. *Evolution* 43: 1683–1689.

1104 Shuster, S. M. 1992. The reproductive behaviour of alpha, beta, and gamma morphs in *Paracerceis sculpta*: A marine isopod crustacean. *Behaviour* 121: 231–258.

1105 Shuster, S. M., and C. A. Sassaman. 1997. Genetic interaction between male mating strategy and sex ratio in a marine isopod. *Nature* 338: 373–377.

1106 Shuster, S. M., and M. J. Wade. 1991. Equal mating success among male reproductive strategies in a marine isopod. *Nature* 350: 608–610.

1107 Shuster, S. M., and M. J. Wade. 2003. *Mating Systems and Strategies*. Princeton, NJ.: Princeton University Press.

1108 Silk, J. B. 1980. Adoption and kinship in Oceania. *American Anthropologist* 82: 799–820.

1109 Silk, J. B., R. M. Seyfarth, and D. L. Cheney. 1999. The structure of social relationships among female savanna baboons in Moremi Reserve, Botswana. *Behaviour* 136: 679–703.

1110 Simmons, L. W. 2001. *Sperm Competition and Its Evolutionary Consequences in the Insects*. Princeton, NJ: Princeton University Press.

1111 Simmons, P., and D. Young. 1999. *Nerve Cells and Animal Behaviour*. Cambridge: Cambridge University Press.

1112 Sinervo, B., and J. Clobert. 2003. Morphs, dispersal behavior, genetic similarity, and the evolution of cooperation. *Science* 300: 1949–1951.

1113 Sinervo, B., D. B. Miles, W. A. Frankino, M. Klukowski, and D. F. DeNardo. 2000. Testosterone, endurance, and darwinian fitness: Natural and sexual selection on the physiological bases of alternative male behaviors in side-blotched lizards. *Hormones and Behavior* 38: 222–233.

1114 Singer, P. 1981. *The Expanding Circle: Ethics and Sociobiology*. New York: Farrar, Straus, and Giroux.

1115 Singh, D. 1993. Adaptive significance of female physical attractiveness: Role of the waist-to-hip ratio. *Journal of Personality and Social Psychology* 65: 293–307.

1116 Singh, D., and P. M. Bronstad. 2001. Female body odour is a potential cue to ovulation. *Proceedings of the Royal Society of London B* 268: 797–801.

1117 Sisneros, J. A., and A. H. Bass. 2003. Seasonal plasticity of peripheral auditory frequency sensitivity. *Journal of Neuroscience* 23: 1049–1058.

1118 Skinner, B. F. 1966. Operant behavior. In *Operant Behavior*, ed. Honig, W. New York: Appleton-Century-Crofts.

1119 Slagsvold, T. 1998. On the origin and rarity of interspecific nest parasitism in birds. *American Naturalist* 152: 264–272.

1120 Slagsvold, T., and B. T. Hansen. 2001. Sexual imprinting and the origin of obligate brood parasitism in birds. *American Naturalist* 158: 354–367.

1121 Slagsvold, T., B. T. Hansen, L. E. Johannessen, and J. T. Lifjeld. 2002. Mate choice and imprinting in birds studied by cross-fostering in the wild. *Proceedings of the Royal Society of London B* 269: 1449–1455.

1122 Smiseth, P. T., C. T. Darwell, and A. J. Moore. 2003. Partial begging: an empirical model for the early evolution of offspring signalling. *Proceedings of the Royal Society of London B* 270: 1773–1777.

1123 Smith, E. A., M. Borgerhoff Mulder, and K. Hill. 2001. Controversies in the evolutionary social sciences: A guide for the perplexed. *Trends in Ecology and Evolution* 16: 128–135.

1124 Smith, H. G. 1991. Nestling American robins compete with siblings by begging. *Behavioral Ecology and Sociobiology* 29: 307–312.

1125 Smith, H. G., and R. D. Montgomerie. 1991. Sexual selection and the tail ornaments of North American barn swallows. *Behavioral Ecology and Sociobiology* 28: 195–201.

1126 Smith, M. S., B. J. Kish, and C. B. Crawford. 1987. Inheritance of wealth as human kin investment. *Ethology and Sociobiology* 8: 171–182.

1127 Smith, R. L. 1997. Evolution of paternal care in giant water bugs (Heteroptera: Belostomatidae). In *Social Competition and Cooperation among Insects and Arachnids, II. Evolution of Sociality*, ed. Choe, J. C., and B. J. Crespi. Cambridge: Cambridge University Press.

1128   Smith, R. L., and E. Larsen. 1993. Egg attendance and brooding by males of the giant water bug *Lethocerus medius* (Guerin) in the field (Heteroptera, Belostomatidae). *Journal of Insect Behavior* 6: 93–106.

1129   Smith, S. M. 1978. The 'underworld' in a territorial species: Adaptive strategy for floaters. *American Naturalist* 112: 571–582.

1130   Smith, T. B., and S. Skulason. 1996. Evolutionary significance of resource polymorphisms in fishes, amphibians, and birds. *Annual Review of Ecology and Systematics* 27: 111–133.

1131   Smuts, B. B., and R. W. Smuts. 1993. Male aggression and sexual coercion of females in nonhuman primates and other mammals: Evidence and theoretical implications. *Advances in the Study of Behavior* 22: 1–63.

1132   Snow, D. W. 1956. Courtship ritual: The dance of the manakins. *Animal Kingdom* 59: 86–91.

1133   Sober, E., and D. S. Wilson. 1998. *Unto Others: The Evolution and Psychology of Unselfish Behavior*. Cambridge, MA: Harvard University Press.

1134   Sockman, K. W., K. B. Sewall, G. F. Ball, and T. P. Hahn. In press. Economy of mate attraction in the Cassin's finch. *Biology Letters*.

1135   Sokolowski, M., and D. Wahlsten. 2001. Gene-environment interaction. In *Methods in Genomic Neuroscience*, ed. Chin, H., and S. O. Moldin. Boca Raton, FL: CRC Press.

1136   Sokolowski, M. B., H. S. Pereira, and K. A. Hughes. 1997. Evolution of foraging behavior in *Drosophila* by density-dependent selection. *Proceedings of the National Academy of Sciences* 94: 7373–7377.

1137   Soler, M., J. J. Soler, J. G. Martinez, and A. P. Møller. 1995. Magpie host manipulation by great spotted cuckoos: Evidence for an avian Mafia? *Evolution* 49: 770–775.

1138   Soma, K. K., A. D. Tramontin, and J. C. Wingfield. 2000. Oestrogen regulates male aggression in the non-breeding season. *Proceedings of the Royal Society of London B* 267: 1089–1092.

1139   Sommer, V. 1994. Infanticide among free-ranging langurs (*Presbytis entellus*) of Jodhpur (Rajasthan/India): Recent observations and a reconsideration of hypotheses. *Primates* 28: 163–197.

1140   Sommer, V. 1994. Infanticide among the langurs of Jodhpur: Testing the sexual selection hypothesis with a long-term record. In *Infanticide and Parental Care*, ed. Parmigiani, S., and F. S. vom Saal. Chur, Switzerland: Harwood Academic Press.

1141   Sorenson, L. G. 1994. Forced extra-pair copulation in the white-cheeked pintail: Male tactics and female responses. *Condor* 96: 400–410.

1142   Sorenson, M. D., and R. B. Payne. 2001. A single ancient origin of brood parasitism in African finches: implications for host-parasite coevolution. *Evolution* 55: 2550–2567.

1143   Soukup, S. S., and C. F. Thompson. 1998. Social mating system and reproductive success in house wrens. *Behavioral Ecology* 9: 43–48.

1144   Spencer, K. A., K. L. Buchanan, A. R. Goldsmith, and C. K. Catchpole. 2003. Song as an honest signal of developmental stress in the zebra finch (*Taeniopygia guttata*). *Hormones and Behavior* 44: 132–139.

1145   Stander, P. E. 1992. Cooperative hunting in lions: The role of the individual. *Behavioral Ecology and Sociobiology* 29: 445–454.

1146   Starks, P. T. 2001. Alternative reproductive tactics in the paper wasp *Polistes dominulus* with specific focus on the sit-and-wait tactic. *Annales Zoologici Fennici* 38: 189–199.

1147   Starks, P. T. 2002. The adaptive significance of stabilimenta in orb-webs: a hierarchical approach. *Annales Zoologici Fennici* 39: 307–315.

1148   Starks, P. T., and C. A. Blackie. 2000. The relationship between serial monogamy and rape in the United States (1960–1995). *Proceedings of the Royal Society of London B* 267: 1259–1263.

1149   Starks, P. T., C. A. Blackie, and T. D. Seeley. 2000. Fever in honeybee colonies. *Naturwissenschaften* 87: 229–231.

1150   Stein, Z., M. Susser, G. Saenger, and F. Marolla. 1972. Nutrition and mental performance. *Science* 178: 708–713.

1151   Stenmark, G., T. Slagsvold, and J. T. Lifjeld. 1988. Polygyny in the pied flycatcher, *Ficedula hypoleuca*: A test of the deception hypothesis. *Animal Behaviour* 36: 1646–1657.

1152   Sterck, E. H. M., D. P. Watts, and C. P. van Schaik. 1997. The evolution of female social relationships in nonhuman primates. *Behavioral Ecology and Sociobiology* 41: 291–310.

1153   Stern, D. L., and W. A. Foster. 1996. The evolution of soldiers in aphids. *Biological Reviews* 71: 27–80.

1154   Stockley, P. 2002. Sperm competition risk and male genital anatomy: comparative evidence for reduced duration of female sexual receptivity in primates with penile spines. *Evolutionary Ecology* 16: 123–137.

1155   Stoutamire, W. P. 1974. Australian terrestrial orchids, thynnid wasps and pseudocopulation. *American Orchid Society Bulletin* 43: 13–18.

1156   Stowers, L., T. E. Holy, M. Meister, C. Dulac, and G. Koentges. 2002. Loss of sex discrimination and male-male aggression in mice deficient for TRP2. *Science* 295: 1493–1500.

1157   Strassmann, J. E., C. R. Hughes, D. C. Queller, S. Turillazzi, R. Cervo, et al. 1989. Genetic relatedness in primitively eusocial wasps. *Nature* 342: 268–269.

1158   Strum, S. C. 1987. *Almost Human*. New York: W.W. Norton.

1159   Stuart-Fox, D. M., A. Moussalli, N. J. Marshall, and I. P. F. Owens. 2003. Conspicuous males suffer higher predation risk: visual modelling and experimental evidence from lizards. *Animal Behaviour* 66: 541–550.

1160   Stumpner, A., and R. Lakes-Harlan. 1996. Auditory interneurons in a hearing fly (*Therobia leonidei*, Ormiini, Tachinidae, Diptera). *Journal of Comparative Physiology A* 178: 227–233.

1161   Stutt, A. D., and M. T. Siva-Jothy. 2001. Traumatic insemination and sexual conflict in the bed bug *Cimex lectularius*. *Proceedings of the National Academy of Sciences* 98: 5683–5687.

1162   Sullivan, J. P., O. Jassim, S. E. Fahrbach, and G. E. Robinson. 2000. Juvenile hormone paces behavioral development in the adult worker honey bee. *Hormones and Behavior* 37: 1–14.

1163   Sumner, S., M. Casiraghi, W. Foster, and J. Field. 2002. High reproductive skew in tropical hover wasps. *Proceedings of the Royal Society of London B* 269: 179–186.

1164   Sündstrom, L. 1994. Sex ratio bias, relatedness asymmetry and queen mating frequency in ants. *Nature* 367: 266–268.

1165   Surlykke, A. 1984. Hearing in notodontid moths: a tympanic organ with a single auditory neuron. *Journal of Experimental Biology* 113: 323–334.

1166   Surlykke, A., and J. H. Fullard. 1989. Hearing of the Australian whistling moth, *Hecatesia thyridion*. *Naturwissenschaften* 76: 132–134.

1167   Suthers, R. A. 1990. Contributions to birdsong from the left and right sides of the intact syrinx. *Nature* 347: 473–477.

1168   Svensson, B. G. 1997. Swarming behavior, sexual dimorphism, and female reproductive status in the sex role-reversed dance fly species *Rhamphomyia marginata*. *Journal of Insect Behavior* 10: 783–804.

1169   Svensson, G. P., C. Löfstedt, and N. Skals. 2004. The odour makes the difference: male moths attracted by sex pheromones ignore the threat by predatory bats. *Oikos* 104: 91–97.

1170   Swaddle, J. P. 1999. Limits to length asymmetry detection in starlings: Implications for biological signalling. *Proceedings of the Royal Society of London B* 266: 1299–1303.

1171   Swaddle, J. P. 2003. Fluctuating asymmetry, animal behavior, and evolution. *Advances in the Study of Behavior* 32: 169–205.

1172   Swaddle, J. P., and G. W. Reierson. 2002. Testosterone increases perceived dominance but not attractiveness in human males. *Proceedings of the Royal Society of London B* 269: 2285–2289.

1173   Swaisgood, R. R., M. P. Rowe, and D. H. Owings. 2003. Antipredator responses of California ground squirrels to rattlesnakes and rattling sounds: the roles of sex, reproductive parity, and offspring age in assessment and decision-making rules. *Behavioral Ecology and Sociobiology* 55: 22–31.

1174   Swan, L. W. 1970. Goose of the Himalayas. *Natural History* 79: 68–75.

1175   Sweeney, B. W., and R. L. Vannote. 1982. Population synchrony in mayflies: A predator satiation hypothesis. *Evolution* 36: 810–821.

1176   Symons, D. 1979. *The Evolution of Human Sexuality*. New York: Oxford University Press.

1177   Taborsky, M. 1994. Sneakers, satellites, and helpers: Parasitic and cooperative behavior in fish reproduction. *Advances in the Study of Behavior* 23: 1–100.

1178   Taborsky, M., and A. Grantnerm. 1998. Behavioural time-energy budgets of cooperatively breeding *Neolamprologus pulcher* (Pisces: Cichlidae). *Animal Behaviour* 56: 1375–1382.

1179   Tallamy, D. W. 2001. Evolution of exclusive paternal care in arthropods. *Annual Review of Entomology* 46: 139–165.

1180   Tarpy, D. R., and 2003. 2003. Genetic diversity within honeybee colonies prevents severe infections and promotes colony growth. *Proceedings of the Royal Society of London B* 270: 99–103.

1181   Tarpy, D. R., and D. I. Nielsen. 2002. Sampling error, effective paternity, and estimating the genetic structure of honey bee colonies (Hymenoptera: Apidae). *Annals of the Entomological Society of America* 95: 513–528.

1182   Temrin, H., and A. Arak. 1989. Polyterritoriality and deception in passerine birds. *Trends in Ecology and Evolution* 4: 106–108.

1183   Temrin, H., and B. S. Tullberg. 1995. A phylogenetic analysis of the evolution of avian mating systems in relation to altricial and precocial young. *Behavioral Ecology* 6: 296–307.

1184   Tewksbury, J. J., T. E. Martin, S. J. Hejl, M. J. Kuehn, and J. W. Jenkins. 2002. Parental care of a cowbird host: caught between the costs of egg-removal and nest predation. *Proceedings of the Royal Society of London B* 269: 423–429.

1185   Thomas, J. A., J. J. Knapp, T. Akino, S. Gerty, S. Wakamura, et al. 2002. Parasitoid secretions provoke ant warfare. *Nature* 417: 505–506.

1186   Thomas, L. K., and A. Manica. 2003. Filial cannibalism in an assassin bug. *Animal Behaviour* 66: 205–210.

1187   Thomas, M. A., K. A. Walsh, M. R. Wolf, B. A. McPheron, and J. H. Marden. 2000. Molecular phylogenetic analysis of evolutionary trends in stonefly wing structure and locomotor behavior. *Proceedings of the National Academy of Sciences* 97: 13178–13183.

1188   Thornhill, R. 1975. Scorpion-flies as kleptoparasites of web-building spiders. *Nature* 258: 709–711.

1189   Thornhill, R. 1976. Sexual selection and nuptial feeding behavior in *Bittacus apicalis* (Insecta: Mecoptera). *American Naturalist* 119: 529–548.

1190   Thornhill, R. 1981. *Panorpa* (Mecoptera: Panorpidae) scorpionflies: Systems for understanding resource-defense polygyny and alternative male reproductive efforts. *Annual Review of Ecology and Systematics* 12: 355–386.

1191   Thornhill, R., and J. Alcock. 1983. *The Evolution of Insect Mating Systems*. Cambridge, MA: Harvard University Press.

1192   Thornhill, R., and S. W. Gangestad. 1999. Facial attractiveness. *Trends in Cognitive Sciences* 3: 452–460.

1193   Thornhill, R., and C. T. Palmer. 2000. *A Natural History of Rape: The Biological Bases of Sexual Coercion*. Cambridge, MA: MIT Press.

1194   Thornhill, R., and N. W. Thornhill. 1983. Human rape: An evolutionary analysis. *Ethology and Sociobiology* 4: 137–173.

1195   Tibbetts, E. A., and H. K. Reeve. 2003. Benefits of foundress associations in the paper wasp *Polistes dominulus*: Increased productivity and survival, but no assurance of fitness returns. *Behavioral Ecology* 14: 510–514.

1196   Tinbergen, N. 1951. *The Study of Instinct*. New York: Oxford University Press.

1197   Tinbergen, N. 1958. *Curious Naturalists*. Garden City, NY: Doubleday.

1198   Tinbergen, N. 1959. Comparative studies of the behaviour of gulls (Laridae): a progress report. *Behaviour* 15: 1–70.

1199   Tinbergen, N. 1960. *The Herring Gull's World*. Garden City, New York: Doubleday.

1200   Tinbergen, N. 1963. On the aims and methods of ethology. *Zeitschrift für Tierpsychologie* 20: 410–433.

1201   Tinbergen, N., and A. C. Perdeck. 1950. On the stimulus situations releasing the begging response in the newly hatched herring gull (*Larus argentatus* Pont.). *Behaviour* 3: 1–39.

1202   Tobias, J. 1997. Asymmetric territorial contests in the European robin: The role of settlement costs. *Animal Behaviour* 54: 9–21.

1203   Tobias, M. C., and G. E. Hill. 1998. A test of sensory bias for long tails in the house finch. *Animal Behaviour* 56: 71–78.

1204   Toh, K. L., C. R. Jones, Y. He, E. J. Eide, W. A. Hinz, et al. 2001. An hPer2 phosphorylation site mutation in familial advanced sleep phase syndrome. *Science* 291: 1040–1043.

1205   Toma, D. P., G. Bloch, D. Moore, and G. E. Robinson. 2000. Changes in *period* mRNA levels in the brain and division of labor in honey bee colonies. *Proceedings of the National Academy of Sciences* 97: 6914–6919.

1206   Tomkins, J. L., and L. W. Simmons. 1998. Female choice and manipulations of forceps size and symmetry in the earwig *Forficula auricularia* L. *Animal Behaviour* 56: 347–356.

1207   Tovée, M. J., D. S. Maisey, J. L. Emery, and P. L. Cornelissen. 1999. Visual cues to female physical attractiveness. *Proceedings of the Royal Society of London B* 266: 211–218.

1208   Townsend, J. M. 1989. Mate selection criteria: A pilot study. *Ethology and Sociobiology* 10: 241–253.

1209   Trainor, B. C., I. M. Bird, N. A. Alday, B. A. Schlinger, and C. A. Marler. 2003. Variation in aromatase activity in the medial preoptic area and plasma progesterone is associated with the onset of paternal behavior. *Neuroendocrinology* 78: 36–44.

1210   Traniello, J. F. A., R. B. Rosengaus, and K. Savoie. 2002. The development of immunity in a social insect: Evidence for the group facilitation of disease resistance. *Proceedings of the National Academy of Sciences* 99: 6838–6842.

1211   Trivers, R. L. 1971. The evolution of reciprocal altruism. *Quarterly Review of Biology* 46: 35–57.

1212   Trivers, R. L. 1972. Parental investment and sexual selection. In *Sexual Selection and the Descent of Man*, ed. Campbell, B. Chicago: Aldine.

1213   Trivers, R. L. 1974. Parent-offspring conflict. *American Zoologist* 14: 249–264.

1214   Trivers, R. L. 1985. *Social Evolution*. Menlo Park, CA: Benjamin Cummings.

1215   Trivers, R. L., and H. Hare. 1976. Haplodiploidy and the evolution of the social insects. *Science* 191: 249–263.

1216   Tsai, M. L., and C. F. Dai. 2003. Cannibalism within mating pairs of the parasitic isopod, *Ichthyoxenus fushanensis*. *Journal of Crustacean Biology* 23: 662–668.

1217   Tschinkel, W. R., and S. D. Porter. 1988. Efficiency of sperm use in queens of the fire ant, *Solenopsis invicta* (Hymenoptera: Formicidae). *Annals of the Entomological Society of America* 81: 777–781.

1218   Tsubaki, Y., R. E. Hooper, and M. T. Siva-Jothy. 1997. Differences in adult and reproductive lifespan in the two male forms of *Mnais pruinosa costalis* Selys (Odonata: Calopterygidae). *Research in Population Ecology* 39: 149–155.

1219   Tuttle, E. M., S. Pruett-Jones, and M. S. Webster. 1996. Cloacal protuberances and extreme sperm production in Australian fairy-wrens. *Proceedings of the Royal Society of London B* 263: 1359–1364.

1220   Tyler, W. A. 1995. The adaptive significance of colonial nesting in a coral-reef fish. *Animal Behaviour* 49: 949–966.

1221   Uetz, G. W., and E. I. Smith. 1999. Asymmetry in a visual signaling character and sexual selection in a wolf spider. *Behavioral Ecology and Sociobiology* 45: 87–94.

1222   Urquhart, F. A. 1960. *The Monarch Butterfly*. Toronto: University of Toronto Press.

1223   Uy, J. A. C., and G. Borgia. 2000. Sexual selection drives rapid divergence in bowerbird display traits. *Evolution* 54: 273–278.

1224   Uy, J. A. C., G. L. Patricelli, and G. Borgia. 2001. Complex mate searching in the satin bowerbird *Ptilonorhynchus violaceus*. *American Naturalist* 158: 530–542.

1225   Václav, R., H. Hoi, and D. Blomqvist. 2003. Food supplementation affects extrapair paternity in house sparrows (*Passer domesticus*). *Behavioral Ecology* 14: 730–735.

1226   Vahed, K. 1998. The function of nuptial feeding in insects: Review of empirical studies. *Biological Reviews* 73: 43–78.

1227   Vallet, E., I. Beme, and M. Kreutzer. 1998. Two-note syllables in canary songs elicit high levels of sexual display. *Animal Behaviour* 55: 291–297.

1228   van Gils, J. A., I. W. Schenk, O. Bos, and T. Piersma. 2003. Incompletely informed shorebirds that face a digestive constraint maximize net energy gain when exploiting patches. *American Naturalist* 161: 777–793.

1229   van Schaik, C. P., and P. M. Kappeler. 1997. Infanticide risk and the evolution of male-female associations in primates. *Proceedings of the Royal Society of London B* 264: 1687–1694.

1230   van Staaden, M. J., and H. Romer. 1998. Evolutionary transition from stretch to hearing organs in ancient grasshoppers. *Nature* 394: 773–778.

1231   Veiga, J. P. 2003. Infanticide by male house sparrows: gaining time or manipulating females? *Proceedings of the Royal Society of London B* 270: S87–S89.

1232   Vieites, D. R., S. Nieto-Román, M. Barluenga, A. Palanca, M. Vences, and A. Meyer. 2004. Post-mating clutch piracy in an amphibian. *Nature* 431: 305–308.

1233   Vincent, A. C. J., A. D. Marsden, K. L. Evans, and L. M. Sadler. 2004. Temporal and spatial opportunities for polygamy in a monogamous seahorse, *Hippocampus whitei*. *Behaviour* 141: 141–156.

1234   Vincent, A. C. J., and L. M. Sadler. 1995. Faithful pair bonds in wild seahorses, *Hippocampus whitei*. *Animal Behaviour* 50: 1557–1569.

1235   Vining, D. R., Jr. 1986. Social versus reproductive success: The central theoretical problem of human sociobiology. *Behavioral and Brain Sciences* 9: 167–187.

1236   Visscher, P. K. 2003. How self-organization evolves. *Nature* 421: 799–800.

1237   Vleck, C. M., and J. L. Brown. 1999. Testosterone and social and reproductive behaviour in *Aphelocoma* jays. *Animal Behaviour* 58: 943–951.

1238   von Engelhardt, N., P. M. Kappeler, and M. Heistermann. 2000. Androgen levels and female social dominance in *Lemur catta*. *Proceedings of the Royal Society of London B* 267: 1533–1539.

1239   von Frisch, K. 1956. *The Dancing Bees*. New York: Harcourt Brace Jovanovich.

1240   von Frisch, K. 1967. *The Dance Language and Orientation of Bees*. Cambridge, MA: Harvard University Press.

1241   Waage, J. K. 1979. Dual function of the damselfly penis: Sperm removal and transfer. *Science* 203: 916–918.

1242   Waage, J. K. 1997. Parental investment—minding the kids or keeping control? In *Feminism and Evolutionary Biology: Boundaries, Interactions, and Frontiers*, ed. Gowaty, P. A. New York: Chapman and Hall.

1243   Wagner, R. H. 1992. Mate guarding by monogamous female razorbills. *Animal Behaviour* 44: 533–538.

1244   Wagner, R. H., F. Helfenstein, and E. Danchin. 2004. Female choice of young sperm in a genetically monogamous bird. *Proceedings of the Royal Society of London B* 271: S134–S137.

1245   Wagner, W. E., Jr. 1992. Deceptive or honest signalling of fighting ability? A test of alternative hypotheses for the function of changes in call dominant frequency by male cricket frogs. *Animal Behaviour* 44: 449–462.

1246   Walcott, C. 1972. Bird navigation. *Natural History* 81: 32–43.

1247   Walcott, C. 1996. Pigeon homing: Observations, experiments and confusions. *Journal of Experimental Biology* 199: 21–27.

1248   Wallraff, H. G. 2003. Olfactory navigation by birds. *Journal fur Ornithologie*. 144: 1–32.

1249   Walther, B. A., and A. G. Gosler. 2001. The effects of food availability and distance to protective cover on the winter foraging behaviour of tits (Aves: *Parus*). *Oecologia* 129: 312–320.

1250   Ward, P., and A. Zahavi. 1973. The importance of certain assemblages of birds as "information-centres" for food finding. *Ibis* 115: 517–534.

1251   Warner, R. R. 1984. Mating behavior and hermaphroditism in coral reef fishes. *American Scientist* 72: 128–136.

1252   Watanabe, T. 1999. Prey attraction as a possible function of the silk decoration of uloborid spider *Octonoba sybotides*. *Behavioral Ecology* 10: 607–611.

1253   Waters, D. A. 2003. Bats and moths: what is there left to learn? *Physiological Entomology* 28: 237–250.

1254   Watson, P. J. 1998. Multi-male mating and female choice increase offspring growth in the spider *Neriene litigiosa* (Linyphiidae). *Animal Behaviour* 55: 387–403.

1255   Watt, P. J., S. F. Nottingham, and S. Young. 1997. Toad tadpole aggregation behaviour: Evidence for a predator avoidance function. *Animal Behaviour* 54: 865–872.

1256   Waynforth, D., and R. I. M. Dunbar. 1995. Conditional mate choice strategies in humans—evidence from lonely hearts advertisements. *Behaviour* 132: 755–779.

1257   Weathers, W. W., and K. A. Sullivan. 1989. Juvenile foraging proficiency, parental effort, and avian reproductive success. *Ecological Monographs* 59: 223–246.

1258   Webster, M. S. 1994. Female-defence polygyny in a Neotropical bird, the Montezuma oropendula. *Animal Behaviour* 48: 779–794.

1259   Wedekind, C., and M. Milinski. 1996. Human cooperation in the simultaneous and the alternating Prisoner's Dilemma: Pavlov versus Generous Tit-for-Tat. *Proceedings of the National Academy of Sciences* 93: 2686–2689.

1260   Wedell, N., and T. Tregenza. 1999. Successful fathers sire successful sons. *Evolution* 53: 620–625.

1261   Wehner, R., M. Lehrer, and W. R. Harvey. 1966. Navigation: Migration and homing. *Journal of Experimental Biology* 199: 1–261.

1262   Wehner, R., and S. Wehner. 1990. Insect navigation: Use of maps or Ariadne's thread? *Ethology, Ecology and Evolution* 2: 27–48.

1263   Weidinger, K. 2000. The breeding performance of blackcap *Sylvia atricapilla* in two types of forest habitat. *Ardea* 88: 225–233.

1264   Weimerskirch, H., J. Martin, Y. Clerquin, P. Alexandre, and S. Jiraskova. 2001. Energy saving in flight formation. *Nature* 413: 697–698.

1265   Weimerskirch, H., M. Salamolard, F. Sarrazin, and P. Jouventin. 1993. Foraging strategy of wandering albatrosses through the breeding season: A study using satellite telemetry. *Auk* 110: 325–342.

1266   Weiss, M. R. 2003. Good housekeeping: why do shelter-dwelling caterpillars fling their frass? *Ecology Letters* 6: 361–370.

1267   Welch, A. M., R. D. Semlitsch, and H. C. Gerhardt. 1998. Call duration as an indicator of genetic quality in male gray tree frogs. *Science* 280: 1928–1930.

1268   Wells, J. C. K. 2003. Parent-offspring conflict theory, signaling of need, and weight gain in early life. *Quarterly Review of Biology* 78: 169–202.

1269   Welty, J. 1982. *The Life of Birds*. Philadelphia, PA: Saunders College Publishing.

1270   Wenner, A. M., and P. Wells. 1990. *Anatomy of a Controversy*. New York: Columbia University Press.

1271   Werner, N. Y., S. Balshine, B. Leach, and A. Lotem. 2003. Helping opportunities and space segregation in cooperatively breeding cichlids. *Behavioral Ecology* 14: 749–756.

1272   West-Eberhard, M. J. 1975. The evolution of social behavior by kin selection. *Quarterly Review of Biology* 50: 1–33.

1273   West-Eberhard, M. J. 2003. *Developmental Plasticity and Evolution*. New York: Oxford University Press.

1274   West, M. J., and A. P. King. 1990. Mozart's starling. *American Scientist* 78: 106–114.

1275 West, P. M., and C. Packer. 2002. Sexual selection, temperature, and the lion's mane. *Science* 297: 1339–1343.

1276 Westcott, D., and J. N. M. Smith. 1997. Lek size variation and its consequences in the ochre-bellied flycatcher, *Mionectes oleagineus*. *Behavioral Ecology* 8: 396–403.

1277 Westneat, D. F. 1992. Do female red-winged blackbirds engage in a mixed mating strategy? *Ethology* 92: 7–28.

1278 Westneat, D. F., and I. R. K. Stewart. 2003. Extra-pair paternity in birds: Causes, correlates, and conflict. *Annual Review of Ecology and Systematics* 34: 365–396.

1279 Wetsman, A., and F. Marlowe. 1999. How universal are preferences for female waist-to-hip ratios? Evidence from the Hazda. *Evolution and Human Behavior* 20: 219–228.

1280 White, S. A., T. Nguyen, and R. D. Fernald. 2002. Social regulation of gonadotropin-releasing hormone. *Journal of Experimental Biology* 205: 2567–2581.

1281 Whiteman, E. A., and I. M. Côté. 2004. Monogamy in marine fishes. *Biological Reviews* 79: 351–375.

1282 Whitfield, C. W., A.-M. Cziko, and G. E. Robinson. 2004. Gene expression profiles in the brain predict behavior in individual honey bees. *Science* 302: 296–299.

1283 Whitfield, D. P. 1990. Individual feeding specializations of wintering turnstone *Arenaria interpres*. *Journal of Animal Ecology* 59: 193–211.

1284 Whitham, T. G. 1979. Habitat selection by *Pemphigus* aphids in response to resource limitation and competition. *Ecology* 59: 1164–1176.

1285 Whitham, T. G. 1979. Territorial defense in a gall aphid. *Nature* 279: 324–325.

1286 Whitham, T. G. 1980. The theory of habitat selection examined and extended using *Pemphigus* aphids. *American Naturalist* 115: 449–466.

1287 Whitham, T. G. 1986. Costs and benefits of territoriality: Behavioral and reproductive release by competing aphids. *Ecology* 67: 139–147.

1288 Whiting, M. J. 1999. When to be neighbourly: Differential agonistic responses in the lizard *Platysaurus broadleyi*. *Behavioral Ecology and Sociobiology* 46: 210–214.

1289 Whittingham, L. A., and P. O. Dunn. 1998. Male parental effort and paternity in a variable mating system. *Animal Behaviour* 55: 629–640.

1290 Whittingham, L. A., P. O. Dunn, and R. D. Macgrath. 1997. Relatedness, polyandry and extra-group paternity in the cooperatively-breeding white-browed scrubwren (*Sericornis frontalis*). *Behavioral Ecology and Sociobiology* 40: 261–270.

1291 Wickler, W. 1968. *Mimicry in Plants and Animals*. London: World University Library.

1292 Wickler, W., and U. Seibt. 1981. Monogamy in Crustacea and man. *Zeitschrift für Tierpsychologie* 57: 215–234.

1293 Widdig, A., F. B. Bercovitch, W. J. Streich, U. Sauermann, P. Nürnberg, and M. Krawczak. 2004. A longitudinal analysis of reproductive skew in male rhesus macaques. *Proceedings of the Royal Society of London B* 271: 819–826.

1294 Widemo, F., and I. P. F. Owens. 1995. Lek size, male mating skew and the evolution of lekking. *Nature* 373: 148–151.

1295 Wiederman, M. W., and E. R. Allgeier. 1992. Gender differences in mate selection criteria: Sociobiological or socioeconomic explanation? *Ethology and Sociobiology* 13: 115–124.

1296 Wiederman, M. W., and E. Kendall. 1999. Evolution, sex, and jealousy: Investigation with a sample from Sweden. *Evolution and Human Behavior* 20: 121–128.

1297 Wiens, J. J. 2001. Widespread loss of sexually selected traits: how the peacock lost its spots. *Trends in Ecology and Evolution* 19: 517–523.

1298 Wigby, S., and T. Chapman. 2004. Female resistance to male harm evolves in response to manipulation of sexual conflict. *Evolution* 58: 1028–1037.

1299 Wikelski, M., and S. Baurle. 1996. Pre-copulatory ejaculation solves time constraints during copulations in marine iguanas. *Proceedings of the Royal Society of London B* 263: 439–444.

1300 Wikelski, M., M. Hau, and J. C. Wingfield. 1999. Social instability increases plasma testosterone in a year-round territorial neotropical bird. *Proceedings of the Royal Society of London B* 266: 551–556.

1301 Wiklund, C., B. Karlsson, and O. Leimar. 2001. Sexual conflict and cooperation in butterfly reproduction: a comparative study of polyandry and female fitness. *Proceedings of the Royal Society of London B* 268: 1661–1667.

1302 Wiklund, C. G., and M. Andersson. 1994. Natural selection of colony size in a passerine bird. *Journal of Animal Ecology* 63: 765–774.

1303 Wilbrecht, L., A. Crionas, and F. Nottebohm. 2002. Experience affects recruitment of new neurons but not adult neuron number. *Journal of Neuroscience* 22: 825–831.

1304 Wilkinson, G. S. 1984. Reciprocal food sharing in the vampire bat. *Nature* 308: 181–184.

1305 Wilkinson, G. S., and G. N. Dodson. 1997. Function and evolution of antlers and eye stalks in flies. In *The Evolution of Mating Systems in Insects and Arachnids*, ed. Choe, J. C., and B. J. Crespi. Cambridge: Cambridge University Press.

1306 Williams, C. K., R. S. Lutz, and R. D. Applegate. 2003. Optimal group size and northern bobwhite coveys. *Animal Behaviour* 66: 377–387.

1307 Williams, G. C. 1966. *Adaptation and Natural Selection*. Princeton, N J: Princeton University Press.

1308 Williams, G. C. 1975. *Sex and Evolution*. Princeton, NJ: Princeton University Press.

1309 Williams, G. C. 1996. *The Pony Fish's Glow*. New York: Basic Books.

1310 Williams, T. C., and J. M. Williams. 1978. An oceanic mass migration of land birds. *Scientific American* 239 (Oct.): 166–176.

1311 Willows, A. O. D. 1971. Giant brain cells in mollusks. *Scientific American* 224 (Feb.): 68–75.

1312 Wilmer, J. W., P. J. Allen, P. P. Pomeroy, S. D. Twiss, and W. Amos. 1999. Where have all the fathers gone? An extensive microsatellite analysis of paternity in the grey seal (*Halichoerus grypus*). *Molecular Ecology* 8: 1417–1429.

1313 Wilson, A. B., I. Ahnesjö, A. C. J. Vincent, and A. Meyer. 2003. The dynamics of male brooding, mating patterns, and sex roles in pipefishes and seahorses (family Syngnathidae). *Evolution* 57: 1374–1386.

1314 Wilson, E. O. 1971. *The Insect Societies*. Cambridge, MA: Harvard University Press.

1315 Wilson, E. O. 1975. *Sociobiology, The New Synthesis*. Cambridge, MA: Harvard University Press.

1316 Wilson, E. O. 1976. Academic vigilantism and the political significance of sociobiology. *BioScience* 26: 187–190.

1317 Wilson, E. O. 1980. Caste and division of labor in leaf-cutter ants (Hymenoptera: Formicidae: *Atta*), II. The ergonomic optimization of leaf cutting. *Behavioral Ecology and Sociobiology* 7: 157–165.

1318 Wilson, E. O. 1983. Caste and division of labor in leaf-cutter ants (Hymenoptera: Formicidae: *Atta*), III. Ergonomic resiliency in foraging by *A. cephalotes*. *Behavioral Ecology and Sociobiology* 14: 47–54.

1319 Wilson, M. I., and M. Daly. 1997. Life expectancy, economic inequality, homicide, and reproductive timing in Chicago neighborhoods. *BMJ.com* 314: 1271.

1320 Wiltschko, R., and W. Wiltschko. 2001. Clock-shift experiments with homing pigeons: a compromise between solar and magnetic information? *Behavioral Ecology and Sociobiology* 49: 393–400.

1321 Wiltschko, R., and W. Wiltschko. 2003. Avian navigation: from historical to modern concepts. *Animal Behaviour* 65: 257–272.

1322 Winfree, R. 1999. Cuckoos, cowbirds and the persistence of brood parasitism. *Trends in Ecology and Evolution* 14: 338–343.

1323 Wingfield, J. C., J. Jacobs, and N. Hillgarth. 1997. Ecological constraints and the evolution of hormone-behavior interrelationships. *Annals of the New York Academy of Sciences* 807: 22–41.

1324 Wingfield, J. C., and M. C. Moore. 1987. Hormonal, social and environmental factors in the reproductive biology of free-living male birds. In *Psychobiology of Reproductive Behavior: An Evolutionary Perspective*, ed. Crews, D. Englewood Cliffs, NJ.: Prentice-Hall.

1325 Wingfield, J. C., and M. Ramenofsky. 1997. Corticosterone and facultative dispersal in response to unpredictable events. *Ardea* 85: 155–166.

1326 Winkler, S. M., and J. Wade. 1998. Aromatase activity and regulation of sexual behaviors in the green anole lizard. *Physiology & Behavior* 64: 723–731.

1327 Winterer, G., and D. Goldman. 2003. Genetics of human prefrontal function. *Brain Research Reviews* 43: 134–163.

1328 Winterhalder, B., and E. A. Smith. 2000. Analyzing adaptive strategies: Human behavioral ecology at twenty-five. *Evolutionary Anthropology* 9: 51–72.

1329 Wise, K. K., M. R. Conover, and F. F. Knowlton. 1999. Response of coyotes to avian distress calls: Testing the startle-predator and predator-attraction hypotheses. *Behaviour* 136: 935–949.

1330 Witte, K., and N. Sawka. 2003. Sexual imprinting on a novel trait in the dimorphic zebra finch: sexes differ. *Animal Behaviour* 65: 195–203.

1331 Wittenberger, J. F. 1981. *Animal Social Behavior*. Boston, MA: Duxbury Press.

1332 Wolanski, E., E. Gereta, M. Borner, and S. Mduma. 1999. Water, migration and the Serengeti ecosystem. *American Scientist* 87: 526–533.

1333 Wolf, N. 1990. *The Beauty Myth*. London: Chatto & Windus.

1334 Wolff, J. O., and D. W. Macdonald. 2004. Promiscuous females protect their offspring. *Trends in Ecology and Evolution* 19: 127–134.

1335 Wolff, J. O., S. G. Mech, A. S. Dunlap, and K. E. Hodges. 2002. Multi-male mating by paired and unpaired female prairie voles (*Microtus ochrogaster*). *Behaviour* 139: 1147–1160.

1336 Woodroffe, R., and A. Vincent. 1994. Mother's little helpers: Patterns of male care in mammals. *Trends in Ecology and Evolution* 9: 294–297.

1337 Woodward, J., and D. Goodstein. 1996. Conduct, misconduct and the structure of science. *American Scientist* 84: 479–490.

1338 Woolfenden, G. E., and J. W. Fitzpatrick. 1984. *The Florida Scrub Jay: Demography of a Cooperative-Breeding Bird*. Princeton, NJ: Princeton University Press.

1339 Woyciechowski, M., L. Kabat, and E. Król. 1994. The function of the mating sign in honey bees, *Apis mellifera* L.: New evidence. *Animal Behaviour* 47: 733–735.

1340 Wrangham, R. W. 1999. Evolution of coalitionary killing. *Yearbook of Physical Anthropology* 42: 1–30.

1341 Wright, J., R. E. Stone, and N. Brown. 2003. Communal roosts as structured information centres in the raven, *Corvus corax*. *Journal of Animal Ecology* 72: 1003–1014.

1342 Wright, R. 1994. *The Moral Animal: Evolutionary Psychology and Everyday Life*. New York: Pantheon.

1343 Wynne-Edwards, V. C. 1962. *Animal Dispersion in Relation to Social Behaviour*. Edinburgh: Oliver & Boyd.

1344 Yack, J. E. 1992. A multiterminal stretch receptor, chordotonal organ, and hair plate at the wing-hinge of *Manduca sexta*: Unravelling the mystery of the noctuid moth ear B cell. *Journal of Comparative Neurology* 324: 500–508.

1345 Yack, J. E., and J. H. Fullard. 1990. The mechanoreceptive origin of insect tympanal organs: A comparative study of similar nerves in tympanate and atympanate moths. *Journal of Comparative Neurology* 300: 523–534.

1346 Yack, J. E., and J. H. Fullard. 2000. Ultrasonic hearing in nocturnal butterflies. *Nature* 403: 265–266.

1347 Yager, D. D., and M. L. May. 1990. Ultrasound-triggered, flight-gated evasive maneuvers in the flying praying mantis, *Parasphendale agrionina*. II. Tethered flight. *Journal of Experimental Biology* 152: 41–58.

1348 Young, L. J., and Z. Wang. 2004. The neurobiology of pair bonding. *Nature Neuroscience* 7: 1048–1054.

1349 Young, M. W. 2000. Marking time for a kingdom. *Science* 288: 451–453.

1350 Yu, D. W., and G. H. Shepard, Jr. 1998. Is beauty in the eye of the beholder? *Nature* 396: 321–322.

1351 Zach, R. 1979. Shell-dropping: Decision-making and optimal foraging in northwestern crows. *Behaviour* 68: 106–117.

1352 Zahavi, A. 1975. Mate selection—A selection for a handicap. *Journal of Theoretical Biology* 53: 205–214.

1353 Zeh, D. W., J. A. Zeh, and E. Bermingham. 1997. Polyandrous, sperm-storing females: Carriers of male genotypes through episodes of adverse selection. *Proceedings of the Royal Society of London B* 264: 119–125.

1354 Zeh, J. A. 1997. Polyandry and enhanced reproductive success in the harlequin beetle-riding pseudoscorpion. *Behavioral Ecology and Sociobiology* 40: 111–118.

1355 Zeh, J. A., S. D. Newcomer, and D. W. Zeh. 1998. Polyandrous females discriminate against previous mates. *Proceedings of the National Academy of Sciences* 95: 13732–13736.

1356 Zeh, J. A., and D. W. Zeh. 1996. The evolution of polyandry I: Intragenomic conflict and genetic incompatibility. *Proceedings of the Royal Society of London B* 263: 1711–1717.

1357 Zeh, J. A., and D. W. Zeh. 1997. The evolution of polyandry II: Post-copulatory defenses against genetic incompatibility. *Proceedings of the Royal Society of London B* 264: 69–75.

1358 Zeh, J. A., and D. W. Zeh. 2003. Toward a new sexual selection paradigm: Polyandry, conflict and incompatibility. *Ethology* 109: 929–950.

1359 Zeifman, D. M. 2001. An ethological analysis of human infant crying: Answering Tinbergen's four questions. *Developmental Psychobiology* 39: 265–285.

1360 Zera, A. J., J. Potts, and K. Kobus. 1998. The physiology of life-history trade-offs: Experimental analysis of a hormonally induced life-history trade-off in *Gryllus assimilis*. *American Naturalist* 152: 7–23.

1361 Zucker, I. 1983. Motivation, biological clocks and temporal organization of behavior. In *Handbook of Behavioral Neurobiology: Motivation*, ed. Satinoff, E., and P. Teitelbaum. New York: Plenum Press.

Illustration Credits

We are grateful to the publishers and photo agencies listed here for their permission to reprint copyrighted material. Full bibliographic information for all illustrations and tables can be found in the Bibliography. Copyrights to the photographs in this textbook are held by the photographers named in the figure captions and by the publishers and photo agencies listed here.

Chapter 1 *Opener*: © Mark Moffett/Minden Pictures. 1.2: © 2004 by Wiley-Liss, Inc., a subsidiary of John Wiley & Sons, Inc. 1.8: © Science Photo Library/Photo Researchers, Inc. 1.11: © S. Nagendra/Photo Researchers, Inc. 1.13: From "The Far Side" by Gary Larson. Reproduced by permission of Chronicle Features, San Francisco. 1.14: © George Schaller/Bruce Coleman, Inc.

Chapter 2 2.2 *sonograms*: © 2002 by the Society for Neuroscience. 2.2 *zebra finch*: Photograph by David McIntyre, courtesy of Atsuko Takahashi. 2.12: © 1976 by the American Association for the Advancement of Science. 2.13: © 2001 by the National Academy of Sciences, U.S.A. 2.16: © 2000 by the Nature Publishing Group. 2.22: © 2005 by Elsevier.

Chapter 3 3.2: © 2004 by the American Association for the Advancement of Science. 3.5: © Nina Leen/Time Life Pictures/Getty Images. 3.15: © 1995 by L.P.P. Ltd. (formerly Laser Pages Publishing Ltd.). 3.27: © Nina Leen/Time Life Pictures/Getty Images. 3.28: © Nina Leen/Time Life Pictures/Getty Images. 3.30: © 1998 by the Psychonomic Society. 3.32D: © Mark Moffett/Minden Pictures.

Chapter 4 4.7: © Ian Wyllie/OSF/photolibrary.com. 4.10: Photograph by John H. Christy, Smithsonian Tropical Research Institute. 4.17: © 2004 by the Company of Biologists. 4.31: © 2002 by the National Academy of Sciences, U.S.A. 4.36: © by MIT Press. 4.38: © 2002 by Oxford University Press. 4.44: © 2004 by Current Biology.

Chapter 5 *Opener*: © François Gohier/Photo Researchers, Inc. 5.26: © 2001 by the Royal Society of London, used with permission of the Perception Lab at the University of St. Andrews.

Chapter 6 6.16 *left*: © Michael Tweedie/Photo Researchers, Inc. 6.24: © 1987 by the American Association for the Advancement of Science. 6.25: © Jen and Des Bartlett/Bruce Coleman, Inc.

Chapter 7 *Opener*: © Kenneth H. Thomas/Photo Researchers, Inc. 7.5: © 2003 by the University of Chicago Press. 7.18: © Mary Beth Angelo/Photo Researchers, Inc. 7.20: © Frans Lanting/Minden Pictures.

Chapter 8 *Opener*: © Frans Lanting/Minden Pictures. 8.1 *left*: © T. Vezo/VIREO. 8.14 *map*: © 2004 by Blackwell Publishing and the Society for Conservation Biology. 8.15 *thrush*: © R. Curtis/VIREO. 8.15 *map*: © 2003 by the American Ornithological Union. 8.17: © 2001 by Nature Publishing Group. 8.20: © 2002 by the Royal Society. 8.24: © 1997 by the Cooper Ornithological Society. 8.30: © P. Craig-Cooper/Peter Arnold, Inc.

Chapter 9 *Opener*: © John Cancalosi/naturepl.com. 9.3: © 1995 by the Nature Publishing Group. 9.4: © Andrew Parkinson/naturepl.com. 9.8: © 2002 by Elsevier. 9.12: © 2000 by the Royal Society. 9.22: © 1987 by Prentice-Hall Inc., Englewood Cliffs, N.J. 9.28: © Roger Wilmshurst/Photo Researchers, Inc. 9.30: © Merlin D. Tuttle/Bat Conservation International/Photo Researchers, Inc. 9.36: © J. H. Robinson/Photo Researchers, Inc.

Chapter 10 10.6: Photograph by David M. Phillips/The Population Council. 10.10: © 2000 by Academic Press Ltd. 10.11 *left*: © Dr. Paul Zahl/Photo Researchers, Inc. 10.11 *right*: © Kelvin Aitken/ANTPhoto.com. 10.13: © 2003 by the Society for the Study of Evolution. 10.14: © 2000 by the American Institute of Biological Sciences. 10.16: © Gregory Dimijian/Photo Researchers, Inc. 10.19: © 1996 by the Royal Society of London. 10.20: © 2002 by the Royal Society of London. 10.22: © 2002 by Oxford University Press. 10.27: © 1979 by the American Association for the Advancement of Science. 10.39: © 2002 by the Royal Society. 10.43 *bowerbird*: © Cyril Webster/ANTPhoto.com. 10.43 *chart*: © 2003 by Oxford University Press. 10.44 *peacock*: © Brian Lightfoot/Nature Picture Library. 10.48A: © Andrew Syred/Photo Researchers, Inc.

Index